Oil Wealth and the Fate of the Forest

Oil production can damage rainforests, but this is just one side of a complicated story about the impact of oil on land use.

This book, a study of eight tropical oil-producing countries, examines the linkages between trade, macroeconomics and policies affecting the environment. In a balanced and comprehensive review, including a detailed assessment of land use in Cameroon, Ecuador, Gabon, Papua New Guinea and Venezuela, Sven Wunder comes up with a counterintuitive suggestion: oil revenues often indirectly come to protect tropical forests.

Oil Wealth and the Fate of the Forest is a topical and accessible book with numerous implications for policy formulation to decide what can be done to diminish deforestation without jeopardising economic growth. This book needs to be read not only by students and academics involved in environmental economics, but also by decision-makers involved in the conservation and sustainable use of forests.

Sven Wunder is Senior Economist at the Center for International Forestry Research (CIFOR), Indonesia.

Routledge explorations in environmental economics
Edited by Nick Hanley
University of Glasgow

1 **Greenhouse Economics**
Value and ethics
Clive L. Spash

2 **Oil Wealth and the Fate of the Forest**
A comparative study of eight tropical countries
Sven Wunder

Oil Wealth and the Fate of the Forest

A comparative study of eight tropical countries

Sven Wunder

Routledge
Taylor & Francis Group
LONDON AND NEW YORK

First published 2003
by Routledge
11 New Fetter Lane, London EC4P 4EE

Simultaneously published in the USA and Canada
by Routledge
29 West 35th Street, New York, NY 10001

Routledge is an imprint of the Taylor & Francis Group

Typeset in Perpetua by
Newgen Imaging Systems (P) Ltd, Chennai, India
Printed and bound in Malta by
Gutenberg Press Ltd, Malta

British Library Cataloguing in Publication Data
A catalogue record for this book is available
from the British Library

Library of Congress Cataloging in Publication Data
A catalog record for this book has been requested

ISBN 0–415–27867–8

Contents

List of figures viii
List of maps ix
List of tables x
List of boxes xii
Foreword xiii
Preface xvi
Abbreviations xviii

Introduction 1

The issue 1
Country sample 2
Theories 3
The hypotheses 4
Forest data and definitions 5
Other concepts and definitions 5
Scope and audience 6
Structure 7
Sources and methods 8
History of the project 9
Acknowledgements 10

1 **The malady of prosperity** 13

Origins of the Dutch Disease 13
To trade or not to trade ... 15
The core-model price mechanism 16
Extensions and modifications 19
Dutch Disease in developing countries 22
Why natural resources? 27

2 **The impact of oil wealth on forests** 30

Exploring the linkages 30
Stating the hypotheses 31

Do mineral rents globally protect tropical forests? 33
Trade, macroeconomics and forests: what do we know? 37
A note on methods 40
A framework for national case studies 42
Impact types: the example of oil extraction 47
A synthesis 51

3 Defining and measuring changing forest conditions 56

How to proceed? 56
Deforestation: what does it mean, and what should it mean? 57
Canopy cover 58
Spatial resolution 60
Sample size 62
Time scale 63
Eliminate the 'forests without trees'! 64
Some examples of deforestation 68
Forest degradation: a residual category 69
Trying to measure the impacts 71
Are we getting any wiser? 75
A way forward 79

4 Gabon 84

Deforestation in Gabon 84
The effect of mineral production on forests 91
The macroeconomic impact of the mineral boom 93
The competitiveness of agriculture and forestry 99
Windfall impacts on government spending 107
Structural changes in income and demand 114
Synthesis and conclusion 122

5 Venezuela 130

Deforestation in Venezuela 130
The effect of mineral production on forests 138
The macroeconomic impact of the mineral boom 140
The competitiveness of land-using sectors 144
Windfall impacts on government spending 154
Structural changes in income and demand 159
Synthesis and conclusion 164

6 Cameroon 170

Deforestation in Cameroon 170
The effect of mineral production on forests 175

The macroeconomic impact of the oil boom 177
The competitiveness of agriculture and forestry 182
Windfall impacts on government spending 195
Structural changes in income and demand 201
Synthesis and conclusion 206

7 Ecuador 214

Deforestation in Ecuador 214
The effect of oil production on forests 220
The macroeconomic impact of the oil boom 224
The competitiveness of agriculture and forestry 227
Windfall impacts on government spending 235
Structural changes in income and demand 238
Synthesis and conclusion 242

8 Papua New Guinea 248

Deforestation in Papua New Guinea 248
The effect of mineral production on forests 256
The macroeconomic impact of the mineral booms 262
The competitiveness of agriculture and forestry 267
Windfall impacts on government spending 280
Structural changes in income and demand 284
Synthesis and conclusion 288

9 More tales of oil wealth and forests: Mexico, Nigeria and Indonesia 296

Mexico 296
Nigeria 303
Indonesia 310
Summary notes 322

10 Comparison, conclusions and recommendations 325

Comparing background variables 325
Comparing the outcomes 331
Conclusions and perspectives: what determines the forest outcome? 359
Policy recommendations 367
Thinking beyond oil 376

Bibliography 383
Name index 415
Subject index 423

Figures

2.1	Linking resource booms to the forest	43
2.2	Example of oil exploration in tropical forests	49
2.3	Example of oil drilling rigs in tropical forests	50
3.1	FAO's classification of forest-cover changes	58
3.2	Forest-cover measurement and spatial resolution	61
3.3	Differences in the tropical forest-cover estimates by TREES, IUCN and FAO	76
4.1	Gabon: capital inflows, petroleum exports and real effective exchange rate	94
4.2	Gabon: timber production and RER	103
4.3	Gabon: age distribution of rural and urban population in 1993	120
5.1	Venezuela: yearly change in cultivated area	137
5.2	Venezuela: capital inflows, petroleum exports and real effective exchange rate	142
5.3	Venezuela: cultivated area, disaggregated into crops and pasture	146
5.4	Venezuela: industrial wood production, agricultural value-added and real effective exchange rate	148
6.1	Cameroon: capital inflows, petroleum exports and RER	178
6.2	Cameroon: industrial wood production, exports and competitiveness	189
7.1	Ecuador: capital inflows, petroleum exports and real effective exchange rate	225
7.2	Ecuador: industrial wood production, agricultural value-added and real effective exchange rate	229
8.1	PNG: capital inflows, mineral exports and real effective exchange rate	263
8.2	PNG: log exports, agricultural exports and real effective exchange rate	270

Maps

4.1	Gabon	85
4.2	Gabon: spatial distribution of rural population in the early 1940s	114
4.3	Gabon: spatial distribution of rural population in 1970	115
5.1	Venezuela	131
6.1	Cameroon	171
7.1	Ecuador	215
8.1	Papua New Guinea	249

Tables

1.1	Dutch Disease in developing countries: an overview	24
2.1	Tropical forest cover of high-mineral exporters	35
2.2	The effect of selected open-economy variables on deforestation	38
2.3	Forest impact types, using the example of oil extraction	48
2.4	Oil wealth and forest decline	52
3.1	Forest-cover and spatial resolution levels	60
3.2	Comparing forest-cover estimates	78
4.1	Gabon: forest cover and deforestation estimates	87
4.2	Gabon: relating oil wealth to relative prices and traded sector production	106
4.3	Gabon: poverty trends	116
4.4	Gabon: oil wealth and deforestation	124
5.1	Venezuela: forest cover and deforestation estimates	134
5.2	Venezuela: relating oil wealth to relative prices and traded sector production	152
5.3	Venezuela: yearly disbursement of agricultural credit during the first oil boom	154
5.4	Venezuela: poverty and inequality	160
5.5	Venezuela: income elasticities of urban household consumption	163
5.6	Venezuela: oil wealth and deforestation	166
6.1	Cameroon: forest cover and deforestation estimates	174
6.2	Cameroon: shifting land-use trends in the humid forest zone after independence	185
6.3	Cameroon: relating oil wealth to relative prices and traded sector production	193
6.4	Cameroon: construction and upgrading of asphalted roads	199
6.5	Cameroon: oil wealth and deforestation	208
7.1	Ecuador: forest-cover and deforestation estimates	217
7.2	Ecuador: agropastoral land-use trends	218
7.3	Ecuador: relating oil wealth to relative prices and traded sector production	233
7.4	Ecuador: oil wealth and deforestation	244
8.1	PNG: forest-cover and deforestation estimates	253
8.2	PNG: average household incomes for landowners affected by mining	260

8.3 PNG: land-use intensity in 1975 and 1996 272
8.4 PNG: relating mineral wealth to relative prices and traded sector production 278
8.5 PNG: mineral wealth and deforestation 290
10.1 Comparison of economic structures in the study countries 326
10.2 Comparison of the forest situation in the study countries 329
10.3 Comparing macroeconomic and deforestation trends 333
10.4 Comparison of forest impacts from mineral production 338
10.5 Comparing petroleum exports and capital inflows' effect on the RER 341
10.6 Comparing RER effects on agricultural production 343
10.7 Comparing RER effects on industrial timber production 346
10.8 Comparing corruption incidence in the study countries 365

Boxes

1.1	A small Dutch Disease dictionary	17
1.2	Expected outcomes from extended Dutch Disease models	23
1.3	Special features of the typical Dutch Disease in developing countries	26
2.1	Different types of booms and busts	34
3.1	Forests according to the FAO	57
3.2	Problems in distinguishing temporary from permanent forest-clearing	66
3.3	Examples of deforestation processes	68
3.4	Examples of forest degradation processes	69
3.5	Some guiding principles for the deforestation country-assessments in this book	80

Foreword

The loss and degradation of tropical forests have been a matter of significant international concern for more than two decades. At first glance, the 'villains' in this story seem obvious – the loggers, miners and ranchers who disturb or remove forests to make money in response to both need and greed. However, in the real world things are much more complicated, and the underlying causes of forest loss and degradation are a complex interaction of factors, many of them operating far away from the forest.

Sven Wunder's book is a valuable contribution to helping us understand the complexity of these forces and the nature of the responses that policy makers might consider as they seek better conservation and development outcomes. This book is about the connection between two things that at first sight would seem to be only marginally related. It explains how oil wealth affects the national forest cover of tropical petroleum-exporting countries. The impact that most people would think about in this respect is the direct one, that is, the picture of oil companies slashing tropical forests to establish drilling platforms, helipads, worker camps and access roads. However, after a meticulous examination of all the direct and indirect effects, the author convincingly shows that the opposed, indirect effects dominate in most cases, so that oil wealth is actually more likely to help protect tropical forests.

This apparently counterintuitive result helps demonstrate the complex interactions that can influence the future of a country's forests, particularly in developing countries. It should alert us to the fact that the indirect environmental effects can sometimes be surprising, especially when significant macroeconomic changes occur. Furthermore, the book indicates that these macro-level changes can often come to have a greater impact on the forest than the things we try to do directly to improve forest management, such as making a new forestry law, fining illegal loggers or implementing a project for planting trees.

How does the author make his somewhat unusual case? The book looks at eight tropical developing countries, with detailed studies of long-term land-use changes in Cameroon, Ecuador, Gabon, Papua New Guinea and Venezuela, as well as summary studies of Indonesia, Mexico and Nigeria. The core macroeconomic mechanism is that oil rents decrease the price-competitiveness of agriculture and logging, which in turn strongly diminishes pressures for deforestation and forest degradation. Gabon is perhaps the showcase of large forests protected by such 'blind' conservation impacts from oil wealth. The economic rents from hard-rock minerals can often have effects that are similar to those of oil.

With this key conclusion, some may feel that this just another overly optimistic perspective on the environment and the oil and mining industries, relying on developing-country economies 'growing their way out' of environmental pressures. Certainly, the book gives a more favourable perspective on these extractive industries, and will challenge those environmental activists who automatically campaign against them. It also indicates that rising national income and employment can sometimes relieve the pressures on forests. However, the book also shows that the type and quality of growth is of crucial importance for the future of forests. In particular, it demonstrates the importance of the particular sectors and technologies that are enabled to advance within different macroeconomic environments.

First and foremost, the book demonstrates that domestic policy responses to oil wealth are vital determinants of forest outcomes. When governments use most oil wealth to upgrade urban infrastructure, this reinforces the core effect by pulling more labour out of land-using and forest-degrading activities. It also warns us, however, that in other cases, when oil revenues are used to finance large road-construction programmes, frontier-colonisation projects or the resettlement of people into forests, the core forest-protective effect of oil wealth can be reversed. Repeated currency devaluations and import protection for heavily land-using domestic sectors also can contribute to increased pressures, leading to forest loss and degradation. Governments in tropical countries should therefore integrate these concerns into their policy planning, especially when forests are strategically important for their macroeconomic development, or where they accommodate a large number of people who depend vitally on forests for their livelihoods.

The policy conclusions coming out of this work have implications beyond the group of tropical oil countries. Other international capital transfers, like bilateral credits, aid or debt relief, can have similar impacts. These measures will thus also alleviate pressures on forests, provided they are not used to bolster specific forest-detrimental policies and practices. In addition, this provides some suggestions on what forest-friendly safeguards can realistically be introduced in the design of macroeconomic policies, considering the potential trade-offs between development and conservation objectives.

The World Bank, the IMF, other multilateral development banks and many bilateral donor agencies are now giving increasing attention to the assessment of the environmental impacts of programmatic lending and macroeconomic adjustment programmes. This is not easy to do because many of these impacts are both complex and indirect and there are frequently confounding effects from a variety of factors not associated with the adjustment lending operation itself. This book shows that some of the impacts of macroeconomic change may be positive while others are likely to be negative. It also demonstrates that, even though it is somewhat difficult to do, one can credibly trace through the impacts of these policies sufficiently to promote the positive changes and minimise the effects of any negative changes better.

The country cases also suggest that, in developing countries with a highly unstable macroeconomic framework, there are few easy solutions and that what might be good development policies may be bad for forests. In these cases, there will be many 'hard' trade-offs between promoting economic development and conserving forests. The future of forests in these circumstances will therefore depend on the conservation community's willingness to pay for forest conservation services on a no-regrets basis for the people and countries where these forests are located. As Sven Wunder suggests in his conclusions, there will be some cases where forest conservation will not be possible unless conservationists can negotiate direct payments for spatially specific forest conservation actions.

To date the level of resources available for direct conservation payments has been extremely limited and, with competing social and economic needs, this is likely to remain the case for the foreseeable future. Solid research is therefore crucially important to illuminate the linkages between macroeconomic change and the environment – a field where, unfortunately, speculative claims have tended to abound. This book is a landmark in adding scientific rigour to the debate, and anybody interested in the topic will benefit greatly from reading it. The country-comparative approach used in these studies has shown itself to be particularly useful in understanding the complexity of these interactions and why different outcomes may result from the same intervention in different countries. The Consultative Group on International Agricultural Research (CGIAR), which the author's institution (CIFOR) is part of, and which the World Bank has been supporting continuously, is especially well-positioned to undertake this type of work. It is an important contribution to global efforts to find improved ways of promoting economic development while safeguarding the local and global environmental and social values of forests.

Kristalina Georgieva
Director, Environment Department
The World Bank

Preface

This is a book by a brilliant Dane, German, sometimes Latin American, written in Indonesia, about the effect of the Dutch Disease on tropical forests. The Dutch Disease is an economic phenomenon that analysts first identified in the Netherlands after the discovery of natural gas in the North Sea. In this case the rapid increase in natural-gas exports brought a deluge of dollars into the country, which strengthened the Dutch currency – the guilder – and caused imports to become cheaper compared to things produced in the Netherlands itself. It also greatly increased the demand for construction and other services that could not be easily imported from abroad, and a lot of the country's resources shifted into these activities. As a result, traditional Dutch exports and economic activities that competed with imports for local markets suffered. The companies involved in these activities had to pay higher prices for their labour and services, faced greater competition from imports, and earned foreign currency that was worth less on the local market. For them it was a real 'disease'.

What in the world does any of this have to do with trees or forests, much less tropical forests? It turns out that a large proportion of the world's tropical forests is in countries that have been affected at one time or another by 'Dutch Disease' type situations. That is to say situations whereby mineral booms or rapid inflows of foreign exchange from foreign assistance, labour remittances or international loans push up the value of local exchange rates, which in turn discourages the production of non-'boom' exports and activities that potentially compete with imported goods. Translated into English, that means that when dollars start pouring into a country because of an oil boom, a government which attracts a lot of donor attention or sizeable loans from abroad, this is generally good for services activities and construction but bad for agriculture and logging.

Still lost? Just think about it for a minute. The Dutch Disease is generally bad for agriculture and logging, particularly if neither of them are the 'boom sectors'. But in most tropical countries these are precisely the sectors that cause most of the deforestation and forest degradation. So if the Dutch Disease is bad for agriculture and logging, it should be good for forests. All you need to do is to spread the disease and you can save the forests.

This is precisely the hypothesis that this book seeks to test. It does so by pulling together the relevant available information about five important tropical countries with large forests that have experienced major oil and mineral booms – Cameroon, Gabon, Ecuador, Papua New Guinea and Venezuela – and looking at how these commodity booms influenced deforestation. By comparing pre-boom, boom, and in some cases, post-boom periods, taking into account possible confounding factors, and looking briefly at three additional cases – Indonesia, Mexico and Nigeria – it manages to pull together a compelling argument about the relations between macro-economic processes and tropical deforestation.

This is very important because over the last few years there has been a lot of debate about how things like economic globalisation, foreign debt and structural adjustment policies affect the environment, but relatively little solid analytical work on the topic. This book goes a long way towards changing that. In so doing, the author has proved that it is, in fact, possible to analyse these impacts seriously, despite the methodological difficulties and often weak statistics. And if this economist could do it, certainly the major international financial institutions such as the World Bank, the International Monetary Fund and the Regional Development Banks should also be able to do it. This means that they can – and probably should – be doing more analytical work allowing them to develop macro-economic, agricultural and infrastructural policies that have the best possible impact on the environment, in particular on tropical forests.

In looking at the evidence, the study does indeed find that mineral and petroleum export booms tend to discourage tropical deforestation and forest degradation. Gabon is probably the clearest case. That country's oil wealth has allowed most people to move out of agriculture and to import the food they eat. As a result there has been very little net forest loss, although logging and hunting for wild meat have caused certain problems. Similar things have occurred, to one extent or another, in the other countries as well.

Unfortunately, though, like most things in life it's not that simple. Sometimes, when countries receive large influxes of money from petroleum or minerals they use some of it to finance activities that promote deforestation like building roads into forested areas or re-settling people there and providing cheap agricultural credit. Households may also use part of their growing incomes to buy agricultural or forest products that come from areas that were previously in 'undisturbed' natural forests. Thus, for example, when incomes went up in Ecuador and Venezuela people started buying more beef, so ranchers went out and converted more forest to pasture. The oil and mining companies may clear some forests themselves to carry out their activities, and the roads they build may open up new areas to farming and logging – although many of the previous claims in that regard have been exaggerated.

So where does that leave us? The book's most obvious message is that any large-scale influx of foreign exchange will tend to reduce forest loss, except in special cases where the money comes to be spent in ways that are particularly harmful to forests. To me at least, however, the book also implies something much more general and profound. To really understand the environmental destruction that threatens the world's natural resources, we need to look beyond just the immediate direct causes and analyse the broader economic and social processes that shape the ways our societies interact with nature. Forests reflect the societies around them. By using the forests as mirrors we can learn a lot about our societies. At the same time, if we really want to keep our forests healthy we may have to change some rather major aspects of our societies. This book provides a lot of good insights into where we might start.

David Kaimowitz
Director General, Center for International Forestry Research

Abbreviations

ABPNG	Agricultural Bank of Papua New Guinea
ACIAR	Australian Centre for International Agricultural Research
AD	Alianza Democrática (Social Democratic Party, Venezuela)
ANU	Australian National University, Canberra
ASEAN	Association of South-East Asian Nations
BAPPENAS	Badan Perencanaan Pembangunan Nasional (National Development Planning Agency, Indonesia)
BNF	Banco Nacional de Fomento (National Development Bank, Ecuador)
CARPE	Central African Regional Program for Environment
CDC	Cameroonian Development Corporation
CDC	Commonwealth Development Corporation
CDR	Centre for Development Research (Denmark)
CEMAC	Communauté Économique et Monétaire de l'Afrique Centrale (Central African Economic and Monetary Community)
CEPAL	Comisión Económica para America Latina y el Caribe (Economic Commission for Latin America and the Caribbean, ECLAC)
CFA	Communauté Financière d'Afrique
CGE	Computable General Equilibrium
CHB	Comptes Hors Budget (Cameroon)
CIFOR	Center for International Forestry Research (Indonesia)
CLIRSEN	Centro de Levantamientos Integrados de Recursos Naturales por Sensores Remotos (Ecuadorian Centre for the Integrated Survey of Natural Resources through Remote Sensing)
CODESUR	Conquista del Sur (Conquest of the South Plan, Venezuela)
CONASUPO	Compañía Nacional de Subsistencias Populares (The National Food Marketing Agency, Mexico)
COPEI	Partido Social Cristiano (Christian Democratic Party, Venezuela)
CPI	Consumer Price Index
CREA	Centro de Reconversión Económica de Azuay, Cañar y Morona Santiago (Centre for the Economic Recovery of Azuay, Cañar and Morona Santiago, Ecuador)
CSIRO	Commonwealth Scientific and Industrial Research Organisation (Australia)

DAL	Department of Agriculture and Livestock (PNG)
Danida	Danish International Development Agency
dbh	diameter above breast height
DEC	Department of Environment and Conservation (PNG)
DGE	Direction Générale de l'Économie (General Department of Economics, Gabon)
DGSEE	Direction Générale de la Statistique et des Études Économique (General Department of Statistics and Economic Studies, Gabon)
ECOFAC	Conservation et utilisation rationelle des ecosystèmes forestiers d'Afrique Centrale (Conservation and rational use of Central African forest ecosystems)
EKC	Environmental Kuznets Curve
ELCOM	Electricity Commission (PNG)
ESA	European Space Agency
EU	European Union
FAO	United Nations Food and Agricultural Organisation
FCFA	Franc de la Communauté Financière d'Afrique
FIM	Forest Inventory Mapping system (PNG)
FIV	Fondo de Inversiones de Venezuela (Venezuelan Investment Fund)
Fob	Free on Board
FORIS	Forest Resources Information System (FAO)
FRA	Forest Resources Assessment (FAO)
GDP	Gross Domestic Product
GFW	Global Forest Watch
GNP	Gross National Product
ha	Hectare(s)
HFZ	Humid Forest Zone (Cameroon)
HIPC	Heavily Indebted Poor Countries
HYVs	High-yielding varieties
IERAC	Instituto Ecuatoriano de Reforma Agraria y Colonización (Ecuadorian Institute for Agrarian Reform and Colonisation)
IFAD	International Fund for Agricultural Development
IGAD	Institut Gabonais d'Appui au Développement (Gabonese Institute to Support Development)
IGBP	International Geosphere Biosphere Program
IIED	International Institute for Environment and Development
IITA	International Institute for Tropical Agriculture
IMF	International Monetary Fund
INC	Instituto Nacional de Colonización (National Institute for Colonisation, Ecuador)
INCRAE	Instituto Nacional de Colonización de la Region Amazónica Ecuatoriana
INDA	Instituto Nacional de Desarrollo Agrario (National Institute for Agrarian Development, Ecuador)
INEC	Instituto Nacional de Estadísticas y Censos (National Institute of Statistics and Census, Ecuador)

INEFAN	Instituto Ecuatoriano Forestal y de Areas Naturales y Vida Silvestre (Ecuadorian Institute for Forestry, Natural Areas and Wildlife)
INPARQUES	Instituto Nacional de Parques (Conservation Agency, Venezuela)
IRF	International Road Federation
ITIC	Technical Institute for Immigration and Colonisation (Venezuela)
ITTO	International Tropical Timber Organisation
IUCN	The World Conservation Union
JATAM	Jaringan Advokasi Tambang (Mining Advocacy Network, Indonesia)
JRC	Joint Research Centre (European Union)
km	Kilometre(s)
LNG	Liquified Natural Gas
MAB	Ministry of Agriculture and Breeding (Venezuela)
MAG	Ministry of Agriculture (various countries)
MARN	Ministerio del Ambiente y de los Recursos Naturales (Ministry of the Environment and Natural Resources, Venezuela)
MARNR	Ministerio del Ambiente y de los Recursos Naturales Renovables (Ministry of Environment and Renewable Natural Resources, Venezuela)
masl	Metres above sea level
MCR	Multi-Country Regression
MFN	Most Favoured Nation
MINEF	Ministry of Environment and Forests (Cameroon)
MoE	Ministry of Environment (various countries)
MoFEC	Ministry of Forestry and Estate Crops (Indonesia)
MoWF	Ministry of Water and Forests (Gabon)
MRSF	Mineral Resources Stabilisation Fund (PNG)
MSS	Multispectral Scanner
NASA	National Aeronautics and Space Administration (USA)
NFA	National Forest Authority (PNG)
NFI	National Forest Inventory
NGO	Non-Governmental Organisation
NNOC	Nigerian National Oil Corporation
NOAA-AVHRR	National Oceanic and Atmospheric Administration's Advanced Very High Resolution Radiometer
NT	Non-Traded
NTFPs	Non Timber Forest Products
OAU	Organization of African Unity
OCISCA	L'Observatoire du Changement et de l'Innovation Sociale au Cameroun (The Observatory of Change and Social Innovation in Cameroon)
OCTRA	L'Office du Chemin de Fer Transgabonais (Gabonese Railway Company)
ONADEF	L'Office National de Développement des Forêts (National Forestry Development Agency, Cameroon)
ONCC	Office National du Café et du Cacao (Cameroon)

ONCPB	Office National de Commercialisation des Produits de Base (Cameroon)
OPEC	Organisation of Petroleum-Exporting Countries
ORSTOM	Office de la Recherche Scientifique et Technique Outre-Mer (France)
OZI	Integrated Operations Zones (Gabon)
PNG	Papua New Guinea
PNGRIS	Papua New Guinea Resource Information System
POSF	Public Officer's Superannuation Fund (PNG)
PREDESSUR	Programa Ecuatoriano de Desarrollo de la Región del Sur (Ecuadcrian Programme for the Development of the South)
PRODESSUR	Sustainable Development Project of the South (Venezuela)
PTC	Post and Telecommunication Corporation (PNG)
RER	Real Exchange Rate
RP	Relative Prices
RWE	Round Wood Equivalents
SAP	Structural Adjustment Programme
SAR	Synthetic Aperture Radar
SEFORVEN	Servicio Forestal de Venezuela (Venezuelan Forest Service)
SNBG	Société Nationale des Bois du Gabon (Gabon's National Woods Company)
SODECAO	Societé de Développement du Cacao (Gabon's Cocoa Development Company)
SOFO	State of the World's Forest (FAO)
SOSUHO	Société sucrière du Haut-Ogooué (Haut-Ogooué's Sugar Company, Gabon)
SUFOREN	Subsecreteriat of Renewable Natural Resources (Ecuador)
T	Traded
t	tons
TM	Thematic Mapper
TREES	Tropical Ecosystem Environment Observations by Satellite
UDEAC	Central African Economic and Custom Union
UNDP	United Nations Development Program
UNEP	United Nations Environment Program
UNESCO	United Nations Educational, Scientific and Cultural Organization
USA	United States of America
USAID	United States Agency for International Development
US$	United States Dollar
WCMC	World Conservation Monitoring Centre
WRI	World Resources Institute
WTO	World Trade Organization
WWF	World Wide Fund for Nature

Introduction

The issue

> Petroleum operations in forest areas inevitably result in both direct and indirect deforestation.
>
> (RAN and Project Underground 1998: 30)

> Badly planned and managed mining operations are posing threats to forests all over the world.
>
> (Finger 1998: 2)

> Oil companies have now leased virtually all remaining tropical forest areas for exploration and/or production, and drilling has caused damage in the Amazon, West Africa, Papua New Guinea, etc.
>
> (WWF and IUCN 1996: 19)

What all these environmentalist observations have in common is that they are focusing narrowly on the site-specific impacts of oil and mineral production on forests. Concerns about the effects of oil and mining operations on forests and the people who inhabit them have been a major focus of global campaigns to save the rainforests. Conservationist efforts to change the behaviour of oil companies and governments have included guidelines on how to minimise oil's environmental impacts in ecologically sensitive production environments (IUCN 1993) and on what sites should not be explored at all because of the risks involved (WWF 2002). But such an approach remains incomplete, because on-site impacts are only one way in which such activity affects forests. Economy-wide impacts related to oil and mineral wealth can – and often do – have dramatic effects on forests. Yet these factors have generally been overlooked in a debate which has focused only on what is apparent to the eye.

How do rising oil prices affect the forest cover of an oil-producing country? Are tropical oil countries more or less likely than non-oil countries to experience high forest loss? Are countries that are growing richer from oil exports clearing more or less forest than before? Are these countries likely to increase deforestation when they are hit with a combination of poor terms of trade, falling export revenues and a heavy debt burden? Overall, in an increasingly globalised environment, how do commodity–price fluctuations ultimately translate into land-use changes?

This book will explore these and other questions in relation to deforestation and forest degradation in tropical developing countries. Its aim is to fill a gap in our understanding of how the fate of tropical forests is linked with foreign trade and domestic macroeconomic policies. Most explanations of deforestation are concerned with direct sources (as in the quotations above), but ignore the powerful indirect effects that emerge in the interaction between international trade, the national economy and the land use resulting from a shifting sectoral structure of production. The direction and magnitude of these *derived* effects are seldom acknowledged. The oil–forest link is probably the strongest of these underlying effects. As will be shown, it is an example where financial transfers originating from international trade can actually help to *slow down* forest conversion and degradation. This is not the scenario frequently portrayed where trade causes new pressures on forest resources, which would otherwise not have materialised. It is rather that a particular kind of trade, the transfer of petroleum rents, limits the pressures that would have occurred in the absence of trade. Any regulatory regime that hopes to be effective in curbing deforestation will have to acknowledge and anticipate this type of influence on forests.

The book focuses on oil countries for two reasons. First, taken together, the two dozen tropical countries that specialise in oil exports have a remarkably high proportion of the world's rain forests. If we exclude Brazil – in many respects a special case – the group of specialised oil and mineral exporters accounts for more than half of the remaining global tropical forests (see Chapter 2). They are therefore certainly worthy of attention in their own right. Second, the economies of these countries often fluctuate dramatically from boom to bust because of their heavy reliance on revenue from a single commodity. Oil countries have also chosen quite different policy responses to boom-and-bust cycles – both over time and across countries. Their 'macroeconomic laboratory' thus offers a good opportunity to identify and study links between external economic changes and the forest. In other words, the selected countries are showcases for the analysis of macroeconomic impacts on the environment.

Country sample

Over the last decade, there has been a growing general interest in the relationship between macroeconomics and the environment, including the links to deforestation. In general, theoretical economic models have been developed, some of which have been applied to single countries. Multi-country statistical models have also been tested at the aggregate, global level (see Chapter 2 for a review). However, none of the studies has combined a country-comparative approach with a detailed case-study analysis of national land-use trends over time. This is the approach that will be taken in this book, and hopefully it will shed new light on a controversial relationship.

Empirically, I will look in detail at five oil-producing countries: Cameroon, Gabon, Ecuador, Venezuela and Papua New Guinea (PNG). Three other countries will be screened more briefly: Indonesia, Mexico and Nigeria. How were the study countries selected? Beyond the aim of a broad geographical representation, the choice of primary countries is intended to cover a variety of basic situations. For instance, PNG is a relatively new oil exporter; Ecuador, Cameroon and Gabon have had major oil exports over the last two to three decades, while Venezuela has been a specialised oil exporter since the 1920s. Two countries are extremely forest-rich (Gabon, PNG), while the other three are in the intermediate range. The majority of the study countries have been registered in various versions of the United Nations Food and Agricultural Organisation's (FAO) Forest Resources Assessment (FRA) as facing low deforestation (PNG, Gabon, Cameroon, Venezuela), but one has faced high forest loss (Ecuador).

All five primary countries are small or medium-sized, which has the advantage that regional heterogeneity is more limited than for the large countries. It thus becomes easier to implement the partial-comparative method that was chosen in this book. However, this also has the drawback that some of the largest tropical mineral exporters, some of which have extensive forest cover, are missing from the sample of primary countries. This is remedied in Chapter 9, where the main outline for one large country from each of the major tropical continents is added: Mexico, Nigeria and Indonesia. These three summary studies will be referred to as 'secondary cases' throughout the book.

Many of the lessons derived from the extreme case of oil-exporting countries will also be applicable to other forested primary commodity exporters, such as coffee- and cocoa-producing countries (e.g. Colombia, Ivory Coast, Kenya), non-oil mineral exporters (e.g. Congo, Central African Republic, Bolivia) or countries with fluctuating foreign-aid inflows (e.g. Ghana, Pakistan). They will also have relevance for global action linking macroeconomics to the environment, such as the likely effect of debt-relief initiatives on forests, or the pros and cons of explicitly building strategic environmental concerns into structural adjustment or poverty-reduction strategy programmes (see Chapter 10).

Theories

The present book brings together two rather different theoretical approaches. On the one hand, we have deforestation theories, which have been developing rapidly over the past decade. Some of these theories draw heavily on the social sciences to explain land-use changes. But naturally, forestry itself and other biophysical sciences claim an important role with regard to deforestation. In general, one can distinguish three different 'schools of thought' on deforestation: the immiseration school, the neoclassical school and the political-ecology school. They distinguish themselves in terms of what they see as the main sources, actors and causes of forest loss; consequently, they also have different relevance in different parts of the world and develop different policy recommendations (Wunder 2000: ch. 2). The macroeconomic causes of forest loss and degradation are discussed in Chapter 2.

The other theoretical approach in this book is macroeconomic, and has come to be known as 'Dutch Disease', a reference to the negative side effects of the natural-gas export boom that the Netherlands experienced in the 1960s and 1970s. The Dutch started to overspend their gas export revenues, thus losing competitiveness, as a result of which Dutch industry lost market share and faced decline. A whole branch of literature on this topic emerged, with both theoretical contributions and case studies (see Chapter 1). When the boom–bust theory of adjustment to trade-induced changes was applied to developing countries like Colombia, Algeria, Nigeria and Indonesia, it became clear that it was agriculture, not industry that suffered from declining competitiveness in these countries. Consequently, the Dutch Disease also had significant land-use impacts, which became the point of entry to this book.

However, 'marrying' the Dutch Disease to deforestation is a formidable conceptual challenge, which is also true of the attempt to bring two different audiences together.

Indeed, when trying to explain the principles of the Dutch Disease to foresters, they tend to change the subject rapidly, and switch opportunistically to talking about 'Dutch *Elm* Disease' instead! Nevertheless, this book makes the basic point that foresters and other forest stakeholders should probably hear out the tale of the *economic* Dutch Disease: it has important things to say about land-use change and the fate of tropical forests.

The hypotheses

One basic, country-comparative hypothesis of the book is that forests in oil-exporting tropical countries tend to face fewer pressures for conversion and degradation than forests in non-oil-producing countries. A related, temporal hypothesis is that these oil countries face lower forest-loss and forest-degradation pressures when there are oil booms than in periods when oil revenues are low and the economy is going through a bust period (see Chapter 3 for a full discussion). Throughout the book, the latter claim will be the central supposition to be tested, and will be called the 'core hypothesis'.

What are the basic mechanisms that are supposed to produce the core-hypothesis outcome? Sizeable foreign exchange earnings from oil raise domestic demand, fuel inflation and cause the exchange rate to appreciate. The oil country's other (non-oil) commodities become more expensive. This loss of competitiveness is a problem for those sectors that compete with foreign substitutes, since they lose market share, profits and growth opportunities. To the extent that (parts of) the agricultural and timber sectors are exposed to foreign competition, an oil boom generally discourages the expansion of agriculture and logging. Moreover, on average, oil countries tend to have a higher proportion of their populations in urban areas, where most of the oil money tends to be spent. All these factors ultimately favour the conservation of forest cover.

The two major oil booms of the 1970s provide a special opportunity to test the core hypothesis. What was the size of foreign-exchange inflows from oil and mining? What was the domestic policy response to shifting external trends? How much redistribution in domestic production (industry, services, agriculture, timber extraction) was triggered by the combination of trade cycles and economic policy response? And were deforestation and forest degradation actually lower during the oil boom than during the pre- and post-boom periods? The type of answer will differ from one country to another. For instance, some countries (Cameroon, Venezuela, Ecuador) had already developed sizeable cash-crop sectors before oil and minerals appeared on the national scene, while others (Gabon, PNG) had not. This has influenced not only the macroeconomic picture, but also the coalitions of interests that shaped domestic policy responses and corresponding land-use trends.

For each country, we will look at a standard set of key variables with potential impact on land use and forests. These factors include: exchange rate movements; public spending on agriculture, road-building, forest regulation and directed settlements; trade policy; urbanisation; poverty; and shifts in the structure of demand. The book illustrates how these factors have varying and contradictory effects on forests. In addition, it will also look at the on-site effects that may occur when oil or mineral deposits are located in forested areas – both in Chapter 2 and in the respective country chapters. However, it should be remembered that these direct physical impacts of oil extraction are not the principal topic of this work.

In exceptional cases, the core hypothesis does not apply because governments spend their oil revenues in a way that is particularly harmful to forests. This underscores the vital importance of in-country policy responses in determining environmental outcomes. In the package of national policies accompanying oil wealth, some of the most forest-damaging instruments identified in this book are spatially explicit deforestation drivers, such as road building, frontier settlement programmes and 'perverse' land-tenure incentives. But there are also general, non-spatial policies in this package, such as repeated currency devaluation, transport subsidies, as well as earmarked credits and protectionist trade policies favouring the most land-extensive sectors, such as cattle-ranching and certain food crops.

Forest data and definitions

In terms of assessing 'the fate of the forest', that is, changes in forest cover and condition, the primary concern of this book is with deforestation, basically defined as the radical removal of tree-crown cover typically followed by land-use change. The second priority is to look at forest degradation, which refers to all other types of change significantly affecting forest structure, for example, logging, hunting and other forest-extraction activities (see Chapter 3 for definitions and data).

Two decades after deforestation emerged on the international agenda, hundreds of models have been constructed to explain it; yet paradoxically the underlying data used to support them remain highly questionable. The documentation of deforestation data over time proved to be extremely unreliable for all the eight study countries. It thus became a subsidiary goal to gather and assess deforestation and land-use data critically, and to come up with alternative best-guess estimates of current deforestation, or at least a likely quantitative range. In some cases, this exercise also changed the initial 'high-low' classification, based on data from the UN's FAO. For instance, the initial plan to include Indonesia among the primary countries had to be abandoned because of the highly complex and contradictory situation regarding existing land-use data.

Other concepts and definitions

The main focus of the book is on the impacts of oil, but some attention will also be paid to rent-producing resources of similar nature. This is because, over certain time periods, they had impacts in the study countries that were quite comparable to the impact of oil. But this makes it particularly necessary to maintain a consistent terminology. In this book, the term 'petroleum' will be used for oil and natural gas, while 'fossil fuels' refer to oil, gas and coal. By 'metals' is meant hard-rock minerals (copper, iron, gold, etc.) that are extracted by the production process generally understood as 'mining'. The terms 'minerals' and 'mineral wealth/rents' jointly refer more generally to both petroleum and metals. Among the oil countries, sometimes reference will be made to 'high-absorbing oil exporters'. This term represents those oil countries that in the long run have been capital-importing. It distinguishes them from the 'low-absorbing' group of population-scarce, high-rent countries that have saved much of their oil revenues abroad, typically some of the Gulf countries like Saudi Arabia or Kuwait, which have a rather different sectoral production structure.

The main theme to be explored in this work is the land-use impact of economic rents. For natural-resource exporters in general and for oil countries in particular, there has been a growing literature on the political economy of rent-generation and rent-seeking. Not only do underground (oil and other mineral) resources generate large economic rents; the same may apply to the timber sector. Illegal logging is an emerging field of research dealing with the appropriation of forestry rents, but this theme is also linked more generally to an enhanced interest in issues of governance and corruption. There are thus strong parallels in the power relations and institutional features that characterise the distribution of income from different high-value natural resources. These political economy factors will be introduced in Chapter 1, and will be evaluated in the concluding chapter.

Obviously, the comparative factor intensity of production in different sectors of the economy is vital for determining the aggregate impact on forests. Production technologies with a high capital content per unit of output will be called 'capital-intensive'. Correspondingly, technologies with a high labour–output ratio will be labelled 'labour-intensive'. On the other hand, those with a high content of land per output unit will be called 'land-*extensive*'. This may seem somewhat illogical, but follows the terminology that has become standard in the literature.

One of the most land-extensive activities with large land-use impacts is the continuous conversion of forests to pastureland. I will generally use the term 'agriculture' as referring to both crop cultivation and the livestock sector. I will use the term 'cattle-ranching' in a broad sense that includes not only the production of meat, offspring or hides, but also dairy farming. This differs from a more narrow definition, often used in the US, where dairying is not seen as an integral part of ranching.

Scope and audience

As mentioned, it is hoped that this work will be valuable to two different groups of readers. The first includes foresters, geographers, scientists and environmentalists interested in the causes of changing agricultural land use, deforestation and forest degradation. The second consists of economists, development planners and social scientists interested in the environmental side effects of macroeconomic fluctuations and policies. Apart from the academic community, the findings are also aimed at influencing the perceptions of development experts, donor agencies, non-governmental organisations (NGOs) and policy-makers. The book should have special appeal in the target countries that have been chosen here as case studies, but through its comparative approach it is also intended to reach a broader public. For this reason, it has generally been written in non-technical style, in the hope of making it accessible to this wider audience.

The highest aggregation level of study in this book is the nation state. This also means that the global impacts of increased oil production on forests are not analysed here. One of the environmental arguments against oil exploration and production is that it helps increase energy supplies, keeping down prices and thus increasing consumption, which will contribute further to the greenhouse gas effect that damages the environment. The current state of knowledge indicates that the specific effects of global warming on forests will vary across regions and are associated with a high level of uncertainty. Temperature and fertiliser effects will probably enhance tree growth in some regions, while in others weather impacts, rising sea levels and increased environmental instability will prove detrimental to forests, their biodiversity and their ecological benefits (IPCC 2001). But to the extent that oil and gas replace the burning of coal, greenhouse gas emissions are reduced because oil and gas have a lower carbon content per energy unit. Obviously, this is too complex an issue to integrate into the analytical framework of the present book.

Another delimitation is necessary in regard to the time perspective. Those looking for the hottest prediction of how forests are affected by the current economic crisis in country X or the new structural adjustment programme in country Y may be disappointed. The main contribution of this work is the identification of long-term structural trends over three to four decades, showing how cyclical factors (e.g. oil-price cycles) have changed these trends along the way. There is thus an important element of economic and land-use history in the present study. The interest is in 'the big picture', rather than in the impact of more isolated recent events. This does not mean that the book has nothing to say about the present. On the contrary, the long-term view should make it possible to make, if not a formal model extrapolation, then at least a consolidated prediction. This may relate to how the extremely volatile world-market oil price is likely to affect land use in the study countries or, say, what effect the recent dollarisation in Ecuador is likely to have on the country's forests (see Chapter 10).

Structure

The book consists of ten chapters. The first presents an overview of the economic theory of oil wealth and the Dutch Disease – its origins, core model, theoretical extensions and application to both developed and developing countries. Chapter 2 links the Dutch Disease to land use and forests, formulates the key hypotheses and explains the basic comparative methodology used for the country case studies. Chapter 3 explains the basic definitions and methods used for the key forest output variables, deforestation and forest degradation. This concludes the introductory part of the book.

The next five chapters look closely at the situations in each of the five primary case-countries, using the comparative methodology developed in Chapter 2. Historical overviews of shifts over time in oil wealth, policies, land use and forest condition will be presented. The order of presentation should aid the reader's understanding of the main mechanisms by starting with the case that most clearly supports the key hypothesis of this book ('oil wealth protects forests'), namely Gabon (Chapter 4). We then move gradually to less conforming cases, introducing additional factors that complicate the land-use picture. Hence, PNG, the last case study (Chapter 8), is the country where the link between macroeconomics and forests is most complicated. Chapter 9 then broadens the perspective to the three secondary, large-country cases: Mexico, Nigeria and Indonesia.

Finally, in Chapter 10 both primary and secondary cases are compared, the key hypotheses from Chapter 2 are revisited, and the different forest outcomes are discussed in terms of country-specific preconditions versus variable domestic policy responses. To anticipate, this comparison will lead to the conclusion that our sample of oil countries had a relatively large room for manoeuvre in domestic policy-making, and that certain domestic policies had a catalytic impact on how much forest protection oil wealth would provide. The book closes with both country-specific and more general policy recommendations.

On the national level, these recommendations identify a few areas where better policies could benefit both economic development and the environment. But opportunities along the trade-off curve between environmental and developmental goals are also discussed, where much could be gained in one domain while sacrificing little in the other. For bilateral donors and multilateral lending agencies, the book should provide a more realistic picture of the likely land use and forest impacts of initiatives designed to revive economic growth and diversify production, such as debt relief and structural adjustment programmes. Drawing attention to the true character of these linkages is the *sine qua non* for developing policies that may succeed in rationalising tropical land use and curbing excessive forest loss.

Sources and methods

Various methods were used to obtain and analyse the data used in this book. Most of the data were gathered during visits to each of the countries, using pre-established key contacts. In some cases, vital information was received in developed-country libraries and universities, such as the Australian National University (ANU) in Canberra for material on PNG or the Walbauinstitut in Freiburg (Germany) for data on Venezuela. In PNG, a counterpart was identified to help with the data collection and the contacts to interview partners. In addition, a three-month consultancy carried out for this project provided a summary report about the contradictory Indonesian deforestation and land-use estimates (Muhamed 2000).

Many interviews were carried out by the author during visits to the five primary study countries (see the next section). Secondary data and statistics were also used. A broad range of information was screened, from deforestation reports and land-use surveys to poverty assessments, consumption surveys, road statistics, forestry and agricultural budgets. Sources included published books, CDs, journal and newspaper articles, maps, international and national statistical volumes, unpublished government reports and ministerial statistics, university theses, donor reports and material from multilateral banks. In addition, a minimum of one but on average 2–3 field sites with ongoing deforestation or forest degradation were visited in each country, in order to have at least some feel for changing land use on the ground.

Processing this wealth of comprehensive but often scattered data included first a systematisation phase. In some cases, long-term time series for key macroeconomic and land-use data were not available, and therefore had to be constructed first. Data analysis included descriptive and graphical methods, but standard quantitative tools such as correlation and regression analysis have also been employed in country Chapters 4–8. For the three secondary cases in Chapter 9, the analysis was limited to descriptive methods.

In assessing the contradictory deforestation and land-use data in the five country chapters, I used the Convergence of Evidence method (see Chapter 3, Box 3.4 for a detailed description). However, it was only after the fourth completed case study that I learned that this was what I was actually doing! This is because 'Convergence of Evidence' is something of a fancy label for the assessment of multiple and contradictory sources, using checks and balances, basic mathematics and, above all, a good deal of common sense, by a single expert – in the case of this book, its humble author. The same method was used by the FAO's FRA group for their country reports underlying the FRA 2000. In the following chapters, I shall explain why, in spite of this methodological coincidence, my deforestation estimates differ from the FAO's for most of the five countries.

History of the project

In March 1997, I visited the Center for International Forestry Research (CIFOR) for the first time, as part of a four-month travel to Southeast Asia. To my surprise, I found that two staff scientists had a marked interest in my old PhD topic, the Dutch Disease, which they planned to link to another topic of my interest, namely deforestation. These scientists were William D. Sunderlin, a rural sociologist and Senior Scientist at CIFOR, and David Kaimowitz, an agricultural economist, at that time Principal Economist, and now Director-General of CIFOR. Both were working in CIFOR's research programme on Underlying Causes of Deforestation. I was eager to establish closer collaboration with them, and found that the trendy term 'a research centre without walls' actually had some tangible meaning to it. A professional partnership developed, which was open-minded, intellectually stimulating, productive and always fair. Over the years, David and William became not only great partners, but also good friends.

Sunderlin and Kaimowitz thus developed the first methodological framework for the project, which I then tested empirically for the pilot case study of Ecuador, the country I knew best from previous work and residence. This occurred while I was affiliated to the Centre for Development Research (CDR) in Copenhagen, Denmark, and resulted in a first case-study report, published as a CDR Working Paper (Wunder 1997). Simultaneously, William Sunderlin took the lead in work that attempted to demonstrate the effect of high oil and mineral exports in curbing deforestation in a pan-tropical cross-country sample. The results from that study, finally published as Sunderlin and Wunder (2000), encouraged us to believe that we had a good case. On average, the occurrence of Dutch Disease seemed to go hand in hand with the greater preservation of forest cover and lower deforestation. But, together with some surprising results from the Ecuador pilot study, it also revealed serious country-specific differences, which could not be accounted for by cross-country regression models. Obviously, a more detailed comparison of selected country cases was called for in order to understand fully the mechanisms at work.

The present book is the result of these comparative case studies. As a consultant to CIFOR, I initially worked on the Venezuela case study in 1998, whereas my two colleagues were to study the remaining four cases. In February 2000, I joined CIFOR as a staff member, based in the headquarters in Bogor, Indonesia. For different reasons, Kaimowitz and Sunderlin both had to withdraw from carrying out the remaining case studies. However, as the 'intellectual fathers' of the project, they continued to provide invaluable inputs. Indeed, this book would not have been possible without them. In particular, I was able to benefit from pre-collected literature, pre-established in-country contacts and their extensive comments on draft chapters. I also profited greatly from the results of previous CIFOR deforestation research in one of the study countries, Cameroon. In spite of all the support, the situation left me with what up to now has proved to be the greatest challenge of my professional career, one that is not to be recommended to people of schizophrenic tendencies, namely to carry out a simultaneous study of five primary and three secondary country cases.

What is my own background for coping with such an ambitious project? I was basically trained as an economist, with specialisation in macroeconomics, and, as already mentioned, my PhD thesis from 1992 was precisely on the Dutch Disease as applied to exports of coffee and drugs from Colombia, and their impact on the national economy (Wunder 1991). My interests in the environment and in forests began during work in Copenhagen for Danida, the development agency under the Danish Ministry of Foreign Affairs. But it was really promoted during 1993–6, when I was affiliated to the Forest Conservation Programme of the World Conservation Union (IUCN), based in the South American Office in Quito, Ecuador. Much of this work was related to deforestation in Ecuador and resulted in the publication of two books, respectively on wood products and forest loss in the highlands (Wunder 1996b) and on deforestation on the national scale (Wunder 2000). Obviously, insights from that period were crucial for the Ecuador chapter of the present book.

What personal presumptions and biases do I carry in my intellectual baggage that may have influenced my analysis? I probably need to confess to the reader at least five preconceptions. First, my developing-country experience is primarily of Latin America, so the contexts of Africa and the Asia-Pacific region are more novel to me, although my work at CIFOR over the last two years has been pan-tropical in scope. Second, I am looking at the issue of land-use change and deforestation through the eyes of an economist – but hopefully not an economist in blinkers. Third, although I have lived and worked in various developing countries, my approach probably remains 'Northern', at least to the extent that I see the global environmental worth of forests as vital, and believe that their conservation must somehow be weighed against and reconciled with local development needs. Fourth, I think that the 'win–win' options for reconciling these two concerns are actually much more limited than is commonly realised. Unfortunately, natural tropical forests tend to have little comparative advantage for poverty alleviation and commodity production (Wunder 2001b). Fifth, my previous work has led me to believe firmly in the strength of extra-sectoral impacts: what happens to tropical forests in terms of deforestation and forest degradation tends to be more a result of factors from *outside* than from inside the forest. The last two preconceptions have been further reinforced in producing this book.

Acknowledgements

At the institutional level, the research for this book was carried out by CIFOR in Bogor, Indonesia. Most of the funding by far came from CIFOR's institutional, core budget. In addition, I am grateful to CDR in Denmark for financing part of the Ecuador work, and the Australian Centre for International Agricultural Research (ACIAR) for financing part of the Venezuela study.

To write a book of this type leaves the author indebted in terms of countless personal favours and help provided by a large number of individuals. This starts with those who provided direct assistance. Library staff at both CIFOR and CDR supported the extensive compilation of literature. Ambar Liano supplied extensive secretarial help. Ahmad Dermawan provided valuable research assistance during the two-year-work in Bogor, while Breno Pietracci assisted in the earlier preparation of the Venezuela and Ecuador drafts. Atie Puntodewo developed the cartographic material, and Widya Prajanthi helped find and enhance the photographs of oil production in forests that were kindly made available by

Caltex-Indonesia. The Tropical Ecosystem Environment Observations by Satellite (TREES) project allowed me to use their maps for publication; the same is true of two Gabon maps from Professor Roland Pourtier of the University of Paris. Thomas K. Rudel (Rutgers University) shared with me his large database of regional deforestation studies in the eight countries that had my interest. Robert Davis, at the time leading FAO's FRA group, gave me access to several unpublished papers prepared for the FRA 2000.

Many people have commented on the draft versions of this book. I am indebted to David Kaimowitz, Robert Parkin, Thomas K. Rudel, William Sunderlin and Stuart White for reading and commenting extensively on the entire book draft. In addition, others have commented on single chapters according to their specific expertise. On the Gabon draft, Jean-Christophe Carret, Robert Nasi and Chris Wilks provided comments. For the Venezuela chapter, I received commentaries from Marta Miranda and Alex Mansutti. For the Cameroon chapter, feedback came from Henriette Bikié, Eric Forni, Benoit Mertens, Robert Nasi and Ousseynou Ndoye. The Ecuador chapter (including the earlier CDR Working Paper) was commented by Douglas Southgate, Marie Bille and the late Phil Raikes. The PNG chapter profited from written remarks made by Michael Bourke, Colin Filer, John McAlpine, Trevor Modowa Gumoi and Luca Tacconi. The introductory chapters (Chapters 1–3) benefited from annotations by Arild Angelsen, Benoit Mertens, Emily Matthews, Robert Nasi and Rona Dennis. Finally, Laura Snook, Marusia Musacchio, Joachim Fiebach and Rona Dennis shared with me their written observations on different sections of Chapter 9.

In addition to these written comments, a large number of people have provided oral information through interviews, and otherwise helped with the in-country data-gathering.

In Gabon, I am grateful to Robert Solem and Nicodeme Tchamou, coordinators of the Central African Regional Program for Environment (CARPE) project, and Robert Nasi, CIFOR, for extensive help in setting up interviews in Libreville during my two visits, in May 2000 and June 2002. I interviewed Clair Mborou, Prosper Obame Ondo, Ousmane Sissoko, Alain Karsenty, Chris Wilks, Stéphane Lombardo, Jean-Philippe Jorez, Filippo Saracco, Raphaél Vinchent, Alfred Ngoye, Sylvain Meye M'eya, Bernard-Henry Voubou, Paul-Henry Nguéma Meye, David Young, Patrice Christy, Gérard Dufoulon, John Bickerton, S. Ziza, Sidi Touré, Rose Ondo, Norbert Gami and Modeste Mfa Obiang. During field visits to Oyem and La Lopé, I talked to Pauwel de Wachter, Victor Ebiang-Ebang, Michel Assoumou Mengué, Mme E. Ekuma, Evane Ndong and Louis Sosthère Ndong-Obiang.

In Venezuela, a special thanks to Elery Cabrera and Samuel Moncada for their help in organising my visit to Caracas. Many other people, in both Caracas and Mérida, were interviewed: Otto Huber, Orlando Ochoa, Raul Huyizzy, Clemencia Rodner, Alfredo Paolilla, Mario Gabaldón, Pedro Delfín, José Betancourt, José Rojas, Euro Segovia, Delfina Rodríguez, Maria Helena Sperandío, Rolando Zamora, Benito Kizer, Armando Torres Lezama, Julio César Centeno, José Miguel Sánchez, Miguel Cabeza, Andrés Peñate, Luis Pedro España, Matías Riutort and Carlos Domingo. At the Venezuela group of the Graduiertenkolleg at the Waldbauinstitut, Freiburg University (Germany), I visited their library and had interviews with researchers Barbara Müller, Martin Lux and Christoph Aicher.

For the Cameroon chapter, Ousseynou Ndoye and his assistants in Yaoundé provided good contacts, information and logistics. Interviews were carried out in Yaoundé with

Frédéric Roger Medjo, Jean-Claude Nguinguiri, Stefan Hauser, Jim Gockowski, Nkamleu Guy Blaise, Henriette-Elise Bikié, Nicodeme Tchamou, David Tchuinou and Urmiah Lynch, as well as with Robert Nasi, William Sunderlin, Chimere Diaw and Benoit Mertens (all Bogor, Indonesia). I also benefited from helpful staff in the libraries of IITA, OCISCA, ORSTOM, the World Bank, the Department of Statistics and National Accounts, the United Nations Regional Information Center and CARPE's Cameroon country office. I am obliged to Mr Ebolo Siriak (IITA) for his patient explanations during a field trip to Ebolowa (South province).

In Ecuador, I received information from a large number of people during my 1993–6 work for IUCN in Quito; space does not allow me to mention them here. During two short-term visits in December 1998–January 1999 and March 2001, I talked to Abel Tobar, Xavier Izko, Remy Ojara, Vilma Salgado, Raul Gaethe, Octavio Recalde, Rodrigo Sierra, Cesar Ajamil, Joseph Vogel, Carlos Larrea, Stuart White, Jorge Meza and Hans Thiel.

For the PNG chapter, Trevor Modowa Gumoi was my local counterpart at the University of PNG in Port Moresby, helping with data collection and in arranging many contacts. Interview partners in Moresby were Goodwill Amos, Paul Barker, Andrew Bond, Dodina Duba, Wep Kanawi, Kilyali Kalit, Ripa Karo, Regina Kiele-Sapak, Joseph Lelang, Leo Mandeakali, Michael Manning, Peter McCrea and Andrew Mika. Dewe Enn and Venatius Muriki accompanied me during a field visit to Madang and the Gogol project area. In Mount Hagen, Sam Imine, Emmanuel Kapanamur, Peter Peng, Philip Senat and staff at Haus Poroman provided me with information on highland agriculture. At the ANU in Canberra, I talked to Michael Bourke, Colin Filer, Debra Grogen, Hartmut Holzknecht and Kenn Mondiai. Finally, in Bogor I received additional information from Mary Milne, Robert Nasi, Ian Rowland and Luca Tacconi.

All these oral and written comments have helped improving this work significantly. Nevertheless, remaining errors are my sole responsibility. Map boundaries shown in this book generally do not reflect any judgement on the legal status of particular territories. The opinions expressed in this book do not necessarily represent CIFOR's official view.

1 The malady of prosperity

The present chapter focusses on oil wealth and the macroeconomics of petroleum-exporting countries. The key concept introduced here is the so-called 'Dutch Disease' – an effect usually caused by large export earnings from a commodity boom. What that term means, why or under what conditions it operates, and what are the likely consequences for the general production structure of an oil-rich developing country is explained in this chapter. This will prepare the ground for the next chapter, where we shall examine how the macroeconomic impact of oil wealth relates to land use and forests.

Origins of the Dutch Disease[1]

It might not appear particularly tactful to name an entire branch of literature after the alleged sickness of one single country, especially when its citizens never actually agreed that their nation had become infected. Nevertheless, this is what has happened *de facto* to the Netherlands, a rich, developed country that experienced a trade-induced commodity boom. In the decades after that experience, the Dutch Disease became a standard framework for the macroeconomic adjustment challenges of mineral-exporting countries, 'mineral' here being defined in a way which, notably, includes oil-producing countries. In the 1990s, poverty has again been brought back to the forefront of international attention, including the various diseases and vicious circles that relate to impoverishment. However, the Dutch Disease describes the opposite phenomenon: it is a disease caused by sudden affluence, more like a severe gastric eructation after an exorbitant and rich meal.

From the late 1960s onwards, the Netherlands faced a significant foreign-exchange boom from the discovery and exploitation of natural gas in the Groningen field. The government used much of the foreign-exchange windfall for additional public spending, which drove up domestic incomes, wages, demand and inflation. This was good for the sheltered, 'non-traded' (NT) sectors of the Dutch economy that produced only for the home market, such as construction, restaurants and hotels. They raised their prices and revenues, profiting from the fact that the boom had given the Dutch people more money to spend. But the exposed, 'traded' (T) sectors, notably industry, came under increasing pressure. As Dutch industrial commodities were in direct competition with foreign goods, industrialists could not just raise their prices, whether in the Netherlands or in export markets: their clients would simply turn to foreign substitutes instead. Thus Dutch industry was squeezed between an appreciating currency and rising wage costs. It could not pass rising costs on to consumers through higher prices. This meant that industrial profitability declined, so that production and employment had to be cut back, or at least did not grow as quickly as before. On the other hand, the sheltered, 'NT' sectors rose with rising domestic prices. The adjustment process thus simply expresses the fact that an economy with extra income and purchasing power cannot satisfy all increased demand through domestic production. It has to concentrate more on those sectors that cannot be traded, while those that are traded will lose competitiveness and lag behind.

The somewhat dramatic label 'Dutch Disease' was first used by *The Economist* (1977) in an article that described the danger of this type of structural change in sectoral production to the British economy, which recently had been benefiting from oil revenues from the North Sea. Industry was generally perceived as the 'leading sector', the long-term engine of developed economies, so 'deindustrialisation' was perceived as being strategically problematic. The Dutch case was painted as an example of how bad things can get if 'easy money' makes the government lose control of fiscal expenditure. As well as being an *ex post facto* assessment of the Dutch case (e.g. Ellman 1981; Kremers 1986), this became a hot policy issue for the UK (Forsyth and Kay 1980; Barker and Brailovski 1981; Eastwood and Venables 1982; Buiter and Purvis 1984; Chrystal 1984). Prior to this, however, similar mechanisms had already been debated in connection with the impact of oil in Norway. The pioneering article by Eide (1973) was followed, for example, by others from Hoel (1981), Baardsen (1987) and Steigum (1989). In Australia, there had already been a long discussion about the 'Gregory effect' (Gregory 1976) of mining revenues on the national economy (see also Snape 1977; Forsyth 1986), including extensive theoretical work by Max Corden (Corden and Neary 1982; Corden 1983, 1984). Mining was also analysed in these terms in Canada (Ansari 1990).

These different developed-country analyses of the possible negative side-effects of positive trade shocks were thus united in the Dutch Disease paradigm. These were later extended to developing countries, where most of the second-generation applications from the mid-1980s onwards are found (see below). Historical analyses diagnosed the Dutch Disease retrospectively in cases such as the impact of the inflow of precious metals from Latin America into seventeenth-century Spain (Forsyth and Nicholas 1983), the first gold discoveries in Australia (Maddock and McLean 1984) and the export of *guano* (a seabird dung used as fertiliser) from nineteenth-century Peru (Topik 1997). Even non-timber forest products can cause Dutch Disease, as in the case of the famous Amazon rubber boom from the end of the nineteenth century (Barham and Coomes 1994). The model can describe a price boom (greater revenue at existing production levels) or a quantity boom (the discovery of new resources and higher production at a constant price). Indeed, as the country chapters below will show, any combination of these two factors may occur, for example, higher oil prices stimulating more exploration and eventually higher oil production. This also means that the duration of such booms can actually differ substantially for different oil countries, because resource discoveries and supply responses differ across booming economies. In Dutch Disease terms, the essential precondition is that the boom brings in significantly higher foreign-exchange earnings.

A major push in Dutch Disease modelling came with the two major hikes in oil prices in the period from 1973 to 1982. However, whether or not such booms should be portrayed as a 'disease' remained controversial (see the discussion in the next section).

The headlines describing the Dutch Disease mechanism ranged from 'How oil revenues can destroy a country' (Attiga 1981) and 'Oil wealth: a very mixed blessing' (Amuzegar 1982) to 'The economics of a lucky country' (a term used for Australia) and, for the actual Dutch experience, 'A case of severe hypochondria' (Rowthorn and Wells 1987). It is obvious that sectoral reallocations and structural changes create winners and losers, and that losers tend to be more vocal in their claims that there is a 'disease' than winners openly boasting about their lucky strike. It is thus necessary to apply consistent observable criteria in using the term 'disease', such as long-term declines in national income and/or employment (Wunder 1991). If we accept this criterion, erratic policies of boom and post-boom management (e.g. anti-growth economic policies, public over-spending, rising corruption and neglect of education) qualify best as candidates for genuine 'disease' scenarios.

To trade or not to trade ...

Let us return briefly to the question of why Dutch industry declined in the booming economy while other sectors were expanding. The Dutch Disease story involves a fundamental difference between 'traded' and 'non-traded' goods. Let us consider a simple example. A Dutchman living in Amsterdam during the natural-gas boom wanted to buy a new sweater. Going to the nearest shop, he could not make up his mind between a domestically produced sweater and an imported sweater. The latter was a 'perfect substitute' for the former, so he would be more inclined to buy it if the former had become unreasonably expensive. But the two sweaters were internationally traded goods, so their prices were kept stable by competitive forces.

However, when the same person went to a restaurant, it would not have been worth his while to drive all the way to Germany just because prices in his favourite Amsterdam restaurant had gone up. When building a new house in the suburbs, he could not just have it prefabricated in, say, Poland, and have it shipped in and erected in the Netherlands without the use of Dutch production factors. Besides possibly violating Dutch laws, the transaction costs of doing so would simply have been too high. Or when hiring a civil servant, the government could not choose to employ a Pakistani citizen, simply because he would accept a lower salary than Dutch applicants for the same job. The services provided by the restaurant, the construction industry and civil servants were all 'non-traded' commodities produced in NT sectors. They expanded unambiguously with the increase in Dutch purchasing power, as they were not threatened by foreign substitutes.

It is normal for NT sectors to grow disproportionately and become relatively more expensive in societies that are becoming richer, a situation that we fully recognise when we visit places like Switzerland or Japan and have to pay dearly for their high-cost services. But it is worth dwelling for a moment on the distinction between T and NT sectors, which is basically determined by the relative importance of transport costs in the final value of a commodity. For instance, the service sector is normally classified across the board as non-tradable. But while this is true for normal services, like those provided by the local hairdresser around the corner, it does not apply for Luciano Pavarotti giving a concert in Amsterdam, or for a high-level computer expert who is brought in by a Dutch firm as a consultant. Their services have a sufficiently high value to become *internationally* traded,[2] which the services provided by the local hairdresser do not. We can recognise the same value difference with regard to the tradability of products. In agriculture, cash crops like coffee and cocoa are, and rice and wheat may be, traded across borders, but voluminous products like plantains or tubers rarely are. In forestry, high-value timbers are often traded goods, low-value species are normally not, and firewood is basically never internationally traded. In addition, cheaper transport costs from, for example, new roads or ports may help to turn non-traded goods into tradable items.

In addition to these 'natural' or cost-determined factors, policies and trade regulations may greatly influence the degree of tradability. For instance, over the past two decades, financial liberalisation has made previously non-traded banking and insurance services a widely traded commodity in developed economies. Some readers may also wonder why the Dutch Disease debate has focused entirely on 'de-industrialisation', when in principle agricultural commodities should be equally affected. The answer is that trade policies in the Netherlands, as elsewhere in the developed world, went to great lengths to shelter Dutch agriculture from external competition. Agriculture was a 'quasi non-traded sector', that is to say, commodities could have been traded, but due to trade policies the sector remained sheltered from foreign competition, so it came to behave much like the genuinely NT sectors.

Some of the key concepts of this and the following sections are summarised in Box 1.1.

Certain commodities may occupy an intermediate position on the tradability scale, because they have had only partial exposure to international competition. These are called 'semi-traded' goods in this book, and may exist for several reasons. First, moderate, non-prohibitive import tariffs may hinder but not prevent the importation of certain goods. Second, technical trade obstacles, like port congestion and slow customs clearance, may *de facto* restrict the tradability of certain commodities. Third, trade policy is often used as a 'stop–go' tool of aggregate demand management, with import liberalisation during booms and reinforced protectionism during crisis periods, thus changing exposure to import competition over time. Fourth, imperfect substitution may prevail. For instance, domestic wheat producers may be protected by prohibitions on wheat imports, but may still be threatened by the free imports of a close substitute (say, rice) that consumers can switch to.

We should thus keep in mind the fact that although the distinction between traded and non-traded goods defines the key sectors in the Dutch Disease adjustment process, the classification of products is neither fixed over time nor, especially, between countries. A T sector in one country may be NT in another, either because of trade policy (quasi NT sectors) or because of the structures that determine relative trade costs. The Dutch Disease is thus basically a sectoral adjustment to higher wealth accruing from a trade windfall and triggering a change in relative prices and production quantities. In the following section, the basic price-adjustment mechanism will be described in general terms, without resorting to a fully formalised mathematical model.[3]

The core-model price mechanism

Relative prices (RP) have a key role to play in the Dutch Disease. RP is defined as the indexed ratio of non-tradable prices (p_{NT}) to tradable goods prices (p_{T}):

$$\mathrm{RP} = p_{\mathrm{NT}} \cdot (p_{\mathrm{T}})^{-1} \qquad (1)$$

Box 1.1 A small Dutch Disease dictionary

Boom/Bonanza – Period of price- or quantity-induced high foreign-exchange earnings from a dominant traded booming sector

Booming sector – A dominant traded sector ('oil') with fluctuating foreign-exchange generation

Bust/Slump – Period of price- or quantity-induced low foreign-exchange earnings from a booming sector

Cost effect – The impact of a different intensity of production factors (labour, capital, importables, energy, etc.) in sectoral cost functions on the profitability of sectors (Dutch Disease model extension)

NT sectors – Sectors that are sheltered from import competition (e.g. construction, private and public services)

'Pure' T sector – Sectors that are fully exposed to foreign competition and prime candidates to be hit by the Dutch Disease, typically non-booming exports

Quasi NT sectors – Potentially tradable sectors that through intervention (typically trade policy) have become *de facto* fully sheltered from imports, and thus behave like NT sectors

Real exchange rate (RER) – Time-series index measuring a country's nominal exchange rate, corrected for the inflation differential with respect to its main trading partners

Relative prices (RP) – Time-series index measuring the price of non-tradables in proportion to that of tradables [p(NT)/p(T)]

Rent, economic – A super-normal profit exceeding the common remuneration of a production factor

Rent-seeking – Economic agents' allocation of resources to the pursuit of economic rents

Resource movement effect – Boom-led impact that makes factors of production move to the booming sector from the rest of the economy

Semi-traded sectors – Sectors that are only partially exposed to import competition (e.g. due to imperfect competing substitutes, non-prohibitive tariffs or shifting trade policies over time)

Spending effect (also 'core effect') – Boom-led stimulation of aggregate demand, featuring a price and quantity redistribution from the T to the NT sectors

T sectors – Sectors producing (non-booming) goods that are exposed to foreign competition, either on domestic or on export markets (e.g. industry, agriculture, forestry or fishery)

We should also remember that domestic tradable prices (p_T) cannot deviate from the cost of foreign (perfect) substitutes, as consumers would simply shift to the cheaper product. This means that the international traded-goods price (p_T^*), measured, for example, in US$, divided by the nominal exchange rate (e), equals the price of traded goods (p_T):

$$p_T \cdot e = p_T^* \tag{2}$$

Economists call relation (2) 'the law of one price'. This means that domestic traded-goods prices can only change if either the nominal exchange rate (e) is de- or revalued, or if world-market commodity prices (p_T^*) change. At the same time, the *RER* is another important price measure, expressing a country's nominal exchange rate corrected for the differential between domestic inflation (measured by p_{NT}) and inflation abroad (measured by p_T^*):

$$\text{RER} = e \cdot p_{NT} \cdot (p_T^*)^{-1} \tag{3}$$

When the RER appreciates (the index goes up),[4] the domestic country loses competitiveness with respect to the rest of the world. In practice, the RER is calculated as a *trade-weighted* index, that is, 'world inflation' (p_T^*) is made up from price indexes in those countries that the domestic country trades with the most (imports and exports). Inserting (2) into (1), we can see that when the 'law of one price' holds (i.e. perfect substitution between foreign and domestic goods), then the two price indicators coincide.

$$\text{RP} = p_{NT} \cdot (p_T)^{-1} = p_{NT} \cdot e \cdot (p_T^*)^{-1} = \text{RER} \Rightarrow \text{RP} = \text{RER} \tag{4}$$

This implies that the RER and RP move in an identical manner, and that we can look at either of the two indicators for a measure of sectoral profitability:[5] a boom creates pressures for the RER to appreciate, and simultaneously for the RP index to turn in favour of non-traded goods (Wood 1987). How is this adjustment achieved in practice? Two potential pathways exist here. First, the nominal exchange rate can be revaluated (in a fixed regime) or appreciate (in a floating regime), responding to the fact that foreign exchange reserves become plentiful with respect to the demand for foreign currency. Second, domestic inflation, as measured by a consumer price index, may grow faster than foreign prices because of excess demand for non-tradable goods. If the country holds exchange rates constant in a regime of fixed exchange rates (as some of our case countries did), then only the price mechanism will be at work.

In this model, which lacks an explicit financial sector, various combinations of currency and price adjustments may occur. What is required is that the RER (and thus the RP) index adjust to a new equilibrium level where the market for non-traded goods is cleared, that is, the supply of non-traded goods has been increased so that it equals higher demand. This must be achieved by reallocating factors of production (such as capital and labour) from the T sector to the NT sector. Similarly, by the end of the boom, when booming-sector revenues and foreign-exchange reserves start to decline, a real depreciation will have to occur, whether through currency devaluation, currency depreciation and/or domestic inflation levels that continue to remain below world inflation levels. In a fully symmetrical way, this will induce factors of production to move back to the T sector. In other words, non-traded production moves pro-cyclically with the fate of the booming sector.

Extensions and modifications

The core Dutch Disease model described above focuses on the spending effects of higher incomes and demand. It leads to unambiguous predictions regarding the impact of a boom on the sectoral structure of the economy:

- Real currency appreciation and an RP shift in favour of non-traded goods.
- Movement of production factors from the T to the NT sector.
- A rise in the domestic production of non-traded goods, at the cost of traded goods.

It is notable that none of the core mechanisms really justifies the idea of a 'disease' in the sense of negative incomes or employment impacts. However, different assumptions and model specifications can change that outcome at the margins. The following modifications focus on those that are particularly relevant for developing countries.[6]

A. *Resource movement effects*. Until now, booming sector income has simply been treated as a transfer payment from abroad ('manna from heaven'), for example, because rising world-market oil prices generate additional foreign exchange out of the same physical quantity of production. That may be an over-simplification when the booming sector is not a pure enclave. For instance, countries that have emerged as new oil exporters (e.g. Ecuador, PNG) allocated resources to the oil sector that had to be diverted from elsewhere in the economy. Resource movement was thus a precondition for the foreign exchange boom. If much labour is needed in the booming sector, this will have to be drawn from the T and NT sectors. This means that, compared to the core effects above, an additional flow of production factors will exacerbate production decline in the T sector, in principle possibly making the outcome for the NT sector ambiguous (e.g. Corden 1984). This modification is particularly valid for booming mining sectors. However, the oil sectors in our countries resemble enclaves with negligible labour demand.[7] Spending effects are thus clearly dominant, and we should not be too concerned about the effect of resource movements in this book.

B. *Relative versus absolute quantities*. The Dutch Disease is not a growth model: it makes predictions on the reallocation of a finite quantity of factors of production with a static technology, operating within a fixed capacity of production. However, in a growing economy we would expect the Dutch Disease to be *superimposed* on existing trends. Thus it would change the composition of Gross Domestic Product (GDP) towards having a *relatively* higher proportion of non-traded production. This means that the T sector may still grow in absolute terms, but at a rate inferior to that of the NT sector. This also includes typical structural changes and transformations in a developing economy, such as the reduction of employment in primary sectors, which are reallocated to secondary and, later on, tertiary sectors. To the extent that the primary sectors are traded and (most of) the other sectors non-traded, we would thus expect the Dutch Disease to *accelerate* the process of structural change (Scherr 1989: 544). The extreme case of an *absolute* decline of the T sector has been observed in some countries of the Organisation of Petroleum-Exporting Countries (OPEC). This phenomenon has been called the 'hyper-Dutch Disease' (Parvin and Dezhbakhsh 1988).

C. *Imperfect substitution and trade policy*. For analytical clarity, the core model assumes that domestic and foreign-traded goods are perfect substitutes, and that 'the law of one price' operates (as in the case of the two sweaters). This assumption may hold for some homogenous commodities, say, iron ore, petroleum or rice, where quality differences determine a fixed premium, but prices tend to move in a parallel manner. However, as discussed above, more heterogeneous commodities (e.g. industrial goods) are often imperfect substitutes. Furthermore, even a fairly homogenous product like rice tends to have positive price cross-elasticities to other foodstuffs that can partially replace it. This is important in understanding the dynamics of food production in many developing countries. Weaker versions of the 'law of one price' may thus be introduced, accepting the notion of imperfect substitution, with domestic traded-goods prices being allowed to differ from foreign ones. This does not mean that competitiveness does not matter. Rather, domestic goods lose *market share* when their prices rise above those of their foreign competitors. This typically leads to the dissolution of the NT–T distinction in favour of a continuous sectoral 'tradability' ranking.[8] Moreover, trade policies have often been used as a 'stop–go' tool in managing aggregate demand, so that 'tradability' may vary considerably over time. Consequently, NT–T relative price and production adjustments become less accentuated, and a sector of 'semi-traded' goods comes to occupy an intermediate position between the two.

D. *Factor markets and cost effects*. The core model assumes that each sector uses only one mobile production factor (e.g. labour). But what if there were an 'input–output' structure in the economy in which different sectors use each other's outputs as inputs? In that case, an oil price boom might negatively affect energy-intensive sectors such as industry. In extreme cases, this may even jeopardise the basic conclusions on RP changes, especially when the resource movement effect (A) is strong (Enders and Herberg 1983; Corden 1984). For our study countries, this has been less relevant. However, importables (machinery, fertilisers, etc.) generally became cheaper there through real currency appreciation, which favoured those sub-sectors that used imported inputs intensively, such as industry and agro-industry. This ultimately also had an effect on land use and forests by favouring 'modern' intensive agriculture (using fertilisers and machinery) over land-extensive traditional farming. In addition, if capital is internationally mobile but labour is not (no immigration into the booming country) and labour intensity differs across sectors, we should expect a significant rise in real wages, making NT, labour-intensive service sectors even more expensive. Real wages indeed rose in most Dutch Disease countries, as did the opportunity costs of labour. As we shall see below, this also reduces the viability of economically marginal forest-degrading activities on the agricultural frontier.

E. *Permanent versus transitory booms*. The core model assumes that economic agents react to a boom as if it were permanent. In the real world few booms are permanent, and people may sometimes predict the transitory nature of a boom from the outset. This is particularly true of price booms. For instance, as the country chapters will show, the first oil-price hike in 1973 was widely perceived as temporary, and most governments reacted cautiously, without over-spending their revenues. However, the second oil-price rise in 1979–80 was more widely perceived as permanent and spending became over-optimistic,[9] so it was a surprise to many when oil prices fell back in 1983 and then plummeted right back down again in 1986. Thus anticipations may or may not be correct, and 'prediction errors' may lead to inefficiencies and higher adjustment costs that can contribute to a real 'disease'. If agents truly perceive a boom as temporary, this should have two different impacts. First, one would probably not want to make lasting resource movements into the booming sector, for example, to plant cocoa trees because of a temporary boom in cocoa prices (see A). Second, those receiving the windfall would be inclined to smooth out consumption over time, so that they consume less now and save a larger portion of their windfall gains for the expected future slump (Bevan *et al.* 1989). To the extent that increased savings translate into higher investments, this would boost imports of, for example, vehicles and machinery. Within the NT sector, temporary boom expectations would trigger less emphasis on consumption (e.g. restaurants, private services) and more on durables, which yield benefits even when the temporary boom is over. Notably, we would expect a boom in construction, the main non-traded investment sector (Bevan *et al.* 1987, 1989; Collier and Gunning 2000c).

F. *Fiscal over-spending, borrowing and rent-seeking*. In a mineral boom, the windfall accrues to the government and the spending effect depends fully on how the government decides to spend the money.[10] Yet many people in the private sector are prepared to 'help' the government in decision-making. The presence of large natural resource rents triggers lobbying and pressures for fiscal expansion (*rent-seeking*) (e.g. Krueger 1978; Gelb and Associates 1988), as well as efforts by public employees to gain control over the allocation of these rents (*rent-seizing*) (Ross 2001). In the case of an oil boom the state normally is the caretaker of the rent, and private individuals try to obtain a share of it. To be successful in competition with others, the rent-seeker needs to invest time and effort that (s)he could have spent on productive activities otherwise. This is why rent-seeking is often seen as socially undesirable. Rent-seeking pressures may change individual mentality and institutional behaviour. Governments may not only spend the entire oil windfall; in some cases they also borrow internationally, thus increasing foreign-exchange inflows further (Mansoorian 1991). Debt accumulation may later cause even higher repayments with interest. If returns to externally financed projects are poor, this also qualifies as a 'disease' element. In many developing countries, the presence of large natural resource rents can weaken institutions and ultimately have a disruptive effect on economic development (see the section on 'Why natural resources?' on the *resource curse* literature). This happens partly because competition in the acquisition of rents implies transaction costs, which are diverted from the entrepreneurial sector in favour of 'directly unproductive profit-seeking activities' (Bhagwati 1983). Another impact is that economic growth and efficiency criteria receive a lower priority in the allocation of resources compared to specific motivations related to politics, ethnic groupings, patronage or prestige (see also Chapter 10). In particular, many boom countries experience increasing corruption and decreasing investment returns over time. Once these features are introduced into the political economy of a country, they may take a considerable time to reverse. This makes them prime candidates for introducing asymmetries into the boom–bust adjustment, triggering effects that may cause a real 'disease' in the form of long-term declines in income or employment.

G. *Time lags, disequilibrium and rigidities.* In the neoclassical world of the Dutch Disease model, production, prices and wages adjust instantaneously to external shocks by reaching new equilibrium levels. This is an oversimplification in several respects. Different model versions have been developed to deal with alternative scenarios of gradual adjustment under disequilibrium regimes. For instance, if the boom occurs in a situation of 'Keynesian unemployment', initial excess capacity may allow the NT sector to expand with less relative changes in prices or factors of production than the core model predicts. On the other hand, prices and wages may be ratcheted, so that, once the boom has increased them, they do not immediately readjust downwards when the economy goes from boom to bust. This may reinforce asymmetries that promote genuine 'disease' scenarios in which the 'productive' T sector has difficulties in coming back after the boom, especially if there are significant adjustment costs involved. Almost all 'disease' scenarios actually predict negative impacts during the post-bonanza *downturn* of the economy, although in this book, Venezuela (Chapter 5), for instance, is an example of a country that was already experiencing crisis symptoms during the oil boom. Another particular feature in developing countries is that non-traded goods may be provided by both a formal and by a low-cost informal sector. The latter tends to expand counter-cyclically, that is, there is a growing 'informalisation' of the economy in crisis periods (Montenegro 1989, 2000). This dichotomy dampens the contraction of non-traded production in conditions of bust.

From these model extensions, we might want to revisit the predictions of the core model made at the beginning of the section. The expected results are shown in Box 1.2. From now on, they should be interpreted as structures that are *superimposed* on pre-existing growth and structural change. For relative prices, we can see that, while real appreciation may be dampened or exacerbated, accelerated or prolonged, *per se* it is indeed expected to occur. Production factors may not be fully mobile in the short term, so it is possible that production adjustments may lag behind price changes. However, the sectors that win or lose out will differ according to the structure of production in the country and the type of boom (recipients, expected duration, etc.). For instance, import-intensive T sectors may gain unexpectedly from an appreciated RER, or a construction boom may occur because a boom is not expected to last so that people favour higher investment over consumption. It is thus not possible to make a general, global classification of T and NT sectors; a country-specific identification is required.

Dutch Disease in developing countries

From the mid-1980s onwards, Dutch Disease analyses moved south. Developed countries seemed to have learned the management lessons of their resource booms. At the end of the day, none of them had actually become 'ill' in the sense of being worse off from their windfalls. It is even doubtful to what extent the fluctuations in the industrial sector should actually be attributed unambiguously to mineral exports in Norway, the UK and the Netherlands (Hutchison 1994). The situation looked quite different for booming resource sectors in developing countries, which experienced more pronounced structural changes. Four of the cases examined in a comparative volume edited by Neary and van Wijnbergen (1986) were developing countries (Colombia, Egypt, Indonesia and Nigeria). Gelb and Associates (1988) compared six developing oil countries, among which were Ecuador and Venezuela. Davis (1983) looked at the fate of exporters benefiting from the boom in

Box 1.2 Expected outcomes from extended Dutch Disease models

- Real currency appreciation and a RP shift in favour of non-traded goods, which
  - may be dampened if the boom is clearly perceived and treated as temporary
  - may be dampened if there is initial excess capacity (e.g. unemployment)
  - may be exacerbated and prolonged by overspending and/or foreign borrowing
  - may be prolonged beyond the boom period by a ratchet on downward price and wage movements.
- Redistribution of production factors from the T to the NT sector, which
  - may be slightly altered if there are resource movement effects (to the B sector)
  - may be lagged in time, due to short-run rigidities in factor mobility
  - is likely to cause real wages to rise, raising the opportunity costs of labour.
- A rise in the domestic production of non-traded goods, at the cost of traded goods, but
  - a temporary boom may set off a non-traded investment boom only ('construction boom')
  - the boom's impact on the sectoral structure may be dampened if
    - it is clearly perceived and treated as temporary
    - substitution is imperfect and there is a 'semi-traded' sector
    - there is initial excess capacity (e.g. large unemployment)
  - import-intensive T sectors (industry, agro-industry) are favoured by lower input costs.

beverage prices in 1975–8. Collier and Gunning (2000a,b) examined twenty-three developing countries that had experienced temporary trade shocks. In addition, many single-country analyses have been carried out. Table 1.1 lists country-case studies that have explicitly used versions of the Dutch Disease model as their framework of reference,[11] grouping them into four main groups: petroleum, hard-rock minerals, export crops and financial transfers.

A quick glance at the table provides several insights. First, with a total of twelve countries (and in the majority of all these studies), petroleum clearly dominates the Dutch Disease literature, thus justifying its emphasis in this book too. In fact, although many more boom–bust stories seem to relate to hard-rock minerals (e.g. Nankani 1979), they have not been analysed using Dutch Disease models. This is perhaps because the core mechanism, the spending effect, is most clearly mirrored by the experience of oil countries: a large-scale transfer of an international economic rent. Second, the range of Dutch Disease sources among the non-oil cases is wide, and it may seem worrying to some that foreign aid is the second-most investigated source, ahead of the beverages, coffee and cocoa.[12]

Table 1.1 Dutch Disease in developing countries: an overview

Product types	*Products*[a]*: Countries*[b]*, References*[c]
I. Petroleum and natural gas	A. **Oil**: Algeria (Gelb with Conway 1988), Cameroon (Benjamin *et al.* 1989; Blandford *et al.* 1994; Amin 1996; Devarajan 2000), Ecuador (Gelb with Marshall-Silva 1988; Larrea 1992; Wunder 1997), Egypt (Martin and van Wijnbergen 1986; Shafik 2000), Gabon (Yates 1996), Indonesia (Warr 1986; Gelb with Glassburner 1988; Scherr 1989; Ahmed and Chhibber 1992; Roemer 1994; Usui 1996), Nigeria (Taylor *et al.* 1986; Collier 1987; Pinto 1987; Gelb with Bienen 1988; Scherr 1989; Struthers 1990; Ezeala-Harrison 1993; Oyejide 2000), Malaysia (Greenaway and Pillay 2000), Mexico (Ros 1987; Scherr 1989; Feltenstein 1992; Collier *et al.* 1994; Gavin 2000), Saudi Arabia (Looney 1992), Trinidad and Tobago (Gelb with Auty 1988), Venezuela (Gelb with Bourguignon 1988; Vaez-Zadeh 1988; Hausman 2000)
	B. **Natural gas**: Bolivia (Morales 2000)
II. Hard-rock minerals	A. **Copper**: Zambia (Kayizzi-Mugerwa 1989)
	B. **Diamonds**: Botswana (Love 1994; Hill and Knight 2000)
	C. **Phosphates**: Senegal (Azam and Chambas 2000)
	D. **Tin**: Bolivia (Morales 2000)
	E. **Uranium**: Niger (Azam 2000)
III. Agricultural commodities	A. **Coffee**: Colombia (Edwards 1986; Kamas 1986; Wunder 1991; Montenegro 2000), Costa Rica (Gonzalez-Vega 2000), Ivory Coast (Ghanem 2000), Kenya (Bevan *et al.* 2000), Zambia (Aron 2000)
	B. **Coca(ine)**: Colombia (Wunder 1991)
	C. **Cocoa**: Ghana (Wetzel 2000), Ivory Coast (Ghanem 2000)
	D. **Cotton and soybeans**: Paraguay (Richards 1994)
	E. **Groundnuts**: Senegal (Azam and Chambas 2000)
	F. **Sugar**: Mauritius (Greenaway and Lamus 2000)
IV. Financial transfers	A. **Foreign aid**: Bangladesh (Azam and Shahabuddin 2000), El Salvador (Paus 1995), Ghana (Younger 1992), Greenland (Paldam 1997), Pakistan (Haque *et al.* 1994)
	B. **Remittances**: Bangladesh (Azam and Shahabuddin 2000), Egypt (de Macedo 1982; Martin and van Wijnbergen 1986; Shafik 2000), El Salvador (Paus 1995)

Notes

a Products/services most often mentioned: oil (twelve countries), aid (five countries), coffee (four countries).

b Geographical country distribution (total: 27): Africa 14 (SSA: 11), Latin America and the Caribbean 8, Asia and Pacific 4, Others 1.

c Only one author reference per product and country.

Third, some cases had more than one 'booming-sector' source of inflows, such as Egypt (oil, remittances), Paraguay (cotton, soybeans), Ghana (aid, cocoa), Senegal (groundnuts, phosphates), Zambia (copper, coffee), and Bangladesh and El Salvador (both aid and remittances). Fourth, the geographical distribution is very skewed, according to the variable degree of reliance on natural resources in different continents. Africa is clearly ahead, with more than half of the cases. This is not due to the oil countries in low-absorbing North Africa (only three representatives), but more to the high-absorbing countries of Sub-Saharan Africa. While Latin America is fairly well represented, apparently only four cases in Asia have been studied, reflecting the fact that on average this continent is less endowed with valuable natural resources. Finally, two individual countries have attracted particular attention, namely Nigeria and Indonesia, which have also been directly compared on several occasions (Pinto 1987; Scherr 1989; Bevan *et al*. 1999). In these comparisons, Indonesia has been seen as a showcase for fairly prudent economic management of oil wealth, whereas Nigeria has been presented as an example of how bad things can get when the state loses control over public spending and economic policy.

The postwar literature on growth and economic development, with its staple theory of growth and concepts of the 'big push' and the development 'take-off', formerly identified capital shortages as the main cause of underdevelopment. This should lead one to believe that a large rent transfer would provide an excellent opportunity to invest in the production capacity of other sectors. Some early Dutch Disease authors also had the impression that a windfall 'seems to present fewer problems for developing countries with few or no other sources of income than to industrial countries where the petroleum sector contrasts with the rest of the economy' (Noreng 1981: 22). Evaluating a decade of resource booms in developing countries, Roemer concluded: 'it has become increasingly clear that a relatively long period of buoyant export booms was *not enough* to stimulate the kind of structural change in the economy that might sustain growth after the export boom' (1984: 3, my emphasis).

In fact, such speculations proved to be completely wrong. With hindsight, one is tempted to reply that the booms might have been *too long* and *too large* to allow for the sort of structural change that Roemer was looking for. Instead of staple-based growth, the notion of a 'staple trap' suggests that 'the rents are transferred from the potentially efficient primary sector into a burgeoning non-tradable sector that includes slow-maturing infant industry and non-productive public sector employment' (Auty 2001: 844). While many economies absorbed their windfalls sensibly during the first years of boom, distortions rose, and both investment returns and the *quality* of windfall spending declined markedly as the boom continued. The body of literature reveals that, contrary to the experiences of the moderate cases of Dutch Disease in the developed world, many developing countries had severe difficulties in managing boom-bust sequences for key primary commodity exports and international transfers, with several true 'disease' cases similar to the Nigerian experience. Box 1.3 reveals some of the outcomes that have proved to be specific to developing countries.

Larger average dependence on natural-resource exports means that the size of booms relative to the rest of the economy has been higher for developing countries. Two complementary features have reinforced this trend over the last three decades (Ross 2001: 11–12). First, since 1970 world market commodity prices have fluctuated more than in earlier decades. Second, the nationalisation of companies and higher royalties have generally given developing nation states access to a much larger share of those commodity rents that previously had been absorbed by Northern trading companies. Paradoxically, this has also increased their exposure to the Dutch Disease. The second bullet point in Box 1.3 states that excessive government-spending sprees have been more pronounced in developing countries, because of a greater institutional exposure to rent-seeking and fiscal-expansion pressures. For publicly owned resources like oil or hard-rock minerals, this is

Box 1.3 Special features of the typical Dutch Disease in developing countries

- Large size of windfalls, compared to size of the economy
- Institutions often too weak to maintain fiscal discipline over several boom years
- A boom may be prolonged or exacerbated by foreign borrowing
- Lower capital mobility between sectors and countries
- Higher labour mobility, including large rural–urban migration to urban NT sectors
- More pronounced trade restrictions (natural or policy-created)
  - create a sheltered, quasi NT industrial sector
  - make agricultural exports the 'pure traded' sector that is most exposed to the Dutch Disease
  - generally cause 'de-agriculturisation' instead of 'de-industrialisation'.

particularly important, as the entire windfall accrues to and is distributed through the government. This contrasts with other cases from Table 1.1, such as private remittances or agricultural crops, although the government may attempt to tax away part of these boom revenues too.[13] Spending pressures may initially be resisted, but this becomes politically less feasible over time. The third bullet point suggests that foreign borrowing may be an integral part of the Dutch Disease picture. Governments may use this option in particular to prolong a spending spree on the cusp of the change from boom to bust, when they find it difficult to default on spending commitments or to disappoint the elevated expectations of powerful interest groups.

Third, although some liberalisation has occurred recently, capital mobility in developing countries has been more restricted than in developed economies. This is true for international capital controls, which affect the private sector's freedom to save part of the windfall abroad if domestic investments give low returns. In addition, the formal financial sector seldom reaches out to all sectors and potential beneficiaries, and capital is thus not fully mobile between sectors. This may imply that saving and investment decisions become directly linked, so that, for example, coffee farmers will tend to invest in their own coffee farms, even when this does not promise high returns in the short or medium term. On the other hand, labour is relatively mobile in developing countries. As many of the NT sectors are both urban and labour-intensive (services, construction, etc.), while the 'losing' agricultural sector is rural and tends also to be labour-intensive, we should expect a commodity boom to cause an enhanced drive towards rural–urban migration and urbanisation.

Finally, trade policy probably affects the economy in a more complex way in developing countries. It is noteworthy that the industrial sectors – the most exposed in the Dutch, British, Norwegian and Australian cases – have traditionally been more protected in developing countries that have tried to apply 'infant industry' tariffs and import prohibitions to achieve import substitution through industrialisation. Industry, including agro-industry, often becomes a quasi NT sector. On the other hand, the most protected sector in the developed world is often the most exposed sector in developing countries, namely agriculture. Many sub-sectors of agriculture become exposed, in particular those that are export-oriented. Others may be partly protected and become semi-traded. Some protection may not directly be policy-induced, but due to the fact that transport infrastructure is less developed, so that high transport costs to and from a given geographical area provide a 'natural' form of protection. On average, resource-abundant economies thus generally become high-price economies, which are not sufficiently competitive to rely on non-booming exports for economic growth (Sachs and Warner 2001). Dutch Disease in developing countries has often been accompanied by 'de-agriculturisation' instead of 'de-industrialisation'. It is important to keep this in mind once we begin to link the Dutch Disease to change in land use and forest cover.

To sum up, the Dutch Disease is a sectoral adjustment to sudden natural-resource wealth, which is particularly prominent in oil-producing countries. Only 'policy failures' can actually make a country truly worse off. There is a reasonable degree of consensus among economists that it would be best to use a publicly accruing windfall to pay off foreign debts and/or to invest a large proportion of it abroad, for subsequent gradual repatriation and absorption into the domestic economy. However, this is not what usually happens in the real world. The political economy of rents goes a long way in explaining why over-spending and other policy failures occur. Besides the Dutch Disease literature proper, the last section of this chapter will deal with the so-called 'resource curse' literature, which can shed additional light on the political aspects of these phenomena.

Why natural resources?

At this stage the attentive reader may ask why the Dutch Disease model has been used almost exclusively for natural resource exports, and not applied to manufacturing. Why is there no 'Japanese', 'German' or 'Swiss' disease to be found in the literature? What is it that makes natural-resource dependency particularly damaging? Most explanations make some reference to the fact that natural resources create economic rents, that is, supra-normal profits that exceed the 'normal' return on factors of production. These rents are claimed to be an unreliable basis for long-term economic development, because of their unequal distribution over time and beneficiaries, introducing incentive structures that are incompatible with sustained long-term development.

The literature on the alleged 'resource curse' overlaps with the literature on the Dutch Disease in part, but it also gives some additional suggestions as to why resource-rich countries underperform resource-poor ones in respect of long-term development. Such underperformance has been disputed for the two decades from the mid-1960s to the mid-1980s, when primary commodity prices were high. Although there are still some sceptics (Davis 1995), evidence for the inferior economic and human development performance of resource-rich countries seems convincingly robust over the long term, even when possible third causal factors are taken into account (Ross 1999; Auty 2001; Gylfason 2001; Sachs and Warner 2001).[14] But there is a range of possible explanations for this paradox. Five factors are briefly discussed in this section: long-term trends in terms of trade, price volatility, the rent-seeking mentality, anti-growth policies and the neglect of education.

First, scepticism about the development potential of natural resources was already raised immediately after the Second World War by Singer (1950) and Prebisch (1950), who argued simultaneously that primary commodities would face a long-term price decline with respect to manufactured goods. This would clearly make industrialisation the way forward for low-income countries, allowing them to diversify out of their dependency on natural resources, which increasingly became subject to an 'unequal exchange' with industrialised countries and kept developing countries in a state of underdevelopment (Amin 1972). However, the hypothesis of the long-term decline in the terms of trade for primary producers has been disputed in more recent studies. For instance, Grilli and Yang (1988) and Bleaney and Greenaway (1993) both find the evidence to be mixed, with conclusions varying greatly according to the choice of study period. This is therefore probably not the best 'resource curse' explanation.

A second argument is that, even if the terms of trade for primary commodities do not decline, their greater fluctuations over time make export earnings unstable, discourage private investment and jeopardise policy-makers' efforts to undertake macroeconomic planning. Again, the evaluation of this hypothesis depends greatly on the time period. Numerous studies found little evidence of higher primary price fluctuations in the 1950s and 1960s (MacBean 1966; Knudsen and Parnes 1975), but found instead that the greater fluctuations in the terms of trade for individual exporters of primary commodities were due to excessive single-commodity export specialisation and insufficient diversification. Some scholars have even suggested that a high degree of instability actually stimulates higher savings, which in the long term may promote higher economic growth (Newbery and Stiglitz 1981). However, since the 1970s the price volatility of primary commodities has increased significantly (Reinhart and Wickham 1994), making it a great challenge to cope with the uncertainties of future commodity prices (Cuddington 1988). This also temporarily produced increased support for, and experiments in, international commodity-stabilisation funds (Maizels 1988). With increasing globalisation and higher international capital mobility, it seems intuitively clear that unpredictable macroeconomic fluctuations are increasingly disadvantageous to a country's development, especially given volatile exchange rates and returns on investments.

A third factor has already been mentioned above in discussing the Dutch Disease literature: the fact that oil and mineral rents normally accrue entirely to governments, rather than to private sector recipients, facilitates non-market-based spending, favouring vested interests and the rise of rent-seeking, which reduces entrepreneurial efforts. Private individuals find it more rewarding to devote efforts to lobbying and 'making the right connections' than finding market opportunities, with the result that remuneration becomes largely divorced from production.

The fourth factor represents the other side of this same coin: policies become less conducive to, or even directly opposed to, economic growth. Prestige and military projects, ideologically and ethnically motivated specific interests, corruption and patronage easily come to govern the agenda of a government that is facing less pressure to 'increase the size of the cake', but more pressure to distribute it in a way that makes its political support groups happy. The result is often that both foreign investors and domestic entrepreneurs find it difficult to see their viewpoints being taken into account in the booming state's policy priorities, which are therefore dominated by distortions (e.g. Karl 1995 on Venezuela; Winters 1996 on Indonesia and Chapter 10).

The fifth point, low investment in human capital, reinforces this rationale. Education is arguably the single most efficient investment for long-run equitable growth, but with large rents being available in the here and now, there is little political reward for such long-term, far-sighted investments. Consequently, the enrolment ratio of specialised oil-producing countries is significantly lower than that of other countries at comparable levels of development (Gylfason 2001: 850–7). In this sense, natural resource wealth may diminish the 'push' towards growth-oriented policies and attitudes. As the same author also expresses it: 'Rich parents sometimes spoil their kids. Mother Nature is no exception' (ibid.: 850).

Notes

1 This first part of the chapter draws heavily on the literature reviews in Wunder (1991) and in Sunderlin and Wunder (2000). I am also most grateful to my colleague William Sunderlin for making available to me an unpublished annotated literature review on more recent work on Dutch Disease and the 'resource curse'.
2 From here onwards, I will use the term 'traded sector' as referring specifically to international trade, irrespective of whether the commodity is traded within the country (regionally or locally) or not.
3 The Dutch Disease spending effect draws considerably on the traditional two-sector trade model developed by Salter (1959). Different core-model versions are described in Corden and Neary (1982), Corden (1984), Neary and van Wijnbergen (1986) or Enders and Herberg (1983).
4 In the literature on RER, some authors use the opposite definition, so that a rising index corresponds to a real *depreciation*. This is simply a question of how the exchange rate has been defined, that is, as number of foreign units per one domestic unit, or vice versa.
5 This is a result of the assumptions that are made. In practice, the two indexes may differ in a number of ways, as discussed in Wunder (1991: 424–30).
6 In addition to the extensions mentioned here, there are therefore others with, in the aggregate, greater relevance for developed economies, such as the explicit modelling of monetary aspects, multi-factor sectoral mobility, positive externalities in the T sector and international factor mobility (see Wunder 1991 for an overview).
7 Financial investments in the oil sector may be considerable, but as long as equipment and services are *imported* through foreign companies, this just corresponds to a reduction of the foreign-exchange windfall. Only when the oil industry *competes* with domestic sectors, for example, drawing labour out of other industries, will the Dutch Disease outcome be affected by resource movements.
8 See Wood (1988) for an empirical investigation of real exchange rate indexes, and Kamas (1986) and Wunder (1991: 205–20) for a discussion of tradability in one particular country, Colombia.
9 It may be difficult to distinguish in practice over-optimism from *prediction errors* (treated in this sub-section) from that of a *politically motivated* spending spree (treated in the next sub-section on p. 14), that is, a scenario where revenues are well-calculated but spending *deliberately* exceed them.
10 However, the situation may be similar when the resources are privately owned but taxed away by the government, for example, by forcing cocoa growers to market their produce through government-controlled boards that do not pass on higher world-market prices to producers (see Cameroon and Gabon chapters).
11 To save space, I have avoided repeating similar analyses of the same countries by the same authors.
12 In fact, following the observations of Dutch Disease symptoms in individual countries, the macroeconomic impact of foreign aid has been subjected also more generally to Dutch Disease analysis (e.g. White 1992).
13 Foreign aid constitutes an intermediate case, as benefits may either accrue directly to the government (e.g. balance-of-payments aid) or to private recipients (e.g. rural development projects).
14 These factors include previous growth (inertia effects) and differences between geographic regions (e.g. Sachs and Warner 2001).

2 The impact of oil wealth on forests

Chapter 1 focussed on oil wealth and its impact on the economy, and defined the concept of the 'Dutch Disease'. This chapter is aimed at bridging from oil and macroeconomics to land use and forests, and thus outlines the key methodologies used in this book. How are land uses and forest conditions affected by oil wealth and its fluctuations over time? From an examination of the literature on the macroeconomic impacts on forests, ten major potential factors of transmission are identified. A framework for the country-case analyses in Chapters 4–9 is developed.

Exploring the linkages

Chapter 1 dealt with the theory of primary commodity export booms and mineral wealth in developing countries and identified a number of causal pathways. As we have seen, the most important macroeconomic consequence was a real appreciation of the booming country's currency (equivalent to a price shift in favour of non-traded goods). NT- and T-protected sectors expand, at the cost of those hit by declining competitiveness. In developing countries, the losing sectors are typically agriculture, fisheries, forestry and mining. The next question is thus: what has all this got to do with forests?

Agriculture

By far the most significant linkage is that Dutch Disease in developing countries causes 'de-agriculturisation'. From the mid-1980s onwards, this was noted for various countries by a number of authors (Roemer 1984; Pollard 1985; Commander and Peek 1986; Scherr 1989; Fardmanesh 1991; Feltenstein 1992). Given the definition of deforestation adopted in this book – removal of forest canopy cover to a level of under 10 per cent – it is also obvious that agricultural expansion, including the establishment of pastures for livestock, is the main source of deforestation. To the extent that the Dutch Disease limits the expansion of agricultural area, deforestation will be reduced in mineral-producing countries (Sunderlin and Wunder 2000: 310–13, and the next section).

Forest product extraction

The second linkage is that RER appreciation is also likely to depress the extraction of products from the forest, to the extent that these are traded goods that are exposed to increased competition. As this is often the case for various timber and non-timber products, real currency appreciation and RP shifts should make these activities less profitable, therefore working towards their decline. If this extraction has a degrading impact on the forest, mineral wealth should thus also help to reduce forest degradation. Of course, the principal argument here refers to timber exports, which may be quite price-elastic. RER appreciation will make buyers go to other countries or products, so that timber-harvesting in the 'expensive' mineral-producing country is reduced. In the deforestation literature, the link between competitiveness and timber harvesting has usually been mentioned for the opposite situation, where crisis-ridden countries with *declining* foreign exchange tend to award cheaper concessions to timber companies (e.g. Sizer and Rice 1995; Sizer and Plouvier 2000). In at least one case, the argument has also been 'reversed' to apply to booming oil wealth. Baardsen (1987) argues that the pulp and paper industry has declined in Norway since the 1970s, probably irreversibly, as a direct result of RER appreciation and higher domestic costs. In some cases, the argument may also apply to non-timber forest products: to the extent that firewood, charcoal, medicinal plants, game, etc. compete with foreign (imperfect) substitutes, their extraction may decline, which may also reduce forest degradation.

Other traded sectors

Third, there is a residual category of other traded, competitively exposed sectors that use either the forest or the land, but which do not belong to the two previous categories (agriculture or forest-product extraction). Non-oil-mining is an obvious candidate for an activity that can cause both deforestation and forest degradation, and is at the same time often a 'pure' T sector producing for the export market. As for timber, the competitiveness argument has usually been made for the opposite situation of a foreign-exchange crisis leading to a drive to the rapid exploitation of natural resources, such as minerals (WWF and IUCN 1996; Miranda *et al.* 1998). There is every reason to believe that an oil boom should lead to a symmetrical *reduction* of mining, at least in those cases where the economic rents are limited, such as certain coal mines or artisan goldminers operating at the margin of profitability. That should take additional pressures off the forests. Another candidate in this residual category is shrimp-farming, which in recent decades has expanded markedly in both South-East (SE) Asia and the Pacific coast of Latin America. Clearing for shrimp ponds has often happened at the cost of mangrove deforestation. Shrimp exports should also be negatively affected by an oil-induced real currency appreciation, thus helping to contain mangrove deforestation (e.g. Parks and Bonifaz 1995). Hydroelectric dams (Rodríguez 1991; Richards 1994) and oil pipelines (see Chapter 6 on Cameroon) are other examples of projects that have been developed to generate or save scarce foreign exchange. If external investment financing is available, these projects may be more acceptable domestically when a country is in crisis than when it is facing an oil bonanza. All these sectors or projects potentially have an expansionary impact on land use, which is likely to accelerate the loss or degradation of forests.

Stating the hypotheses

Spatial scale

The validity and theoretical underpinning of these three principal linkages – less forest conversion to agriculture, less degradation from timber extraction and other reductions in forest interventions – will be discussed in detail in the rest of this chapter. But first it may be worth while formally stating the theoretical expectations as hypotheses. An initial question here is at what *spatial scale* we would expect the land-use linkages to operate. The Dutch Disease is basically a national-scale phenomenon. Oil and minerals are publicly owned underground resources. As we saw in the last chapter, the spending effect of a boom depends entirely on the government's budget priorities. Nominal currency exchange-rate appreciation would apply to an entire country, as the same currency is used over its entire territory. Yet *real* currency appreciation may not be identical nationwide, as inflation rates may differ across regions. There are examples of specific regional booms, either because of strong resource movement effects (e.g. workers are offered higher wages to work in the region's oilfields) or because of a geographic concentration of revenue spending (e.g. governments spending most of their oil money in the capital).[1] Accordingly, land-use effects will also depend on where the main agricultural areas are located, and which crops are hit by competitiveness declines. In other words, our hypothesis should be applied on the national scale, but taking into account the basic geography of the country, that is, without losing sight of the fact that impacts may differ across regions.

Time scale

A second *a priori* concern is the *temporal* dimension. As mentioned in the Introduction, land-use changes are unlikely to operate fully, or to be measurable in practice, on a year-to-year basis. A main concern of this book is therefore with *long-term* changes, analysing to what extent the *emergence* of oil wealth has made a structural difference to land use and forests in the study countries. Another question is whether, in the *medium term*, change in the *amount* of oil wealth that causes either bonanzas or crises has an impact on land use. This impact of a relative shortage or alternatively abundance of foreign exchange also has clear implications for non-oil-producing countries. Let us start with the long-term hypothesis:

> **Hypothesis 1**: *Tropical forested countries specialising in oil and other high-rent mineral exports tend to face lower deforestation and forest degradation than other tropical forested countries.*

The long-term differential forest conservation outcome is difficult to measure as a time-series hypothesis. It does make sense for certain long-established oil exporters, such as Venezuela in our sample. This country experienced a historical transformation from a land-expansive agrarian nation to a largely urbanised economy with low deforestation pressures – a process that fully coincided with the oil boom. As we can also demonstrate the oil-induced factors that were behind this change in land use, it becomes plausible to attribute a major role in Venezuela's low degree of forest loss to oil. However, the argument necessarily involves implicit long-term assumptions about hypothetical 'non-oil' development paths, which for some countries may be controversial. It would thus be preferable to carry out supplementary comparative tests at the country level. This could be a comparison between groups of oil/mineral versus non-oil/non-mineral countries (see next section). Within the sample range covered in this book, we could even compare countries with a large per-capita oil windfall (Gabon or Venezuela) with regional neighbours with similar preconditions, but less oil wealth that fluctuates more (Cameroon or Ecuador – see Chapter 10).

> **Hypothesis 2**: *Tropical forested countries specialising in oil and other high-rent mineral exports will, over time, experience lower deforestation and forest degradation during boom periods than during bust periods.*

The second hypothesis is fully internal to the group of oil- and metal-exporting countries, and could thus in principle be tested exhaustively at the time-series level: when national oil wealth goes up, deforestation and forest degradation are supposed to go down, or even be reversed. For countries that have become oil exporters more recently, such as Cameroon or PNG, one might in principle be able to compare the 'before' and 'after' situations. Otherwise, however, unlike the previous hypothesis, this one would not involve an 'oil' versus 'non-oil' dichotomy, but rather the impact of a shifting, measurable *degree* of oil wealth over time. This makes Hypothesis 2 the most relevant for application beyond the sample of oil-producing countries, allowing comparisons with other fluctuating macroeconomic variables, such as foreign debt, financial crisis, boom and bust in foreign investments, etc. In a very oil-rich country facing a bust with an economic crisis (e.g. Venezuela), we would not expect to find land-use structures that are fully identical with those of non-oil countries. We would rather expect to see *changes at the margin*, with some additional pressures on forests, making that country somewhat more similar to the group of non-oil-producing countries.

Boom and bust types

The popular terms 'boom' and 'bust' can cover a variety of situations that prove to be significantly different in terms of their intensity, (a)symmetry and duration. This is illustrated graphically by the plotted examples of boom-revenue time-profiles in Box 2.1. One factor of distinction is whether we are dealing with a price boom, typically with revenues that fluctuate from year to year, or a quantity boom, for example, the discovery of a new oilfield, often with a longer duration. As indicated, in the 1970s and 1980s many high-absorbing oil and metals exporters used the increased creditworthiness of their booming economies to borrow more funds on international capital markets. Where that pattern is applicable, foreign indebtedness should thus be analysed as an integral part of the country's boom–bust adjustment, since credits were only made possible by the resource boom in the first place. Ultimately, Hypothesis 2 about the land-use impact of oil-induced foreign-exchange fluctuations should thus be investigated formally, preferably using statistical methods, such as regression analysis.[2]

Do mineral rents globally protect tropical forests?

The basic argument implicit to both hypotheses, namely that oil and mineral production help to protect tropical forests, is likely to be greeted with some disbelief, to put it mildly. After all, oil and mining companies have been a main target of numerous international campaigns to save the rainforests. Even if this book could show that Hypothesis 2 holds

Box 2.1 Different types of booms and busts

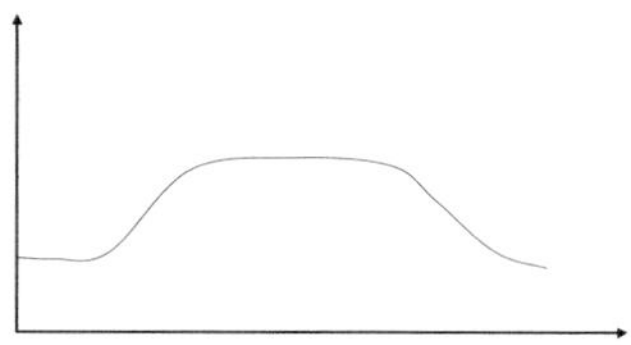

(a) Temporary boom

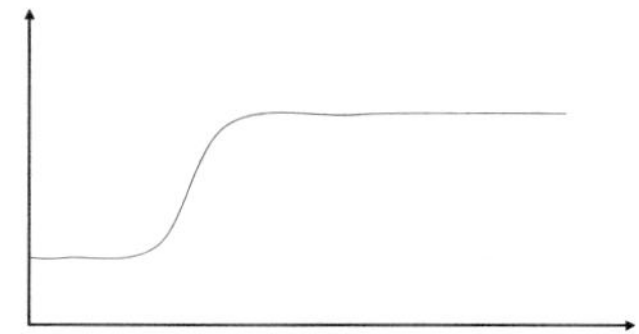

(b) Permanent boom

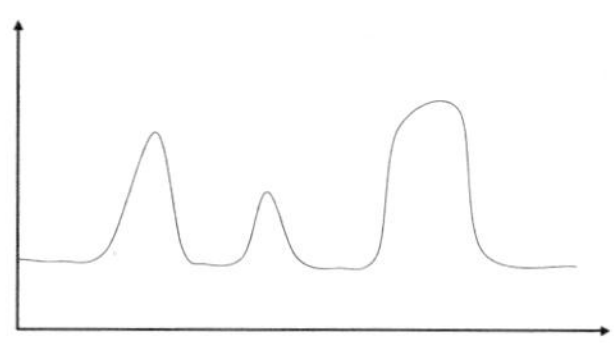

(c) Repeated mini-booms

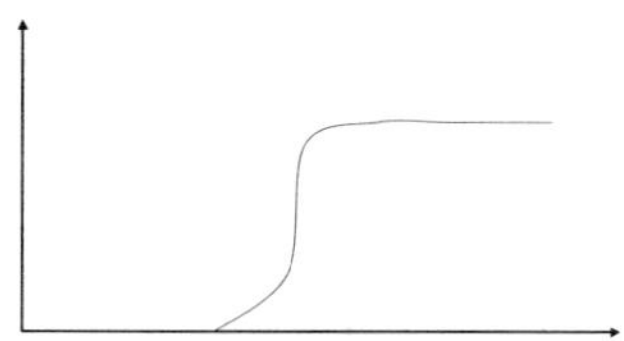

(d) Booming new resource exporter

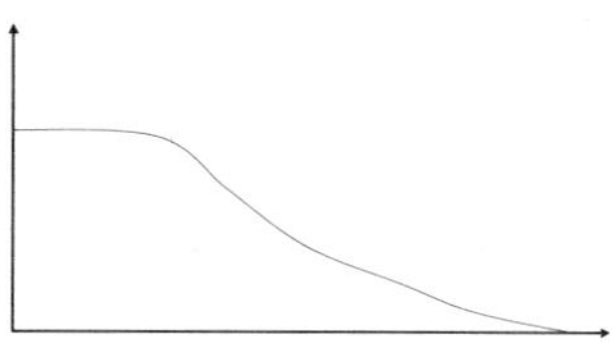

(e) Resource-exhaustion bust

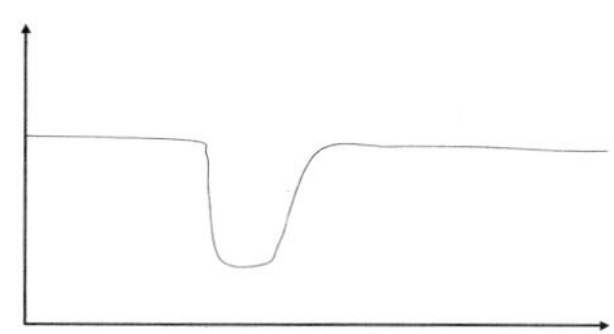

(f) Mini-bust

Notes
X-axis: Time.
Y-axis: Foreign-exchange inflows from a booming sector.

true for all (or most) of the five study countries, the sceptical reader may reject that approach *a priori* by claiming that a small sample of five countries cannot be representative. In the spirit of persuading such sceptics not to put the book aside at this early stage of reading it, we shall thus start by presenting some aggregate cross-country findings that may be considered to support Hypothesis 1.

An initial overview exercise is to see how much forest tropical mineral exporters control compared to other tropical countries. The forest cover of twenty-three specialised exporters of minerals (oil and metals)[3] is given in Table 2.1, using data from the last two FRA by the FAO – two alternative though not necessarily compatible sources.[4] For 1995, we can see that the relatively small number of mineral-exporting countries control almost half of all remaining tropical forests in Africa and in the Asia-Pacific region. In the FRA for 2000, this share is even higher (63.3 per cent in Africa, and 57.1 per cent in Asia and Oceania). Due to the dominance of a single country, Brazil, the 1995 figure is only

Table 2.1 Tropical forest cover of high-mineral exporters, 1995 and 2000

	Forest area ('000 ha)		2000	
	1995	*2000*	*Land area* ('000 ha)	*% Forest cover of land area*
Africa				
Angola	22,200	69,756	124,671	56
Cameroon	19,598	23,858	47,544	50
Congo	18,537	22,060	34,200	65
Dem Rep of the Congo	109,245	135,207	226,705	60
Gabon	17,859	21,826	26,767	82
Guinea	6,367	6,929	24,586	28
Liberia	4,507	3,481	11,137	31
Nigeria	13,780	13,517	91,283	15
Togo	1,245	510	5,679	9
Zambia	31,398	31,246	74,135	42
All mineral exporters	244,736	328,390	666,707	49
All tropical forests*	504,901	519,000	n.a.	n.a.
Forest share of mineral exporters	48.5%	63.3%	—	—
Asia/Oceanic				
Brunei Darussalam	434	442	577	77
Indonesia	107,791	104,986	190,457	55
Malaysia	15,471	19,292	32,975	59
Papua New Guinea	36,939	30,601	46,284	66
All mineral exporters	160,635	155,321	270,293	57
All tropical forests*	321,669	272,000	n.a.	n.a.
Forest share of mineral exporters	49.9%	57.1%	—	—
The Americas				
Jamaica	175	325	1,097	30
Mexico	55,387	55,205	195,820	28
Trinidad and Tobago	161	259	513	50
Bolivia	48,310	53,068	109,858	48
Ecuador	11,137	10,557	28,356	37
Guyana	18,577	16,879	21,498	79
Peru	67,562	65,215	128,522	51
Suriname	14,721	14,113	16,327	86
Venezuela	43,995	49,506	91,206	54
All mineral exporters	260,025	265,127	593,197	45
All tropical forests*	907,389	780,000	n.a.	n.a.
Forest share of mineral exporters	28.7%	34.0%	—	—
Global				
Mineral exporters	665,396	748,838	1,530,197	49
All tropical forests*	1,733,959	1,571,000	n.a.	n.a.
All tropical forests (excluding Brazil)*	1,182,820	1,038,519	n.a.	n.a.
Forest share of mineral exporters	38.4%	47.7%	—	—
Forest share of mineral exporters (excluding Brazil)	56.3%	72.1%	—	—

Source: FAO (1997, 2002).

Note
* Figures are not based on national statistics, but estimated from the FAO's pan-tropical remote-sensing survey.

28.7 per cent for the Americas (34 per cent in 2000). The share of land area covered by forest is also high in mineral-exporting countries: the weighted average is 49 per cent in Africa, 57 per cent in Asia and Oceania and 45 per cent in the Americas. Globally, the twenty-three mineral exporters controlled 47.7 per cent of tropical forest cover in 2000 (38.4 per cent in 1995). If Brazil, historically and at the federal-state level itself an important mineral exporter, is excluded from the calculus, the forest-cover share of mineral exporters in 2000 is no less than 72.1 per cent (56.3 per cent in 1995).

Mineral exporters' share of remaining tropical forest stocks is thus extraordinarily high, but one should also look at the speed of current forest loss in these countries. Sunderlin and Wunder (2000: 315–18) examined this for the same twenty-three countries, using the binary classification of high versus low deforestation from Rudel and Roper (1997a), for both the 1976–80 and the 1981–90 periods. This deforestation measure is not based on the FAO's figures, but on a series of reports about forest-loss trends at the sub-national level. A chi-square test showed a significant relationship between high mineral exports and low deforestation for both periods. But, not surprisingly, the relationship was more significant (at the 1 per cent level) for 1975–80, a mineral boom period characterised by high international oil and metals prices, than for 1981–90, when the boom turned into bust in most of the countries concerned. Mainardi (1998), using forest-cover data from the FAO's FRA for 1990, also finds that sixteen developing-country mineral exporters had a lower deforestation rate during the 1980s than those countries not specialising in minerals.

At this point, one might rightly object that a bivariate, positive correlation between a country's high mineral exports and its high remaining forest cover and low deforestation rates may be a deceptive indicator. We know from the deforestation literature that many different factors cause forest loss. Hence, in a fully developed multivariate model in which the impact of these other variables is controlled for, the partial forest-protecting effect of high mineral exports might simply vanish. However, this suspicion does not hold true. Sunderlin and Wunder (2000: 320–2) replicated a consolidated cross-country model in the study by Rudel and Roper (1997), mentioned earlier, and added mineral exports to the model as an additional explanatory variable. In a sample of sixty-six countries, mineral exports proved to be an important predictor of deforestation rates, with the expected negative sign: high mineral exports cause lower deforestation rates, even when other variables, such as road-building, topography, foreign debt and population growth, are controlled for.

Similar multivariate results by Mainardi (1998: 39–40, 47) also confirm the general curbing effect on forest loss of the bias towards mineral exports. In his regression model, a dummy variable for the sixteen high mineral exporters in a sample of forty-two tropical countries is estimated with a positive, significant figure. However, the opposite is true for a smaller subset of five high deforestation countries, where large open-pit mines or alluvial deposits are located specifically in forested areas. This indicates that while mineral *rents* may protect forests, site-specific production activities in forests are obviously detrimental. A statistical test of independent means (Sunderlin and Wunder 2000: 318–20) gives us some indications as to why mineral rents in general help to maintain forests: mineral-rich countries have a significantly smaller proportion of their GDP, labour and land in agriculture, and a larger proportion of urban population. Again, the multivariate coefficient for mineral exports is more significant (at the 5 per cent level) for the 1976–80 mineral boom period than for the 1981–90 bust period (α = 10.97 per cent).

To sum up, this section has presented evidence that tropical countries specialising in mineral exports have more forest left, and lose this remaining forest at a significantly slower rate, than other tropical countries. This result is fairly robust for different measures of forest cover, deforestation and mineral export specialisation, as well as for different (bivariate and multivariate) model types. There are preliminary indications that mineral rents indirectly protect forests through an 'urban bias', that is, by encouraging more rapid urbanisation and discouraging agricultural development. However, the significance of this forest-safeguarding effect is somewhat variable both over time and between countries. This further underscores the relevance of explicitly testing Hypothesis 2 on the land-use adjustment over time of mineral-exporting countries, which is the main objective of this book.

Trade, macroeconomics and forests: what do we know?

Hypotheses 1 and 2 both presuppose a macroeconomic link between mineral rents and RP, changing land use and forest conditions. But what is the current 'state of the art' in this field? This is not the place for a complete review of the macroeconomic drivers of deforestation.[5] But before we develop a framework for oil-wealth impacts on forests, we should take a closer look at findings from other studies into external and trade-induced factors, since they are closely related to the oil-wealth situation. Variables of particular interest would be foreign-exchange in- and out-flows, foreign indebtedness, terms-of-trade fluctuations, and a number of policy variables mediating the effect of external changes on RPs and structural adjustment (e.g. devaluation, trade liberalisation, etc.).

In deforestation analyses, it has become increasingly acceptable to distinguish between three different causal levels (Kaimowitz and Angelsen 1998): deforestation *sources* (e.g. mining or agricultural expansion), *immediate causes* (e.g. road-building or new agricultural technologies) and *underlying causes* (e.g. trade liberalisation or population growth). Obviously, the factors relating to external trade and macroeconomics all belong to the last group, that is, to the underlying causes of forest loss (Contreras-Hermosilla 2000). The effects of these factors on forests have been investigated in three types of model. First, non-empirical, analytical models contribute exclusively to the development of theory. Second, computable general equilibrium (CGE) models aim to quantify complex links and multi-sectoral spillover effects by building a dataset for single countries. Third, cross-country regression models, like those discussed in the previous section, employ structural differences between countries over a given period to simulate stylised patterns of change over time. These methods will be discussed further, but perhaps the single most important common finding is the complexity and variability of the linkage between the macroeconomic level and the forest. Table 2.2 reproduces selected results from a global review of about 150 economic deforestation models (Kaimowitz and Angelsen 1998), focusing specifically on three open-economy factors: devaluation, external debt and trade liberalisation. Only empirical models (cross-country and CGEs) are considered. Among the three factors, devaluation has an unambiguous impact in terms of accelerating deforestation, especially in the short run. Basically, this is due to improving RPs for agriculture, which leads to expanding agricultural production and forest conversion and to improved incentives for timber harvesting. Trade liberalisation is analysed in some of the CGE

Table 2.2 The effect of selected open-economy variables on deforestation

Type	*Study*	*Country*	*Independent variables*		
			Devaluation	*Trade liberalisation*	*Foreign debt*
CGE	Aune *et al*. (1996)	Tanzania	Increase		
	Barbier and Burgess (1996)	Mexico		Increase	
	Cruz and Repetto (1992)	Philippines	Increase	Increase	
	Lopez (1993)	Ghana		Increase	
	Mwanawina and Sankhayan (1996)	Zambia	Increase		
	Thiele and Wiebelt (1994)	Cameroon		Reduction	
	Wiebelt (1994)	Brazil	Increase		
MCR	Bawa and Dayanandan (1997)				Increase
	Burgess (1991)				Increase
	Capistrano (1990)			Increase	Mixed
	Gullison and Losos (1993)				No effect
	Inman (1993)				Mixed
	Inman (1990)				Mixed
	Kahn and McDonald (1995)				Increase
	Kahn and McDonald (1994)				Increase
	Kant and Redantz (1997)				Increase
	Kimsey (1991)				No effect
	Mainardi (1998)		Increase		Increase
	Rudel and Roper (1997a)				No effect
	Rudel and Roper (1997b)				Mixed
	Rudel and Roper (1996)				Increase
	Shafik (1994a,b)		Increase		No effect

Source: Kaimowitz and Angelsen (1998: 64, 85).

Note
Blank spaces indicate that the variable was not analysed in the particular study. These are results from multi-country regression (MCR) and computable general equilibrium (CGE) models.

models, and three out of four models show an effect promoting forest loss, again by promoting primary traded sectors using land and forests. On the other hand, overvalued exchange rates and protectionist policies that harm agriculture and timber exports clearly tend to slow down deforestation (ibid.: 97).

The most relevant variable for our purposes is external debt, because continuous debt service payments represent net outflows of foreign exchange, that is, the *opposite* situation to an oil or metals boom. Table 2.2 shows that seven out of fifteen cross-country models predict an increase in deforestation from higher debt, four find the impact to be mixed, while another four identify no significant impact. The signs and significance of the debt variable are highly inconsistent across studies. This is attributable to different debt and deforestation indicators, but different time periods and model specifications also play a role. For instance, Kahn and McDonald (1995) concluded, from a sample of deforestation in sixty-eight tropical countries from 1981 to 1985, that higher debt triggered significantly higher forest loss. Higher debt service payments allegedly marginalised many people and caused poverty-led deforestation, while also forcing the indebted countries to adopt myopic 'resource-mining' strategies in order to expand foreign-exchange earnings from timber and agricultural exports. Rudel and Roper (1997) analysed debt impacts in the context of a sixty-seven-country dataset with two sub-periods, the 1970s and 1980s. Foreign debt per capita had a significant accelerating impact on deforestation in all the regressions. In particular, debt is confirmed to be a push factor for deforestation in countries with smaller, more fragmented forests. On the other hand, a similar cross-country study by Capistrano and Kiker (1995), using data from forty-five developing countries over the 1967–85 period, reached the opposite conclusion. Debt service was not significant, except for the 1972–5 sub-period, where its impact was actually negative, that is, higher debt service caused *less* deforestation. This may result from the high availability of new international credit during this sub-period, which would more than counter higher debt service payments. Thus a measure of net capital inflow would probably be a better indicator. In terms of bivariate correlation coefficients, Angelsen and Culas (1996: 15–17) found no significant connection between debt and deforestation at either the global level or for sub-samples according to income levels or periods. It seems fair to say that most cross-country multivariate regressions indicate a conditional relationship of more debt causing more deforestation, but they do not provide us with much knowledge about the nature of the link between capital outflows and deforestation.

A more reliable route is thus to look at national time-series, if these are available. In these studies, debt-crisis impacts prove highly variable from country to country, which also foreshadows the variability of oil-wealth effects, to be analysed in the following chapters. In Brazil, for example, it is widely acknowledged that the debt crisis of the 1980s curbed Amazon deforestation, mainly by reducing the availability of public finance for road-building and settlement programmes, and by limiting private agricultural investments (Young 1995: 19). Higher debt thus went hand in hand with less deforestation. In the WWF's comparative studies on structural adjustment and the environment, it proved hard to identify unambiguous relationships between debt and deforestation (Reed 1992: 143–6). For example, Ivory Coast had the highest deforestation rate in the world, but this could not be linked to large levels of foreign indebtedness. As in the cases of Thailand and Mexico, forest loss was associated rather with specific land-extensive national development strategies. These root causes of deforestation were already at work in the 1970s, before a high foreign debt had been accumulated and was being serviced under the soaring real interest rates of the 1980s.

To sum up, devaluation usually promotes higher levels of deforestation, and trade liberalisation in many cases has the same effect. However, the latter will ultimately depend on the specific sectors from which negative or positive trade biases are removed, and how these sectors are linked to land use. On the other hand, it can be concluded that 'the link between debt and deforestation rates is tenuous' (Angelsen and Culas 1996: 19). In most cases, foreign-exchange outflows increase pressures to devalue and to increase exports, but much will depend on how existing policies and development strategies are affected. In some forest-rich countries, wasteful land-use projects may, for example, be curbed by shortages of foreign exchange. To analyse these effects, comparative case studies are probably better instruments than CGE models. Things are even worse with global regression models that implicitly assume a large degree of homogeneity in a country's development and land-use trajectories.

A note on methods

Opposed effects

The challenge is thus to make explicit how the macroeconomics are linked to forests over time in mineral-rich developing countries. From the above, we already have an idea of the complexity of these linkages. In developing a framework for the case studies, it is important to recognise contradictory effects. In a recent analogous analysis of the root causes of biodiversity loss, micro-level environmental changes in ten regional study areas were carefully linked to trade and macro-policy changes (Wood *et al.* 2000). But it seemed surprising that all the underlying linkages that were identified were supposed to contribute unambiguously to biodiversity loss: there were no 'stabilisers' in the system. As will be shown in what follows, it is more realistic to expect some underlying factors to work towards higher environmental pressure while others work against it, the net effect depending on the balance between them.

Let us take an example. Higher income from an oil boom has a 'core' spending effect that favours urban non-tradables, causes 'de-agriculturisation' and thus reduces deforestation. But what if higher incomes also change food habits, for example, towards increased beef consumption, thus promoting a land-hungry livestock sector? Or what if the government uses a large share of its oil money to open up new settlement frontiers, for example, to finance the colonisation of remote forests? In both cases, stronger forces may overpower the core effect of the resource boom in curbing deforestation. Various ambiguities may apply, so it is necessary to count positive and negative effects alike, determining their relative weight and net impact.

Partial versus general approach

A fundamental choice in describing the relations between the macroeconomic level and forests is whether to use a partial or a general equilibrium approach. The big advantage of the CGE approach (see Table 2.2 for study references) is in the analysis of high levels of complexity. It can cope with simultaneous systems, feedback loops and weighting effects that go in opposite directions, and thus ultimately detect counterintuitive impacts that may differ from those one gets by simply adding up partial impacts.

Yet CGEs also have considerable weaknesses. First, deforestation is often concentrated in certain regions and places where specific dynamics, neighbour-effects and interactions between local deforestation agents apply. Although some models try to distinguish macro-regions (e.g. Cattaneo 2001), the spatial specificity of deforestation seems almost impossible to capture in CGE models. Second, CGEs are heavily data-intensive, presupposing knowledge of a range of variables and parameters, which in practice is unavailable in most developing countries. In the process of economic modelling, parameters are thus often tacitly copied from other models or countries, and behaviour is assumed to take certain functional forms that turn out to drive key model outputs (Kaimowitz and Angelsen 1998: 67). This is even more problematic for the land-use component of such models, where data gaps and insecurities are overwhelming (see Chapter 1). New economic history has shown that good theoretic modelling can sometimes help to bridge gaps arising from bad or absent data, but the assumptions made and the influence they have need to be fully transparent. CGE models are anything but transparent. To the non-specialised reader, and sometimes even to fellow modellers, CGEs are 'black boxes' delivering an output that is impossible to trace back to its different causal components and in which the importance of a myriad of assumptions made by the model builder are very difficult to check.

Faced with these methodological difficulties, the approach adopted in this book is a partial one. The advantage of this method is that it provides a flexible framework for pulling together a patchwork of multiple factors, in spite of incomplete data coverage and different spatial scenarios. It does so in a fully transparent manner that does not 'bridge' knowledge gaps by concealed model assumptions. The disadvantage is that a 'recursive' (hierarchical) system is used that does not allow for multi-factor feedback loops on 'high-level' variables (e.g. from higher deforestation back to national income or consumption). It is also only a semi-quantitative method, so it may not always fully resolve ambiguities relating to the weighting of partial impacts with opposite numerical signs. In these cases, a partial method must also draw on the *ad hoc* judgements of the analyst and his readers.

Which baseline is relevant?

Finally, in discussing Hypothesis 2 above, it was mentioned that the main interest was in the long- and medium-term adjustment to changing oil and mineral wealth. The choice of not constructing a full dynamic model thus has consequences for the baseline chosen for comparison, in the sense that Dutch Disease adjustment to a mineral boom is *superimposed* on common trends of growth and structural change in a developing country's economy (see Chapter 1). Some of the most common trends are:

- economic growth and population growth;
- a gradual shift from primary to secondary and later to tertiary sectors (GDP and labour share);
- long-term urbanisation;
- reduced elasticity of food (especially staples) consumption over growing household incomes;
- long-term expansion of cultivated area.

Obviously, there will also be country-specific trends and events that are unrelated to oil wealth (natural disasters, civil war, abrupt political changes, etc.), but significantly affect both the macroeconomic picture and land-use changes. This means that it is only in the most dramatic cases of changing mineral wealth that we may see our hypotheses confirmed in *absolute* terms. However, in other cases where mineral wealth was not the single dictating development feature but one among many, we may still find our expectations confirmed *at the margin of* other tendencies in the economy, that is, in terms of trend changes or an altered *rate* of change. For instance, a lower rate of deforestation during an oil boom, compared to the pre- and post-boom periods, is clearly a relative confirmation of our core hypothesis, given that other exogenous trends (such as population and consumption growth) continue a general drive towards expansionary land use. But obviously, even stronger confirmation would be provided if forest cover actually started to *increase* in absolute terms because rural cultivation areas were being abandoned directly as a response to the mineral boom. It is thus important to distinguish between absolute and relative assessments to enable the reader to know when a hypothesis is to be confirmed absolutely or relatively, and when it is to be rejected outright.

A framework for national case studies

From underlying causes to sources: avoiding hasty conclusions

Over the past decade, the deforestation debate has reached an increasing level of consensus about the dominant role of 'underlying causes'. As expressed in the WWF's and IUCN's *Forests for Life* policy book: 'the real causes of forest destruction can occur far away from the forest itself. Key issues include: current consumption levels, international debt and structural adjustment, pressure for trade and development' (WWF and IUCN 1996: 12). On the other hand, there is a temptation to draw up relatively simple causal links, as when the same source goes on to say that '[o]f the 17 most indebted countries, 14 have tropical forests. ... In practice, debt servicing is often achieved by cashing in natural resources such as timber' (ibid.: 13). A set of interlinked misconceptions is commonly derived from this type of direct underlying causal explanation. One view is that forest decline is unambiguously and uni-directionally linked to all the 'big problems' of society, such as poverty, inequality, globalisation, population growth, excess consumption and the foreign debt crisis. A second is that the environmental decline of forests and other biome is 'over-determined' by these factors (e.g. Wood *et al.* 2000: 33–5), so that it cannot be halted unless most or all of these factors are addressed simultaneously. A third is that proximate agents of forest decline should be absolved from any responsibility for environmental destruction: people with chainsaws and axes in their hands are merely 'victims' or 'instruments' being driven deterministically by the distant underlying evils of modern society.

These may be overly pessimistic conclusions concerning the actual options for forest conservation. The graphical framework presented below uses a step-wise recursive system, moving gradually from 'underlying' to 'proximate' causes and 'sources' of deforestation. This procedure is meant to avoid the common confusion over which factors are the 'most underlying' ones in a complex causal chain (Contreras-Hermosilla 2000: 5–6). Along the way, we shall encounter very few uni-directional causalities, but rather a mix of factors drawing the forest outcome in opposite directions: some of these underlying factors put additional pressures on forests, while others alleviate them. This underscores the fact that, unless the analysis 'follows through all the steps' from the supposed underlying causes to the sources and agents of deforestation, there is a severe risk of oversimplification and of producing inadequate policy recommendations. Key factors of leverage for the forest outcome of a mineral boom will be identified.[6] Figure 2.1 (see left-hand side) is divided into five different levels of analysis:

1 External (foreign-exchange inflows from minerals and foreign borrowing).
2 Macroeconomic (national income, demand structure, wages and RPs).
3 Sectoral (traded versus non-traded production, and urban versus rural sectors).
4 Land use (sectoral area expansion or contraction).
5 Forests (degradation and deforestation – environmental impacts).

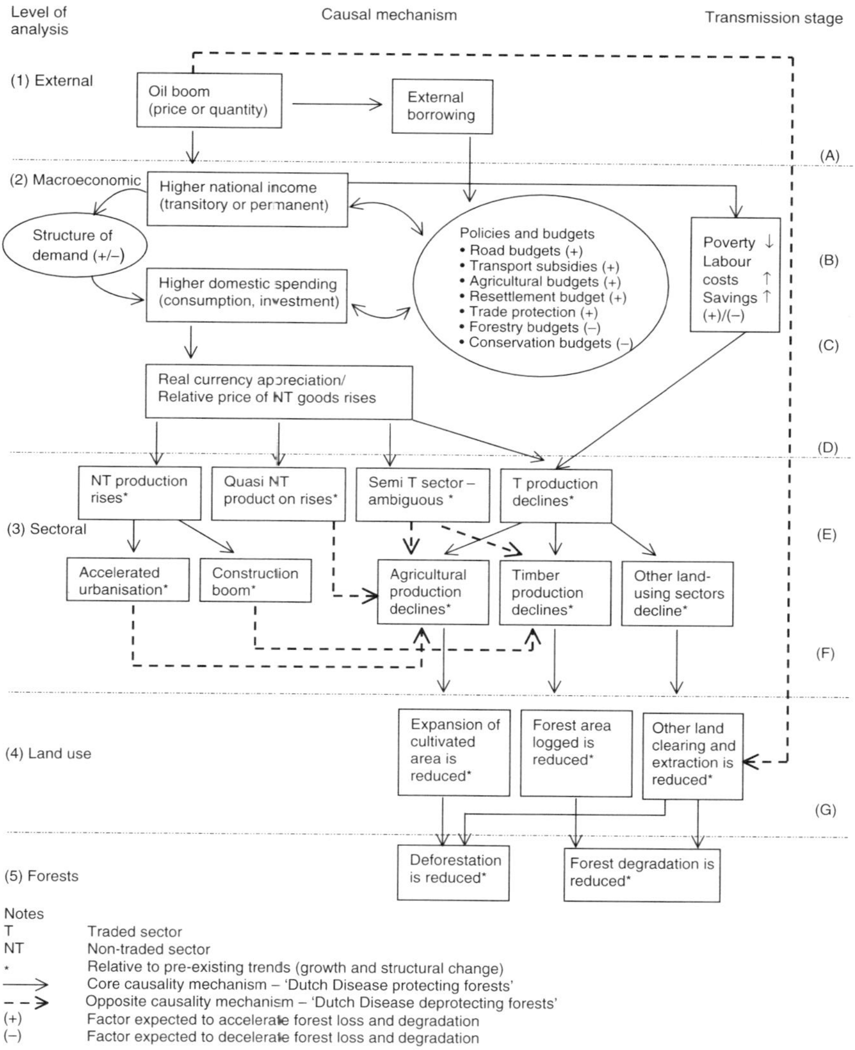

Figure 2.1 Linking resource booms to the forest.

A graphic illustration

From top to bottom, the stages of transmission in Figure 2.1 have also been marked (A–G) for easier reference (right-hand side), as each of them involves additional suppositions. The mainstream effect, that is, the boom alleviating pressures on forests, is expressed throughout the figure by fully drawn lines; corresponding policy and budget effects are marked by '(−)'. Contrary effects working towards higher forest clearing are shown by dotted lines;

the policy and budgetary aspects that are expected to have this effect are marked '(+)'. Note that, from the third level downwards, predictions of stock variables (production, land use, forest size, etc.) are relative, that is, changes are superimposed on pre-existing trends. Here are some explanations for each analytical level (with literature references in the endnotes):

(1) *External level*. As explained in Chapter 1, shifting mineral export prices (e.g. a price boom) and mineral production quantities (e.g. new resource discoveries, or the exhaustion of existing resources) together determine the size of the foreign-exchange inflows for the 'booming sector'. Mineral revenues feed directly into current public budgets, but absorption effects may be mediated by commodity stabilisation funds (as in PNG) or by extra-budgetary savings accounts abroad (as happened in Cameroon). Access to foreign borrowing may be directly related to the boom, justifying the explicit inclusion of this variable.

(2) *Macroeconomic level*. High capital inflows raise both national income and consumption, but for publicly owned underground resources like petroleum and metals, this mainly happens through the public sector (transmission stages A and B). Mineral revenues and public foreign borrowing are generally absorbed through budget allocations, that is, public wages, consumption, investments and transfer payments, with only few minor exceptions.[7] The impact on the rest of the economy is thus directly contingent upon a set of budgetary priorities and policies. Higher public spending has some general effects on the economy and land use, but the literature on deforestation shows that selected spending categories are likely to trigger the following policy-led impacts on land use and forests:

- *Roads* are the single most likely instrument increasing deforestation and forest degradation. This is mainly the case if new roads are built through forested areas, less so if existing roads are improved.[8] The road effect is spatially explicit (opening up specific new locations) and long term (lasting beyond the boom period). Note that rail or waterway constructions or improvements can have similar effects by lowering transport costs.
- *Transport and energy subsidies*. Lower private transport costs through subsidies will also increase forest degradation and forest loss in remote frontier forest areas by linking them more to markets for forest and agricultural products. As a minor alleviating effect, urban consumers may switch more rapidly from firewood to subsidised fossil fuels, reducing any potential over-harvesting pressures.[9]
- *Agricultural budgets*. Higher spending on agricultural projects, investments in parastatal enterprises, better rural extension services, higher input subsidies and lower export taxes are all measures that can increase deforestation. This happens if they effectively increase total demand for cultivated land (but *not* if they overwhelmingly cause intensification that limits aggregate land use), and if this additional land is taken from the forests.[10]
- *Forestry budgets*. Higher spending on forestry agencies may reduce forest degradation, in particular if the additional resources help improve the enforcement of forestry laws and practices on the ground.[11]

- *Conservation budgets*. Higher spending on protected areas (equipment, park rangers, etc.) and other forest conservation initiatives reduces deforestation and degradation, but only if the money is spent in a way that effectively reaches forest conservation objectives.[12]
- *Directed settlement*. Government-sponsored resettlement programmes, like *Transmigrasi* in Indonesia, or geopolitically motivated frontier settlement, as on the disputed forested border between Ecuador and Peru, promote deforestation if people are moved into forests to clear new land.[13]
- *Trade protection and subsidies*. Governments often choose to spend some boom revenues to subsidise agricultural sectors that are hit by declining competitiveness. This limits the decline in cultivated area, and may even increase deforestation and degradation, if some land- or forest-using sectors are fully protected by import quotas or prohibitive tariffs (*quasi non-tradables*).

The land-use impact of a mineral boom thus depends crucially on how these selected components are represented and weighted in the accompanying policy package. In each case, it is necessary to find out which budgets and policies in particular benefited from mineral booms or from the long-term presence of oil wealth. The second task is then to see how they interacted with specific economic sectors, and to what extent they fed through to forest cover and quality.

However, there are also some more generalised spending effects. Increased demand always causes real currency appreciation, through some combination of rising prices and an appreciating nominal exchange rate, shifting RPs and production incentives in favour of non-tradables (i.e. the 'core effect' discussed in Chapter 1). Income distribution and demand structures change, which, as indicated in Figure 2.1, may have contradictory impacts. First, if poverty is reduced, and real wages and the opportunity costs of labour rise, this will raise costs and squeeze profits in T sectors. Some highly labour-intensive but economically marginal activities, such as slash-and-burn farming and firewood gathering, will typically decline. On the other hand, reduced poverty may also free up new resources for savings and investment in rural land-using and forest-converting activities. The net impact of poverty reduction on forests is thus *a priori* ambiguous: it depends on the weight of different factors in a specific situation.

Second, many studies have shown that higher incomes change the structure of aggregate demand. Luxury goods increase their share of household consumption, inferior goods reduce their share, and some goods decline even in absolute terms.[14] Changes in food demand often have an impact on land use. For instance, dairy and meat products are often luxury goods. If a higher income causes a disproportionate rise in demand for them, and if, because of import restrictions, that demand is mostly satisfied domestically, this may increase forest loss. This is because animal foodstuffs require much higher biomass inputs per calorie produced than plant sources. But preference changes connected with higher incomes may also induce a shift away from local staples to exotic ones like rice and wheat. This may reduce deforestation if the former come from land-extensive slash-and-burn systems while the latter are either imported or produced in more land-intensive, permanent systems. Again, the aggregate effect of a changed demand structure on forests is product- and case-dependent.

(3) *Sectoral level*. As explained in Chapter 2, the change in RPs increases both non-traded production (private and public services, construction) and quasi non-traded production (import-protected manufacturing and other selected sectors). Semi-traded sectors experience shifting levels of import protection over time or across sub-sectors. This often applies to agricultural sub-sectors, and the outcome is ambiguous. T sectors decline, among which are always the export sectors (e.g. coffee, shrimp-farming, timber or mining) and that part of home-market production in agriculture, forestry, fishery, etc. that competes with imports. As most NT and quasi NT sectors are urban, the mineral boom will accelerate rural–urban migration. Overall, urbanisation is expected to alleviate pressures on forests. Although pressures on periurban forests may increase in order to supply crops, livestock and dairy products for the growing cities (see F), some urban food consumption will typically be imported. Domestic crop production for urban markets will also tend to be more intensive and use less land than if the same population was living in rural areas with farming as their core livelihood activity. A construction boom can raise domestic demand for timber (for construction poles, furniture, etc. – see F), also counteracting the core reduction effect. These contrary effects will be stronger the more imported-protected these sectors are, allowing them to rise concurrently with domestic purchasing power.

(4) *Land-use level*. Curtailing aggregate agricultural production, whether in absolute terms or relative to previous trends, should also decrease the demand for expanding cultivated areas, compared to what would have happened without a mineral boom (see F). Likewise, declining extraction of timber and other forest products should reduce forest degradation pressures. If the overall *land-use intensity* of agriculture or forest extraction changes, for example, due to product or technology changes affecting aggregate production functions, this will of course change the effect. For instance, appreciated RERs will explicitly favour the use of imported inputs such as fertilisers, pesticides or machinery, which could lead to intensification and a reduction in area *beyond* what is implied by production trends alone. Similarly, if timber extraction under the new high-cost scenario focuses more on a few high-value species (less quantity extraction per hectare), logged-over areas would decline *less* than in proportion to timber extraction. Finally, oil and metals extraction normally also directly affects land use, for example, through open-pit mines, oil-drilling platforms, access roads, use of timber for extraction, etc. This direct effect on land use is expressed by the right-hand dotted line from 'oil boom' to 'other land-clearing and extraction'.

(5) *Forest level*. Lastly, the transmission (G) involves a distinction between 'affected area' and 'affected forest area'. A main question is to what extent land-use effects overlap spatially with forested regions. For instance, if all of a country's currently exploited oil reserves are located off-shore, the on-site area impacts of oil production will not affect forests at all, although off-site pipelines and roads may go through forests (see the next section). Likewise, if a new sugar plant triggers an expansion in sugar plantations solely into natural grasslands, this land-use impact would also not translate into an on-site forest impact, although again off-site effects may occur.[15] This indicates that, although in the tropics natural forests generally constitute the main 'land reserve' from which additional land uses are drawn, one must make an 'overlap check' to determine to what extent this applies to specific land-use changes. For non-converting forest-degradation processes, one must

also make some basic checks. If a mineral boom significantly reduces, say, commercial firewood collection and selective timber extraction, this would be expected to result in lower pressures on forests and increased forest quality. But this actually presupposes that both firewood collection and timber extraction were 'unsustainable' prior to the boom, in the sense that they caused noticeable forest degradation, which may or may not be true. In other words, one needs to scrutinise land-use impacts for both their relevance to the forests and the type and degree of their environmental impact.

Impact types: the example of oil extraction

A last conceptual clarification concerns what is meant by different types of 'impact'. Table 2.3 outlines a general analytical structure. Here it is used specifically for the case of oil extraction from forested areas. This will also give us an opportunity to identify forest impacts relating to oil production, determine the sign and expected numerical size of deforestation and forest degradation effects respectively, and become familiar with certain technicalities before coming to the country chapters.

Direct impacts

Direct impacts on forests from oil extraction are caused by the exploration, extraction and transportation processes *per se*. Some of them occur on or adjacent to the spatially defined production locality (on-site impacts), others affect areas further away (off-site impacts). They are 'direct' in the sense that they are exclusively biophysical in their nature; we need not look at their interaction with social systems to determine their environmental effect, as we have to do for both *indirect* and *derived* impacts (see section on 'Indirect impacts').

In the tropics, there is a relatively large spatial correlation between forests and petroleum deposits (RAN and Project Underground 1998). By their very nature, direct impacts from oil interventions are always to some extent detrimental to forest cover and quality. However, aggregate direct clearing for oil and gas operations in forested areas in developed countries has been small. For instance, a case study in Louisiana (US) found that only 0.32 per cent of a study area of 2,674 acres was cleared (Newbold *et al*. 1988: 331). This is because 'conventional' oil and gas deposits are extracted through wells from deep underground reservoirs, an operation that does not demand much space (RAN and Project Underground 1998: 4). On the other hand, some 'unconventional' oil production processes have the potential to cause extensive direct impacts on forests.[16] Special risks occur when the activities are carried out in ecologically sensitive ecosystems such as mangroves, in particular in terms of pollution effects, water-flow disruptions, etc. (IUCN 1993).

What specific processes are we talking about? Oil exploration through seismic surveys is the first direct impact of oil activities (see Figure 2.2). With the input of energy, acoustic waves are measured in space to collect geological information about subterranean rock formations, in order to determine the most likely location of hydrocarbon deposits. In water, air or water guns are employed; land-based seismic surveys are either done by the Vibroseis or the explosive-source method (IUCN 1993: 13–17). Seismic arrays are grouped in straight lines and need to be cut across all types of vegetation. The Vibroseis method needs

Table 2.3 Forest impact types, using the example of oil extraction

Impact type	*Description*	*Examples*	*Deforestation impacts*	*Forest degradation impacts*
Direct, on site	Physical site impact caused by production, wastage and infrastructure requirements	• Forest clearing for platforms, seismic exploration, workers' camps and helipads • Toxic pollution • Dredging in mangroves • Timber extraction for building platforms	+ small	+ significant
Direct, off-site	Other local physical impacts caused by production, wastage and infrastructure requirements	• Forest clearing for access roads • Downstream river pollution • Oil pipelines (clearing, spills)	+ normally small	+ significant
Indirect	Effects stimulated by locally accruing oil incomes, benefits – and changing access conditions	• Forest clearing adjacent to access roads • Bushmeat hunting by oil workers • More in-migration of labour/less out-migration of surplus labour due to better incomes, schooling, health, etc. • More local food, services, etc. sold to company and its workers	+/− can be significant	+/− significant
Derived (economy-wide)	All other (economic) effects from oil income – at a higher scale of aggregation	• Urban-biased spending sprees • Agricultural competitiveness loss • Timber competitiveness loss	−/+ dominant	−/+ significant

Notes
+ Increases forest degradation/deforestation;
− Reduces forest degradation/deforestation;
+/− Either reduces or increases forest degradation/deforestation.

Figure 2.2 Example of oil exploration in tropical forests.

Source: Caltex, Indonesia.

roads or paths 2.5–4 m wide along the lines, to transport the seismic laboratory on a vehicle to the sites. In tropical forests, this method was often used in the 1950s to the late 1970s and had a relatively large impact on canopy opening. Recently, for exploration the explosives method is being applied more to rugged areas of difficult access, including tropical forests (T. Barr, personal e-communication, 6 September 2002). This method uses helicopters for access and only needs an opening to accommodate a maximum line strip 1.5–2 m wide cut in the undervegetation, without canopy opening. This is the process we can see in Figure 2.2. So-called 3D and 4D seismic technologies enable companies to obtain both a better picture of the underground reservoir and a prediction of how it will be affected by the production process (RAN and Underground Project 1998: 10). Here, deforestation proper occurs only where helipads and 'drop zones' are established, typically at 300–400 m intervals along the seismic lines. In Gabon, the use of the 'lighter' method reduced clearing to 0.06–0.2 ha/km of seismic line, compared to 0.3–0.4 ha/km using the old method (Wilks 1992: 35). In other words, a more careful choice of method over time has typically also reduced the vegetation impact of oil exploration.

The second direct impact concerns actual oil drilling for further exploration and, later, for production. At the production stage, the largest deforestation impact results from the establishment of infrastructure like roads, pipelines, drilling platforms and camps for workers (see Figure 2.3). Again, the size of forest clearing for these purposes has been reduced over time; it is now typically 5–7 acres for each drilling rig (T. Barr, personal e-communication, 6 September 2002). The rigs are used both for exploration and production wells, but once a well is on production they are removed. An additional effect may be

Figure 2.3 Example of oil drilling rigs in tropical forests.

Source: Caltex, Indonesia.

the exploitation of timber and sand and laterite quarries in the construction of roads, platforms, etc. For example, Amigransa *et al.* (1997: 88) estimate that about 4,000 timber boards are needed for an oil platform, which are often extracted from the surrounding forests. Some observers stress that these effects can still add up to significant amounts. For instance, the Quito-based NGO Oilwatch estimated that in Ecuador, the construction of helipads for seismic exploration has led to the deforestation of 54,000 ha since oil activities started (Amigransa *et al.* 1997: 87). While that estimate may be high in absolute terms, two additional caveats apply. First, this clearing is spread over a period of at least twenty-five years; the yearly figure (2,160 ha) thus corresponds to only 0.5–1.5 per cent of Ecuadorian deforestation (see Chapter 7 on Ecuador). Second, according to the same source, the size of each clearing is about 0.5 ha. Under normal levels of disturbance, an isolated opening of that size in a tropical forest that is left alone will grow back into forest within a relatively short period. Admittedly, the process still represents transitory deforestation, but counting it in an *accumulative* manner would be like adding up what shifting cultivators have cleared over the decades, without taking into account any forest regrowth. The relevant figure to look at is how much *net deforestation* (clearing minus regrowth) the oil industry causes. As we will see, this figure is also in the tropics likely to be quite small.

However, direct forest-degradation impacts from oil may be more severe than deforestation effects, and have a larger off-site reach. Again, it all depends on how modern (and costly) a technology is employed. In Ecuador's Amazon forests, where rudimentary technology and practices were used in the early periods of production, there was a large on-site discharge of production wastewater (toxic brine). Pollution impacts also included toxic drilling-mud, the overflow from rudimentary waste pits and the burning of natural gas. Oil production seriously affected fish and terrestrial wildlife populations and had devastating nutritional and health impacts on indigenous forest people (Kimerling 1991: 77–84; Amigransa *et al.* 1997: 75–6). Nowadays, modern technologies and improved practices can minimise this damage, for example, by re-injecting toxic wastewater into the ground (see Chapter 8 on PNG).

Indirect impacts

Indirect effects are sub-national (local or regional) 'development' impacts that are not a direct consequence of production activities, but rather of local socioeconomic changes – pressures and opportunities – triggered by the presence and activities of oil companies. Spatially these impacts are off-site local or regional, and may include forest degradation as well as deforestation. They usually act against forests, although there are exceptions where they help locally to conserve forests (see Chapter 8 on PNG). The agents that act upon them can be local (e.g. farmers increasing their cash-crop activities) or come from the outside (e.g. loggers harvesting along oil roads). Their size is determined by parameters such as the length, importance and durability of access roads, the number of local employees or the value of locally purchased goods. Indirect effects will often be more powerful than direct ones. Just as in the case of direct impacts, company practices *vis-à-vis* indirect effects have also changed over time. To improve their environmental record, many oil companies try to control the land-grabbing, forest conversion, timber or bush-meat extraction that follow in the wake of new oil roads and trigger the most radical forest impacts.

Derived impacts

The *derived* or *economy-wide* impacts of an oil boom form a residual category, which covers all the national effects that oil rents cause: Dutch Disease, changes in policies and budgets, sectoral structure, etc. (see Figure 2.1). Spatially, these effects are not restricted to any particular site in a country.[17] Normally the oil-rent effects are more powerful than the direct and indirect effects combined. For hard-rock mining, for example, some open-pit mining in forests, the proportions may be different, due to the larger physical impact at the local and regional level (larger direct effects), the lower element of economic rent (less derived effect). Due to their assumed dominance, the derived oil effects is the key subject of analysis in this book.

A synthesis

Based on the factors and mechanisms described in this chapter, Table 2.4 provides a summary of the partial-effect comparative framework that will be used in the five country

Table 2.4 Oil wealth and forest decline – overview of potential country impacts

Economic and productive impacts			*Links to forest land use*		*Forest impact*	
No.	*Type*	*Intensity*	*Type*	*Strength*	*Type*	*Intensity*
1	Land- and forest-using sectors lose competitiveness	Weak/Medium/Strong ●●●	Export and food crops/forest extraction decline or stagnate	→ ⇨ ➔	Less forest loss and degradation (absolute or relative)	Weak/Medium/Strong ▼▼▼
2	Higher urban labour absorption accelerates urbanisation	Weak/Medium/Strong ●●●	Less (expansion of) cultivated area and forest extraction	→ ⇨ ➔	Forest regrowth/less deforestation and degradation	Weak/Medium/Strong ▼▼▼
3	Higher budgets of forestry and conservation agencies	Weak/Medium/Strong ●●●	Improves on-the-ground forest management	→ ⇨ ➔	Less forest encroachment and, usually, less illegal extraction	Weak/Medium/Strong ▼▼▼
4	Oil and mining production increases	Weak/Medium/Strong ●●●	Local (on- and off-site) direct and indirect land uses/impacts	→ ⇨ ➔	Forest clearing and degradation (mainly pollution)	Weak/Medium/Strong ▲▲▲
5	Higher budgets for agricultural development	Weak/Medium/Strong ●●●	More (rapid) expansion of cultivated area	→ ⇨ ➔	More (rapid) deforestation	Weak/Medium/Strong ▲▲▲

6	Roads/rail construction/ improvement; transport subsidies	Weak/Medium/Strong ●●●	Reduced transport costs stimulate product trade	→ ⇨ ➔	More deforestation and degradation	Weak/Medium/Strong ▲▲▲
7	Agricultural and timber trade protection increases	Weak/Medium/Strong ●●●	More (rapid) expansion of cultivated area/logged area	→ ⇨ ➔	More (rapid) deforestation and degradation	Weak/Medium/Strong ▲▲▲
8	More public directed settlement programmes	Weak/Medium/Strong ●●●	Forest clearing for new farms and more forest extraction	→ ⇨ ➔	Mainly more deforestation, some degradation	Weak/Medium/Strong ▲▲▲
9	Higher urban incomes shifts food demand	Weak/Medium/Strong ●●●	Substitution of staple crops, fuels; more animal foodstuffs	→ ⇨ ➔	Various opposite effects	Weak/Medium/Strong
10	Reduced absolute poverty	Weak/Medium/Strong ●●●	Higher labour (opportunity) costs; more savings	→ ⇨ ➔	Various opposite effects	Weak/Medium/Strong

Notes
1–3 field – reduces forest decline.
4–8 field – increases forest decline.
9 and 10 field – ambiguous.

chapters that follow. The table shows ten major impacts on forests from a mineral boom or period of rapid foreign exchange inflows. It is a gross list of the factors that are potentially important, but not all of them have to be present in each country. Factors 1–3 in Table 2.4 imply that the partial mechanism is expected to cause *an improvement* in forest cover and/or quality, either in absolute terms or relative to pre-established trends – for example, a slow-down in deforestation or in the expansion of unsustainable logging. Accordingly, factors 4–8 indicate that the effect is a *forest decline*, while factors 9–10 are ambiguous – they could cause either improvement or decline. The numerical sign of individual effects may vary between cases: for example, in one country agricultural trade may be liberalised, while in another land-using home-market sectors may be more protected, causing opposite forest impacts.

In the country chapters, the different factors will be presented in the table according to their supposed importance, as shown by the ranking given in column 1. This is based on their respective intensity, which is described in the last column (the triangular symbols). Intensity is determined as a product of two factors: the significance of the underlying economic and productive impact in column 3 (e.g. 'how much more urbanisation?', 'how much more money for the forestry agency?', 'how many new roads?') (the bullet-point symbols); and the strength of the link from production in that sector to forest land use in column 5 (e.g. 'how much forest regrowth in abandoned agricultural areas?', 'how much improvement in on-the-ground forest management?', 'how much clearing around new roads cutting through forests?') (the arrow symbols). The weight and distribution of factors causing forest decline or improvement will determine the aggregate effect of oil and mineral wealth on forests, to be analysed in the country chapters.

Notes

1 See, for instance, Collier (1998) on the impact of an oil boom on the oil-producing federal state of Chiapas in Mexico. Regional effects are more common in the case of agricultural commodity booms, where at least some of the revenues go to local producers rather than the state. See, for example, Wunder (1991: 321–57) on the local impacts of the Colombian coffee boom and ibid.: 362–87 on the local effects of the drug boom.

2 Regression analysis is a data-analysis tool comparing *marginal* changes, fitting a line of minimum deviation through a scattered plot of two (bivariate) or more variables (multivariate). It will be employed in the country chapters explicitly to test relations between the partial variables identified in this chapter.

3 For the sake of comparability, we follow Sunderlin and Wunder (2000: 314; 329–30) in classifying countries that had an above-average mineral export share during 1979–81 as high mineral exporters.

4 More precisely, this is the 1995 update of FRA 1990 and the figures from the recent FRA 2000. As discussed in Chapter 1, due to different sources and assumptions, these forest-cover assessments differ greatly in numerical size (see, e.g. the differences for Angola and the Democratic Republic of the Congo), which is a reason to look at them both simultaneously.

5 Instead, the reader is referred to Brown and Pierce (1994) and Kaimowitz and Angelsen (1998). For an attempt to distinguish different approaches to or 'schools of thought' on deforestation, see Wunder (2000: ch. 2).

6 The main causal factors were identified in the pioneer study by Wunder (1997), juxtaposed with the factors from the global review of deforestation models by Kaimowitz and Angelsen (1998), and refined in Sunderlin and Wunder (2000). These are thus not necessarily supposed to

be the *only* explanatory factors, but in our experience and by our criteria they are the most important ones.

7 For instance, in PNG part of the mineral windfall (metals and oil) was channelled to local landowners in the form of compensation payments and royalties.

8 The main role of roads in promoting deforestation has been demonstrated forcefully in spatial regression models, such as Chomitz and Gray (1996) or Mertens and Lambin (1997). But the impact of roads has also been confirmed in regional and cross-country regression models (e.g. Andersen *et al*. 1996; Rudel and Roper 1997a; Mainardi 1998) as well as in comparative case studies (e.g. Jones *et al*. 1990 on Central America). Kaimowitz and Angelsen (1998), Geist and Lambin (2001) give more literature references.

9 An oil-price hike will, in *the absence* of subsidies, increase pressures for harvesting more firewood as a substitute. There may also be incentives to build large hydroelectric dams in forested areas to generate energy from non-fossil sources (Rodríguez 1991). However, both energy subsidies and the income effects from a boom (richer consumers and producers, more urbanisation) are likely to more than cancel out the effects in oil-exporting developing countries.

10 See Kaimowitz and Angelsen (1998: 19, 31, 58, 64, 81) for overviews of different types of models predicting higher deforestation from changed agricultural incentives and policies.

11 In the literature, the argument has been applied more to the reverse situation. When structural adjustment and state 'modernisation' programmes cut back the budgets and staff of forestry and national park agencies, this may limit their ability to enforce forest laws on the ground (e.g. Reed 1996; Sunderlin and Rodríguez 1996). There seems to be no *a priori* reason why the opposite should not apply symmetrically in a situation of foreign exchange abundance and increased budgetary allocations.

12 The arguments here are also normally 'reversed' (e.g. WWF and IUCN 1996), just like those for the effect of fiscal austerity on forestry agencies (see note 11 on Forestry budgets).

13 On *Transmigrasi* and deforestation, see, for example, FWI/GFW (2002: 48–51). On geopolitical motives for deforestation, an illustrative case is Brazil (Hecht and Cockburn 1990).

14 These are called 'Giffen goods' in economic theory, and in developed countries have included goods like margarine (heavily replaced by butter when income rises) or potatoes (replaced by a variety of more costly staples).

15 Yet imagine that these grasslands were previously occupied by, say, cattle-ranchers who are bought out by the sugar-planters. Ranchers may reinvest the money to clear the forest elsewhere, moving their ranches. In other words, even if there were no direct overlap between additional land demand and forests, there may be an indirect effect on demand for land. For example, this was very important when soybeans advanced into non-forested areas in Brazil, causing smallholders to move increasingly into the Amazon (Kaimowitz and Smith 2001).

16 Contreras-Hermosilla (2000: 4) reports that although extensive oil sands in the boreal forests of Alberta (Canada) are currently not being exploited, if production was started here, this would affect large forest areas in the US Pacific Northwest and in western Canada.

17 An exceptional case at the borderline between 'indirect' and 'derived' effects occur when mineral rents are being distributed not only nationally but also with direct allocations to the mineral producing federal state or province (for instance, see Chapter 8 on PNG).

3 Defining and measuring changing forest conditions

Before starting the country analysis, it is vital to deal with the concepts and measurement of the central variables in this book: deforestation and forest degradation. After examining different current definitions of these two phenomena, four critical dimensions are discussed: canopy cover, spatial resolution, sample size and time-scale. It is argued that any unbiased forest definition should build on measurable tree-cover rather than on assumed land-usage; there should be no 'forests without trees'. Unfortunately, for our five primary study countries, knowledge about forest-cover and -loss remains scattered and contradictory. As long as we have no direct measurements of deforestation, analyses of other (mainly agricultural) land uses can give important additional insights. Likely ranges of deforestation estimates in the country chapters are given and compared with the FAO's. Some proposals as to what it would take to produce a truly consolidated forest-cover information system conclude the chapter.

How to proceed?

In setting a baseline for each of the country cases, the condition of forests must first be considered from the historical perspective. Has climate change over time naturally caused forest cover to shrink or to expand, for example, at the expense of savannah and grasslands? How long has the country's territory been occupied by humans? How large was the 'original' or 'default' forest cover prior to that occupation? What early anthropogenic impacts on forests have occurred, over time and in different regions? What changes were introduced during the colonial period? What land uses have the independent nation states tried to develop? It is often tempting to assess current forest-cover trends in comparison to 'original' forest cover,[1] in order to have an idea of humans' accumulated impact on the current vegetation climax. In many cases, this widely used term is ambiguous. For example, countries like Gabon or Cameroon have faced considerable historical fluctuations in forest cover, due to changing climate conditions – even during periods when the land was inhabited by humans (see Chapters 4 and 6). Thus it matters a lot what time-period is selected as 'original'. In this book, 'original' forest cover will generally refer to the early Holocene (the post-Pleistocene glacial period), about 8,000 years ago, when:

1 The climate was essentially the present-day one.
2 Vegetation had had time to adjust to post-glacial conditions.
3 'Benign' human subsistence activities (hunting, gathering, later shifting cultivation at low population densities) had already become established.

A second step is to provide a snapshot of forests in the present. What forest types are there, and how rich are they in products, services and biological diversity? Who, and how many people, live in and around these forests? How are they distributed geographically *vis-à-vis* national centres of commodity production? Having answered this second set of questions, we can turn to an examination of ongoing changes in forest conditions. However, a number of methodological issues must be clarified first. For the purposes of this discussion, it is convenient to distinguish between two types of forest intervention, namely 'deforestation' and 'forest degradation'. In the following, they will each be defined, discussed and illustrated with reference to specific examples.

The main preoccupation of this book is deforestation, the second priority being a number of processes that, taken together, are categorised as forest degradation. As a term, 'deforestation' continues to have highly negative connotations. To the extent that it is considered an 'undesirable land-use change', what change processes should be included? Can measurement and desirability criteria possibly go hand in hand? The chapter ends with a suggestion concerning what deforestation *should* mean, that is, tree removal below a threshold level that can be objectively measured, which may or may not be desirable, and which does not hinge on futile speculations of expected future land use.

Deforestation: what does it mean, and what should it mean?

Any global definition of 'forests' and 'deforestation' must strike a compromise between, on the one hand what may conceptually appear an appropriate threshold between 'forests' and 'non-forests', and on the other the pragmatic criteria that are already used in measurements worldwide. Box 3.1 reproduces the current forest and forest-change definitions used by the UN's FAO (with my added keywords in capitals), as applied in the most recent FRA (FRA 2000). It first provides a definition of 'forests', the meaning of 'deforestation' being a logical extension of this definition.

Forests are thus defined by both a sufficient presence of trees (the crown-cover criterion) and by the absence of other dominant forms of production (the usage criterion).

Box 3.1 Forests according to the FAO

Forests are:

1 lands of more than 0 5 ha [UNIT SIZE];
2 with a tree-canopy cover of more than 10 per cent [CROWN-COVER];
3 which are not primarily under agricultural or urban use [USAGE].

Deforestation is:

1 the conversion of forests to another land use [USAGE]; or
2 the long-term reduction of tree canopy cover below the 10 per cent threshold [CROWN-COVER].

Source: FAO (2000a: 7,10).

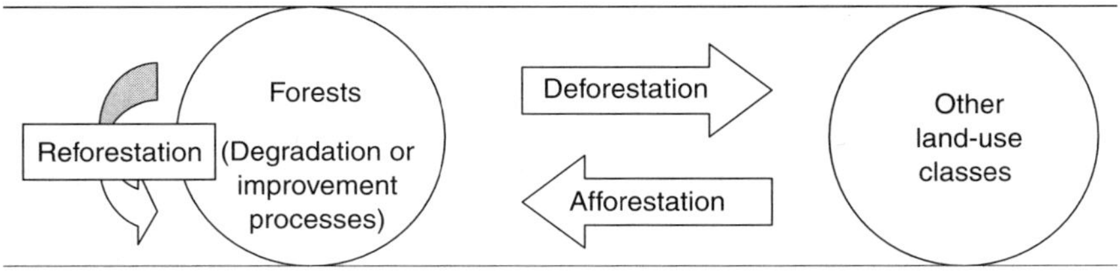

Figure 3.1 FAO's classification of forest-cover changes.

Source: FAO (2001a: 24).

For instance, areas under agroforestry are to be regarded as non-forested, as they mainly produce agricultural outputs, even though their tree-crown cover may actually exceed 10 per cent. There are a number of important caveats and clarifications. Trees should have a minimum height of 5 m. Tree plantations are forests (including, e.g. rubber and cork oaks), unless they produce agricultural outputs (fruit, estate crops, etc.). Temporarily unstocked zones, such as tree nurseries and clear-felled or recently reforested areas, are still counted as forests. The 'long-term' reduction of tree-canopy cover is delimited to 'more than ten years', while less than this counts as a 'temporary' reduction (FAO 2001a: 24). In this sense, non-agricultural tree-covered areas with temporary reductions in their tree-crown cover to below 10 per cent are categorised as being *reforested*, without losing their status as forests in the period when they are not covered by trees. According to the FAO's definition, reforestation refers to the re-establishment of forest formations after a temporary condition with a crown-cover density of less than 10 per cent. The removal of trees on areas intended for later reforestation is thus an internal cycle that is not counted as 'deforestation'. By contrast, areas that lose their minimum 10 per cent tree cover for more than ten years and/or are cropped are considered *deforested*. The (natural or man-made) stocking of non-forested or deforested areas with trees is called *afforestation*. Figure 3.1 conceptualises the relationship between the different terms. Deforestation as measured in most statistics, as well as in this book, refers implicitly to *net* deforestation, that is, to gross deforestation minus afforestation. Afforestation thus reduces net deforestation, while reforestation, as a fully internal forestry process, does not.

I generally follow the FAO definition in this book, but try to distinguish between plantations and natural forests whenever possible. However, four key aspects of the FAO definition will also be critically reviewed in this chapter: canopy-cover criteria, spatial resolution, sample size and time perspective.

Canopy cover

The 10 per cent canopy-cover threshold in the FAO's definition has been criticised by many of those concerned with the biological integrity of forests. They believe that too many heavily degraded areas are being counted as forest, even though they have lost up to 90 per cent of their originally full canopy-crown cover, as well as many of their environmental functions as forest. It should certainly be kept in mind that the FAO uses a 'broad' definition of forests, and therefore a 'narrow' definition of deforestation. It takes drastic intervention into a closed forest to bring average tree-crown cover below 10 per cent. In particular, selective logging does not cause this. This 'narrow' definition thus tends to view deforestation as a conversion process involving a drastic intervention causing land-use change, although there are exceptions to this rule.[2] This criterion is favoured by foresters, geographers, economists and land-use planners. On the other hand, ecologists, biologists and environmental campaigners tend to focus more on the quality and integrity of forest functions. They have thus preferred a 'broad' definition of deforestation that includes forest degradation (logging, over-hunting, over-grazing, etc.) as deforestation processes. A 'broad' definition of deforestation tends to produce larger aggregate estimates (e.g. Myers 1994) and thus tends to alert the environmentally sensitive public more than the FAO figures (Wunder 2000: 9–11). The validity of different definitions of deforestation is thus fully dependent on the analyst's ultimate goal. There is probably a growing general consensus that the simple observation of selected product over-harvesting and disturbed forest functions should count as forest degradation, whereas a quantitative reduction-criterion for crown-cover is needed to talk about deforestation.

However, whether that should be the FAO's 10 per cent threshold is highly debatable. Foresters, government agencies and ministries in charge of forest administration have an interest in seeing as large an area as possible designated as 'forested' or as 'potential forest'. They are therefore likely to resist raising the 10 per cent canopy-cover threshold, which would reduce the forest area potentially under their mandate and control. Other international organisations have chosen to adopt a more differentiated criterion, and to use several thresholds. For instance, the TREES project which is financed by the European Union (EU) has, in simplified terms, classified forest cover greater than 70 per cent as 'dense forest' and the 40–70 per cent range as 'fragmented forest' (e.g. Eva *et al.* 1999: 12–13).[3] The International Geosphere Biosphere Program (IGBP) has used a 60 per cent threshold for forests and 40–60 per cent for woodlands (Defries *et al.* 2000: 249), while UNEP (2001) uses 40 per cent for closed forests and 10–40 per cent for open or fragmented forests. Although the FAO's 10 per cent criterion is thus situated at one end of the spectrum of definitions, at least it has been used consistently since the 1960s for tropical forests. The scientific argument has been that 10 per cent of crown cover is 'the minimum canopy density where naturally occurring formations of trees exist as communities ... as opposed to areas where trees exist scattered in the landscape or in rows' (FAO 2000a: 5). However, arguments for a higher cut-off may seem at least as valid.[4]

The choice of a cut-off criterion can make a large difference in landscapes with open forest formations, scrub land or forest–savannah transition zones. The FAO formerly applied a 20 per cent crown-cover threshold to countries outside the tropical zone that belonged to the group of developed countries. In the FRA 2000, this was lowered to 10 per cent in order to obtain a unified global criterion. For a country like Australia, with a lot of open vegetation formations, this reclassification meant that an additional 118 million ha of forest were suddenly 'discovered'.[5] In developing countries with a high proportion of fragmented forests, the differences can also be large. For instance, for India national forest cover using the 10 per cent criterion for crown cover was estimated at 19 per cent of land area, while by using a 40 per cent criterion reduced this figure to just over half (11 per cent of land area; see FSI 1997, cited in UNEP 2001: 4). Obviously, changing crown-cover criteria can severely limit the comparability of forest assessments over time. So the first question one has to pose for each deforestation figure is: Which cut-off criterion was used for canopy cover?

Spatial resolution

Second, one must ask a related question: At what spatial resolution was forest cover measured? The country studies in the following chapters show that comparisons between national forest-cover estimates over time are often invalidated by different map or assessment scales with different resolution levels. Table 3.1 provides a numerical illustration of the problem, based on numbers that were derived from the Gabon data.[6] Let us first look at two patches of moist forest, as indicated by an imaginary set of satellite photos or aerial photographs. One image shows a primary, unlogged forest; the other forest has recently been logged. We are examining vegetation cover at five different pixel sizes that increase stepwise: 1 m^2, 10 m^2, 100 m^2, 1,000 m^2 and 10 km^2. Since we are only interested in distinguishing the state of 'forested' from 'non-forested' land, we use an algorithm that represents the FAO criterion of 10 per cent minimum forest cover.

The unlogged, primary forest includes natural tree-fall gaps. At an extremely detailed scale of 1 m^2 pixel size, these gaps cover approximately 5 per cent, so that forest cover is 95 per cent. At the next pixel size (10 m^2), they are reduced to 1 per cent, not being registered at all at lower resolution levels: the small openings are 'averaged out', and the area will simply appear to be completely covered by tree-canopy cover. For the logged-over forest, there will be more canopy openings due to tree-fall gaps, skidder trails, secondary roads and principal roads with wide deforested borders. At 1 m^2 pixel size, 20 per cent of the area appears to be without forest cover, but that figure diminishes steadily with resolution levels, because the openings become relatively smaller. At 10 km^2 pixel size, resolution is too coarse to register any logging activity: the area appears to be fully forested. The numbers from the logged–unlogged forest comparison are also illustrated graphically in Figure 3.2. Note that the difference between the two forest types, which can be interpreted as deforestation caused by logging, shrinks from 15 per cent points to 12, 7, 2 per cent points, and finally to zero. So, as can be seen in Figure 3.2, estimated deforestation is inversely proportional to pixel size. Different thresholds of detectability produce different estimates of deforestation.

Table 3.1 Forest-cover and spatial resolution levels. A hypothetical example with three landscape types

Landscape type	*Pixel size* (%)				
	1 m^2	10 m^2	100 m^2	1 km^2	10 km^2
1 Unlogged primary moist forest	95	99	100	100	100
2 Logged forest ('cleared islands in a sea of forests')	80	87	93	98	100
3 Forest patches in a savannah area ('forest islands in a sea of open space')	15	13	7	2	0

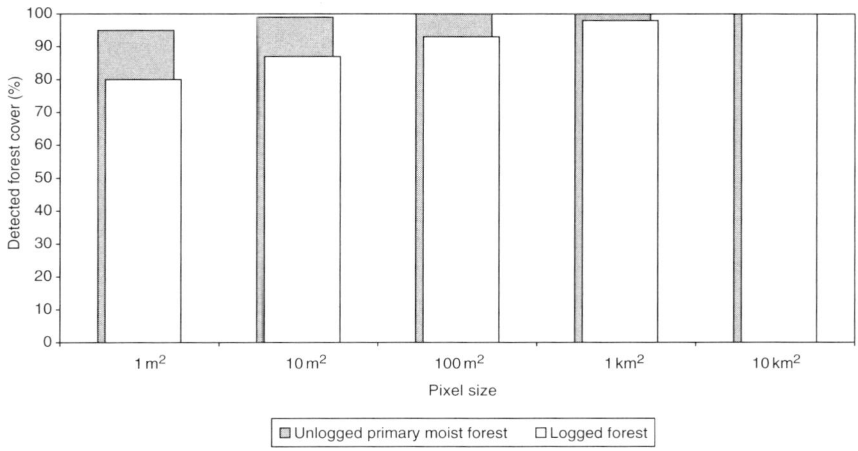

Figure 3.2 Forest-cover measurement and spatial resolution. Primary and logged forests compared.

Source: Table 3.1.

Note
Differences between unlogged primary moist forest and logged forest shows deforestation.

The example of logged forest was just one of several options in what one might more generally call a landscape of 'cleared islands in a sea of forests'. Alternatively, one might have chosen patchy shifting cultivation, small-scale artisan miners, or oil exploration and drilling in an otherwise undisturbed, closed forest. But, besides the primary and logged forests, let us also take a look at a third type of landscape, which we can call the 'forest islands in a sea of open space'. This could represent, for instance, isolated forest patches in an area of savannah (as found in our case countries of Cameroon, Gabon and Venezuela), or forest remnants in a highland region dominated by agriculture (as occurs in Ecuador). Note that the relation between pixel size and forest cover is precisely the opposite in this case: the coarser the resolution level, the less our vegetation screening captures forest fragments. So, at 10 km^2 pixel size, none of the forest fragments is large enough to occupy 10 per cent of the pixel. Hence, there would appear to be no forest cover. Estimated deforestation of our forest fragments is positively proportional to pixel size.

What does this small example imply for deforestation comparisons over time in our country cases? Many of the nation-wide historical forest maps available from the 1950s, 1960s and 1970s are of a coarse-resolution type, for example, at the 1 : 500,000 or 1 : 1,000,000 scale. Colonial governments often produced more detailed maps from aerial photos, for example, at a 1 : 50,000 or 1 : 100,000 map scale, but only for selected sub-national sites. Since the 1970s, technology to produce nation-wide imagery has improved, and political pressure has increased to monitor changes in national forest cover. Hence, for the last decade or two, we have more nation-wide high-resolution maps available. But these are not directly comparable to the old, coarse-resolution ones. In countries with a predominance of the 'forest islands in a sea of non-forests' situation (e.g. Ecuador or Venezuela), these newer forest maps have revised national forest cover significantly upwards. This is because many previously ignored forest fragments are now being detected. In countries where the 'cleared islands in a sea of forests' are more common – for example, because of a lot of shifting cultivation (see Chapter 6 on Cameroon) – the new forest-cover estimates have been revised downwards. With other types of deforestation patterns and geometrical shapes, the relationship between spatial resolution and estimated deforestation may be different.[7]

The point is also important for the comparison of different types of satellite imagery. At one end of the spectrum, we have the US National Oceanic and Atmospheric Administration's Advanced Very High Resolution Radiometer (NOAA-AVHRR), with a low resolution of 1.1 km^2. They thus provide images at the left end of Table 3.1. Their main advantages have been wide-area coverage, free availability and daily frequency, which allows for screening that selects the best images (e.g. without smoke or cloud cover), and have thus been preferred for the mapping of large eco-regions. On the other hand, we have medium-high resolution images (10–80 m^2), such as those from Landsat (Multispectral Scanner, MSS; Thematic Mapper, TM), SPOT (Malingreau 1993), or the new generation of high-resolution satellites such as Ikonos and Quickbird producing imagery from 2.8–4 m^2 in multispectral mode up to 0.70 cm^2–1 m^2 in panchromatic mode resolution (R. Dennis, personal e-communication, 14 July 2002). These types of imagery provide the base for a more precise classification of vegetation, including in the context of forest-change processes. But until now they have been expensive and are still available only at a frequency where they often show non-classifiable areas, for example, due to cloud cover. For Brazil, it has been shown that cloud cover can introduce significant errors into Landsat-based assessments (Andersen 1996; Mayaux *et al.* 1998: table 8). On the other hand, AVHRR-based fire imagery for 1987 produced a large overestimate of deforestation in the country. The exceptional number of satellite-based deforestation estimates in Brazil allowed a gradual consensus to be reached about methods (Downton 1995).[8] More recent products, such as the Synthetic Aperture Radar (SAR) images are not affected by cloud or smoke cover (e.g. Kuntz and Siegert 1999), but their potential for producing forest-cover estimates on a wider scale is not established (B. Mertens, personal communication, July 2001).

The TREES project has used a methodology based on low-resolution imagery but combines it with high-resolution products, that is, validating NOAA-AVHRR-based assessments with Landsat TM using a correction procedure (Malingreau 1993; Mayaux *et al.* 1998). This seems to use some of the advantages of both types of product. On the other hand, it has been argued that the correlation between the two is not sufficiently strong, and that the results of the integration efforts are not very satisfactory (Päivinen *et al.* 2000: 12). Variations in deforestation estimates due to resolution gaps thus continue to exist, and should be kept in mind when different sources are compared. This is true particularly for fragmented landscapes, and will become more important as humans penetrate more into tropical forests and continue to fragment them.

Sample size

A third critical point concerns the size of selected study areas, in particular the question of *spatial sampling* versus *wall-to-wall* measurement. For cost reasons, it has not yet been possible to measure the entire world's forest cover at an equivalent level of detail. Some observers have argued that this can be effectively overcome by careful spatial sampling. For instance, in the FAO's global forest assessment, 117 sites representing a 10 per cent sample of tropical forests are analysed using remote-sensing images. The results are then used to extrapolate forest-cover changes for different forest types. The TREES project has now produced new deforestation estimates for the humid tropics, revising FAO's estimates downward by 23 per cent, based on 100 sites with high deforestation that represent 6.5 per cent of tropical land area (Achard *et al.* 2002). A recent study has even argued that the sub-national level is more appropriate than national data for the study of deforestation processes and causes (Geist and Lambin 2001), but some of their sub-national regions, such as the Brazilian Amazon, are actually larger and more heterogeneous than most tropical nation states.

Measurement options used in extrapolating sub-national results on to the national level depend very much on the type of landscape and land use under analysis. This is clearly demonstrated by the country sample in this book. For a country like Cameroon, an economic crisis induced an increase in food-crop production and subsequent acceleration of forest-cover loss from the mid-1980s onwards. Different regional studies from the forest zone all confirmed the same trend, although this varied with closeness to the market. In Cameroon, a stratified sample would give a fairly good reflection of the countrywide picture. But for Ecuador, the spurt in deforestation was much more regional, because it hinged on shifting areas being opened up sequentially for settlement by new roads, oil production, specific crop booms, etc. In a newly opened area, clearing followed a quite distinct land-use pattern than in an area that represented an 'old frontier'. In other words, predicting national deforestation in Ecuador based exclusively on samples from the Amazon, the highlands or the coastal region would have given severely biased results. In line with an observation by Matthews (2001: 6), in general this should caution us against sample-based deforestation assessments, including the extended sample-based World Forest Survey proposed by the FAO for future assessments (Päivinen *et al.* 2000: 13). For many countries with heterogeneous land use and multiple sources and fronts of deforestation, only 'wall-to-wall' methods can provide us with the correct answers.

Time scale

Finally, the time perspective relating to deforestation matters a great deal. The FAO explicitly refers to *long-run* (or permanent) loss of forest cover. The term 'excludes areas where the trees have been removed, due, for example, to harvesting or logging, and where the forest is expected to regenerate naturally or with the aid of silvicultural measures within the long-term' (FAO 2001a: 25). In a way, this represents the typical forester's view in a developed country setting of what 'his' domain is within land-use planning: an area set aside for forest does not cease to be 'forested' just because there are no trees on it. For instance, such an area may have been temporarily clear-cut to harvest the trees, as an integrated part of a rotation scheme. The area only truly leaves the forester's domain once a planned, long-term decision is taken to put it to an alternative use, for example, to enlarge a residential area or to crop it. In that case, the land will obviously not be *reforested*, and may thus safely be counted as *deforested*.

It is not only foresters who stick to the classification of temporary versus permanent clearing. Environmentalists are also deeply concerned about permanent deforestation, while often finding the temporary loss of forest acceptable, especially if it is being carried out by small subsistence farmers who need to do so in order to survive. The principal fear is that tropical forests are *disappearing*, rather than that they are being cut for a certain period within the framework of dynamic land-use systems, before returning at some stage to forest cover. There is thus also a strong ideological argument that justifies developing an international focus on permanent deforestation, which would urge us to make a clear distinction between temporary and permanent conversion.

However, applying this distinction to the reality of the tropics runs into severe problems. First, land use in the tropics tends to be highly unpredictable, because of both volatile economies and insecurity of land tenure. For instance, squatters may occupy areas intended as plantations to grow annual crops or, vice versa, plantation companies may crowd out local farmers. To predict on the ground whether an area that is cleared now will be tree-covered again in ten years is a formidable task. Certainly, I would not like to be in the shoes of the forest statistician who is asked to make that prediction. Furthermore, from a biodiversity point of view, it is not at all clear why, for instance, a large area that has been clear-cut by a pulp mill should *not* be considered deforested, just because it is left alone to regenerate, or maybe is intended for reforestation by someone sometime in the future. Conversely, why should a clearly tree-dominated area that is extensively used for slash-and-burn agriculture with long periods of fallow be considered deforested, just because it produces crops? Unlike the FAO criterion, in this view long-fallow shifting-cultivation systems would *not* be regarded as deforestation.

Eliminate the 'forests without trees'!

Should shifting cultivation thus be exempted from accusations of deforestation, using the argument that it tends to be just as temporary a removal of forest-cover as certain silvicultural operations? Actually, much of the ambiguity concerning the time dimension of deforestation surfaces in the discussion about shifting cultivation, which often has clear political undertones. For instance, in the debate over deforestation in Indonesia, shifting cultivation has been seen both as the primary cause of forest loss and as a key means of sustainable forest use. The validity of the two arguments depends entirely at what end of the 'forest–farming continuum' one is standing (Sunderlin 1997). A long-fallow subsistence system at low population densities can sustain cyclically regenerative forest cover, but a short-fallow rotation in areas near markets with growing population pressures often leads to forest degradation and, ultimately, to deforestation. Moreover, as will be shown in this book, shifting economic conditions nation-wide can also lead to comprehensive changes in the extent and intensity of shifting cultivation (see especially Chapter 6 on Cameroon).

Consequently, I think it would be wrong to absolve shifting cultivation, even long-fallow temporary clearing, from all responsibility for causing deforestation. First, it actually is a significant source of forest clearing. A recent estimate is that more than 10 million ha of tropical rain forest are *temporarily* cleared each year for shifting cultivation (IFAD *et al.* 2001: 45). This is more than the FAO's most recent estimate of global, net *permanent* deforestation during the 1990s (FAO 2001a). Second, there is no clear boundary between 'sustainable' systems of shifting cultivation and wasteful land-degrading systems of 'nutrient-mining', which everybody agrees should be defined as deforestation (see below): it is impossible to come up with a generally acceptable criterion for dividing the two. Third, the same insecurity about future land uses that surrounds forestry also applies to shifting cultivation (see section on 'Some examples of deforestation').

Personally, I believe that one should go in the opposite direction. *All* forest-cover removal at a given point in time should count as 'deforestation', including areas that have been clear-felled for allegedly temporary silvicultural purposes. In terms of the terminology in Figure 3.1, the proposal is thus not to try to distinguish between 'temporary' and 'permanent' forest clearing, nor between 'reforestation' and 'afforestation'. I use the wording 'try to', because much of what is reported as deforested from national governments through to the FAO today probably already fails to make this distinction in practice. In other words, my proposal is ultimately to delete the 'usage criterion' in Box 3.1 and *use exclusively a 'tree-cover criterion' to define deforestation*. The justification for this proposal lies in two interrelated claims:

1 The difference between 'temporary' and 'permanent' forest-cover removal cannot be measured objectively, other than, at best, historically and retrospectively. For the purpose of monitoring forest loss, 'temporary' forest removal is thus not scientifically verifiable in any consistent manner.
2 Even if, hypothetically, this distinction could be objectively measured, any such temporal cut-off criteria would tend to be arbitrary, and would include 'desirable' and 'undesirable' land-use processes of change on both sides of the selected borderline.

Regarding the first claim, the examples in Box 3.2 show that one can only safely distinguish between temporary and permanent deforestation *ex post facto*. In principle, one can only assess whether the initial clearing has proved to be temporary or not ten years after it occurred. The first two examples in Box 3.2 show that the empirical measurement of temporary forest-clearing makes little sense in societies that are subject to a high degree of structural change. This applies whether one uses the FAO's criterion that exonerates forestry operations (the first example), or the alternative that seeks to exonerate shifting cultivators (the second example). In both cases, some *intended* temporary clearing inevitably becomes permanent. The two clearing types are better seen as part of a temporal continuum. The third example is meant to substantiate the second claim: many activities that qualify as a 'temporary clearing' norm are undesirable land-use changes by almost any criterion. There are other examples than the banana case described in Box 3.2. The burning of the Amazon forest and the temporary conversion of forest to low-productive, land-extensive pasture (see the Ecuador and Venezuela case studies) is one of them: some forests may well come back before the ten-year limit, but most people would intuitively count this as deforestation. On the other hand, converting forests permanently to a stable, land-intensive, high-return form of land use may in many cases be more 'desirable' (potentially for both the people and the environment) than a shifting, low return, technologically static form of land use that expands monotonously with ever-growing product demand. So, the exclusive ideological focus on permanent changes in land use may simply be misdirected.

Box 3.2 Problems in distinguishing temporary from permanent forest-clearing: three examples

Example 1: Silvicultural clearing is allegedly temporary, but ...

A pulp mill clear-cuts a natural forest concession area of 1,000 ha. It intends to reforest it with fast-growing species within the next five years, in order to supply the factory with raw materials in the future. The area is thus temporarily without trees, but not *deforested* according to current FAO criteria. Consider two scenarios:

Scenario A. A local community has long claimed rights to the land the company harvested. A shift in government now strongly promotes decentralisation. The newly empowered community decides to occupy the treeless land and begins to plant annual and perennial crops, both to obtain an income and to consolidate their claim of ownership to the land. Permanent deforestation, after all, due to conflicting land uses?

Scenario B. The country experiences a financial crisis. The pulp mill runs into severe financial difficulties and also has little success with its plantation programme. Consequently, it postpones its plans for reforestation until further notice. Some regrowth has occurred on the area of cleared forest, but the recurrent burning of adjacent agricultural fields spreads and burns the area on several occasions. Eventually, persistent *alang-alang* grasses (*Imperata cylindrica*) invade it. Permanent deforestation, after all, by default?

Example 2: Clearing for shifting cultivation is allegedly temporary, but ...

A forest plot is cleared by a small farmer for shifting cultivation of, say, plantains. The plantains are only supposed to be cultivated for *X* years. This falls below our predefined cut-off of *Y* years for temporary forest clearing, which we have fixed in order to accommodate the fact that sustainable long-fallow swidden systems may help to conserve forests. So the felling and burning of trees is thus not deforestation, and the field of plantains is actually a forest. Consider two change scenarios when year *X* arrives:

Scenario A. The government receives an oil windfall, and to help farmers it chooses to subsidise imported fertiliser. At the same time, road-building has reached the nearest village, so the farmer can now sell plantains in the provincial capital. He decides to use fertilisers to prolong cultivation of the plot beyond year *Y*. Should his original clearing be reclassified in retrospect? Who is going to keep account of the changes?

Scenario B. Crop prices change and the farmer considers eliminating the already relatively unproductive plantains to plant a more profitable perennial. Does the act of planting perennials thus count as deforestation? Which remote sensing technique allows us to detect such a change? The farmer has also received an offer from a cattle rancher, who wants to buy the plot to put it into pasture. Does the latter then become the agent of deforestation, while the original felling of the trees was only 'forest degradation'?

Example 3: Is the temporary clearing of forest preferable to permanent deforestation?

After the Second World War, Ecuador faced a 'banana fever': a new capitalist export sector emerged, which indirectly helped to open up a large, previously forested region on the Pacific coast for settlement (Wunder 2001a). For more than a decade, the rudimentary cultivation technique was based on the nutrient mining of hilly slopes, which also caused a lot of erosion: slash, burn, cultivate, exhaust the soils and move on to the next site, within a few years. Huge areas were affected. But, as for shifting cultivation, the banana cultivation period *per se* would be too short to call (permanent) deforestation. Consider the following scenarios:

Scenario A. In some cases, the land actually went back into forest, within a period of less than *X* years. Does this mean that the clearing and nutrient-mining for the commercial banana crop was not deforestation?

Scenario B. In some places, banana workers stayed on the abandoned banana farms to cultivate food crops beyond year *X*. Would this then be deforestation?

Scenario C. In some cases, the hilly land became too eroded by nutrient mining for forests to come back. Does this mean that the same technique would constitute deforestation in scenario C but not in scenario A? Who can make that prediction and classification at the time of the initial clearing?

Scenario D. At the end of the 1970s, stationary capital-intensive banana-growing techniques were introduced, so that the bananas were now cultivated on small but permanent plots. If these areas had been previously forested, this would count as deforestation, whereas the forest-wasteful nutrient-mining 'move-on' technology in scenario A would not. Is this a useful classification?

Until now, the FAO has retained its emphasis on a temporal distinction, arguing that land *use* is eventually more important than land *cover* (FAO 2001a: 7). Others have also preferred land-use data (e.g. agricultural census figures) to satellite imagery, though not necessarily with a ten-year horizon. But the scope for ambiguity and arbitrary manipulation in classifying areas of 'forests without trees' seems insurmountable. This also distracts attention from the fact that we know far too little about tropical land *cover*. The case studies in this book will make it painfully clear that an insufficient measurement of vegetation cover ultimately also leaves us in the dark about land uses, and that 'ground truthing' alone is not going to help us out of this dilemma. Consequently, areas that have less canopy-cover than a pre-defined cut-off, measured by remote sensing and other supplementary tools, should unambiguously count as 'de-forested' or 'non-forested'. Preferably, the FAO's 10 per cent canopy-cover threshold for deforestation should be raised, reserving the 10 per cent and some intermediate canopy-cover category exclusively for open forests and fragmented areas.

Some examples of deforestation

Box 3.3 lists what types of intervention would typically count as deforestation, according to the above adopted 'narrow' definition, and where in the tropics these processes are likely to predominate. Conversion to crops and pasture are obvious global candidates. Mangrove expansion into shrimp ponds (e.g. in SE Asia) or the general expansion of cities, roads and other infrastructure are other examples of changing land-uses that clearly comply with the FAO's criterion. But as we move down the list, cases may appear more contentious. Few would object to large-scale open-pit mining being called deforestation, although subsequent land use may be uncertain. Yet shifting cultivation should also be counted as deforestation, to the extent that forest cover within an area passes under the 10 per cent canopy-cover threshold. Such deforested areas should include those that are fully cleared at any single point in time, plus regenerating fallow areas that are not (yet) tree-covered as judged by the FAO's 5-m height-criterion. It should also include areas that have been clear-cut for wood harvesting, for example, for pulp and paper or for charcoal production. Note that different factors in Box 3.3 can apply sequentially, for example, charcoal clear-cutting followed by agricultural conversion. Finally, by the logic of an exclusively tree-cover-based definition, natural and semi-natural processes should not be ignored. This would include large-scale windfalls, or areas that have been (repeatedly) burnt, to the extent that they no longer satisfy the canopy-cover criterion. This has been the case for large areas on Borneo (e.g. Kuntz and Siegert 1999; Siegert and Hoffmann 2000). Some of these cases would probably be hard for many foresters to accept as deforestation but, as argued above, such a classification makes sense from the point of view of measurability and consistency.

Although the list in Box 3.3 might be controversial, many of the deforestation studies reported throughout this book already use similar criteria. A crown-cover criterion is easier to measure on a fairly objective and transparent basis. It is also frequently more relevant with respect to forest ecological services, such as carbon storage, to which the global

Box 3.3 Examples of deforestation processes

Type of intervention	*Where does it mainly occur?*
Forest conversion for permanent agriculture	Pan-tropical
Forest conversion for permanent pasture	Pan-tropical, esp. Latin America
Shifting cultivation (intensive)	Pan-tropical, esp. Africa, South Asia
Mangrove clearing for shrimp ponds	Especially Asia and Latin America
Infrastructure expansion (roads, rail, etc.)	Pan-tropical
Urban/residential area expansion	Pan-tropical
Dam construction floods forested areas	Pan-tropical
Open-pit mining	Pan-tropical
Clear-cutting for pulp and paper-harvesting	Especially in Asia
Clear-cutting for charcoal-making	Especially Africa (artisan); Brazil (industrial)
Hurricanes, canopy-eliminating fires, etc.	Pan-tropical

society is attaching more and more weight. Neither satellite images nor 'ground-truthing' reports come with labels that indicate what will happen to an area in the future. Distinctions between 'permanent' and 'temporary' forest removal thus tend to be highly speculative. Furthermore, our different preconceptions about 'good' and 'bad' deforestation should interfere as little as possible with our measurements of deforestation. In what follows, the terms 'forest loss', 'forest-clearing' and 'deforestation' will therefore be used as synonyms.

Forest degradation: a residual category

There are many other human interventions into forests that do not satisfy the deforestation criterion of (temporary or permanent) forest-canopy removal below a 10 per cent threshold. These processes are labelled 'forest degradation' in this book: they change forest structure considerably, but do not induce complete forest loss. Degradation is frequently the first step in a chain of events that may subsequently lead to deforestation, and the relationship is often causal. The FAO defines degradation as 'a reduction of the canopy cover or stocking within a forest . . provided that canopy cover stays above 10 per cent', which is generally to be understood as 'the long-term reduction of the overall potential supply of benefits from the forest' (FAO 2001a: 26). Again, the temporal aspect ('long term') in that definition may be ambiguous: talking about 'overall potential supply' may be like comparing apples to oranges (see the next section), and the term 'stocking' may generally be associated more with wood than, for instance, the disruption of environmental services. However, the key feature of forest degradation is that it is defined residually and therefore includes a rather mixed menu of dishes (see Box 3.4). Furthermore, in some cases we can measure the reduction in crown cover, for example, through remote-sensing techniques and/or forest inventories. Unfortunately, it is much more difficult to count the stocking of animals or medicinal plants in a forest. The second part of the FAO's definition of

Box 3.4 Examples of forest degradation processes

Type of intervention	*Where does it mainly degrade?*
Selective logging	Pan-tropical, esp. in SE Asia
Shifting cultivation (small-scale, extensive)	Pan-tropical, esp. Africa
Small-scale clearing for artisan mining	Pan-tropical
Recurrent forest fires	Pan-tropical
Oil pollution from drilling	Pan-tropical
Mercury pollution from gold miners	Pan-tropical
Firewood over-extraction	Especially in dry forests and periurban areas
Over-grazing	Dry forests (e.g. Sahel, India)
Over-hunting ('defaunation')	Pan-tropical, esp. Central Africa
Other NTFP over-extraction (e.g. medicinal plants)	Pan-tropical
Deterioration of regeneration processes	Pan-tropical

degradation thus becomes extremely hard to quantify. Nevertheless, reduced stocking is a vital aspect of degradation in some countries, and it will be referred to more qualitatively in this book.

Selective logging, that is, timber extraction that does not involve clear-felling, is the prime candidate and most measurable intervention within the forest-degradation category. Although this type of logging does not normally eliminate canopy cover to less than 10 per cent (and thus does not constitute deforestation), it can easily affect more than half the tree cover. This may depend, for instance, on the ecological sensitivity of the area, the extraction technology used or the number of trees harvested. Impacts are particularly high in the dipterocarp forest of SE Asia, where the density of commercial timber species is high. Vegetation, soils and animal populations can all be negatively affected by logging. However, as the discussion in some of the country chapters will show, there is much debate about the 'sustainability' of logging. Other clear-cut examples of degradation include pollution by oil firms (from installation spills, mud made toxic by drilling, waste pits, etc.) or by artisan gold-miners using mercury.

However, degradation ambiguity applies to many of the categories in Box 3.4 too. Clearing small forest patches for extensive shifting cultivation or artisan mining, on a scale that does not qualify for deforestation, may or may not have damaging impacts. Increases in forest fires have been linked convincingly to logging (Mackie 1984; Woods 1989; Cochrane *et al.* 1999; Nepstad *et al.* 1999; Siegert *et al.* 2001), but it may be difficult to distinguish this phenomenon clearly from naturally occurring fires, or from other types of anthropogenically induced fires. Among the over-harvesting of non-timber forest products (NTFPs), over-hunting ('defaunation') is perhaps the most widespread. Firewood extraction may affect dry forests negatively but have negligible impacts in wood-abundant moist forests with low human population densities. Seasonal livestock overgrazing is a phenomenon restricted to dry forests.

Like 'deforestation', 'forest degradation' also has clearly negative connotations. As expressed in the FAO definition above, degradation is clearly equated with 'non-sustainable forest use', implying that future forest harvests are being compromised. As one distinguished ecology expert has expressed it, 'almost any type of resource extraction conducted in tropical forests will have an ecological impact' (Peters 1996: 54). As indicated at the bottom of the list in Box 3.4, this may include interventions that many non-ecologists would tend not to classify as degradation, such as the over-harvesting of single species (e.g. medicinal plants), the population reduction of key pollinator or seed-dispersal species, or the over-collection of Brazil nuts (*Bertholletia excelsa*) on the forest floor, gradually inducing genetic impoverishment (ibid.: 43–6).

From a social scientist's viewpoint, a definition of degradation that includes just any imaginable disruption to biological processes must have many caveats. Also, a forester will find that much of what (s)he calls 'forest management' is degradation by a strict ecological criterion. From the perspective of human benefits, degrading one species may sometimes increase the supply of another, whether as an unexpected by-product or as a conscious effort to manipulate nature. For instance, one might mention:

- selective logging to promote the growth of rattan (*calamus*) or bamboo (*bambusa*);
- periodic burning of the forest to help produce the biomass needed for herbivore game species;

- frequent burning to promote the regrowth of fire-resistant commercial palm species; and
- temporary forest clearings increasing the presence of valuable timber species (such as *okoumé* [*Aucoumea klaineana*] – see Chapter 4).

Of course, this is not to say that most forest degradation is good for human needs, or that it cannot lead to forest loss. For instance, recent Amazon research shows that traditional logging is affecting larger areas than those that are being deforested annually, produces cumulative changes in vegetational structure ('cryptic forest impoverishment') and increases forest flammability significantly. Changes from logging and unintentional burning thus favour future burning and further degradation (Nepstad *et al.* 1999).

Much of the confusion has to do with the fuzziness of 'sustainability' as a concept. What ultimately has to be sustained over time: the land-user's income, the total value of forest produce harvested, the physical quantities of wood or non-wood production, the provision of ecological services or the conservation of biodiversity? For each of these layers, management answers to questions of sustainability are bound to differ. Criteria vary not only according to type of scientific analysis, but also forest beneficiaries. For instance, some observers would claim that declining harvest values from the forest over time (reduced 'natural capital') may be defensible, as long as forest users are able to accumulate savings and channel them into more remunerative assets outside the forest (e.g. product-processing equipment, means of transport or children's education).[9] We should thus keep in mind that what has been listed as forest degradation in Box 3.4 is not necessarily either 'unsustainable' or 'undesirable' to all forest stakeholders. This book will not attempt to address the discussion of what types of forest intervention are either undesirable or unsustainable. It will refer to the different country-specific pressures on resources from case to case, but mostly leave it to the reader to evaluate the normative aspects of that process.

Finally, 'deforestation' not only diverges from 'forest degradation' by criteria concerning canopy-cover threshold: the two are also likely to differ as to their underlying economic nature. To the extent that deforestation represents a conscious choice to convert forests to alternative land uses (which holds true for the majority of examples in Box 3.3), forest clearing *per se* has an economic cost, which is only justifiable to the land-use decision-maker in terms of future returns. In other words, deliberate land-clearing is basically an *investment* in future alternative land uses. For comparison, forest degradation through the over-harvesting of forest commodities seldom requires large-scale investments and is more likely to involve the cashing-in of economic rents. Consequently, the two phenomena are also likely to be linked in a fundamentally different way to the impacts of oil wealth. This constitutes an additional reason for distinguishing the two clearly in the case-study analyses.

Trying to measure the impacts

The preceding sections gave an outline of how deforestation and forest-degradation processes should be defined and assessed. The next step is to measure these processes empirically, which would then allow us, first to compare how they have evolved over time in the five study countries, and then to relate them to potential explanatory factors. Unfortunately, the poor data available make this a major and somewhat frustrating undertaking, especially in developing countries with restricted institutional capacity, where the size and condition of forests has never been a high-priority issue. In addition, as we shall see in the country chapters, national forest agencies have often had an incentive to underestimate deforestation levels in order to claim a mandate over larger areas of land and relieve international political pressure. Country forest estimates are thus scarce, heterogeneous, often biased, and frequently not comparable over time or between countries. This makes it a gargantuan task for any international agency to gather together the pieces of the puzzle into a consistent picture. A recent appraisal by the World Resources Institute (WRI) states: 'At the beginning of the twenty-first century, official intergovernmental processes do not produce consistent and replicable estimates of the world's forested area' (Matthews 2001: 1). As will be shown below, this general characteristic definitely holds true for our five study countries.

Starting with forest degradation, or significant human-induced changes in forest structure, some attention has been given to measuring logging. What indicators are available? The country descriptions in this book will show that the expansion of concession areas is normally *not* a good indicator in quantifying impacts on forests. The appropriation of control over concessions is a process that is driven by legal, infrastructural and anticipated market factors and not very closely correlated to actually logged-over area. Timber production figures are probably the best widely available substitute, and these will be used for each of the five countries. However, uncertainties include the size of illegal and informal logging and trade, especially for an often significant domestic market. Production from plantations could not be separated consistently from production from natural forests, so total production has been used instead. Parameters from the International Tropical Timber Organisation (ITTO) and other sources allowed for the conversion of different wood products to Round Wood Equivalents (RWE). This will, where relevant, facilitate comparisons over time and between countries.

Forest area *affected* by logging may be a more direct indicator of logging impacts. Even so, logging impacts depend greatly on the number of species harvested, which often varies over time and with the preferences in different buyers' markets. These changes are obviously more adequately accounted for by production figures than area figures, as they refer to the amount of wood actually extracted. The WRI has published some statistics on the annual logging of closed broadleaf forests in the past (e.g. WRI 1994), but these figures were imprecise and have been discontinued in more recent versions of the same publication series (e.g. WRI 1998). The WRI is now aiming to produce specific estimates for selected countries (E. Matthews, personal communication, Bogor, July 2001). Remote sensing techniques do allow for the monitoring not only of logging, but also other interventions affecting biodiversity, notably fire impacts. For instance, from Landsat TM images it is possible to detect these impacts if they are measured within about a year of their occurrence (Nepstad *et al.* 1999: 505). This may become an important management tool in the future. However, in what follows I shall argue that at the national level, producing reliable deforestation statistics should have the first priority.

For deforestation, the FAO has been the most widely cited source in recent decades, combining national government and project data (reports, maps, satellite images, etc.) with the FAO's own various reporting and modelling methodologies. Every ten years, the FAO has carried out a global FRA and published new forest-cover and deforestation estimates for the previous decade. The results of the FRA for 1980 basically came from an evaluation of the view of selected forest experts (Lanly 1982). The FRA 1990 (FAO 1993), updated in the mid-term publication *State of the World's Forests*, SOFO (FAO 1997a),[10] has been used extensively in the country chapters below. The FRA 1990 used two methods in the Forest Resources Information System (FORIS) to process different statistical and spatial data (Singh 1993). First, forest loss between 1980 and 1990 was measured at 117 selected tropical forest sites, representing a 10 per cent global sample of different forest types. Second, a mathematical model was used to intra- and extrapolate national deforestation that could not be measured for lack of data (FAO 1993). The second method proved decisive for the determination of deforestation rates in the 1990 FRA (Grainger 1996).

To measure deforestation, one needs at least two (comparable) forest-cover estimates at different points in time. To compare the evolution of deforestation over time, one needs estimates for at least three points. The problem is that, among the 90 tropical countries surveyed in the 1990 FRA, only 21 (23 per cent) actually reported more than one national forest assessment to FAO. Furthermore, for those that had just one assessment (66), including all our 5 study countries, more than half (39) were from before 1980 (including 3 of our 5 cases), that is, they fell partly outside the period supposedly being examined, namely 1980–90 (FAO 1993: 5; table 1b). For instance, for Gabon the most recent forest-cover measurement had been a sample-based survey that dates back to 1970. Forest cover thus first had to be projected forward to 1980, and then model-predicted further for the decade up to 1990. Even the best imaginable model is bound to be inaccurate for such a long-term extrapolation of a single point estimate.

For none of our five cases did the FRA 1990 present more than one *measured* forest-cover estimate, so the forest-loss figures were necessarily *model-driven* for all the five countries. The FAO's controversial deforestation model is disaggregated according to forest type (wet, dry, very dry). Its dynamic is driven by population growth that changes the population density within these different forest types, interpreting deforestation as a logistical growth function (FAO 1993: 11–13; Singh 1993). Humans slowly start to 'eat their way' into a still complete forest stock, then the process accelerates, but levels off in the late stages, when little forest cover is left (the 'S-curve'). This also meant that, when population growth in the tropics generally showed a marginal slow-down in the early 1990s, the FAO's predicted global deforestation showed a similar reduction (FAO 1997a). Some researchers erroneously used the FAO figure to test for causes of deforestation that included demographic change, thus involving an obvious circularity, given that population growth had been used to generate the data in the first place (see Rudel and Roper 1997a: 54–5). Notwithstanding this criticism, the 1990 FRA was a comprehensive effort to measure global forest change, which for a decade the FAO was sufficiently confident of to promote it decisively to its clients. As a senior FAO-FRA official expressed it at the time: 'The FRA 1990 gives a fairly good, and in any case the best ever, description of the present forest area and of the changes between 1980 and 1990' (Janz 1993: 7).

Whether or not FRA (1990) and SOFO (1997) reflect global reality or correctly captured 'the norm' of tropical countries will not be discussed here. But, as will be shown in the country chapters, our sample of tropical oil countries clearly deviated from 'the norm', and the deforestation estimates in the 1990 FRA were therefore inadequate for our selected countries. The most glaring example in this book is Gabon, where the FAO's model, based on forest stock, forest type and population growth, predicted a yearly net deforestation of 116,000 ha for 1980–90 and 91,000 ha for 1990–5. However, Gabon's considerable oil wealth over that period meant that people did not 'eat their way' into the forest by following a logistical S-curve. Rather than clearing the 1.62 million ha over fifteen years as predicted by the model, in net terms they probably did not clear any forest at all! Gabon's deforestation was an FAO 'modelling mirage', but still it found its way into many national and international publications.

Fortunately, this has been acknowledged in the FAO's new (not model-based) FRA assessment for 2000. Gabon's annual deforestation has been revised downwards to a mere 10,000 ha (0.0 per cent) (FAO 2001a: 31), an adjustment which an earlier draft of the Gabon chapter in this book contributed to significantly.[11] For Venezuela, the estimates of the FAO model for deforestation were also excessive. For Ecuador and Cameroon, the estimates may have been within an adequate range of long-term averages over two decades, but this conceals important fluctuations between sub-periods. In other words, even if the FAO's model of deforestation may have been good at predicting forest loss in 'typical' tropical country, it certainly did not predict it well for the five countries in this book, simply because none of them was sufficiently 'typical'.

A number of other sources on forest-cover change have also been used by international deforestation researchers and will therefore be presented in the country chapters. One has been published by FAO's Forestry Unit itself, in the form of the deforestation figures in the FAO *Production Yearbook*. These data come exclusively from forestry agencies reporting to the FAO, but the forest definition used is not the same. Forest cover includes scrub growth and areas that are *intended* for reforestation 'in the foreseeable future' (my emphasis). On the other hand, estimates exclude 'forests used only for recreation purposes' (FAO 1996: viii, note 6). The idea here is obviously to define forests more narrowly for foresters interested in harvesting timber. But for our five countries, the reported figures do not make any sense at all. They are often static over decades, or show even a slight net reforestation in countries that have undoubtedly suffered forest loss. For the sake of completeness, the figures will be presented in the chapters below, but they cannot be recommended even for narrow wood-production purposes.

A second alternative source has been the Conservation Atlas, produced jointly by the IUCN, the World Conservation Monitoring Centre (WCMC) and CIFOR, with one volume published for each tropical continent (Collins *et al.* 1991; Sayer *et al.* 1992; Harcourt and Sayer 1996). This is a useful additional source for forest stocks, but no independent deforestation estimates were produced. Some of the maps' underlying estimates of forest cover are rather old and in some (though not all) cases, their coarse resolution means that they are not very precise in their quantitative assessment. Consequently, they run into some of the resolution comparability problems described earlier in this chapter.

Probably the best current forest-cover estimates are those derived from new satellite-imagery analysis, especially in two pan-tropical projects: the European Joint Research Centre's (JRC) TREES project, and the Pathfinder project financed by the US National Aeronautics and Space Administration (NASA). Other assessments have been made by the US Woods Hole Research Center and by the United Nations Environment Programme (UNEP). Some of the TREES estimates, published in Mayaux *et al.* (1998), Laporte *et al.* (1995) and based on maps of the tropical Americas (Eva *et al.* 1999), Africa (Mayaux *et al.* 1997) and Asia-Pacific region (Stibig *et al.* 2002), have been used in this book. However, except for a couple of cases (see below), these projects are not yet at the stage where they are producing national deforestation estimates. In fact, for political reasons they may be reluctant to enter into open competition with the FAO, the organisation with the clearest international mandate to produce such estimates. The Pathfinder project has decided to focus most on the tropical moist-forest areas. It will produce some national deforestation estimates, but this will not cover all tropical countries.[12]

Among other international sources, the WRI publication *World Resources* (e.g. WRI 1998) reproduces the FAO's FRA estimates. The problem with earlier volumes (WRI 1994, 1995) is that the WRI quoted, simultaneously with the FRA 1990 figures for 1981–90, another series for the 1981–5 sub-period, without specifying the source (e.g. WRI 1995: table 19.1; 312–13). For the two series to be consistent, as I show below, would often entail extreme shifts in deforestation trends between the 1981–5 and 1986–90 sub-periods. Most likely, the two series should never have been compared. By uncritically reprinting these estimates, there is a danger of engaging in *data laundering* – lending authority to estimates that may not deserve it.[13] As part of its Frontier Forest project, WRI has also published figures of the 'original' forest cover for individual countries' (Bryant *et al.* 1997), which was discussed above. Useful country-specific data were also exchanged with researchers from the WRI-headed consortium Global Forest Watch (GFW), which prepared country reports for three of the five cases in this book (Cameroon, Gabon and Venezuela).

There are a number of additional statistical sources on forest-cover change. Sometimes particular cross-ministerial or task-oriented studies on forest-cover change are carried out relating, for example, to climate change or protected areas. There may also be national forestry agency or ministerial reports, with variable definitions, coverage, sources and quality. Donors may finance historical land-use assessments of particular geographical areas for development or conservation projects. University scholars may select particular sites to test out new methodologies. A petroleum company may want to document what local land-use impact its activities have caused. Sub-national reports on forest loss in particular regions may add to the national picture, but one should remember the caveats about spatial sampling mentioned above. Different efforts to produce global systematisations and analyses of sub-national case studies exist (Rudel and Roper 1996; Geist and Lambin 2001).

Are we getting any wiser?

In the new FRA for 2000, the FAO has stopped using its controversial deforestation model to 'fill the holes', but it still employs its sample-based model to detect trends in different forest types. In a new approach, national annotated literature reviews were prepared to capture in-country 'grey' literature. Based on an array of sources, an FRA expert then weighted and combined the information to provide a best guess (Convergence of Evidence technique). In the next stage, this product was subjected to an iterative evaluation by a small group of experts (Delphi technique) (Päivinen *et al.* 2000). Judging from the deforestation results, government-reported forest statistics have in some cases regained their primacy. In many ways, the FRA 2000 methodology is much closer to FRA 1980 than FRA 1990.

At the time of writing (July 2001), preliminary figures for FRA 2000 have been presented, but these are not yet official, and often reach dramatically different results by retrospective re-estimating the 1990 baseline. For some countries (e.g. Gabon), these preliminary results are definite improvements; for others (e.g. Venezuela), they seem to add to the confusion. Consequently, I have chosen to present the FRA 2000 results *ad hoc* in this overview chapter, but not in the respective country chapters, which might have proved too confusing for the reader. As a first step in analysis, Figure 3.3 gives a comparison of different forest-stock estimates from different sources for each of the five countries. The 'baseline' (represented by the vertical axis of zero deviation) in the figure is the TREES estimates in Mayaux *et al.* (1998), which employ a combination of NOAA-AVHRR (low-resolution) data validated by an index correction factor derived from Landsat (high-resolution) imagery. Except for the very high estimate for Ecuador, I would personally regard these stock estimates to be the most reliable currently available. The horizontal bars show the relative deviation (in per cent) of other pan-tropical sources from these TREES estimates.

An observation which extends to the full sample of tropical countries (Mayaux *et al.* 1998: 46–7) is that there is generally more agreement between the TREES and the IUCN/WCMC/CIFOR classification than between TREES and the two FAO assessments. As already observed by others, this may have to do with the FAO's choices regarding what forest types to include (Grainger 1996; Mayaux *et al.* 1998). One obvious explanation is the very different canopy-cover criterion. That probably explains much of the great

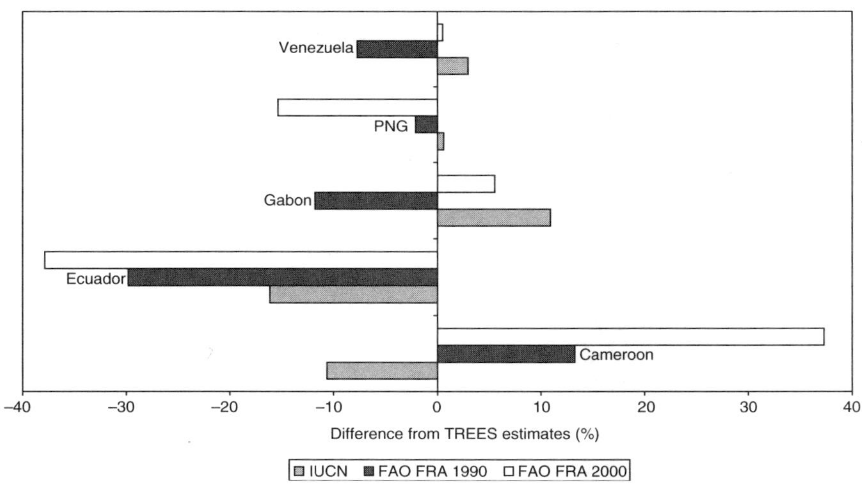

Figure 3.3 Differences in the tropical forest-cover estimates by TREES,[a] IUCN[b] and FAO.[c,d]

Notes

a TREES figures from Mayaux *et al.* (1998).

b IUCN, WCMC and CIFOR's Conservation Atlas.

c FAO FRA 1990 figures from FORIS (wet, moist and montane forest).

d FAO FRA 2000 figures are preliminary, from FAO (2001a).

difference for a country with sizeable forest–savannah transition zones, such as Cameroon. The FAO's more inclusive 10 per cent cut-off counts large areas as forests which TREES views as open forests or woodlands. However, this cannot explain why the FAO's estimates of forest stocks are much *lower* than the two other sources for PNG and Ecuador, where we should rather expect the opposite deviation. It is also surprising that, in comparing FRA 1990 to FRA 2000, for three of the five countries the gap between the FAO and TREES has widened rather than narrowed, quite strikingly so for PNG and Cameroon. The time difference (i.e. deforestation between 1990 and 2000) should only account for a minor share of that difference. But we should obviously have a closer look at the underlying changes between FRA 1990 and FRA 2000.

Table 3.2 reproduces in detail the FRA 2000 results (columns 1–4) and the FRA 1990 ones (columns 5–7) for the five countries. A quick glance at the two columns with yearly loss rates (3 and 7) would seem to indicate that deforestation rates have decreased dramatically from the 1980s to the 1990s in Venezuela (from 599,000 to 218,000 ha, or from 1.2 to 0.4 per cent), Gabon (from 116,000 to 0,000 ha, or from 0.6 to 0.0 per cent) and Ecuador (from 238,000 to 137,000 ha, or from 1.8 to 1.2 per cent). There are increases for Cameroon (from 0.6 to 0.9 per cent) and marginally PNG (from 0.3 to 0.4 per cent).

But things are not that simple. Some of the key country data sources are unpublished documents, so it would be helpful to receive key additional information (e.g. coverage, source resolution and canopy-cover criteria). Nevertheless, the footnotes from the FAO's web page do reveal that for three of the countries (Cameroon, Gabon and PNG), the underlying linear extrapolations that have produced forest-loss figures go back to map data from the 1970s. For the two other countries, extrapolations actually cover half of the 1980s (Venezuela 1985–95) or even more years from the 1980s than the 1990s (Ecuador 1985–92). It may thus seem generally more appropriate to label the alleged 1990–2000 deforestation figures in FRA 2000 (2, 3) 'past deforestation', as they appear to be at least as much influenced by trends in the 1970s and 1980s as by what happened specifically over the decade after 1990. They definitely do not seem to be able to detect trend changes between the 1980s and the 1990s, although this is probably exactly the interpretation that many of the readers of the FRA for 2000 will give them.

In comparing deforestation rates with my own 'best guess' (11), it is actually likely that deforestation rose in Cameroon and declined in Ecuador, but it is very unlikely that forest loss was reduced to one-third in Venezuela or suddenly eliminated in Gabon. The main underlying problem is that the FRA 2000 baseline figures for 1990 (4) conflict with the FRA 1990 figures for 1990 (5), which the 1980–90 forest-loss figures were based on in the first place. In the comparative part of the table, (8) compares the 1990 baselines from FRA 2000 and FRA 1990. It shows that 1990 forest stocks were revised upwards by as much as 28 per cent in Cameroon, 20 per cent in Gabon and 12 per cent in Venezuela. The PNG estimate is 12 per cent lower and the Ecuador estimate 1 per cent lower than in the FRA for 1990. How much does that stock readjustment make up, compared to current forest loss? Column (9) shows how many years of deforestation (by FRA 2000 standards, 3) the stock readjustment corresponds to. For Ecuador (−1yr), it is negligible, but for three other countries, the result is in the range of three decades. For Gabon, the situation is extreme: at the low forest-loss speed given in the 2000 FRA, the Gabonese could still keep

Table 3.2 Comparing forest-cover estimates. FRA 2000, FRA 1990 and own estimates (in '000 hectare)

Countries	*FRA 2000*				*FRA 1990*			*Comparison*		*Own 'best guess'*	
	Total forests 2000[a]	*Yearly deforestation 1990–2000*	*Yearly deforestation rate 1990–2000 (%)*	*Baseline 1990*[a]	*Total forests 1990*[a]	*Yearly deforestation 1980–90*	*Yearly deforestation rate 1980–90 (%)*	*1990 forest area difference: FRA 2000/ FRA 1990 (%)*	*1990 forest area difference, expressed as no. of deforestation years (FRA 2000)*	*Forest cover (~1990)*[a,b]	*Deforestation (~1990s)*
	(1)	(2)	(3)	(4)	(5)	(6)	(7)	(8)	(9)	(10)	(11)
Cameroon	23,858	222[c]	0.9	26,078	20,372	122	0.6	28.0	26	17,000–24,000	150–200
Ecuador	10,557	137[d]	1.2	11,927	12,026	238	1.8	÷0.9	÷1	15,500 (11,500–17,000)	>180
Gabon	21,862	10[e]	0.0	21,962	18,265	116	0.6	20.2	370	20,500–23,000	0
PNG	30,601	113[f]	0.4	31,731	36,043	113	0.3	÷12.0	÷38	32,000–37,000	50–70
Venezuela	49,506	218[g]	0.4	51,686	46,052	599	1.2	12.2	26	49,000 (44,000–49,000)	250–400

Sources: FAO (1993, 2001a), http://www.fao.org/forestry/fo/country/index.jsp (accessed 31 July 2001); own estimates.

Notes
a Including plantations.
b Preferred point estimate and/or likely range under variable assumptions/definitions.
c Cameroon: 1975–99 extrapolation; Faure (1989) (cited by FAO) estimate of 200,000 ha/yr for agricultural conversion.
d Ecuador: 1985–92 extrapolation; two remote-sensing national forest maps (latter unpublished).
e Gabon: 1970–99 extrapolation; Nguépi 1999 (cited by FAO) 15,000 ha/yr; Wunder 2000.
f PNG: 1975–2000 intrapolation; 1975–85 zero deforestation *assumed*; Forest Inventory and Mapping System.
g Venezuela: 1985–95 extrapolation; comparison of two Carrero maps (the latter unpublished).

on deforesting for 370 years before their forests were reduced to the level that FRA 1990 had predicted them to be.

This further underlines another fact pointed out by Matthews (2001): changes in the definitions and baselines in the 2000 FRA come to dominate strongly over changes to forest cover in the real world. Of course, the international public is craving for the latter, rather than the technicalities of the former. But the degree of baseline reassessment and the uncertainties involved should be made transparent to the reader at the individual country level. Even at the global level, it may sound surprising when the FAO states that 'net deforestation [in the 1990s] has likely decreased since the 1980s' (FAO 2001a: 8, my addition), considering that just one page later the conclusion drawn from the pan-tropical remote-sensing survey is that 'the reduction in deforestation rates between the two decades is not significant' (ibid.: 9).

The drastic adjustment of the FRA 1990 baseline and deforestation rates may come as a surprise to those among the international public who truly believed in the previous optimistic FRA self-appraisal, as expressed in the words of Klaus Janz, quoted above. It may in particular be worrying to clients who have forest-stock baselines high on their agenda, such as the carbon-sequestration community. If the FAO really thinks that the 2000 FRA has produced much more consolidated estimates than the 1990 FRA, it would be well advised to acknowledge openly that both forest stock and deforestation estimates in the 1990 FRA were erratic, and by a margin that is anything but trivial. Such baseline corrections would then, of course, be urgent and highly desirable. Gabon is an excellent case in point where an important improvement has been made. However, corrections should then also be made backwards in time, that is, re-estimating deforestation from 1980 to 1990. But an overarching question is: Has the FRA 2000 really provided us with consolidated knowledge that allows us to make once-and-for-all baseline adjustments, which we then can solidly rely on in our daily work, without fearing new and equally drastic adjustments in the future?

A way forward

In trying to answer this question, the final two columns of Table 3.2 (10, 11) show the 'best-guess' ranges or point estimates from the country analyses in this book. The forest-cover estimates for 1990 (or around that year) adopted in this book are summarised in column 10. Where ranges are given, this implies that different forest definitions, cut-off criteria and resolution levels have been applied in the underlying sources, which makes a big difference, especially for those countries where there are a lot of forest fragments (e.g. Cameroon, Ecuador). The presentation of intervals may also be a better way to deal with the inherent insecurities than an obsession with publishing point estimates without the expected range of error.

For deforestation in the study countries, the FRA for 2000 seems to be more closely correlated to my estimates than that for 1990. An exception, however, is Venezuela. From an excessively high figure of 599,000 ha in the FRA (1990), the estimate has now been cut drastically to just over one-third of that figure (218,000 ha). This is now somewhat below the range I calculate (250,000–400,000 ha). It also means that the FAO has conceded the low deforestation estimates that the government has been claiming for years, which was a major source of discussion in Venezuela. This will inevitably create an erratic image, and a wrong political signal, that Venezuelan deforestation has been declining in the 1990s: it is actually more likely to have increased (see Chapter 5 on Venezuela).

From this characterisation of 'official' deforestation data, it is probably evident that, for the five case countries, I have slightly higher confidence in my own deforestation estimates (points or intervals) than in those in the FRA (2000). In principle, both build in their first stage on the 'Convergence of Evidence' technique, using common sense and basic checks to make the best use of contradictory evidence. FRA (2000) then validated the numbers by an expert panel (Delphi technique), although the question then is, for how many tropical countries can a sufficient number of qualified experts actually be identified for this method to work in the intended way? It is also not always sufficient to 'eliminate outlier opinions' in order to land at some consensus number. From my reading of four of the FRA (2000) background reports,[14] I would point to three distinct attributes of the country chapters in this book, which are described in Box 3.5.

First, many of the data, especially perhaps those provided by government agencies and ministries, have to be assessed very critically for their origin, coverage and definitions, and sometimes even to be sure that the numbers in a column add up (see, e.g. Chapter 5 on Venezuela). If biases exist, one then has to determine their size and direction, and to judge whether certain corrections would make the numbers more comparable to other

Box 3.5 Some guiding principles for the deforestation country-assessments in this book

1 Critical review of sources

- Check geographical coverage, scale/resolution, cut-off criteria, time period
- Determine errors and reliability of source
- Evaluate biases and comparability.

2 Inclusion of sub-national results

- Screen case studies available in-country
- Screen other systemised databases for sub-regional studies
- Check pre-identified deforestation 'hot spots'
- Evaluate the representativeness of the sample of sub-national results.

3 Cross-check with other land-use data

- Identify the main forest-replacement land uses on the national scale
- Screen data on these land uses (e.g. agricultural censuses, mining surveys, etc.)
- Screen data on land-use substitutes (e.g. production quantities)
- Check consistency of deforestation figures with sum of land-use changes.

information. Second, in the absence of a fully consolidated national picture, regional case studies can often provide important additional clues about the size and timing of forest clearing. Nevertheless, as discussed above, there are also severe limitations on the prediction of national deforestation from a restricted sample of sites. It is thus important to make an effort to try to obtain all the studies available. In this book, searches were made for regional case studies, both during visits to the case countries and in existing databases. The latter also included other scholars' systemised collections of regional deforestation case studies, such as the unpublished database maintained by Tom Rudel of Rutgers University (USA) and, as a check-up at a late stage in the writing, the case-study collection in the recent book by Geist and Lambin (2001). An additional source of pan-tropical information is the mapping of deforestation 'hot spots', which were identified and described by a panel of international experts in 1997 (Achard *et al.* 1998). They provide an idea of where major forest loss has occurred in the recent past, and what specific pressures are responsible for it.

Third, and perhaps most importantly, governments tend to have relatively little interest in deforestation. They often care more to monitor non-forest land uses that they assume to be more productive, such as agriculture. On the other hand, changing land use is a zero-sum game (a country's total land area is normally fixed), so expanding cropland or pasture has to occur at the expense of other land cover. Numbers not only have to provide a consensus among experts – they also have to add up! In principle, one should be able to disaggregate an annual deforestation estimate into the elements listed in Box 3.2 to find out what alternative land uses the converted forest has been put to, such as permanent crop expansion, ranching, mining, shifting cultivation (crops and fallow periods), etc. In practice, the full range of land uses may not be available. In the absence of land-use data, one can often obtain additional clues from data on agricultural production or stocks, which can be converted to land-use data if reasonable land-productivity assumptions can be made. For instance, pastures are the main end-use of deforested land in Venezuela, but ceased to be surveyed at the end of the 1970s. Consequently, the growth in areas used for pasture could be extrapolated from the growth in head of cattle (see Chapter 5 on Venezuela). This method has some caveats, but it provided an absolutely vital understanding of land-use changes and forest loss over the last two decades, an understanding that could never have been reached by merely comparing a couple of national forest maps.

Thus if, for instance, all our estimates of land-use elements in Box 3.2 for a given year total 400,000 ha,[15] and the official deforestation estimate is only 200,000 ha, then the latter is probably wrong. To check deforestation estimates by agricultural census, survey and other land-use data can be a powerful 'insurance' against adopting forest-loss figures that are wide of the mark. As other scholars have recognised, figures for expansion of cultivated area provide substitute indicators in the analysis of deforestation processes (e.g. Houghton *et al.* 1991; Southgate 1994; Barbier 2001). Agricultural data are particularly relevant for those countries where forests constitute the main 'default' land use (e.g. Gabon, Ecuador), so that any pasture or cropland expansion can only be at the expense of forest stocks. In countries with important non-forest types of natural vegetation, the calculation is more complicated, as one first has to determine what type of area expansion occurs specifically in the country's forested zone (e.g. tubers, plantain, coffee and cocoa in the case of

Cameroon). Whichever of the two cases applies, this constitutes valuable information that has been underutilised in all the previous FRAs.

On aggregate, the FRA (2000) country-level deforestation estimates may or may not be better than those in FRA 1990. They are certainly more transparent in a number of ways in their methodology and sources, and information is now much more easily available through the FAO's FRA web homepage. Yet the methods applied – contradictory evidence validated by an expert, and then subjected to a consensus-seeking panel – are in themselves still sufficiently tentative to ensure that many estimates will have to be revised again in the future. In principle, that may also be true for the figures produced in this book. The bottom line is that nothing can replace the comprehensive *measurement* of forest-change processes using the best available technology. At present, these types of data are not available for any of the five study countries on a national scale. Remote-sensing should play a much more prominent role in accomplishing that global task by combining low- and high-resolution data with 'ground-truthing' efforts. The FAO should take the lead in undertaking a comprehensive task of this nature, but it cannot complete it alone. The following recommendations[16] may provide a way forward:

- intergovernmental fora must fully recognise the weaknesses of our current forest-cover knowledge;
- intergovernmental fora and donors should provide the FAO with a mandate and the resources to form a consortium of the best experts and organisations to produce a more consolidated FRA, drawing on the considerable but currently dispersed international expertise and information on forest cover;
- 'wall-to-wall' measurement should be one of the guiding principles for the forest-cover component of a future FRA, combining different resolution levels and different source types;
- a single 'state-of-the art' methodology should be agreed centrally among the participants of this consortium and applied globally to forest-cover assessment;
- baseline establishment should have first priority, enabling historical forest-change analyses;
- forests (and deforestation) should be defined basically by canopy-cover criteria, eliminating arbitrary use criteria. Crown-cover criteria should be more highly differentiated, including at least one cut-off point above the current 10 per cent (e.g. 50 or 70 per cent canopy cover);
- a comprehensive monitoring system should produce forest-cover indicators more frequently than once a decade, accommodating the recurrent need for timely information on a rapidly advancing process;
- the centrally agreed-upon methodology should be promoted by building up capacity in tropical countries, enabling in the medium term both a more consistent classification and more reliable government reports to international fora, but leaving room for other assessment needs on a lower scale;
- the growing scientific recognition of the role of forests in the global carbon cycle and the problem of global warming should underscore the urgent need to raise the sophistication and frequency of forest-cover assessments.

Notes

1 I owe the suggestion in this paragraph, and the three criteria below, to Dr Stuart White, one of the reviewers of this book.

2 Just felling a couple of trees in an open forest patch with a tree-canopy cover of 10.1 per cent could bring that cover down to 9.9 per cent, and thus count as 'narrow' (FAO) deforestation (A. Angelsen, personal communication, Bogor 25 April 2001). However, the drastic intervention in a more closed forest represents the more typical situation.

3 Actually, the TREES classification refers to both forest cover (percentage of forest area) and canopy cover (density of tree canopy). For instance, for the TREES map of SE Asia, the category 'closed high density forest' corresponded to >70 per cent forest cover and 10–40 per cent canopy cover (R. Dennis, personal e-communication, 14 July 2002).

4 For instance, UNESCO (1973), cited in UNEP (2001: 6), argues for the 40 per cent threshold because 'when the coverage of the trees is 40 per cent the distance between two tree crowns [equals] the mean radius of a tree crown'.

5 This figure is the difference between the 1990 figure published in the FRA 1990 (40 million ha) and the revised 1990 figure in the FRA 2000 (158 million ha) (Matthews 2001: 2).

6 I am much obliged to Chris Wilks, Africa Forest, based in Libreville, Gabon, who formalised this hypothetical example in his valuable comments on my country chapter. He provided the figures in the two first rows, as an approximation from his extensive work with remotely sensed data in Gabon.

7 There are, of course, other clearing types than these two, which, in the classification by Geist and Lambin (2001: 66–72), are called 'patchy forests' and 'diffuse clearing'. These authors distinguish four additional types of clearing: geometric (modern-sector large-scale clearing), corridor (e.g. roadside colonisation), fishbone (the classic pattern of resettlement schemes) and island (e.g. circular clearing in periurban areas).

8 Downton (1995: 235) notes that, even when using the same satellite (Landsat), the studies by the Brazilian Space Agency and by Skole and Tucker reached results that differed by 50 per cent. Apparently, this was because of the different type of images (1: 250,000 versus 1: 500,000 scale and use of coloured versus black-and-white images), as well as subjective differences in the interpretation of images.

9 This principle is sometimes referred to as 'weak sustainability' (Pierce *et al.* 1990: 1–9), with a high degree of 'sustainability' between natural and man-made capital (Costanza *et al.* 1997: 96–107).

10 The SOFO deforestation update was mainly achieved through extrapolation of the mathematical model in FORIS. However, new forest inventories were integrated for nine countries, among which is our example of PNG.

11 The draft chapter for Gabon was passed to FAO-FRA in December 2000, and is cited three times in the FAO-FRA 2000 Gabon section on forest-cover changes, at http://www/fao.org/forestry/fo/country/index.jsp (accessed 31 July 2000).

12 M. Steininger, NASA/GSFC, personal communication (e-mail), March 2001.

13 I owe this colourful metaphor to Tom K. Rudel, one of the reviewers of this book.

14 These annotated bibliographies, for example, Ortiz-Chour (1999) for Nepal, Corrales and Kleinn (2000) for Ecuador, or Kleinn and Corrales (2000) for Venezuela, are available on the Internet, at http://www.fao.org/FORESTRY/FO/FRA/index.jsp (under 'Publications').

15 Remembering that the deforestation factors in Box 3.2 were not fully exclusive (e.g. wood-harvesting may be combined with agricultural conversion), such a calculation must be careful not to *double-count* items.

16 Many of these coincide with or were inspired by Matthews (2001), Grainger (1996) or Malingreau (1993).

4 Gabon

Gabon is the textbook case of a country where, in line with the core hypothesis, forests have been decisively protected by oil rents, attracting people to urban areas through government spending, grossly reducing agricultural competitiveness and production, and providing strong general disincentives for deforestation and forest degradation. In this respect, the Gabon story will provide a baseline which can be used to compare the other countries' land-use stories with in the following chapters.

Deforestation in Gabon

Vegetation history

Gabon is today one of the most forested countries in Africa. Bordered by Cameroon and Equatorial Guinea in the north, Congo in the east and south, and the Atlantic Ocean in the west, the country is part of the Congo Basin. Of its land area of 267,665 km^2, tropical forests currently cover more than 80 per cent, complemented by transition zones and savannahs (see Map 4.1). Its forests fall into three categories: the broad group of coastal basin forest, the more homogeneous forests of central Gabon, and the north-eastern forests that share characteristics with semi-deciduous forests (Drouineau and Nasi 1999). Gabon's small population retains a forest cover close to 20 ha for each inhabitant, an average reached by no other African country.[1] For Gabon, the interesting comparative questions thus relate to why deforestation did *not* occur. For instance, why was forest clearing reduced to point impacts? Why is net forest cover likely to have regenerated over the past three decades? What role has oil wealth had to play in this?

How has vegetation cover changed historically? Climatic change played an important role. Twenty-five thousand years ago, savannahs covered most of Gabon's land area (Clist 1995), but during sub-periods (20,000–15,000 BP and 2,000–2,800 BP), only micro-refuges of forest survived (Maley 1998: 1). Human presence in Gabon at least dates back to hunter-gatherers in the Middle Stone Age, 100,000–40,000 BP. They probably practised extraction from rivers, forests and savannahs. In the Iron Age, the development of agriculture also facilitated settlement expansion, but population density remained low, at around 0.5 inhabitants/km^2 (Clist 1995: 198). Human impacts on the forests only become noteworthy after 10,000 BP, especially through the use of fire. The forest–savannah distribution has thus historically been shaped jointly by climatic change and by humans. For example, the south-western savannahs have evolved with repeated burning by Bantu tribes, who migrated into the area in the fifteenth and sixteenth centuries.[2]

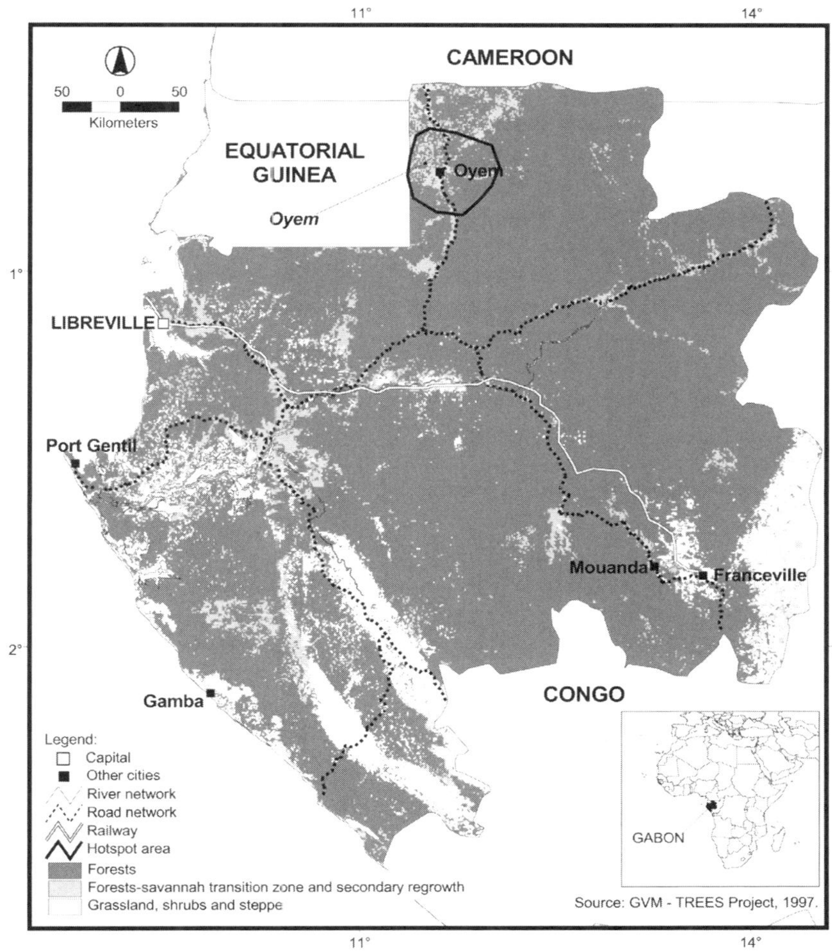

Map 4.1 Gabon.

Colonial settlement by the Portuguese and French initially concentrated in Estuary Province. Penetration into the interior only occurred later, in 1850–80, along the Ogooué river (Ropivia and Djeki 1995). Trade with the Portuguese, Dutch, English, German and French focused on extractive products (ivory, palm oil, timber, honey and beeswax), much more than agricultural commodities (NEA 1994: 19–25). These early trade flows provided no direct incentive for additional forest conversion to crops. On the contrary, the emerging slave trade from Gabon, fuelled by colonial labour demands for overseas plantations, depressed population densities. Much later, between 1911 and 1933, French and German systems of forced labour led to a number of famines in villages, further reducing population density (Adams and McShane 1996: 212). This limited human presence and intervention has greatly favoured forest preservation in Gabon.

The country never developed a strong agricultural tradition. It has been claimed that, at independence, between 80 (Richard and Léonard 1993: 230) and 86 per cent

(Yates 1996: 64) of the labour force worked in agriculture. But one should rather see this rural population as a 'forest people' practising complementary subsistence cropping. Some observers claim that agricultural neglect is rooted in the extremely adverse natural conditions, that is, 'the predominance of swamps and dense forests' (EIU 1997: 22). Gabon does suffer from soil limitations and a less diverse agricultural potential than, for example, neighbouring Cameroon (Richard and Léonard 1993: 58–63). Yet, as French Gabon expert Roland Pourtier notes, biophysical conditions do not differ much from other central African countries where agriculture is far more important (Pourtier 1989a: 35).

Gabon's abundance of forests and agricultural underdevelopment is best explained by its particular socio-economic and historical circumstances (ibid.: 146). For instance, at the beginning of the twentieth century, efforts were made to promote oil-palm plantations in Gabon, but these failed because of the small labour force, which was unaccustomed to working in plantations and clearly preferred extractive activities (Pourtier 1989b: 141–3). Following independence in 1960, foreign trade gradually shifted towards mining resources (manganese, uranium and petroleum), which also had little or no direct forest impact (see section on 'The effect of oil and mineral production on forests'). Actually, only one externally traded commodity directly affected forests in Gabon: the selective logging of *okoumé* (*Aucoumea klaineana*), a valuable timber species that grows in most of the country (see section on 'The competitiveness of agriculture and forestry'). Until the 1980s, exploitation occurred in easily accessible coastal areas, especially Estuary Province (Raponda-Walker and Sillans 1961: 28). In recent decades, progress in transport and extraction technologies has rapidly expanded production into the interior (Droineau and Nasi 1999: 8–13).

On the whole, a combination of historical, demographic and socio-cultural factors have meant that an extraordinarily large part of Gabon's land area has preserved its natural forest cover. The most important, and interrelated, reasons for this seem to be the country's low population density, the negligible trade-led incentives for the development of cash crops and the economy's continuous reliance on extracted rather than cultivated resources. This is the background against which the more recent macroeconomic changes have to be compared.

Current forest loss

Gabon is extraordinarily rich in forests, but there is no consensus on the extent of forest cover or its change over time. Table 4.1 gives an overview of the estimates. The most frequently used source is FAO's FRA for 1990 (FAO 1997a), which estimates total forest cover as having been 18,314,000 ha in 1990 and 17,859,000 ha in 1995. For 1995, this would correspond to only 69.3 per cent of Gabon's land area. Almost all of this is natural forest: in 1995, plantations accounted for a mere 21,000 ha. The FRA reported an annual deforestation of 116,000 ha throughout the 1980s and slightly less (91,000 ha) in the first half of the 1990s.

The model-based FRA methodology is particularly critical in the case of Gabon. The FRA 1990 figures are extrapolated from just a single forest assessment, dating back to 1970. Hence, the FAO rightly classified the reliability of both the forest stock and change estimates as 'low' (FAO 1993: annex 1). Yet, as will be argued, the FAO-FRA figures are not only unreliable for Gabon, they are totally misleading. They underestimate forest cover and

Table 4.1 Gabon: forest cover and deforestation estimates

Author	*Forest cover* (in ha)	*Year*	*Annual deforest.* (in ha)	*Relative decline* (%)	*Period*	*Source type*	*Coverage notes*
FAO (1997a) FRA	18,314,000 17,859,000	1990 1995	91,000	0.5	1990–5	Model estimate	Total forests, >10% tree cover
FAO (1993) FRA	18,235,000	1990	116,000	0.6	1980–90	Model estimate	Total forests, >10% tree cover
TREES – Mayaux *et al.* (1998)	20,677,000	1991–5	—	—	—	NOAA-AVHRR satellite images	Evergreen and semi-deciduous forests, >70% tree cover
TREES, cited in Collomb *et al.* (2000)	21,338,900	1991–3	—	—	—	NOAA-AVHRR satellite images	Unspecified
TREES, cited in Wilks (2000)	22,957,000 1,839,000	Not specified	—	—	—	NOAA-AVHRR satellite images	All forests (>10% tree cover) Fragments (10–70% tree cover)
WRI (1998)	19,411,000 18,314,000 17,859,000	1980 1990 1995	109,700 91,000	0.6 0.5	1980–90 1990–5	FAO and ITTO	All forests (incl. plantations)
FAO (1996) *Production Yearbook*	20,030,000 19,960,000 19,966,000 19,900,000	1979 1984 1989 1994	14,000 −1,200 13,200	0.01 −0.0 0.01	1979–84 1984–9 1989–94	Forestry agency reporting	Prod.forests + other categ.[a]
IUCN (1990)	22,531,000	1987	—	—	—	Map G.Caballé (Edicef)	Total forest (>10% tree cover)
Sayer *et al.* (1992)	23,544,500*	1987	—	—	—	Maps IGN Paris, INC Libreville	Rain forests[b]
Myers (1994)	22,000,000 16,400,000	'Original' 1989	230,000	1.4	1989	Not specified	Tropical moist forests
Collomb *et al.* (2000)	25,767,000 22,000,000	'Original' unspecified	—	—	—	WRI MoWF	Unspecified Unspecified

Notes

a Production forests + other wooded land + intended reforestation ÷ recreation forests.

b According to the source, forest cover is overestimated because forest fragmentation is underestimated.

* Regional estimates.

severely overestimate deforestation: there is no reason at all to believe that about 100,000 ha of forest are lost each year, since there is no alternative land use that could explain conversion on this scale.

A direct forest stock measurement from NOAA-AVHRR satellite images has been undertaken by the TREES project under the EU-financed Joint Research Centre. TREES takes into account evergreen and semi-deciduous dense forests, with a minimum of 70 per cent tree-cover, compared to the FAO's 10 per cent criterion (Mayaux *et al.* 1998). This excludes highly fragmented forests and forest–savannah transition zones, a definition that *ceteris paribus* should make the TREES estimate smaller than the FAO's. However, the order of magnitude is the reverse: the TREES estimate of 20,677,000 ha (80.2 per cent of land area) for the early 1990s is 15.8 per cent higher than the FAO's model-derived figure for 1995.[3] Wilks (2000: chs 1–2) gives a slightly different interpretation of the TREES data and stresses the fuzzy character of the distinction between the forest and non-forest categories. Including fragmented forests (10–70 per cent forest cover), which is closer to the FAO's definitions, Gabon's total forest area would thus be 23,957,000 ha (of which 21,188,000 ha is dense forest), corresponding to no less than 89.5 per cent of total land area. These discrepancies stress the outdated empirical basis of the FAO data.

While the TREES figures represent the best currently available knowledge on forest stock, most other estimates shown in Table 4.1 build on a mixture of sources, many of which are of dubious quality. WRI (1994) combines the FAO data with other (unspecified) deforestation sources for 1981–5. As 1981–5 forest loss figures are much lower (0.1 per cent yearly loss), it would require an unlikely acceleration of deforestation during 1986–90 (to 1.1 per cent/yr) for this to remain consistent with the 0.6 per cent average for the whole of the 1980s. Data in the *FAO Production Yearbook* (FAO 1996) show random variation around a forest stock of 20,000,000 ha, and are as unreliable for Gabon as for the other countries analysed in the present book. The Gabon Report of the WRI Global Forest Watch initiative (Collomb *et al.* 2000: 34) cites an estimate of 22,000,000 ha from Gabon's Ministry of Water and Forests (MoWF) for the mid-1990s. The same source publishes an 'original' forest estimate from the WRI's Frontier Forest initiative that corresponds to 97 per cent of Gabon's land area (25,767,000 ha), but scientific evidence shows that, on the contrary, historically there was *less* forest in Gabon.[4] Finally, the 23,544,500 ha reported in the IUCN Conservation Atlas[5] is, as admitted in the source, a very high figure, due to the rough scale of the underlying maps that cause both small clearings within the forest and forest fragments within transition areas to be overlooked.

The TREES data thus provide a reliable current estimate of forest stock, but no deforestation figures over time. One indicative exercise is to calculate the FAO-FRA's wrongly extrapolated figures from the 1990 estimate (18,235,000 ha) back to their base year, that is, forest cover in the original map of 1970. With an alleged 0.6 per cent yearly deforestation during 1970–90, 1970 forest cover must have been 20,567,338 ha. Although map methodologies and resolutions are probably different, it is noticeable that this is marginally *lower* than the recent TREES estimate for 1991–3 (20,677,000 ha). The comparison would indicate that forest cover has basically remained unchanged since 1970, probably with marginal net *reforestation*.[6]

Recent changes in forest cover over time thus appear to be minor, and not well documented. Logging may be the main influence on forests. As Wilks (2002) indicates, this would cause deforestation of 14,500 ha if measured at a very refined resolution level (see above), and forest modification of an estimated 350,000 ha/yr. However, we shall assume that logging causes degradation rather than deforestation, thus using a coarser scale of resolution that is more common for national forest assessments (see section on

'The competitiveness of agriculture and forestry' and Chapter 1). The TREES classification of deforestation hotspots (Achard *et al.* 1998) also includes only one single site, the Oyem area, which traditionally has been the most agricultural region in Gabon (see Map 4.1).

Can additional clues as to land use be gained from other sources, such as agricultural censuses? Unfortunately, agricultural sources are of little help in Gabon. An agricultural survey was carried out in 1960–1, and a proper census in 1974–5 as part of an FAO project. After 1975, no agricultural census exists for Gabon, reflecting the generally low priority of land-use data.[7] Some recent figures are available, based on yearly reports from regional agricultural offices, but there is clear evidence that these figures grossly overestimated cropped area. Even for the 1961 and 1975 point estimates, methodologies may differ.[8] Still, the comparison is interesting for our purposes: in 1961, Gabon was not yet an oil-dependent economy (the *avant-pétrole* period), while 1975 was in the midst of the first international oil price boom (MAEDR 1975: 56–61). The number of cultivating households increased marginally (1.8 per cent), from 71,440 in 1960–1 to 72,734 in 1974–5. Yet cultivated area declined by a stunning 33,542 ha, from 106,280 to 72,738 ha (31.6 per cent). The average age of the cultivating household head had also risen remarkably: in 1960–1 the subgroup over fifty years made up 30.4 per cent, while in 1974–5 this share was 52.2 per cent. This reflects the fact that the younger generation migrated to urban areas for better job opportunities (see section on 'Structural changes in income and demand'), leaving behind an ageing rural population, including retired return migrants. As is confirmed by case studies, these old or retired farmers tend to cultivate much smaller plots.[9]

Can any changing land-use trends be identified from the sub-national analyses? A series of regional deforestation studies have focused on land-use changes in and around urban areas (see section on 'Structural changes in income and demand'). This concerns urban areas: Libreville (Magrin 1994; BDPA 1998; Trefon 1999), Franceville (Wolff *et al.* n.d.; IGAD 1997; Rippert 1997) and Oyem (Wolff *et al.* n.d.). In periurban areas, there has been an expansion of land colonisation, and even garden plots in intra-urban areas are cultivated. Plantains, tubers and legumes are the main crops in intensified, semi-permanent production systems (Trefon 1999: 45). This food supply only partly satisfies increasing urban demand, and most foodstuffs consumed in Gabon are imported. For a town like Oyem, satellite analysis shows that cumulative forest clearing (cultivation and fallow areas) reached 5,015 ha from 1961 to 1990, corresponding to about 25 per cent of the study area (Wolff *et al.* n.d.: 22). For Franceville, 1,699 ha of forest (about 30 per cent) were cleared or degraded for cultivation from 1953 to 1994, and 1,489 ha of savannah integrated (ibid.: 26–7). Around the capital, Libreville, a recent study commissioned by the Institut Gabonais d'Appui au Développement (IGAD) (BPDA 1998: 18–26) shows that the currently cultivated area is 9,000–10,000 ha, while the total area affected (including fallow and secondary regrowth) is 55,000–65,000 ha (ibid.: 23). Periurban cropped areas may perhaps make up about 50,000 ha in the whole of Gabon.[10]

However, periurban deforestation only represents a partial, quite specific phenomenon, which can in no way be generalised to the entire country. Higher forest loss around the cities is likely to have been more than counterbalanced by abandoned slash-and-burn activities and reduced periods of fallow in rural and semi-rural areas. There are many examples of this process of rural abandonment. Pourtier (1989b: 270, 277) documents a sharp contraction of cultivated area around the small towns of Makokou and Lastoursville. Studies of rural areas by the Libreville-based consultants Africa Forest comparing aerial photographs from the 1950s and the 1990s find a substantial decline in cropped area. For instance, the population of the village of Nyonyie (southern Estuary, near Libreville) declined from 500 to only twenty people, and a 4 km strip of coastal plantations has now regenerated as secondary forest (Christy *et al.* 1990a). Another coastal village, Oyan, was reduced from 100 to 10 people, with a similar decline in traditional agriculture (Christy *et al.* 1990b). Similarly, at least 13.5 per cent of the forest had historically been cleared for agriculture in the Avocette area prior to regeneration, which is ten times the impact of local oil-producing activities (Wilks 1992: 34). Adams and McShane (1992: 207–13) describe rural abandonment and forest regrowth in north-east Gabon. Overall, it is estimated that only about 50,000 families are still practising traditional agriculture in Gabon (Marchés Tropicaux 1998: 20). While current rural population levels may be stagnant rather than declining, the continuous ageing of farmers probably still causes some net abandonment of cultivated areas in the country's interior.

It thus seems probable that Gabon experienced some net *reforestation* over the past three decades, with forest regeneration outstripping deforestation. Extensive traditional agriculture in the interior of the country has been reduced massively by the rural exodus, and more intensive systems near the urban areas have expanded only moderately. From the TREES data, Wilks (2000: 13–14) estimates that industrial plantations in Gabon cover 23,000 ha (0.1 per cent), traditional agriculture about 90,000 ha (0.3 per cent), while the mosaic of post-agricultural vegetation at different stages of regeneration occupies nineteen times the area cropped, that is, 1,718,000 ha (6.7 per cent).

For the sake of a dynamic interpretation, let us make the conservative assumption that about 90 per cent of this regenerating vegetation is secondary forest (1,538,000 ha), while non-forest fallows make up 10 per cent (180,000 ha). Total deforestation from traditional agriculture would thus as a rule of thumb be three times the area cropped; in the present hypothesis, 90,000 ha times three equals 270,000 ha. Imagine now that the area under slash-and-burn in 1961 was twice as large as it is currently (180,000 ha), that fallow length has not changed, and that there were no industrial plantations.[11] This would have caused a deforested area in 1961 of three times the cropped area, that is, 540,000 ha. More than 270,000 ha today would thus have been deforested in 1961 by slash-and-burn agriculture combined with long fallows. This is more than ten times the size of the 23,000 ha of industrial plantations that have been created up to the present day. Net forest cover would thus have been 247,000 ha less in 1961. Although this exercise remains speculative, it indicates that food-crop cultivation is the decisive driver of deforestation, simply because of its extensive land requirements. Hence, the combination of agricultural decline and the shift to sedentary, intensive systems is likely to have permitted a sizeable expansion in forest cover.

It is equally clear that the FAO-FRA 1990 model-derived figure of a current yearly forest loss of about 100,000 ha was completely unrealistic. It is indicative that the total accumulated zone under cultivation is roughly in the range of 100,000–150,000 ha, which makes an alleged yearly net increment of almost the same size meaningless. As will be shown from the analysis of other deforestation sources, there is no motivation to clear so much land in Gabon. Indeed, the following section will demonstrate that the forest-loss impact of the oil and mining sectors is very limited.

The effect of mineral production on forests

Direct oil impacts

The presence of oil in Gabon was known as early as 1928, but promising reserves had only been found by 1955 (Shell Gabon 1996). It was thus only after independence in 1960 that oil production started to develop on a significant scale. From 1.4 million tonnes (t) in 1966, production expanded to 5 million t in 1969 and 11.3 million t in 1975 (EDIAFRIC 1985: 80). Originally, the bulk of oil reserves were exploited off-shore, mostly by the French firm Elf Aquitaine. But with the discovery of the coastal fields of Rabi-Kounga, Gamba-Ivinga and Echira by Shell at the end of the 1980s, on-shore production has come to play the most important part in the industry.

On-shore exploration has been carried out along the entire coast and its hinterland, but today production is concentrated on the coastal stretch between Port Gentil and Gambá (southern Gabon). Two oil export terminals have been built, connected by a pipeline, in Gambá and Cap Lopez. Production zones coincide with areas of wetlands, mangroves, rainforest, savannah and savannah transition zones. In on-shore areas, oil production therefore has potential direct and indirect deforestation impacts. According to the scenario presented by the international NGO Oilwatch, oil impacts appear dramatic: the on-shore exploration area in Gabon is said to be 46,073 km^2, and exploration is seen to be 'the principal cause' of the FAO-cited 0.6 per cent (116,000 ha) yearly deforestation in Gabon (Amigransa *et al.* 1997: 34). As will be argued from environmental impact assessments of both the oil companies and independent sources, this is a highly misleading picture: both direct and indirect deforestation impacts are actually extremely limited.

The consultancy firm Africa Forest has carried out a series of site-specific environmental impact studies in this area.[12] Seismic exploration according to the old 'heavy' method needed roads to be cut along the seismic lines, but this was replaced by a 'light' method in the late 1970s (see Chapter 3). On one site where both methods had been applied, the 'light' method reduced the area affected by 50–75 per cent, but even the *combined* impact of both methods affected only 13 ha of forest, corresponding to 0.5 per cent of the Ofoubou study area (Basquin *et al.* 1991: 67).

The second impact concerns actual oil-drilling for exploration and production. Here, the largest deforestation impact results from the establishment of roads, pipelines, drilling platforms and other infrastructure. A minor impact is caused by the exploitation of sand and laterite quarries, used for the construction of roads, platforms, etc. Shell Gabon has attempted to estimate the forest loss impact of its oil activities in the Rabi field, which in the mid-1990s produced 55 per cent of Gabon's total oil output. The Rabi Phase I project still included 'heavy' methods, such as clearing a 30 m safety zone around platforms, which implied a larger forest loss than modern and environmentally more friendly techniques. According to the study, the direct oil-related deforestation impacts of Rabi I amounted to 1,405 ha. Half of this (735 ha) occurred inside the proper Rabi Field Permit, that is, there was forest loss corresponding to 5.7 per cent of the permit area (Shell Gabon 1996: 37). Roads and pipelines accounted for half the on-site effects. Off-site effects were 670 ha, all of which was due to infrastructure. Quarries consumed a minor 50 ha, platforms 74 ha and other infrastructure (stations, airport, camps, etc.) 101 ha, while non-forest regrowth areas covered 117 ha.[13] Roads and pipelines are thus clearly the main space occupants, causing the largest deforestation. But even for Rabi I, an older and not environmentally up-to-date operation, forest loss remains relatively limited: the 1,405 ha correspond to only 0.06 per cent of the total Setté Cama Permit Area, and to 0.0066 per cent of Gabon's estimated total forest area.

The observation of very limited direct impacts of oil on forest cover is confirmed by several studies in oil-concession areas along Gabon's coast. In forest–savannah transition zones and in old logging areas, oil companies may (and normally do) take advantage of pre-existing open areas to install their infrastructure without clearing too much forest. Typically, the forest-loss impact of oil operations lies between 0.2 and 0.5 per cent of the corresponding study area's forest cover. This is much lower than the agricultural conversion and logging pressures that these areas have faced historically, before they became depopulated during the last couple of decades (see section on 'Rural–urban migration'). In terms of the record of historical land-use changes, oil thus forms a minor part (see the Africa forest studies cited above). Also, new techniques to facilitate forest regrowth imply that clearing figures actually do not represent *permanently* deforested areas.[14]

Can one derive a credible figure for the overall impact of oil production on the national scale from the various case-based forest-loss estimates? Adding up the presumed impact at Rabi I (1,405 ha), Rabi II (70 ha), Gambá-Ivinga (2,810 ha) and Setté Cama (782 ha), one source arrives at a total of 5,067 ha for the Gambá complex, which is the most important on-shore production area (Blaney *et al.* 1998: 97). However, as an unpublished map of the area distribution of oil concessions shows,[15] there are actually many minor on-shore exploration sites along the entire coastal strip of Gabon. Considering size, production frequency and technology, a best guess would be to double the impact found in the Gambá complex (J. Bickerton, personal communication, 8 June 2000) to a national direct forest loss from oil of about 10,000 ha. This *accumulated* forest loss represents a mere 0.04 per cent of Gabon's land area, a figure that is quite different from the estimates issued by the environmental campaign mentioned at the start of this section.

Indirect oil impacts

Although oil is a typical enclave sector with limited employment generation,[16] the indirect forest-loss impacts may sometimes be more important than the direct ones. For instance, the southern town of Gambá basically arose as a settlement of oil workers on Shell's concessions. Many of the immigrant women have started diversified slash-and-burn cultivation of food crops for both subsistence use and local sale. A WWF survey of the Gambá area found that households cultivate on average 1.2 ha. For all Gambá households, this adds up to a total forest conversion of 6,224 ha. This is thus higher than the direct oil impacts of 5,067 ha (Blaney *et al.* 1998: 97). Fortunately, low population pressure in Gabon limits these indirect impacts to local needs: there is no wave of immigrants or additional colonisation following the oil companies.

Environmental recommendations to avoid road construction have not always been followed by the oil companies, because the local people insisted on roads being built to improve their transport access (C. Wilks, personal communication, 29 May 2000). This indicates that indirect forest-loss impacts may sometimes be desirable from a local point of view. In parenthesis, the same consideration also frequently applies to road-building by logging companies that may be desired by local communities (Simons 1996). Oil roads almost always provide better access for hunting, including for the transport of bushmeat to urban markets. Local villagers, temporary oil workers and urban-based hunting crews may all be active in poaching.[17] Declines in wildlife are greatest in those areas that are closest to urban bush-meat markets (see section on 'The structure of consumption').

Other mining impacts

Gabon also has other export mining sectors, namely manganese and uranium: do they cause either direct or indirect deforestation impacts? This can be answered in the negative. Sizeable manganese deposits exist near Moanda (50 km from Franceville, in the interior). They were originally transported by a 76 km long cable car and the Congolese railway to the port of Pointe-Noire. With the completion of the Trans-Gabonese railway to the Estuary port of Owendo, near Libreville, manganese production increased (Dautuille 1996: 13–14). Uranium deposits are also found in the same region, the most important ones near Mounana. Some extraction has been by open-pit mining, and some was subterraneous. Uranium exports have been declining heavily since the 1970s (ibid.: 15–16). However, the bottom line in deforestation terms is that these non-renewable resources are both mainly deposited in savannah areas, where they occupy limited space and where their exploitation does not compromise forested areas (IUCN 1990: 37–8). Nonetheless, changes might occur in the future. Several South African firms are currently prospecting for gold in forested areas, and open-cast mining would cause forest loss (C. Wilks, personal communication, 6 October 2000).

On the whole, the oil and mining sectors thus have a negligible deforestation impact in Gabon. Oil, the uncontested mainstay of the Gabonese economy, has a very limited forest impact, both in direct (oil-related) and indirect (access-providing) terms. Indirect impacts tend to be limited to oil workers' food crops and to hunting pressures. The indirect colonisation impacts would probably have been much larger had it not been for the general long-term bias against rural development and small-scale agriculture in Gabon, to be described in the following sections.

The macroeconomic impact of the mineral boom

Over the last three decades, the rise in oil exports and the high inflow of foreign exchange per capita have created an economy that has been called the 'African Emirates' – the ultimate *rentier* state depending heavily on one single, wealth-generating export commodity. Throughout the history of Gabon, rent-generating extractive sectors have always dominated, such as *okoumé*, uranium and manganese. But oil increased rents significantly. Figure 4.1 shows long-term trends for three key macroeconomic indicators: oil exports, capital inflows and RPs.

The value of oil exports remained limited in the 1960s and early 1970s but rose dramatically from 1974 onwards, as a combined effect of the oil-price boom on the world market and of rising Gabonese oil production. Inflow peaked in absolute terms with the second oil price boom in 1980, when export revenues reached a nominal value of

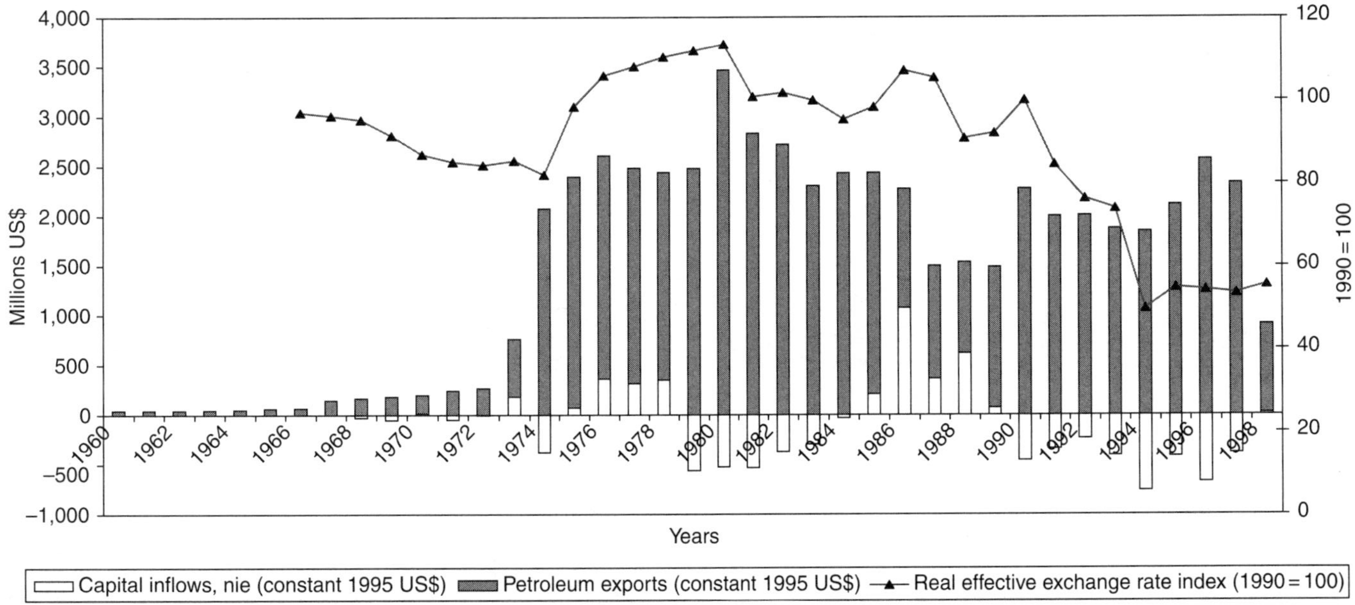

Figure 4.1 Gabon: capital inflows, petroleum exports and real effective exchange rate.

Sources: DGE (1998, 1999); DGSEE (1997); IMF (1990, 1999, 1999b); World Bank (1999a, 2000a); Zomo Yebe (1993).

Notes

1 Real effective exchange rate, 1966–77: 1970 weights; 1978–9: 1985 weights; 1980–98: 1990 weights.

2 Capital inflows, not including exceptional financing, 1968–88: Other capital nie; 1989–98: Financial account nie.

3 Petroleum exports, 1960–85: Crude petroleum exports, 1986–98 Petroleum exports.

US$1.9 billion (US$3.5 billion in constant 1995 prices). What is remarkable compared to other oil exporters is that increased oil production partly compensated for declining oil prices by the end of the 1980s. The discovery of important new fields, such as Rabi-Kounga on the coast, pushed export production levels up from 7.9 million t in 1987 to 14.7 million t in 1991 and 18.2 million t in 1995 (DGSEE 1997: 19). This means that, in spite of less favourable prices during the 1990s, real oil revenues only declined by about 25 per cent, compared to the peak from the mid-1970s to 1985. In other words, the quantity element of oil discoveries gave a semi-permanent character to Gabon's version of the oil boom.

Oil prices have fluctuated greatly. The peak of US$22.25/barrel for Gabon's Mandji crude in October 1996 (EIU 1999a: 17) was followed by a severe price slide below US$10/ barrel in December 1998 (DGE 1999: 15). This brought down real export revenues to almost one-third: in constant 1995 prices, oil revenues fell from US$2.6 billion in 1996 to US$0.9 billion in 1998 (see Figure 4.1). As oil accounts for more than half of public revenues, this directly triggered corresponding declines in the government's budget. During 1998, falling oil prices meant that the fiscal budget had to be revised downwards by 15.1 per cent (DGE 1999: 15). Yet, during 1999–2001 world-market oil-prices have again skyrocketed to above US$30/barrel (not shown in Figure 4.1), more than offsetting the decline in Gabon's production quantities in the short term. The balance of payments showed a surplus of around 20 per cent of GDP in 1999–2001 (Söderling 2002: 3).

A second element in capital inflows has been foreign borrowing, but in *relative* terms indebtedness has been less pronounced than for most high-absorbing oil countries. Gabon has accumulated some of the highest foreign debt per capita in Africa, but because of high income levels, total debt size currently corresponds to less than one years' GNP (95.7 per cent), and the ratio of debt service to exports was only 13.1 per cent in 1997 (EIU 1999a: 30). As can be seen from Figure 4.1, some borrowing against oil revenues occurred during 1975–8 and especially 1986–8, compensating for the abrupt drop in oil prices. Gabon adopted several structural adjustment programmes with the World Bank and the IMF, but frequently failed to service its debts and accumulated arrears, especially in the 1988–93 period (Statistisches Bundesamt 1994: 88). Gabon's capital account has a negative structural element, because of oil-related financial transactions (EIU 1999a: 29). This element implies that there are few years with significant net capital inflows. But in terms of year-to-year fluctuations, Gabon's borrowing path was clearly counter-cyclical, as the country used foreign capital to smooth the unpredictable fluctuations in international oil prices.[18]

As Gabon has an extremely open economy, after 1973 this steady cash-inflow pattern was directly translated into growth and fluctuations in domestic demand. Following moderate but stable economic growth in the 1960s, higher oil production and the two oil-price booms suddenly made Gabon an affluent country. This happened from 1972 to 1980, when the US$ value of per capita GDP rose almost *eightfold*, from US$803 to US$6,193. After 1983, the then highly oil-dependent economy did not, as in Cameroon, follow a sustained pattern of crisis and dramatically falling per capita GDP, but rather suffered year-to-year oscillations, still at a highly elevated level of per capita GDP (US$4,200–6,000). Variations over time have basically been determined by oil-price and US$-exchange rate fluctuations *vis-à-vis* the French franc. In Gabon, the entire quarter of a century after 1973 should thus be seen as one continuous boom period, even though oil-price fluctuations created budgetary shortfalls and a series of 'mini-crises'. A crisis and post-boom adjustment comparable to that of other Sub-Saharan countries should only be expected once oil reserves run dry during the decades to come.

These radical long-term changes in the structure of the Gabonese economy are also reflected in RP changes, as indicated by the real trade-weighted exchange-rate index in Figure 4.1. As in Cameroon, the CFA franc (FCFA) is Gabon's national currency. It was for a long time held constant at 50 : 1 to the French franc, until it was devalued to 100 : 1 in 1994. The nominal exchange rate *vis-à-vis* the dollar zone is thus determined by changes in the dollar–franc exchange rate – from 2002 onwards by the dollar–euro rate. Compared to other CFA countries, changes in competitiveness are set by inflation differentials. Gabon thus cannot independently devalue its currency in response to a bust in oil exports. This limits the degree of freedom in economic policy-making, and tends to keep the RER overvalued in bust situations.

From 1970 to 1980, the oil boom caused a real appreciation of about 75 per cent against the dollar: domestic inflation continuously exceeded US$ inflation, and the nominal exchange rate of the French franc appreciated by 35 per cent.[19] In the late 1980s oil revenues declined, while in the early 1990s the RER depreciated by almost 50 per cent. Note that this process had already got underway with the deflationary period of 1990–3 (IMF 1999b: 447, line 64), that is, well before the nominal 50 per cent devaluation of the FCFA against the French franc. Devaluation was followed by an instantaneous one-time inflationary surge to 35 per cent in 1995, cancelling out much of the RP effect (CIA 1999a: 5).

What were the criteria and priorities of the government of Omar Bongo – the man in power since 1967 – in distributing the country's oil wealth? A number of main areas can be outlined:

- an increase in public employment, and in the salaries and benefits of public employees;
- investment in transport infrastructure (Transgabonese railway, Owendo port);
- investment in urban infrastructure (construction, health, education);
- investment in and subsidies to large-scale parastatals.

First and foremost, government bureaucracy has swelled after independence, in a manner that would have been impossible without oil and mining rents. The number of public employees (excluding the para-public sector) grew from 3,842 in 1965 to 9,800 in 1970, 15,400 in 1975, 35,479 in 1980 and 42,664 in 1983 (Pourtier 1989b: 205). The trend clearly underlines the link between rising oil income and higher public employment. Until the IMF austerity plan of 1987, civil servants also received fringe benefits like free cars and housing (Yates 1996: 207). During 1985–7, when oil prices fell sharply, total labour costs in the budget were reduced by 16.2 per cent (Zomo Yebe 1993: 37), while in 1989–90 fixed employment was reduced by 912 persons (Statistisches Bundesamt 1994: 45). Yet these cutbacks were of short duration. Unpublished official data from the Direction Générale du Budget (not fully comparable to Pourtier's figures)[20] show that, even during the mini-crisis of the 1990s, public employment grew every single year. In 1994–5 the nominal public payroll grew less than FCFA inflation (IMF 1999a: 30). But, with the 1999 recovery of oil prices, public salaries were raised significantly, so that the total public payroll, including indemnification payments, grew from FCFA 149 billion in 1998 to FCFA 212 billion in 1999 (République Gabonaise 2000: 16). In the revised 1999 budget, this corresponded to 34.4 per cent of the budget (ibid.).

Between 1975 and 1990, public employment thus grew by 111 per cent while private and para-public employment was reduced by 31 per cent (Statistisches Bundesamt 1994: 45). There are clear indications of over-staffing, and absenteeism is widespread among public employees.[21] Giving new employment opportunities to people from different regions and ethnic groups may generally be seen as a way of continuously buying balanced political support within an overall situation that Yates (1996) calls the 'allocation state'. In 1995, the public and para-public sectors employed no less than 70 per cent of wage earners (Poupart and Pilichowski 1997: 49–50). But the granting of oil-financed benefits to rent-seeking groups goes beyond the budgetary sphere. According to Yates (1996: 209–11), institutionalised corruption extends to luxury goods and personal shareholdings that the local political elite has acquired in Gabonese corporations as returns for favours. Zomo Yebe (1993: 62) calls this the rise of a 'kleptocracy'. The business environment in Gabon remains closed, singularly dominated by French interests (Barnes 1992: 71–5; Yates 1996: 187; EIU 1999a: 10) and linked to personal ties. The borderline between 'public' and 'private' spheres is extremely fluid. The problem seems to have accelerated severely over time, as illustrated by the incredible number of false invoices circulating in the public sector.[22]

Second, a large share of the oil money went into what was at the time one of the largest construction projects in the world: the *Transgabonais* (Transgabonese) railway, linking Libreville's port of Owendo in the north-west to Franceville in the south-east, near the border with the Congo. Estimated total costs over the 1973–86 period vary from US$3 billion (EIU 1999a: 13) to US$4 billion (Yates 1996: 181), the latter figure being close to the size of Gabon's total foreign debt. The main rationale of the project was to open up Gabon's interior to development and increase the extraction of natural resources, such as timber, manganese and iron ore.[23] However, the World Bank declined to support it because of a lack of profitability. Only the oil boom, combined with credits from mainly bilateral sources, made it financially possible for President Bongo to fulfil this personal ambition (ibid.: 177–8; Pourtier 1982: 116). Mismanagement and an absence of competitive tendering caused large cost overruns, which have indeed driven up investment to a level that cannot be justified by the modest current returns (Statistisches Bundesamt 1994: 73). This has led one observer to call the Transgabonais railway 'a costly infrastructure through an empty space' (Pourtier 1982: 127), while another commentator has called it 'a train that leads nowhere … an iron path to more rentierism and primary extraction' (Yates 1996: 183). In spite of some undeniable benefits, the average share of 19.1 per cent of public development budgets absorbed by the railway during 1972–85 appears extremely high.[24]

A third boom-spending category refers to investments in urban infrastructure. In general, the bulk of oil revenues were consumed in the cities, especially in the capital. This included a series of urban prestige projects. The most extravagant were the works undertaken prior to the summit of the Organisation of African Unity (OAU), held in Libreville in 1977. This included new hotels, theatres, convention centres, etc. (Richard and Léonard 1993: 157). The number of employees in the construction sector grew from 17,252 in 1970 to 37,754 in 1975 (Statistisches Bundesamt 1994: 45). During 1977, as much as half of the state's annual budget may have been committed to construction costs, contributing heavily to its fiscal crisis and the forced adoption of the first IMF austerity programme in 1978 (Yates 1996: 196–8). Other lavish constructions of this type included the second presidential palace and several luxurious ministerial buildings.

Much was also invested in urban social infrastructure, such as schools, hospitals, etc. By African standards, Gabon has achieved good increases in school enrolment ratios and hospital bed capacity, although life expectancy and literacy ratios remain close to average (Statistisches Bundesamt 1994: 18). In spite of stagnant per capita GDP, certain social indicators improved further in the 1990s (UNDP 1999b: 21–2). This reflects both higher social spending and improved priorities as part of the IMF-sponsored structural adjustment programme (EIU 1999a: 11–12). Yet, within the sub-group of higher middle-income countries, Gabon still lags behind in terms of three of the four indicators of the 'human development diamond' applied in World Bank (2000a). In spite of high investments in the social sectors, inefficiencies have impeded better human-development results (S. Meyé, UNDP, personal communication, 30 May 2000).

The fourth major oil-rent recipient has been Gabon's sixty-five parastatal companies in industry, services, transport and agriculture, which have been recipients of both investments and subsidies to cover running deficits. In 1982, twenty-four of these companies alone generated a publicly subsidised deficit of FCFA 21 billion (US$56.6 million – Zomo Yebe 1993: 75; IMF 1999b: 446). Agro-industrial parastatals also have a direct potential for deforestation, which will be discussed in the next section. For now, it may suffice to say that much of the parastatal investment was diverted or simply wasted, 'due to massive political interference in management, high operating costs, and gross overstaffing' (World Bank, cited in Yates 1996: 210). This is also witnessed by the current poor financial state of most companies, some of them being forced into privatisation, others into bankruptcy (UNDP 1996: 16; Marchés Tropicaux 1998: 13–14).

On the whole, the combined price and production increases gave Gabon's oil boom a semi-permanent feature, with substantial though unequally distributed wealth. Due to mining revenues (uranium, manganese), the RER had already appreciated in the 1960s, but this was greatly accentuated from 1974 to 1982 by petroleum revenues. Only in the early 1990s was there a substantial real currency depreciation. The Bongo government also indulged in foreign borrowing, especially for the construction of the Transgabonais railway. But borrowing remained relatively low and tended to smooth out fluctuations in oil revenues. In spite of elaborate multi-year development plans, macroeconomic policy has been dominated by discontinuous and reactive stop–go policies that have been held hostage to fluctuating oil prices.

Yet, a worrying recent trend is that Gabon's economy has not grown in real terms during the last couple of years, in spite of a huge oil boom. After a deep recession in 1999 (GDP fell 9.6 per cent in real terms), the economy did not recover in 2000, with a further 1.9 per cent decline in real GDP. This coincides with a record-high share of oil in GDP, reaching 47.6 per cent in 2000 and also a 8.2 per cent real growth in wood production (DGE 2001: 90–1). In other words, the rent sectors grew, along with the service sectors that depend on these rents (+5.2 per cent in 2000), but the rest of the economy declined so much that the net result was negative. This is a typical symptom of over-saturation in the absorption of rents, known from other specialised oil countries (see Chapter 5 on Venezuela), which indicates an extreme inefficiency in the economy.

On the whole, market factors seldom operate independently in Gabon; the state is an omnipresent catalyst. Oil wealth was generally distributed with a strong urban bias, favouring prestige projects in urban construction, infrastructure, parastatal companies and social sectors. Except for the Transgabonese railway, all major public projects were 'forest-neutral' in their direct effects. Their indirect impact was to pull labour out of rural areas on a massive scale towards more remunerative employment options in the cities, in the civil service, in parastatal companies, private services or construction (section on 'Structural changes in income and demand'). As described in the following section, this urban development bias and the corresponding neglect of agriculture, have greatly reduced pressures on the forests.

The competitiveness of agriculture and forestry

Agriculture

Prior to the oil boom, Gabon's agricultural base was already weak, due mainly to a combination of labour scarcity and extremely rich extractive resources. From ivory to slaves, from *okoumé* to manganese, uranium and oil cycles, the economic prospects of establishing a cash-crop plantation economy could never compete with the here-and-now pay-off from the extractive commodity that predominated at the time. However, oil rents brought foreign exchange inflows that were much greater than in any previous extractive cycle. Adverse RP effects further squeezed the agricultural sector. But the impact was different in each of the four sub-sectors of subsistence farming, import-competing food crops, export crops and parastatal agroindustries.

First, subsistence farmers in isolated rural areas mainly cultivate plantains and cassava, and to a lesser extent bananas and yams, using shifting cultivation. They probably suffered less direct impacts from RP changes, as their produce was less exposed to foreign competition. A scattered population, restricted demand and bad road connections meant that many of these rural areas were poorly integrated into local markets, so that their products remained semi-sheltered from competing imports. An exception were those food items where consumer prices in the interior were subsidised (on trade policy, see section on 'Trade policy impacts'). But the decline of the subsistence sub-sector was due more to (forced) rural resettlement (section on 'Windfall impacts on government spending') and to (voluntary) urban migration (section on 'Structural changes in income and demand'). Rising wages in the urban NT sectors drove up labour opportunity costs, inducing many to leave the countryside.

The import-competing food sector was directly hurt by the oil-led loss of competitiveness. In 1960, Gabon was still self-sufficient in most foodstuffs; by 1988, about two-thirds of food items were being imported (H. F. Henner, cited in Zomo Yebe 1993: 12). Rising domestic costs reduced the competitiveness of Gabonese enterprises in a number of ways. Zomo Yebe (1993: 79–84) points to three main cost elements that hold back domestic production: high salaries, high transport costs and – particularly interesting from our perspective – the increased costs of the deforestation needed to expand agriculture.[25] Food imports multiplied by a factor of eight during the 1970s oil boom, from US$12 million in 1970 to US$102 million in 1978 (ibid.: 58). Even during 1987–93, they still grew at an impressive 8 per cent per year in real terms (Marchés Tropicaux 1998). The EU and South Africa have been two main suppliers, especially of meat (section on 'Structural changes in income and demand'). In addition, there are large informal imports of food crops and their derivatives (plantains, cassava flour, fruit, legumes, etc.), by ship or truck, from Equatorial Guinea, Congo and especially southern Cameroon.[26] Unfortunately, statistical evidence on food crops is extremely sketchy. Official reports give current production figures for cassava (250,000 t), plantains (240,000 t), yams (100,000 t), peanuts (10,000 t), rice (10,000 t) and bananas (9,000 t – see Dauthuille 1996: 39–46). As mentioned above, even these small figures are probably overestimates, but their marginal expansion during the last decade, led by a slightly more competitive RER following the 1994 devaluation, may have caused some deforestation.

The third sub-sector, the 'purely traded' export crops, cocoa and coffee, was the one that was most severely hit by declining competitiveness. In the 1950s, the northern Woleu–Ntem 'cocoa triangle' (Oyem–Bitam–Minvoul) experienced a period of prosperity. However, production stagnated with the rise of mining sectors and the corresponding increase in labour costs and out-migration. Disease also played an increasing role. During the oil boom of the 1970s, this decline was aggravated, in spite of the fact that international coffee and cocoa prices experienced a price hike of their own. Many peasants simply stopped cultivating cocoa: from 5,500 t in 1974, production fell to 3,000 t in 1977 and 1,600 t in 1987 (Pourtier 1989b: 294). During the past two decades, these two export crops have been completely wiped out. Cocoa exports in 1996–7 were only 627 t, coffee exports a tiny 42 t (Marchés Tropicaux 1998: 25). The combined planted area in the country today is reported to be less than 1,000 ha (DGE 1999: 32). This veritable economic extinction also implies that the two crops do not at present exercise any pressure on forests. Had it not been for the impact of the massive oil rents, the Woleu–Ntem province would probably have embarked on land-use trends comparable to those in the adjacent Humid Forest Zone of southern Cameroon. This area has a similar natural endowment, and much forest was lost to cocoa and food crops.[27]

Only the fourth agricultural sub-sector profited from the oil boom, namely agribusiness. Large-scale parastatal chicken farms, oil palm, sugar and rubber plantations, cattle ranches, etc. all received generous credits and subsidies from the government after 1975. As explained in section on 'Windfall impacts on government spending', these companies were sheltered from the oil-induced competitiveness squeeze, as oil money was used to cover both high investment costs and recurrent deficits. In addition, this highly capital-intensive sector profited on the cost side from the appreciation of the RER, lowering expenditures for the imported inputs it depended on. But the sector was highly inefficient, which implies that it failed to boost output – or to expand cropped areas. Where it did succeed, much of the incorporated land was taken from savannahs. Hence, little forest was cleared for agro-industry; the TREES estimate cited in Wilks (2000: 13–14) is 23,000 ha (0.1 per cent of Gabon's land area).

On the whole, agriculture was the main loser from Gabon's oil bonanza, although it also declined from an already weak pre-boom base. Agriculture's share of the economically active population plummeted from 35 per cent in 1969 to 15 per cent in 1985 (Zomo Yebe 1993: 56). Its share of GDP fell even more dramatically, from 32.2 per cent in 1960 to 18.6 per cent in 1970 and 6.5 per cent in 1980. In the 1980s, the emphasis on agribusiness raised the share of GDP to 11 per cent in 1987, before it fell continuously to 7.5 per cent in 1998 (World Bank 1999a), 4.9 per cent in 1999 and only 3.9 per cent in 2000 (DGE 2001: 31). A number of factors worked against agriculture. State agencies' cocoa and coffee prices became increasingly adverse to producers (see section on 'Trade policy impacts'). Inadequate road infrastructure impeded crop commercialisation (Pourtier 1984; section on 'Windfall impacts on government spending'). And, most of all, massive public spending pulled labour out of the countryside and into urban areas (section on 'Structural changes in income and demand').

Forestry

If deforestation is currently a negligible issue in Gabon, what about forest degradation? Has it been linked in any way to the oil cycles and macroeconomic changes that occurred? Timber extraction is the most obvious candidate. With its vast forest area and valuable species, Gabon has a clear comparative advantage for forestry. It is estimated that more than 90 per cent of its forest area can be commercially exploited, the highest ratio in Africa (Marchés Tropicaux 1998: 30). As explained above, one species, *okoumé* (*Aucoumea klaineana*), was historically Gabon's main export commodity. In 1960, timber made up almost three-quarters of Gabon's exports, but with the expansion of oil exports, this share was reduced to less than 10 per cent by 1980 (Pourtier 1989a: 191). *Okoumé* is used mainly for plywood, and Gabon is the principal producer of raw logs. It is found in about 70 per cent of the forested area, that is, excluding the east and northeast regions. *Okoumé* still accounts for 60–75 per cent of timber export value, complemented by *ozigo* (*Dacryodes buettneri*) and what are normally referred to as 'miscellaneous species' (*bois divers*). Only 4.3 per cent of the extracted logs were processed in 1998 (DGE 1999: 27, 41), but the government has recently changed policies so as to enforce existing legislation that should raise that share to 30 per cent (EIU 2001). The development of value-added industries has been hindered by high domestic costs, as well as the fact that the largest logging operators preferred to feed their pre-installed mills overseas (R. Nasi, personal communication, Bogor, 21 August 2000).

By our definitions, timber extraction in Gabon does not cause deforestation. *Okoumé* is usually found at a density of 1–3 trees/ha sized above 70 cm diameter above breast height (dbh), the legal minimum (Larivière 1996: 162–3), so that logging remains highly selective. It opens up 5–20 per cent of the canopy, depending on the size and route of logging roads (Wilks 2000: 11–13); the average may be 10 per cent (IUCN 1990: 29). Using the FAO's criterion of 90 per cent crown-cover loss, it is thus only at very fine resolution levels (e.g. 1–10 m pixels) that deforestation from logging would be recorded. Using this unusually detailed scale, logging operations would cause a yearly forest loss of 5,000–15,000 ha.[28]

Okoumé is known to regenerate fairly well in forest openings, for example, in previously cropped areas.[29] Logging inevitably changes the forest structure; up to 50 per cent of the canopy may be ecologically altered (Collomb *et al.* 2000: 10–11). In the coastal zone, easy access has meant that many forests have already been logged several times, but, as indicated by case studies of vegetation cover, this has not impeded forest regrowth. Rather, the genetic and commercial value of these forests may be diminished when the tallest and straightest trees are cut repeatedly.[30] Since 1957, two-thirds of Gabon's forests have been logged at least once, and concessions have increased sevenfold. A marked expansion occurred from 1994 to 1999 (Collomb *et al.* 2000: 13). Whether logging causes forest degradation thus ultimately depends not only on the type and frequency of harvesting, but also on the criteria applied to determine forest quality. *Okoumé* regrowth may be favoured by long-cycle harvesting. Some observers suspect that minor disturbances in the Gabonese forest from selective logging may also benefit large mammals such as elephants and favour a higher local biodiversity in the mosaic of regenerating secondary vegetation (Adams and McShane 1996: 217).

Indirect forest impacts from logging do exist, since it eases access and raises demand for other forms of extraction, notably for hunting by logging workers, commercial hunters and local communities (Sayer *et al.* 1992: 172; Simons 1996; Gami and Nasi 2000). Logging may also reduce the availability of non-timber forest products that are traditionally harvested by local people, such as fruits (Simons 1996: 56). But with Gabon's low-population density and an absence of agricultural pressures, there is no indirect 'opening-up' effect permitting further deforestation by land-hungry squatters. This distinguishes Gabon from many other countries in Africa. On the whole, logging causes forest degradation in some but not all cases, 'defaunisation' from the bushmeat trade being the most harmful of these impacts.

Concessions have expanded sharply during the past decade, but this does not give a reliable indication of changing activity levels, as many concessions remain underused (Simons 1996: 37). It is more relevant to look at production data. Figure 4.2 shows the development of logging production over the last four decades (prior to 1986, only data for *okoumé* are available). Domestic demand is minor; over the past decade, export quantity shares were 95 per cent for *okoumé* (1987–98) and 93.4 per cent (1987–96) for all timber (Collomb *et al.* 2000: 37). After expansion in the 1960s, in the oil era *okoumé* extraction fell by more than one-third in 1975. Until the early 1990s, *okoumé* production then fluctuated around the much lower level of 1 million m^3/yr; total timber production was around 1.4 million m^3/yr.

Several factors appear to explain the decline and later stagnation. One was the crisis in international markets for Gabon's timber (Brunck *et al.* 1990: 94). In addition, Gabon's coastal stocks gradually became exhausted, and no road construction was carried out to make new areas in the interior accessible (Barret-feuvre and Dufoulon 1979: 15, 33). Until 1998, the Société Nationale des Bois du Gabon (SNBG) had a monopoly of *okoumé* and *ozigo* timber exports. It applied high commercialisation margins, which made it less rewarding for private companies to expand exports (EDIAFRIC 1985: 124–7; EIU 1999a: 24). In the wake of the Asian crisis, severe financial and management problems forced the SNBG to liberalise exports to Asia. But our core effect of oil-led rises in costs and lower competitiveness with an appreciated RER was also very important (Barnes 1992: 78), coinciding with the period of lower timber exports (Figure 4.2).

The renewed expansion in timber exports from 1992 to 1997 was the result of a mixture of factors. The most important was the rapid rise of Asian markets. Exports to France, traditionally the largest buyer of *okoumé* and *ozigo*, remained rather stagnant after 1990, but China's imports rose exponentially from a tiny 12,300 m^3 in 1991 to 1,015,800 m^3 in 1997 (Collomb *et al.* 2000: 38). Asian destinations accounted for 62 per cent of Gabon's

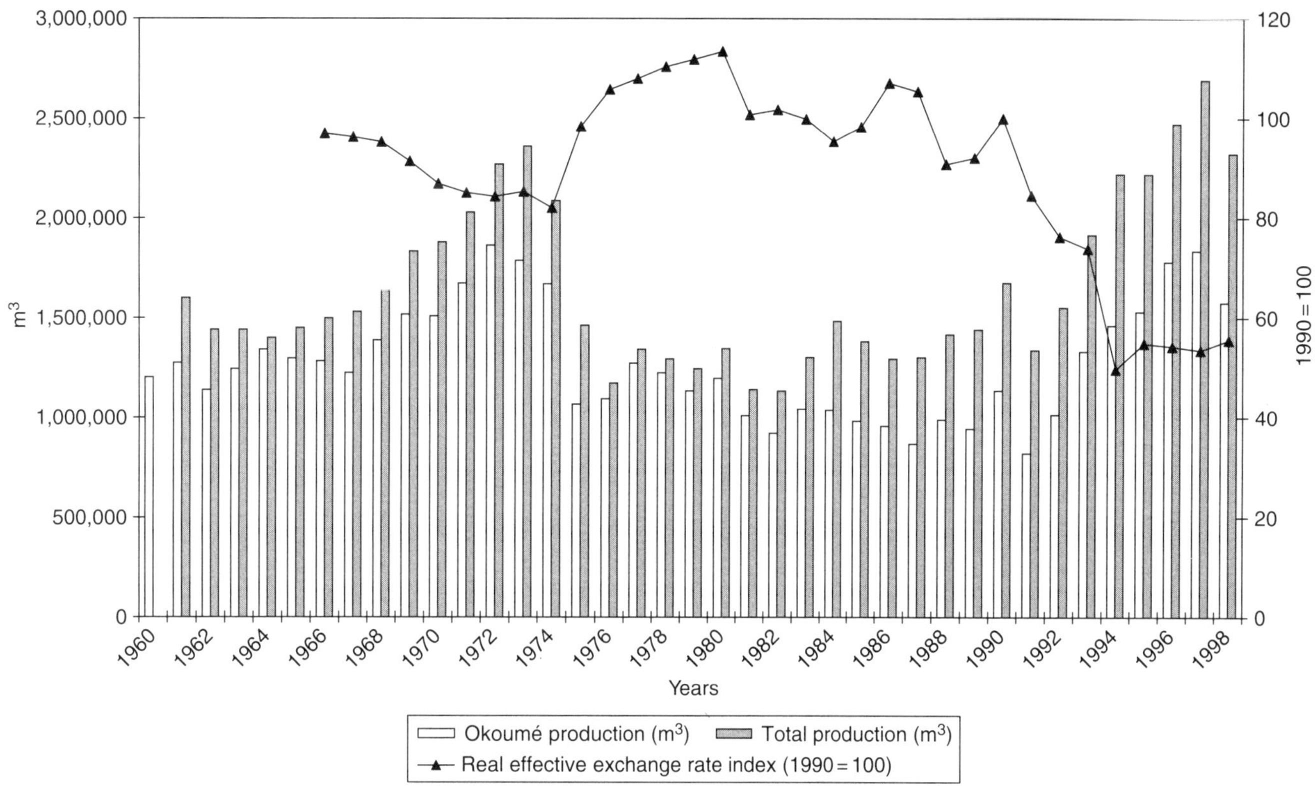

Figure 4.2 Gabon: timber production and RER, 1960–98.

Sources: Brunck *et al*. (1990); FAOSTAT (2000b); Collomb *et al*. (2000).

Note
Okoumé production 1960–74 was converted from metric tons to m^3 by using the 1974 conversion factor in Brunck *et al*. (1990: 92).

total timber exports in 1997 (Marchés Tropicaux 1998: 36). The completion of the Transgabonais railway in 1986 was a key enabling factor, opening up new extraction areas in the interior with the help of cheap transport costs (section on 'Windfall impacts on government spending'). Last not least, the sharp real depreciation in 1990–4, induced by declining oil revenues (see Figure 4.1), made Gabon's timber exports more competitive (UNDP 1996: 5) and encouraged sizeable investments by Asian firms (Marchés Tropicaux 1998: 30–1). Dutch Disease factors were thus highly instrumental in explaining both the stagnation and rise of Gabonese timber production.

Trade policy impacts

Potentially, oil-country governments may apply price and trade controls (e.g. import quotas, tariffs or export subsidies) to try to protect domestic sectors that are exposed to the Dutch Disease. In Gabon, the situation since the 1970s was mostly the opposite: trade-policy and price-control interventions exposed private producers in the T sectors even further to competition from abroad, especially in agriculture. Although trade policies were not the decisive factor, they further revealed the low priority of agriculture in government policy. There is a degree of pride in the statement that 'Gabon has the means to provision abroad' (Richard and Léonard 1993: 215): rising agricultural imports are seen as an achievement rather than a problem.

The most prominent examples of discriminatory trade policies are Gabon's two export crops, coffee and cocoa. During the oil boom, international beverage prices also faced a boom, but private producers received little benefit. Government-controlled marketing agencies, mostly the Caisse de Stabilisation et de Péréquation, siphoned off the bulk of the profits. When prices turned around, these profits were not recycled back to producers. Rather than 'stabilising' producer prices, therefore, producers were further penalised by the scheme (Pourtier 1989b: 294–5). In 1979, cocoa producer prices were higher in Cameroon than in Gabon, and this gap widened during the 1980s (Zomo Yebe 1993: 87).

Other interventions also had a bias against agriculture. Price controls, for example on foodstuffs, were used both to control inflation and to ensure 'fair' prices for consumers in the interior of the country. The latter meant that certain processed food items would cost the same in a remote interior village as in Libreville or Port Gentil, where they were produced or imported (Zomo Yebe 1993: 51). Obviously, by reducing prices, locally produced foodstuffs lost the *de facto* protection which high domestic transport costs for urban goods would otherwise have provided.

The government continuously used protectionist measures for larger processing industries, such as sugar, cement and bottled water (D. Young, personal communication, US Embassy, Libreville, 2 June 2000). However, helping these parastatal industries to stay alive or expand did not usually have any forest impact. The only protected sector to use land was sugar. Zomo Yebe (1993: 76) reports that the sugar price of the Société sucrière du Haut-Ogooué (SOSUHO) *before* subsidies was 280 FCFA/kg, while the corresponding price on the European market was only 100 FCFA/kg, thus illustrating the vital role of subsidies. Sugar plantations were established in the savannah areas near Franceville, thus causing savannah conversion but not deforestation.

With the recent round of trade liberalisation in the Central African UDEAC, Gabon's options in using protectionist measures have been greatly reduced. However, in the past informal imports of low-value food crops (plantains, manioc, etc.) from neighbouring Cameroon, Equatorial Guinea and Congo have also been *de facto* free of duty. Higher value food products, notably from France and South Africa, have paid tariffs, but not at a level that significantly restricted imports of them. On the whole, Gabon's fairly liberal trade policy has thus further discouraged domestic agriculture. Indirectly, this has provided an additional obstacle to the expansion of cultivated areas, thus contributing to the preservation of forest cover.

A quantitative view

How important is 'price competitiveness' to the long-term performance of Gabon's logging and agricultural sectors? Did rising costs and an appreciating RER seriously hamper the development of timber and agricultural exports? Did oil wealth thus indirectly protect, or at least delay, the loss and degradation of forests? And how effective was nominal devaluation in reviving growth in these sectors growth? These are some of the key questions.

As emphasised in this section, competitiveness was a key factor. Zomo Yebe (1993: 52–7) has estimated the typical competitiveness effects for the 1969–85 period econometrically. He finds that higher oil-export revenues reduced both RPs for farm goods and agricultural employment. Also, food imports significantly rose with higher incomes and the appreciation of the RER. In other words, the oil boom caused the RP and 'de-agriculturisation' effects predicted by the theory. The new regressions in Table 4.2 use an approach similar to Zomo Yebe's, but for a longer period (including the period of real depreciation) and for different variables. For those readers that are not accustomed to regression analysis, it is a statistical method to analyse the relation between one or more independent variables (here: oil revenues and capital inflows) and a dependent variable (here: the RER). By using pairs or groups of these variables that correspond in time, we can estimate the correlation between changes in the independent and the dependent variable. How good a fit does our model provide? How much of the changes in the latter can be *explained* by the former, compared to what has to be attributed to other factors not captured by our theoretically underpinned model? In simplified terms, the higher the *model-explained* share (expressed by the *F*-test and the indicator R^2) and the more *significant* the individual independent variables (as measured by a *T*-test), the more are our theoretical expectations confirmed.[31]

As expected, RER appreciation during 1966–97 was positively influenced by oil export revenues and capital inflows (both expressed in 1995 US$). Both coefficients are significant, at the 5 and 1 per cent levels respectively (Regression 1). As an independent variable, the RER has the expected negative impact on agricultural output (in fixed 1995 US$; Regression 2) and timber production (in cubic metres; Regression 3). Both coefficients are significant at the 1 per cent level, but the competitiveness variable explains much less of the variation in agricultural output ($R^2 = 44.26$ per cent) than in timber production ($R^2 = 71.61$ per cent). This fits well with our observations above, that the decline of agriculture is a more complex phenomenon that does not depend only on RPs.

As regards timber, a qualification must be made for *okoumé*, a species that has no perfect substitutes on the world market.[32] As mentioned above, SNBG's monopoly in marketing *okoumé* has also hampered competitive supply responses on behalf of producers. As shown in the disaggregated estimate for *okoumé* (Regression 4), just about one-third of the

Table 4.2 Gabon: relating oil wealth to relative prices and traded sector production. Regression results, 1966–98

Independent variables/ dependent variables	*Petroleum exports (millions 1995 US$)*	*Capital inflows (millions 1995 US$)*	*RER (1990 = 100)*	R^2	*F-value*	*Years*
RER (1990 = 100)						
Coefficient	0.008	0.027		0.3103	6.3	1968–98
T-value	2.324**	3.432***				
Agricultural value added (constant 1995 US$)						
Coefficient			−2286191.1	0.4426	23.822	1966–97
T-value			−4.88082***			
Total timber production (m³)						
Coefficient			−20506.032	0.7161	78.186	1966–98
T-value			−8.842***			
Okoumé production (m³)						
Coefficient			−10287.36	0.3644	17.78	1966–98
T-value			−4.216***			
Miscellaneous timber production (m³)						
Coefficient			−10218.67	0.7699	103.713	1966–98
T-value			−10.184***			

Notes

** Parameter *T*-value significant at the 5 per cent level.

*** Parameter *T*-value significant at the 1 per cent level.

variation in production of the species is explained by the RER (R^2 = 36.44 per cent). For the miscellaneous category (Regression 5), that share is much higher (R^2 = 76.99 per cent).[33] The similar size of the two marginal coefficients (both are highly significant at a level of 1 per cent) implies that a real currency depreciation of 1 per cent causes *okoumé* and the miscellaneous category to rise equally (10,287 m^3 and 10,218 m^3, respectively). However, the quantity of *okoumé* exports in 1997 was almost triple that of miscellaneous woods. *Okoumé* is thus somewhat less price-elastic, probably because it contains a greater rent element than the miscellaneous category.

Where there are rents, rent-seeking behaviour is usually common too. Forestry rents are currently much less taxed in Gabon than in Cameroon.[34] Forestry taxes have lost their real value, as they have not been inflation-adjusted since the mid-1970s, that is, precisely since the rise in oil rents boosted state revenues. The abundance of oil rents has thus reduced efforts to use taxation as an instrument to cut into private-sector timber rents. Hence, a good part of the rising private-sector rents have been captured by a limited number of logging companies. SNBG allocates production quotas to the companies and then buys back the wood at a fixed price, a system that easily lends itself to a lack of transparency. SNBG also captured rents, which have frequently been diverted to political campaigns among other things. The timber sector thus adopted some of the oil economy's rent-seeking features (A. Karsenty, personal communication, Libreville, 26 May 2000).

On the whole, declining competitiveness from oil wealth was a prime factor in the performance of primary sectors. Policies tended to accentuate market factors rather than stabilise them, due especially to a pronounced anti-agricultural bias that hit small producers. Our regression results show that price competitiveness was even more important for logging, which is almost entirely an export sector, though there are elasticity differentials between timber species. But the overall results clearly confirm the core hypothesis of this book: oil wealth has protected forests in Gabon from conversion and exploitation by shifting RPs against these activities.

Windfall impacts on government spending

Agriculture and forestry

Land use is not only affected by RPs, but also by other policies that encourage or discourage it. In relation to agriculture and forestry, an additional question is thus whether oil wealth brought more resources to public regulatory and development agencies and, if so, what policies they implemented that had relevance for land use. In other words, did increased financing for agricultural and forestry agencies also increase their capacity to change the course of events on the ground, for example, through technical support to expanding land uses, subsidised credits to small-scale farmers, or a better control of forestry practices that reduced degradation?

Following independence, increasing mineral wealth (manganese, uranium and oil) initially did not provide any more funding for agriculture: in fact the sector became marginalised in development planning. In the 1966–70 and 1971–5 plans, agriculture had a total budget of 5.3 billion FCFA; for comparison, 100 billion FCFA were allocated to the mining sector (Zomo Yebe 1993: 69). There was little rural investment, most of the emphasis being on what Yates (1996: 151) calls 'rural "welfare" policy', that is, distributing oil wealth to public consumption, employment and social spending, mainly in order to equalise benefits between different regions and ethnic groups.

When oil revenues multiplied in the mid-1970s, policy-makers become more ambitious about the productive potential of rural areas. The overall idea in the 1977 Agricultural Master Plan was to create a modern sector that would be the complete opposite of traditional slash-and-burn agriculture in being 'sedentary, intensive, mechanised [and] salaried' (cited in Pourtier 1984: 449), and, one could add, 'large-scale and state-controlled'. Capital-intensive technologies were meant to overcome rural backwardness, chronic labour shortages and deficiencies in food supplies to urban markets. Implementation of the strategy was halted by the financial crisis of 1978, but brought back on to the agenda when state revenues soared in the second oil-price hike of 1979.

The revised plans for the 1980s included a multitude of regions and products, such as cattle-ranches, vegetable gardens, ricefields, integrated chicken farms and plantations of cocoa, coffee, sugar, plantains, banana, rubber and oil-palm. Many of these projects required new land, and the potential land-take involved in implementing the first project round would have been an area of 12,650 ha.[35] Some crops, such as plantains, rice and cocoa, would indeed have required the forests to be cleared, but others (ranching, sugarcane) would have drawn mainly on savannah areas, while tree crops (rubber, oil palm) would have involved replacing the natural forest with tree-crop plantations.

Subsidised credits were a main instrument used in channelling resources to agriculture: the sector's share in overall public investment rose from 1.2 per cent in 1966–75 to 4.5 per cent in 1976–80 and to about 10 per cent in 1980–8, an impressive rise in absolute terms. During the 1990s, agricultural investment was cut back. Following the downswing in oil prices in 1986, the agricultural development strategy also became more selective in terms of sectors and regions, based on a more rigorous assessment of profitability. Across-the-board rural welfare subsidies were being faded out (Yates 1996: 167–70). Yet throughout the post-1970 period, the agro-industrial parastatal sector absorbed the lion's share of agricultural funding, including credits from abroad (Pourtier 1989b: 281–2; Marchés Tropicaux 1998: 20). The government also compelled oil companies to recycle some of their profits into agro-industries (Yates 1996: 71).[36]

However, the master plan to turn the agricultural sector upside down by implementing a top-down technocratic strategy turned out to be a failure, mainly because of mismanagement in the parastatal companies (R. Vinchent, personal communication, 30 May 2000). Over-staffing and rent-seeking flourished in the agro-industries, behind which a single clan of influential families concentrated more and more power and resources (Pourtier 1989b: 292). Much funding was swallowed by the bureaucracy itself: 'these administrative and para-administrative offices consumed the vast majority of public monies earmarked for so-called "rural" development' (Yates 1996: 147). Funds were wasted, operations never reached their planned scale, and high running costs caused extreme deficits, such as for coffee[37] and cocoa.[38] Many agro-industrial parastatals are currently close to bankruptcy (UNDP 1996: 16). Rural labour was successfully drawn out of the peasant sector, but was often insufficiently qualified for work on mechanised plantations.[39] There were political pressures to invest in rapidly created projects, where crucial feasibility factors like soil suitability or water scarcity had not been thoroughly evaluated beforehand. In other cases, vested interests deliberately sacrificed social profitability to private benefits or political and ethnic pressures.

There are thus two fundamental reasons why the policy-induced emphasis on agro-industry did not cause any notable deforestation. One was the capital- and land-intensive character of the planned production compared to traditional land-extensive agriculture. The second was the failure actually to implement the plans. Although some industries fared better than others,[40] most plantations lagged far behind the originally planned scale. It is estimated that all agro-industrial plantations covered about 13,000 ha in 1987 and 23,000 ha in the mid-1990s (TREES estimates from 1998; see Wilks 2000: 13–14). The latter barely corresponds to 0.1 per cent of Gabon's land area.

Oil-induced funding for agricultural development has thus been limited, unstable and ill employed over the last four decades; it proved highly inefficient in promoting production. As a consequence, it also caused little deforestation. The next question is whether the reverse effect occurred: that is, was oil money made available to improve forestry regulation and to finance protected areas, thus helping to protect forests more efficiently?

For both areas, the question can be answered in the negative. Forestry regulation in Gabon remained extremely weak until 1996, when a process of legal reform was initiated, mostly due to external pressures. This has now resulted in the adoption of a new Forestry Code (per 31 December 2001), providing an important legal step towards sustainable forest management.[41] It can be argued that soaring oil revenues for a long period encouraged the government to neglect the regulation of the forestry tax and resource base. Oil money helped to improve human resources in forestry, for example, by financing the establishment of a National Forestry School (École Nationale des Eaux et Forêts). But post-colonial public forestry administration fell increasingly into the hands of a centralised bureaucracy, which maintained little presence or activity in the field (G. Dufoulon, personal communication, Libreville, 30 May 2000). Of the 310 agents employed by the Ministry of Water and Forests, more than half are Libreville-based. Two-thirds of the Ministry's vehicles are used in the capital; in 1997, only eighteen cars were deployed outside Estuary Province (Collomb *et al.* 2000: 21–2, 39–40). Poaching erodes wildlife resources near the most populated and accessible areas, but it has only been regulated half-heartedly (Simons 1996: 41–2), and existing laws have generally not been enforced (IUCN 1990: 3).

The situation concerning protected area management is similar: forest conservation has not been a priority in Gabon, either for policy-makers or for civil society (Adams and McShane 1996: ch. 11; Simons 1996: 28–9). This is quite understandable, given the enormous forest wealth in Gabon. Most of the protected areas described by Brugière (1999) were created before or shortly after independence, many of them as hunting reserves. In the total protected area of 1.045 million ha (3.9 per cent of land area), savannahs are thus overrepresented and lowland forest under-represented (Sournia 1998: 105–6). A large share of the total still consists of 'hunting grounds' (*domaines de chasse*) (Brugière 1999: table 4.1). Except for the recently created Minkebé Reserve, Ipassa (near Makokou) and the southern part of La Lopé, in general protected forest zones have been logged over at least once (IUCN 1990: 40; C. Wilks, personal communication, Libreville, 6 October 2000).

Oil money has done little to improve public budgets for protecting forests. As in Cameroon, most finance for conservation has been left to international donors.[42] Recently, the government has shown more interest in actively conserving its tremendous biological heritage, but conflicts concerning protected areas were exemplified by the controversies surrounding logging in the biodiversity-rich Lopé Reserve, Gabon's oldest and most famous protected area (White and Oslisly 1998). The regional EU-financed ECOFAC project that assists Gabon's wildlife service in the management of Lopé at one point threatened to withdraw, as contradictory legal texts permitted the allocation of new logging concessions within the Reserve's boundary (Durieu *et al.* 2000). Yet, in financial terms the ECOFAC project, with US$520,000, has a yearly budget almost double the size of the entire Wildlife Department's operating budget of US$280,000 (Adams and McShane 1996: 224). This illustrates how little budgetary impact the country's oil wealth has had on protected area management in Gabon.

Roads and rail

Roads

A rentier economy has the fiscal means to improve transport infrastructure, in order to facilitate the private sector's commercialisation of agricultural and other commodities. Nevertheless, this was not the policy that was chosen in Gabon; on the contrary, roads were widely neglected there, as they have always been.[43] By 1935, a much more elaborate road system already existed in southern Cameroon and western Congo. Northern Gabon was connected by road to southern Cameroon for the marketing of cocoa and coffee, but basically the rest of the country had no roads, and Libreville was an island surrounded by the rainforest. River transport served to evacuate *okoumé* logs, the country's principal export commodity. Prior to independence, roads therefore had a low priority. In the 1960s, it became a principal aim to make Libreville a true capital in the economic sense too. Many bridges were built to replace ferries; in 1960, it was still necessary to take nine ferries to make the journey from Libreville to Franceville (G. Dufoulon, personal communication, Libreville, 31 May 2000).

Despite some improvements, the road network remains severely underdeveloped and poorly maintained. While Gabon had the densest air-transport network in Africa, with fifty-two airports in 1998, in 1996 the road network covered only 7,670 km, of which only 629 km were surfaced (CIA 1999a: 7). This situation has basically not changed since 1991, when the total was 7,518 km, of which 614 km were surfaced (IRF 1994: 14). The current Gabonese road density of 0.03 km/km^2 is among the lowest in the world (IRF 2000: 11). Unpaved roads are often impassable during the rains, thus making the transport and marketing of products highly unpredictable. Lately, the government has given greater emphasis to roads and drawn up a plan to have 3,580 km of asphalt roads in the long term (EIU 1999a: 12–13). Though donors have been slow to respond, there are several ongoing projects, some with support from the EU.

It is thus clear that road construction has in no way been a national priority, nor has it been favoured by oil wealth. As far as passenger transport is concerned, oil wealth has made air transport the preferred option. For goods transport, river and rail have been better alternatives, while the generally limited interest in agriculture means that there is little lobbying for road construction (B. H. Voubou, UNDP, personal communication, Libreville, 30 May 2000). Limited competition in transport and marketing may also have driven up middlemen margins, providing further disincentives to rural producers who are receiving lower prices. Transport and distribution costs combined are thus reported to make up to 50 per cent of the retail price of agricultural produce in Libreville, and as much as 80 per cent of the retail price of staple crops like plantains and manioc (Poupart and Pilichowski 1997: 69).

Will a greater future emphasis on road construction and improvement thus translate into more forest loss? The quantitative analyses of deforestation over the last three decades in the periurban zone of Oyem and Franceville shows clearly that forest-clearing tends to occur in a 2–4 km range around roads, which are thus a significant spatial land-use determinant of forest loss (Wolff n.d.: 29, 48, 53). Near the capital, the paving of roads has certainly eased agricultural marketing and expansion along the Libreville–Lambaréné axis, but not much along the stretch from Libreville to Ndjolé (A. N'Goye, IRAF, personal communication, Libreville, 29 May 2000). Remoter asphalt stretches, for example, between Libreville and Oyem, have until now not brought about any land-use changes. They may actually have facilitated the import of foodstuff from Cameroon to these areas, and thus retarded local agricultural development (own observation, 1 June 2000; P. de Wachter, personal communication, Libreville, 26 June 2002).

It thus appears that road construction and improvement is a necessary but not a sufficient condition for forests to be cleared. Closeness to markets, soil suitability and labour costs and availability are also vital preconditions. However, if declining oil revenues and a true economic crisis were simultaneously to reduce labour costs in Gabon, then improved market access by surfaced all-weather roads are likely to lead to forest-clearing for commercial agriculture. Although some newly built roads have not had a deforestation impact until now, the existence of these roads may well enable forest-clearing in the future.

The Transgabonese railway

What Gabon did not invest in roads, it certainly put into railways. The single largest absorber of oil wealth was the US$ 3–4 billion invested into the Transgabonese railway running from Libreville's port of Owendo to Franceville in the southeast (see Map 4.1 above). The economic implications of this project have already been discussed in the section on 'The macroeconomic impact of the mineral boom', but what were its impacts on the forests? To answer this question, we should distinguish between direct impacts, that is, the forest-clearing associated with the physical railway construction itself, and the indirect ones, that is, the forest-related changes in human activities that the railway made possible.

Historical photographs of the railway construction (1973–86) leave a dramatic impression of bulldozers navigating their way through the mud, stripping forest cover from large areas. How much forest disappeared as a direct impact of the railway? No thorough studies seem to have been made, but a thesis from Omar Bongo University deals with the impact of the first 182 km of rail, from Owendo to N'djolé (Ella Nguema Rolly 1979). This stretch mostly passed through flat, secondary forest areas that had previously been submitted to logging and shifting cultivation. The author estimated that the clearing of forest along the track had been in the range of 60–100 m, depending on topography. Using a simple average of 80 m, which would include some additional clearing for stations, workers' camps, etc. and multiplying this with a length of 182 km yields a cleared area of 1,456 ha.[44]

The total length of the railway is 650 km, but forest-clearing on some stretches was under the average of 80 m, right down to zero in some places. This is mainly because the railway also crosses savannahs and other non-forested or transitional forest areas. Using an average of 60 m for the full 650 km length of the railway, the total cleared area would be 3,900 ha. Adding some deforestation impact from the selective logging of wood used as sleepers, a 'best guess' would be a direct deforestation impact of 4,000–4,500 ha, or about 0.02 per cent of Gabon's land area. This accumulated impact was distributed over the construction period of fourteen years. However, not even this small figure represents *permanent* deforestation. As can currently be observed by travelling on the Transgabonais, there has been a remarkable degree of forest regrowth, almost covering the track in some areas. Using an overall average (including zero for previously cleared or non-forested areas) of 30 m (own observation), permanent direct forest-clearing would only be about half of the figure given above, that is, 2,000–2,250 ha.

Indirect deforestation impacts refer to the 'opening up' of new areas for settlement and clearing, such as an expansion of agricultural production destined for the Libreville market. There is surprisingly little impact of this type. Some settlement has been encouraged around the train stations. Some food is sold to travellers, and some agricultural produce is transported to Libreville. But this seems to be the exception rather than the rule. Most cultivated products are destined for local consumption. Not even the area close to the railway around Libreville has experienced any major cash-crop development (Ella Nguema Rolly 1979: 54–7). This confirms the impression from the road section above that the provision of physical infrastructure is insufficient on its own to encourage commercial agriculture in Gabon, as long as better-remunerated activities are available.

Direct and indirect deforestation from the railway is thus negligible, but the impact on forest *structure* is significant. One effect of the railway has been to increase bush-meat supplies from areas made accessible to urban markets, notably Libreville (Trefon 1999: 47). But more important is the fact that it has increased timber extraction significantly. Logging was from the outset one of the main economic rationales for building the Transgabonais. As the Minister of State remarked, in a contemporary internal document on the expected economic returns from the railway: '[the transport] of wood in particular will make up a substantial part' (Boumah 1975: 2, my translation from French). *Okoumé* forests in the area near the coast had been over-harvested, and the railroad cut through new *okoumé* zones in the interior that would open up 3 million ha to exploitation (G. Dufoulon, personal communication, 30 May 2000). Transport costs would be much reduced compared to road transport (Pourtier 1982: 125). Today, 50 per cent of the railway's revenue is generated by timber transport (EIU 1999a: 33). It is significant here that the recent tendering to privatise OCTRA, the company operating the railway, was won by a consortium with a strong representation of forestry companies (EIU 1999a: 13). However, as shown in Figure 4.2, logging only really increased in the 1990s, when there was the *combined* impact of higher competitiveness, increased demand in Asia and cheaper transport from new concession areas to the port of Owendo.

Directed settlement

The resettlement (*regroupement*) of villages was an important historical factor in shaping the current unequal spatial distribution of Gabon's rural population (Richard and Léonard 1993: 120–1). Both the stylised map in Raponda-Walker and Sillans (1961: 8) and the population-density maps from the 1920s and 1940s reproduced in Pourtier (1989b: 106–7, 122) still show a rather equal distribution of people in the rural, overwhelmingly forested area (see Map 4.2). A somewhat higher human density only occurred in selected regions: Estuary Province, the north (Woleu–Ntem), the southeast (Franceville area) and around the Ogooué river. The contrast with the subsequent map from the 1970s is startling (Map 4.3). Basically all the population was now concentrated along pre-existing roads and rivers, in what Pourtier (1989b: 102) describes as a pattern of 'linear space'.

The policy of both colonial and post-colonial administrations was not to bring people from overpopulated regions into the forest to colonise new space, but, on the contrary, to get them *out of the forest* in order to concentrate them in larger sedentary agglomerations.[45] The aim of this top-down strategy was both to promote development objectives (providing social infrastructure, attaining economies of scale, increasing food production, etc.),[46] and to increase political control. As one might expect, these attempts to disrupt local livelihoods fundamentally often met with fierce resistance, though sometimes resettlement was negotiated. The colonial state embarked upon this policy from the 1930s onwards, but a nation-wide and efficient linear concentration of the rural population was only achieved after independence, notably in the early 1970s. The means used to achieve this goal included both sticks (e.g. burning villages) and carrots (e.g. providing schools, transport facilities and housing).

To what extent was this policy related to oil wealth? Obviously, it formed part of a vision of 'modernisation' that already existed before the oil era. It might be argued that, in indirect terms, oil revenues strengthened the ability of the state to implement this vision, providing the resources required both to force people to move and to build the social infrastructure in the newly created villages. However, most of this process was completed in the early 1970s, that is, when oil revenues were only just beginning to increase. This implies that the link with oil is probably a weak one: revenues from timber and other mining sectors (uranium, manganese) may have contributed more.

What impact did rural resettlement have on land use and forests? Of course, whenever it was necessary to clear forests to create the new 'linear' villages, this caused immediate and permanent deforestation. But the semi-nomadic traditional lifestyle of rural Gabon[47] had also implied many (voluntary) resettlements. A more interesting question is to what extent the shift to sedentary settlement was accompanied by lasting changes in production patterns. Pourtier (1989b: 114) suggests that there was little change in productive strategies, though agriculture became more intensified by shortening fallow lengths and sometimes shifting to perennial crops. Intensification reduced the clearing of forests over time. As an important economic side-effect, abandoned cropped areas may have facilitated valuable *okoumé* regrowth (R. Nasi, personal communication, Bogor, 21 August 2000). But perhaps most important of all was one particular unintended impact of the resettlement policy: relocation increased contact with urban areas and thus facilitated the rural exodus, especially of young people (see next section). This reduced both the rural population and

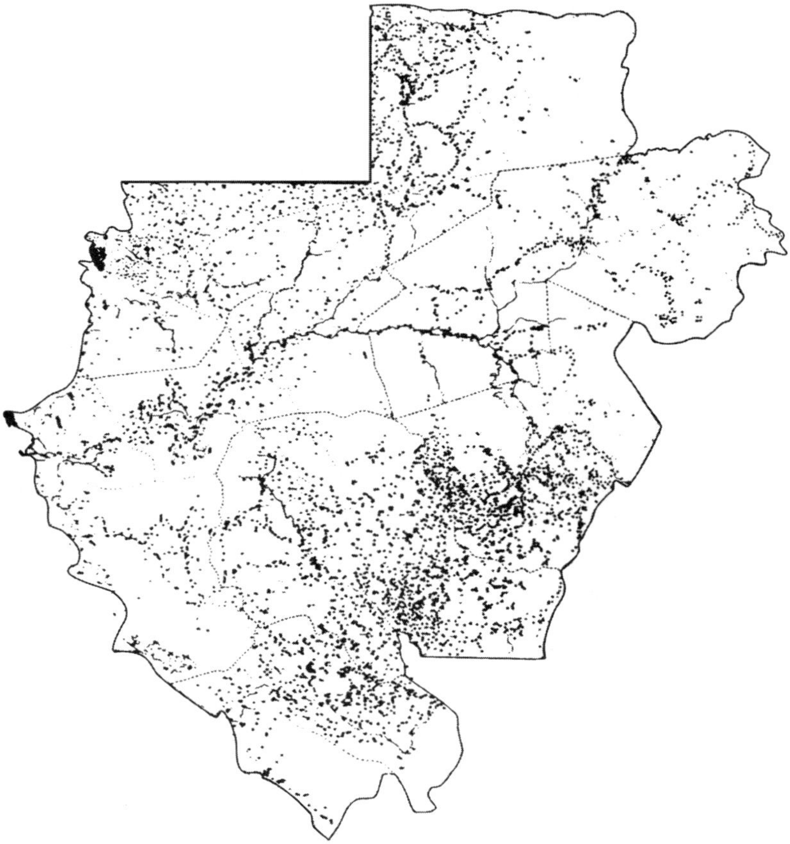

Map 4.2 Gabon: spatial distribution of rural population in the early 1940s.

Source: Pourtier (1989a: 116).

the average size of cultivated plots. Therefore, resettlement policies evidently contributed to net forest regeneration, though oil revenues had a minor role in their implementation.

Structural changes in income and demand

In the previous sections, it has been argued that Gabon's transformation to a high-rent oil economy was accompanied by a loss of competitiveness and by a series of policy interventions that both favoured the conservation of forests. However, a third type of impact may occur through the rise in incomes, causing structural changes in income distribution, in aggregate demand, and in the spatial distribution of the population. These changes could potentially work for or against forests. For instance, in Latin America there has been a tendency for a richer and more urbanised population to consume more meat and dairy

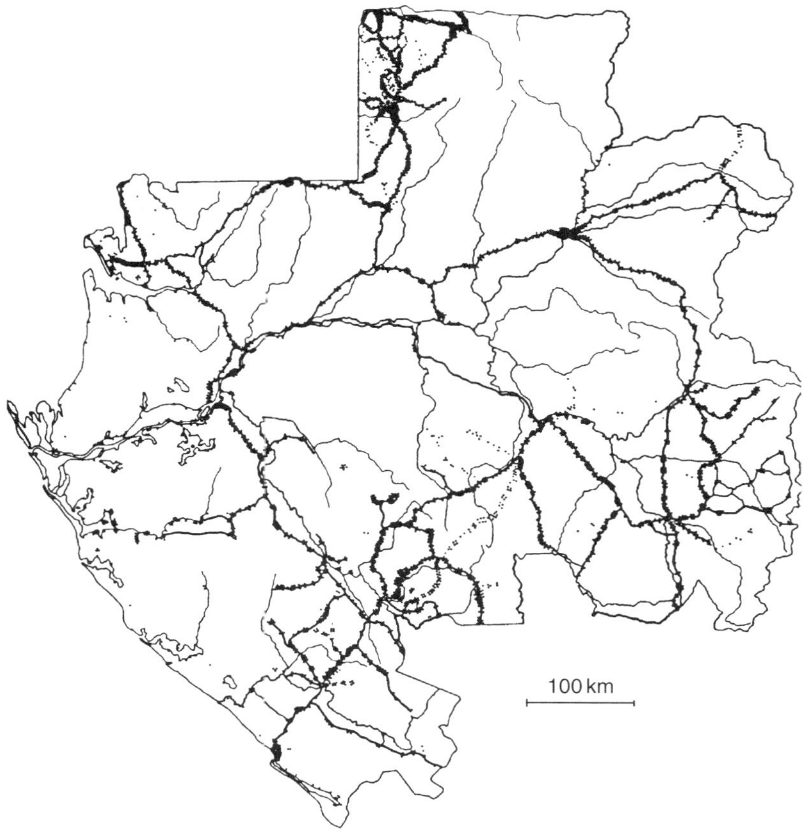

Map 4.3 Gabon: spatial distribution of rural population in 1970.

Source: Pourtier (1989a: 117).

Note
Each dot represents village of 100 residents.

products, causing a rise in land-extensive cattle-ranching that promotes deforestation. In this section, we shall investigate these types of changes in Gabon, starting with changes in income distribution and poverty levels.

Poverty alleviation

In Gabon's dual economy, with its capital-intensive export enclave and high concentration of power and assets, income distribution has remained extremely unequal over the last four decades. On the other hand, absolute poverty has been conspicuously reduced. Statistical sources on poverty and income distribution are scattered, and for rural areas mostly non-existent. But Table 4.3 reproduces World Bank estimates from Poupart and Pilichowski (1997), which combine household expenditure and survey data with the national accounts.

Table 4.3 Gabon: poverty trends 1960–94

Indicators/years	*1960*	*1968*	*1975*	*1985*	*1990*	*1993*
Minimum wage						
Incidence	87	83	86	85	84	81
Intensity	60	61	54	53	51	51
2/3 of average consumption						
Incidence	68	67	62	62	62	62
Intensity	39	35	30	30	30	30
Dollar per day						
Incidence	66	57	26	15	14	11
Intensity	38	27	8	4	3	2
Human Development Index	0.26	0.38[a]	n.a	0.47[b]	0.53[c]	0.56

Sources: Poupert and Pilichowski (1997:17). UNDP (1999b:22).

Notes
a 1965.
b 1980.
c 1992.

The two first lines show measures of relative poverty, the third of absolute poverty. Each of these gives both the percentage of population living at or below the respective poverty line ('incidence'), and an indicator of the gap between the poorest and this line ('intensity'). For both the minimum-wage and the two-thirds-of-average-consumption lines, until the mid-1970s poverty incidence fell slightly, while its intensity was fairly reduced. After 1975, there was little change in any of the relative poverty measures. However, the incidence of absolute poverty, indicated by the percentage of people earning less than US$1/day (in fixed 1985 prices), fell dramatically and continuously from 66 per cent in 1960 to 11 per cent in 1993. This fall was clearly strongest during the peak oil-boom years. The intensity of absolute poverty was also much reduced. For comparison, the UNDP's Human Development Index, which takes into account both GDP and social indicators, also shows a continuous improvement over the same period (see Table 4.3, fourth indicator). On these grounds, the World Bank report seems far too hasty in concluding that 'the link between economic growth and poverty reduction is tenuous' (Poupart and Pilichowski 1997: 19). Indeed, although income inequality is reproducing itself in the Gabonese economy,[48] it is doing so at a higher income and welfare level, where the number of absolute poor has experienced an extraordinary and sustained decline due to significant trickle-down effects.

This long-term decline has occurred overwhelmingly in urban areas: the main route for poverty reduction has been to move out of rural areas, as well as out of the traditional and into the modern sector. Although lagging behind urban remuneration, rural wages have also risen, due to remittances from urban relatives, government projects and employment in rural areas, and some marketing of crops and forest products in urban markets. Clearing forests in Gabon remains a highly labour-intensive activity, which in 90 per cent of rural households is carried out by axe and machete, while only 10 per cent possess a saw (ibid.: 67). High labour costs are an important impediment to projects involving manual forest-clearing (Zomo Yebe 1993: 80, 83–4). Poverty alleviation thus raised the opportunity costs of rural labour, which *ceteris paribus* reduced forest conversion.

Urban unemployment emerged only after 1986, due to downswings in the oil cycle and structural adjustment involving cuts in public sector spending. In 1998, unemployment was 20 per cent, affecting mostly young people (UNDP 1999a: 20). Private transfers play an important role in their survival strategies (DGSEE 1994a: 10). Unemployment often reflects an increasing mismatch between labour demand and a Gabonese workforce that is unaccustomed to the private sector. Job-seekers often lack skills, enterprise and a willingness to accept low-paid or informal jobs (ibid.: 53–6; EIU 1999a: 23). This helps explain how an unemployment rate of 20 per cent can coexist with an immigrant population that makes up 15 per cent of residents (DGSEE 1993: 18–19) and that continues to dominate key sectors like commerce, transport and fisheries (ibid.: 44).

Unlike countries with a more limited oil wealth, such as Cameroon, people seldom return to the countryside as a response to rising urban unemployment: the gap between urban and rural remuneration probably remains too high. Zomo Yebe (1993: 82) found that in the 1980s an urban self-employed person in the informal sector typically had three times the salary of a rural poor person, and 17 per cent more than even an independent agricultural producer. On the other hand, an urban salaried worker earned 37 per cent more than a rural one. Although the proportions may have changed since, the figures make it clear why return migration to rural forested areas has not been a common response by people who have become marginalised in urban labour markets.

A more rational survival strategy may be to maintain urban residence while diversifying into periurban 'weekend farming'. Survey results indicate that 5 per cent of households in Libreville and 27 per cent in secondary centres have partly followed this path (Poupart and Pilichowski 1997: annex B, p. 6). Another response to the mini-crises may have been to increase the extraction of open-access forest resources, such as game, timber, charcoal and non-timber forest products (Trefon 1999; Wolff *et al.* n.d.: 48–52). Except for bushmeat (to be further discussed in the next section), such mini-crisis impacts do not seem to have been widespread, nor to have caused anything more than 'point impacts' on forests in periurban areas.

In sum, the long-term reduction in absolute urban poverty and massive urban labour absorption suggests that rural–urban migration has been a main (and predominantly successful) poverty-alleviation strategy, which at the same time has held back deforestation. This main mechanism of adjusting to oil wealth will now be analysed further. The lagging rural sector remained poor, with an increasing gap *vis-à-vis* the urban economy. Higher rural wages not only reduced poverty but also generally raised labour costs, adding to the disincentives against clearing forests.

Rural–urban migration

The single most important transformation of Gabonese society during the last half-century has been the accelerated urbanisation of a forest people. In 1950, Gabon still had the largest share of rural population among the six central African countries, far more than Congo or Cameroon. Yet since 1975 it has been the second most urbanised country after the Congo (Wolff n.d.: 5, World Bank 1999a: table 3.10). Urbanisation estimates vary greatly according to how urban areas are defined. Gabon's population-census criterion has been to consider agglomerations of more than 3,000 inhabitants as 'urban' (Richard and Léonard 1993: 152–3). Officially, urbanisation was thus 73 per cent in the 1993 census (DGSEE: 1993: 2). International comparative sources like the World Bank (1999a) yield the more conservative estimate of 52.1 per cent for 1997. More interesting than variable static definitions are changes over time. Pourtier (1989b: 237) reports a rise from 20 per cent in 1960 to about 60 per cent in 1985; Moriconi-Ebrard (1993, cited in Wolff n.d.: 5) from less than 5 per cent in 1950 to about 45 per cent in 1990, with a marked increase during the main oil boom years (1970–80). The World Bank (1999a) figures are 17.4 per cent in 1960 and 52.14 per cent in 1997, showing an almost linear urbanisation trend over the entire period.

The accelerated process of urbanisation is closely related to the creation of an independent state, strengthened gradually by the rise in rents from mining and especially oil. As Pourtier expresses it, cities and towns are 'the daughters of the state'. At the same time, they are also the most tangible direct product of Gabon's mineral extraction cycles: 'The rent has been converted into towns' (Pourtier 1989b: 237). This is most obvious for the exponential growth of Libreville, the modern capital, where the power to distribute publicly accruing rents is concentrated. Until recently, the city doubled its population every decade: 12,500 (1950), 31,000 (1960), 77,000 (1970), 185,000 (1980), 338,000 (1990), 450,000 (1998).[49] Port-Gentil, the second-largest town and main port, has benefited from the export of both timber and oil. Finally, the urban triangle of Moanda–Mounana–Franceville in the southeast has been created by uranium and manganese mining, as well as by generous public projects that President Bongo has initiated in what is his home region.

Even in smaller administrative outposts in the interior – isolated 'forest towns' with little production and few trade links – the state has often become the 'leading sector', civil servants' salaries, public social infrastructure and construction projects proving decisive in stimulating a local urban economy (Richard and Léonard: 264–74). Sometimes the false impression prevails that this public spending has occurred in a geographically highly skewed manner. In fact, spending priorities seem to have been regionally fine-tuned (ibid.: 269), in order to mirror the distribution of population, and of votes.[50] In the 1980s, about half of this urban population, notably women, still practised urban or periurban agriculture, in spite of the rising density of the urban population: lack of space led to longer transport times to the fields and, in particular, a shortening of fallows. The latter was achieved, *inter alia*, by shifting to crops like manioc, which are more tolerant of reduced fallows (ibid.: 271). In other words, rising crop production in and around the towns went hand in hand with intensified cropping. At the same time, imported foodstuffs penetrated most urban areas massively, thus limiting the demand for periurban products (see next section).

The net result was that, even where periurban cropped area expanded at the expense of forests, as happened around larger agglomerations such as Libreville, Franceville and Oyem, this sort of expansion and forest loss was proportionately far below the rise in population. As mentioned in the section on 'Deforestation in Gabon', from 1961 to 1990, although 5,015 ha of forests were cleared near Oyem (25 per cent of the area studied by Wolff *et al.* n.d.: 17, 22), population increased sevenfold from 3,000 to 22,404 (DGSEE

1993: 88). In Franceville, 1,699 ha of forest (about 30 per cent) were cleared from 1953 to 1994 (Wolff *et al.* n.d.: 18, 26–7), but just from 1972 to 1993 population rose sixfold, from 5,000 to 31,183 (DGSEE 1993: 86). Around Libreville, 9,000–10,000 ha are currently being cultivated (BPDA 1998: 18–26), for a population of as much as 450,000 (Marchés Tropicaux 1998: 17). This corresponds to 0.02 ha per inhabitant.[51] There can therefore be little doubt that urbanisation has been a highly 'land-preserving' undertaking on the whole.

Where did the urban migrants come from, and why did they move? Inter-provincial migration flows up to 1970 (Pourtier 1989b: 260) and 1993 (DGSEE 1993: 62) document a continuous rural exodus from the interior to Libreville and, secondarily, to Port Gentil and the Franceville triangle. Up to 1970, Libreville-bound migrants originated mainly from Ngounié and Nyanga Provinces in the south and Woleu–Ntem in the north. By 1993, there had also been a large migration flow to Libreville from Haut-Ogooué Province in the southwest, as well as from abroad. 'Pull' motives such as urban job opportunities were the main motive for migration; there were no demographic or other pressures worth mentioning in the areas of origin (Pourtier 1989b: 266). In some cases, such as Oyem in the mid-1970s, there was a direct causal link between oil-boom revenues, new urban and labour-intensive projects being undertaken by the state, rural exodus and the abandonment of cropped areas.[52]

How were land use and forests in the sending areas affected? In the rural areas, population has either declined or stagnated since 1960. No historical rural census figures seem to be available at the provincial level. Yet, most rural provinces, like Ogooué–Ivindo, Ogooué–Lolo, Nyanga and Ngounié, have maintained their low 1960 densities of around 1–2 (rural and urban) inhabitants/km^2; for others, there has been a moderate rise to 2.3 (Moyen–Ogooué) and 2.5 (Woleu–Ntem). More importantly, the age and sex structure has changed dramatically in the countryside (see Figure 4.3). Rural areas have built up a deficit in the most productive age groups (20–45), especially of men, who have migrated to the towns to find better job opportunities. Rural areas are thus rapidly ageing, with a lower share of economically active persons. This is perhaps the main factor behind the abandonment of cultivated areas and the increases in forest regeneration.

The structure of consumption

In a country where per capita GDP rises eightfold in eight years, and remains for three decades at 4–5 times its inflation-corrected pre-boom level (section on 'The macroeconomic impact of the mineral boom'), significant changes in the composition of demand are bound to occur. As a society grows richer, different goods are consumed, even if this wealth is distributed unequally. Production of especially higher calorie-intake food products may potentially lead to an extensification of land use and accelerate forest-clearing. It will be argued that demand structure in Gabon indeed changed dramatically, but that this had unimportant impacts on land use because food imports grew spectacularly.

From 1961 to 1973, per-capita daily calorie consumption grew moderately from 1,959 (208 from animal products) to 2,141 (267), but petroleum wealth accelerated this increase to 2,626 (425) in 1984 (FAO 2000b). This reflects the fact that rising purchasing power also stimulated food sales, especially of animal products such as beef. Bovine meat

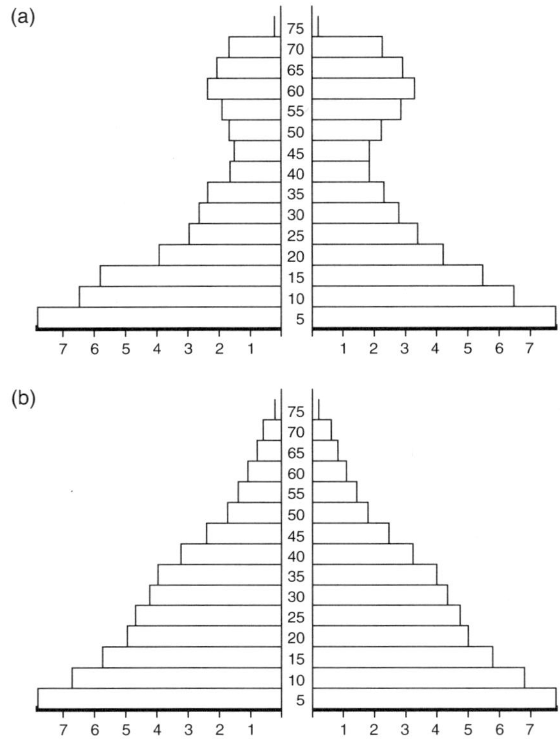

Figure 4.3 Gabon: age distribution of (a) rural and (b) urban population in 1993.

Source: DGSEE (1993: 8).

consumption had already increased from 1,459 t in 1961 to 2,297 t in 1973, but with the oil boom it went up sevenfold to 13,474 t in 1984. The domestic share of beef production was 12 per cent for both 1961 and 1973, and although production expanded by a meagre 105 t up to 1984, its relative share plummeted to 3 per cent. Meat consumption levels stayed high until 1995, after which there was a sharp decline to 5,094 t in 1998. For the first time since independence, the domestic market share has now reached 21 per cent, though only because imports have declined sharply. Absolute domestic production levels have been stagnant since 1990. For comparison, Gabon's bovine production corresponds to 2 per cent of the Central African Republic's, 1.2 per cent of Cameroon's and 0.2 per cent of South Africa's production in 1998 (ibid.).

In other words, basically all the rapidly rising demand for bovine meat from 1973 to 1984 was satisfied by imports, mainly from the EU and South Africa. In the second half of the 1980s, agro-industrial investment in cattle increased domestic production. The main ranches are at Nyanga, Ngounié (both in the south) and Lekabi (near Franceville). However, these are almost exclusively situated in savannah areas (Dauthuille 1996: 51; DGE 1999: 34–5). With sizes of 100,000, 65,000 and 50,000 ha respectively, all of them are under-utilised in spatial terms, and even a fair rise in domestic production would probably not make it worth while to clear forest for pasture expansion. The deforestation impact of cattle-ranching in Gabon is thus currently zero.

With oil wealth, Gabon also experienced an increasing 'Westernisation' of food habits. At the luxury end, Gabon apparently has the world's highest per-capita consumption of champagne, which was already known as *jus d'okoumé* in the heyday of timber exports. More importantly, the basic diet changed. In the coastal towns, maize flour and French breads replaced traditional tuber staples such as starchy maniocs (Zomo Yebe 1993: 56; Trefon 1999: 43). Consumption of 'luxury' foods, such as fruit, vegetables, meat and cereals, generally increased (Marchés Tropicaux 1998: 17). One source estimates that, during the early 1980s, Gabon imported as much as 96 per cent of its food consumption value (*The Economist*, cited in Yates 1996: 213). Hence, consumption shifts further reduced incentives for agricultural production, and with them, potential sources of forest loss.

In the late 1980s, agriculture bounced back somewhat. There are indications that some tuber staples consumed in urban areas now increasingly come from Gabonese sources, while other foods are still imported.[53] On the other hand, new trends are emerging, such as the massive penetration of Asian rice as a staple crop in Libreville and Port Gentil. In the towns of the interior, high transport costs from the ports have hitherto protected the predominant place of tubers and plantains (DGSEE 1994a: 249–54). However, in the two main towns on the coast, rice and cereals together now provide 36 per cent of calorie consumption, while tubers only make up 17 per cent (DGSEE 1994b: 20).

If a growing, richer and more urban population did not cause much land conversion, did it extract more forest products in a way that decreased forest quality? As mentioned, domestic demand for wood products (firewood, charcoal, timber) is unlikely to cause more than point impacts in a country with so much forest per inhabitant, especially for wood energy uses that tend to decline in urban areas. Bushmeat is the prime candidate for a form of extraction that diminishes forest *quality*.[54] Historically, hunting has provided the bulk of meat supplies in Gabon. As shown by beef consumption above, meat is also a foodstuff with a relatively high income elasticity (0.45 in Gabon's urban centres),[55] implying that it expands fairly rapidly when a society grows richer.

Bushmeat figures are very tentative, given the decentralised and semi-legal nature of its exploitation. Indjieley (1998: 2) quotes a WWF report to the effect that sales in three Libreville markets over twelve months (1992–3) equalled 500 t, corresponding to a yearly consumption of 1.2 kg/inhabitant. Steel (1994) estimates that, for all six urban markets in Gabon, yearly sales are 1,105 t. But when extra-market roadside sources are included, extraction reaches 6,900 t. Rural use is estimated at 11,000 t. The national result is a yearly bushmeat consumption of 17,900 t (17.65 kg/person). This is 57 per cent higher than the 11,381 t estimate by Wilkie and Carpenter (1998: 5), and it is claimed to be double the size of beef consumption (Steel 1994: summary). Also, urban bushmeat use is said to be rising over time, as a cheap open-access resource that becomes attractive under conditions of economic crisis (Trefon 1999: 46).

However, there seem to be several flaws in these arguments. An initial note of caution is that roadside sales and subsistence uses are extremely difficult to extrapolate on to the national scale from a limited set of observations, yet these uses make up 94 per cent of Steel's estimates (1994). Even if the absolute figures reflected reality, their relative size is still highly exaggerated. It is not correct to say that bushmeat makes up double the quantity of beef consumption: the 17,900 t represent less than one-third of total meat supply.[56] Also, urbanisation does not increase, but reduces bushmeat consumption. Using the population census estimates from 1993 (742,296 urban and 272,680 rural inhabitants; DGSEE 1993: 6), Steel's figures imply an average urban consumption of 25 g/ person/day, compared to 110 g in the countryside. The 'urban crisis' argument also seems dubious, as urban bushmeat is not very cheap, whether for suppliers to hunt, store, transport and sell, or for consumers to buy.[57] Urban consumers keep eating bushmeat to a limited but regular extent out of preference and cultural tradition, not as a specific response to economic crisis.

In general, a picture emerges in which urbanisation and changing consumption patterns have tended to favour food imports. This has helped to protect Gabon's forests by moving the 'ecological footprint' of swelling urban areas outside the country's borders. Of course, this factor works in tandem with the loss of competitiveness of Gabonese agriculture. This trend was extremely pronounced from 1973 to 1985. Recent changes at the margin of these trends, which are related to urban crisis in spite of oil wealth, produced only a vague revival of domestic food crops.

Synthesis and conclusion

Among the cases analysed in this book, Gabon is the country that most clearly confirms the core hypothesis. Following independence in 1960, Gabon was gradually transformed from a timber-exporting into a mineral-extracting economy, a structural change that, together with the limited population density, also created conditions extremely favourable to forest preservation. The oil boom was the driving force in the creation of a rentier state. During the price hike of the 1970s, GDP per capita rose sevenfold. In the mid-1980s prices dropped, but the rise in Gabon's oil production meant that since 1973 Gabon has remained a considerably richer country than before. In spite of unequal distribution, absolute poverty has been reduced dramatically. The government's spending of oil wealth has generally focused on the following areas:

- more public employment, higher salaries and benefits;
- transport infrastructure (railway, ports, but *not* roads);
- urban infrastructure (construction, health, education);
- large-scale parastatals.

The direct impact of swelling public employment and urban infrastructure was 'forest-neutral', but in indirect terms this triggered a rural exodus to urban areas, especially of young people in the most productive age groups. An ageing rural population thus lived more and more from private transfers and public projects, and reduced or abandoned crop cultivation. Together with iterative state policies of rural resettlement, this led to forest regrowth in many increasingly 'empty' parts of the interior. Several case studies of changes in vegetation cover show that cutbacks in extensive slash-and-burn systems led to a significant expansion of forest area. A quote from a village chief in northeast Gabon is illustrative of a situation in which a reduced human presence went hand in hand with forest rehabilitation: 'Nobody lives here any more. ... The young are leaving, and the elephants and gorillas run freely through our gardens, destroying what little we grow to eat'.[58]

Generally speaking, there are no motivations for large-scale deforestation in Gabon; rather, the decline in land-using sectors since the start of the oil era is likely to have triggered natural net *reforestation* over the past three decades. This evidently belies the FAO's FRA 1990, which suggested an annual rate of deforestation of around 100,000 ha. Current net deforestation is probably either zero or negligible. Land use has generally declined, and because of the greater concentration of the population, it has also become more intensive. Table 4.4 summarises how oil wealth has triggered a series of market- and policy-induced changes. Ten partial pathways are classified according to their economic intensity, and to the strength with which they are linked to forests. These two criteria jointly determine the intensity with which deforestation is either curbed or accelerated by that pathway.

Factors reducing forest loss made up most of the pathways (seven out of ten), particularly the strongest ones. Urban labour absorption (1) was clearly the most important. This worked together with poverty reduction (3), mostly achieved in urban areas, and rural resettlement (9) towards the goal of rapid urbanisation. Jointly, these factors liquidated a fair share of traditional agriculture. A second group of factors clusters around demand substitution. This included RER appreciation (2), increasing exposure to it caused by adverse trade and pricing policies (8), and shifting urban consumption preferences induced by higher incomes (4). Domestic production could not compete within this scenario and was increasingly replaced by imports. This put further pressure on agriculture, which declined dramatically in terms of production, employment and cropped area.

The Dutch Disease thus induced a long-term loss of competitiveness, which hurt agriculture and other non-mineral T sectors. Timber was one of them. Exports have proved highly price-elastic in recent decades, as implied by the regression results above. The extensive but highly selective type of logging in Gabon implies that it has an impact on forest structure but does not cause deforestation. For agriculture, the export crops, coffee and cocoa, were subjected to real economic extermination; import-competing domestic food crops were also hit and have only recently experienced a slight revival in periurban areas, as a response to urban mini-crises. As regards the acceleration of forest loss, impacts remained weak. The agro-industrial sector (8), the state pet designed to replace traditional agriculture from the 1980s onwards, was much less land-extensive and, above all, too inefficient to reach its planned output. Road-building (5) was entirely neglected, while much money was put into the Transgabonais railway. This expanded logging, which may have accelerated forest degradation, but its impact in terms of deforestation has been negligible. And the booming oil sector (7), although being mostly on-shore based, had an equally minor (and often exaggerated) impact on deforestation.

Gabon's development strategy in recent decades therefore looks in many ways like a sophisticated conspiracy against agriculture, which coincidentally has proved to be an involuntary but highly efficient forest-conservation strategy! It seems to matter little that

Table 4.4 Gabon: oil wealth and deforestation – an overview of the long-term impacts

Economic and productive impacts			*Links to deforestation*		*Deforestation impact*	
No.	*Type*	*Intensity*	*Type*	*Strength*	*Type*	*Intensity*
1	Urban labour absorption in (para-)public sectors and private services	Very strong ●●●	Rural exodus; cultivated areas are massively abandoned	Close ➔	Forest regrowth in the country's interior	Very strong ▼▼▼
2	Long-run loss of traded sector competitiveness	Strong ●●	Export and food crops shrink, reduced timber extraction	Close ➔	Less forest conversion	Strong ▼▼
3	Reduced absolute poverty	Strong ●●	Higher opportunity costs of rural labour	Medium ⇨	Less forest conversion	Medium ▼
4	Higher urban incomes shifts food demand	Medium ●	Imports go up, substituting local staples	Medium ⇨	Less forest conversion	Medium ▼
5	New road and rail construction	Weak ❍	Open up frontiers for product trade	Close ➔	Maybe more degradation, but no deforestation	Weak Δ
6	Higher budgets of development agencies	Medium, unstable ●	Capital- and land-intensive agro-industries	Weak →	Deforestation point impacts	Weak Δ
7	Oil and mining production	Medium ●	On-site direct and indirect forest loss	Weak →	Deforestation point impacts	Weak Δ
8	Agricultural protectionism and subsidies	Close to zero	De-protection of food crops and ill-priced export crops	Medium ⇨	Reinforced decline in cultivated area	Weak ∇
9	Rural resettlement	Weak link to oil ○	Spatial concentration of rural population	Medium ⇨	Reduced fallow length	Weak ∇
10	Higher budgets of forestry/ park agencies	Close to zero	Weak domestic policy priority	Medium ⇨	Improved law enforcement?	Close to zero

Notes
1–4 and 8–10 area – effect reduces deforestation; 5–7 area – effect increases deforestation.

the agencies responsible for publicly protected areas remain hopelessly under-funded or that forestry regulation is weak (10), as long as one follows 'the Gabonese recipe' in achieving maximum forest conservation:

- draw as much labour as possible out of rural areas by spending your money in the cities;
- let your exchange rate appreciate, over-tax export agriculture and favour food imports;
- neglect demands for new road-building in the interior;
- force people to settle in concentrated roadside agglomerations;
- deny credit and technical assistance to small-scale agricultural producers;

- waste most of your agricultural budget on agro-industrial 'white elephants';
- nourish a rent-seeking environment in which few people find it worthwhile to produce.

Not only did the state discriminate against agriculture; the sector also has a generally compromised status among the Gabonese people themselves. Forest-clearing for crops is carried out by men, but cultivation is exclusively relegated to women (Pourtier 1989a: 194), and most commercial agriculture is left to foreigners. For instance, almost 75 per cent of crop producers around Libreville are foreigners (BPDA 1998: 32). However, once oil revenues decline the situation will gradually change. Even most observers who are sympathetic to forest conservation would probably admit that some rise in agricultural expansion will be inevitable in terms of Gabon's need to adjust to lower oil revenues, although it will trigger some deforestation.

How fast will the decline in oil revenues occur?[59] Some forecasts have foreseen a very rapid decline, based on a linear extrapolation of currently known reserves. For 2005, a production level of only about 6 million t has been projected, with a probable complete exhaustion of all oil resources in 2010 (Marchés Tropicaux 2000: 1437). Correspondingly, the CGE model of Söderling (2002) assumes a halving of oil production between 2000 and 2007, from 13.6 to 6.1 million t, triggering massive macroeconomic adjustment in the near term (ibid.: 15). These decline scenarios are highly relevant for the medium term, but in the short run they are too pessimistic. Gabon is likely to remain a significant oil exporter for at least another fifteen years. This is both because existing fields are declining more slowly than earlier predicted, and because the minor new discoveries have been more important than earlier believed:

- The decline phase of existing fields is not linear; rather, the curve levels off over time, due to geological and physical characteristics.
- New discoveries, systematically related to the larger exploration efforts, which are triggered by high current oil prices[60] and better conditions offered to the foreign oil companies by the government,[61] have partly compensated for the decline in the old fields.

Consequently, the most grim oil scenario for Gabon seems to be one of declining international prices, reduced exploration and a production decline somewhat lower than over the past three years, with a marked decline of revenues especially after 5–10 years. The most optimistic scenario would be the maintenance of fairly high prices, a stabilisation of production quantities over the next five years, followed by a gradual decline in both production and revenues. Still, this outlook buys more time in order to adopt rational policies preparing for a future without oil.

Will that decline translate into greater pressures to occupy land for agriculture, at the cost of higher deforestation? The development of agriculture has long been a planning goal,[62] but it may only happen once oil income seriously declines and forces the necessary shifts in RPs. Eventually, agriculture and land pressures, especially from expanding low-return food crops, would be a crisis-led 'default strategy' to embark on. That would be necessary to the extent that Gabon is unsuccessful in developing other rent sectors (e.g. gold-mining or natural gas) and value-added production activities that do not require land (e.g. manufacturing or wood-processing).[63] The economy generally needs to become less interventionist, reducing '*l'étatisation*'; a domestic entrepreneurial class would need to develop. In Gabonese society, consumption is still highly divorced from production, a mentality that has been created by its rich extractive resources. In this sense, a quote from the classical work of Raponda-Walker and Sillans (1961: 31) is still valid: 'Why bother to cultivate in a country where everything is at hand?' While forests can contribute positively to economic development, it seems also quite obvious that they have constituted a default condition of 'non-development' in the countryside.

Notes

1 Using FAO's conservative 1995 forest-cover estimate (see section on 'Current forest loss'), combined with the rather high 1995 population projection of 1.3 million, forest-cover per capita is 13.5 ha (FAO 1997a: 182), almost ten times higher than in Cameroon (1.5 ha per capita). Using the more cautious population figure of 1.15 million from the last census, combined with the more credible forest-cover estimates from TREES, the figure rises to as much as 20 ha forest-cover per capita.

2 Descoing (1974) and Fontes (1978), cited in Sayer *et al.* (1992: 169).

3 Collomb *et al.* (2000: 34) cite a higher TREES estimate (21,338,900 ha – see Table 4.1), which is the figure prior to the correction procedure applied by TREES and thus less accurate.

4 See Wilks (2000: 5–10), and the section on 'Vegetation history'.

5 Sayer *et al.* (1992); updated French version by Larivière (1996).

6 This tentative conclusion is not only supported by the 109,662 ha (0.5 per cent) higher 1970 estimate compared to the TREES estimate for 1990. Deforested areas in 1991–3 were more concentrated (periurban clearing, agro-industrial plantations) than in 1970 (slash-and-burn plots dispersed evenly in the interior). This implies that 1970 forest cover was probably overestimated, as the map scale was not sufficiently detailed to register small clearings. The same argument is used to caution against the high forest-cover estimates in the IUCN Conservation Atlas (Sayer *et al.* 1992).

7 Personal communications from R. Vinchent (IGAD) and S. Touré (Direction d'Agriculture), 30 May and 3 June 2000, Libreville, respectively.

8 Pourtier (1989a: 278) notes that 'the statistical services of the Ministry of Agriculture are so to say non-existent, or they are limited to vague estimations' (my translation from the French). This certainly seems true for the 1999 figures provided by Mr S. Touré (Direction d'Agriculture, personal communication, 3 June 2000). According to reports from the provincial offices for agriculture, cultivated areas in Gabon add up to the high (and suspiciously round) figure of 170,000 ha. However, validating these figures for Estuary Province, which has been studied in detail by BPDA (1998), it turns out that the official figure is 70–80 per cent higher: the official report is 16,150 ha, while BPDA finds a cultivated area of 9,000–10,000 ha (ibid.: 23).

9 For instance, see Rippert's description (1997: 11) for the area around Franceville.

10 This excludes fallows. The total for the three towns may be 40,000 ha, to which one should add other towns such as Port Gentil, Lambaréné and Lastoursville.

11 These assumptions are also rather conservative. Fallow length was probably longer in 1961, as agriculture near urban areas has intensified over time. Area decline (45,000 ha) in our calculation is comparable to the difference in cropped area between the 1961 and 1975 censuses, but may actually be much larger, as indicated by the Africa Forest case studies.

12 See Christy *et al.* (1990a,b), Basquin *et al.* (1991), Christy *et al.* (1991a,b) and Wilks (1992).

13 There were also about 53 ha with dead trees, due to disturbances from erosion etc. Shell Gabon called this 'indirect deforestation' (ibid.: 37), a term that should rather be reserved for the non-oil impacts that are facilitated by the oil operations (e.g. hunting and agriculture; see section on 'Indirect oil impacts').

14 Compared to slash-and-burn agriculture, however, it is true that logging and oil can create impacts that sometimes, though quantitatively fewer, are *longer lasting*, because of processes of soil compactation, erosion, laterite deposits, etc.

15 I am grateful to John Bickerton of Amerada Hess for providing me with this information.

16 UNDP (1999b: 52) reports a direct employment of only 1,690 people for 1998.

17 See Christy *et al.* (1990a: 27–8), Basquin *et al.* (1991: 68–9), Christy *et al.* (1991a: 35) and Blaney *et al.* (1998: 83).

18 This claim is supported by the Pearson correlation coefficient between real oil revenues and real capital inflows (−0.44); the negative coefficient is significant at the 5 per cent level.

19 In the 1960s, RER levels appreciated relatively, *inter alia* due to other mining exports (manganese, uranium). These might be analysed as another source of Dutch Disease wealth, but for the sake of simplicity only oil and financial inflows were included.

20 Apparently, the classification of non-permanent employees and military personnel varies. I am grateful to Mr S. Ziza (Ministry of Finance) for making these unpublished figures available to me.

21 Yates (1996: 212), and my own observations.

22 In a 1999 audit of government liabilities, 18 per cent had to be cancelled due to double, triple or fictitious invoicing, while another 29 per cent was debt paid but still recorded as due (Söderling 2002: 5).

23 Originally, it was planned to connect the port of Owendo to Booué and to the iron-ore deposits of Bélinga. With the poor world-market prospects for iron ores, the Bélinga link was postponed in favour of the connection with Franceville in the south-east. Formerly, the manganese exploited in nearby Moanda had been transported by cable car and rail to the Congo for shipping (Richard and Léonard 1993: 207). The *Transgabonais* now brings the manganese directly to the port of Owendo. In addition, Franceville is in President Bongo's home region, and the rerouting of the railway considerably strengthened his political position internally (Yates 1996: 180).

24 Calculated from yearly shares given by Alexandre Barro-Chambrier, and cited in Yates (1996: 182). On the first stretch from Owendo to Booué, construction costs came out at twice the budget (ibid.: 179). The World Bank has claimed that the expenses of the railway amounted to six times the comparable international standard costs (ibid.: 182).

25 The author compares the cost of traditional and bulldozer-operated forest clearing. Due to the high labour costs, the latter tends to be a cheaper option, but high fixed costs hinder its adoption by small-scale farmers (ibid.: 83–4).

26 M. Mfa Obiang (Director of Macroeconomics, DGE), Libreville, 8 June 2000, personal communication, and own observations.

27 See Chapter 6 on Cameroon for a detailed analysis.

28 Chris Wilks of ECOFAC (Libreville) estimates that 2 m pixels would yield a total of 15,000 ha of net yearly deforestation (canopy opening minus natural regrowth) while for 20 m pixels the figure would be 5,000 ha (C. Wilks, personal communication, 6 October 2000).

29 As a light-loving species, it seems to grow well in previously cultivated plots, where it can develop high densities in the early stages of regrowth, especially on sandy soils on the coast, where it has few competitors (Fuhr *et al.* 1998). This ability also explains why locally *okoumé* has often been called 'the son of manioc' (Aubreville 1948).

30 See the Africa Forest case studies referred to above, such as Christy *et al.* (1990a,b), Basquin *et al.* (1991), Christy *et al.* (1991a,b) and Wilks (1992).

31 For the multiple regression models in this book, we use only linear model specifications. In the interpretation, I will only refer to the indicator for 'explanatory power', R^2 (which varies between 0 and 100 per cent), not to the F-test. The T-test is a test for the significance of individual variables, which allows us to see which variables contribute most to the general model fit. Generally, the higher the significance, the *lower* the probability that the variable can be statistically deemed random in its correlation to our dependent variable. This means that if a variable is significant, for example, at the 10 per cent level only, that is less supportive of its contribution to the model than if it is significant at the 5 per cent or even at the 1 per cent level.

32 Apparently, the Asian *meranti* and Cameroon's *ayous* species are the closest substitutes.

33 For the sake of simplicity, *ozigo* was included in the miscellaneous woods category.

34 Taxes have not been inflation-adjusted over time and thus have lost their real value. With US$30.8 million, Gabon's forestry tax revenue is only half that of Cameroon's (Collomb *et al.* 2000: 26).

35 Estimated by adding up the individual project figures given in Yates (1996: 164–7).

36 For instance, the IGAD receives part of its funding from Elf Gabon; the same was true for the now failed Sogacel project to clear land for eucalyptus and pine plantations, and the legume-producing Agripog project (R. Vinchent, IGAD and C. Wilks, Africa Forest, personal communications, Libreville, 30–31 May 2000).

37 Coffee plantations were projected to achieve a production target of 10,000 t in 1978–9, but actual production was only 234 t (Yates 1996: 161).

38 The parastatal for coffee and cocoa, *Socagab*, produced cocoa on 635 ha and coffee on 95 ha in 1998, but its annual costs are three times the size of sales revenues, and it relies on state subsidies to cover its recurrent deficits (Marchés Tropicaux 1998: 25).

39 For instance, this was a main problem for the Société sucrière du Haut-Ogooué (SOSUHU) near Franceville. Public investments in SOSUHO ran to around 17 billion CFAF in 1975–81. A shortage of skilled labour encouraged increases in mechanisation and in the proportion of highly paid foreign employees (Pourtier 1989b: 284–5). A 6,000 ha sugar plantation was planned for an over-sized processing plant of 60,000 t/yr, but production never reached even half the intended production (24,000 t in 1997 – Marchés Tropicaux 1998: 26).

40 The most successful agro-industry has been rubber-producing Hévégab, with a reported rubber plantation area of 8,396 ha in 1997 in the centre-north, mainly near the processing plant in Mitzic (Marchés Tropicaux 1998: 25; Ovono-Edzang 2001). But even Hévégab was recently hit by severe mismanagement; production plummeted from 10,963 t in 1998 to 2,363 t in 2000 (DGE 2001: 38).

41 See *HEBDO Informations*, no. 452, 16 March 2002, for a reprint of the official legal text.

42 C. Wilks (Africa Forest) and P.O. Ondo (WWF), personal communications, Libreville, 28 and 31 May 2000.

43 This paragraph draws on Pourtier (1989b: 219–28).

44 Using the same parameters, the author reaches a total deforestation of 15,000 ha, that is, ten times my figure. This seems to be an error produced by placing a comma in the wrong place (ibid.: 32).

45 This paragraph draws on Pourtier (1989b: 102–22).

46 In the 1980s, efforts to combine resettlement with agro-industrial complexes were made in the so-called Integrated Operations Zones (OZI), but the idea was later abandoned (Poupart and Pilichowski 1997: 68).

47 Pourtier (1989a: 230–7) analyses the frequent shift of village sites in traditional rural systems in Gabon. He finds that there is no agricultural motivation for high mobility, that the exhaustion of forest products and disease may provide partial explanations, but that many movements are seemingly irrational ways of responding to the opportunities provided by open, unoccupied space.

48 For instance, inequality of urban consumption in Libreville and Port Gentil basically remained unchanged between the censuses of 1962 and 1994 (DGSEE 1994a: 156).

49 See Richard and Léonard (1993: 152, 156); Marchés Tropicaux (1998: 17).

50 For instance, P. Michaud (cited in Yates 1996: 206) puts forward the inequality argument that, by the mid-1980s, no less than two-thirds of public spending occurred in three provinces (Estuaire, Ogooué-Maritime and Haut-Ogooué). However, this distribution mirrors almost perfectly their corresponding share of the national population (665,401 people and 65.6 per cent – DGSEE 1993: 5).

51 In spite of the short transport distance, high labour costs reduce product competitiveness *vis-à-vis* imports from Cameroon, thus limiting the scale of crop production (ICRA and IGAD 1996).

52 Prior to the independence celebrations in 1978, public construction projects in the capital of Woleu–Ntem Province seemed to be paying such high salaries that many people abandoned their cocoa plots moved permanently into town (Pourtier 1989b: 294).

53 In the 1993 household survey for Libreville and Port Gentil, it is estimated that 76 per cent of the total volume of road-transported staple crops (40,000 t) comes from Gabon, a share that seems high. Cameroon remains the biggest supplier of vegetables, fruit (except for bananas) and manioc flour, that is, products with a higher value per weight unit (DGSEE 1994a: 275–93).

54 See Wilkie and Carpenter (1998) for a summary of the ecological impacts of bushmeat extraction.

55 The figure is from the 1993 household survey for Libreville and Port Gentil (DGSEE 1994a: 234). This linear elasticity coefficient tells us that, with a 1 per cent rise in household income, meat consumption rises on average by 0.45 per cent.

56 For the year in question (1993), the FAO (2000b) reports a bovine meet ('beef') consumption of 19,207 t, which is larger than the alleged 17,900 t of bushmeat consumption. The FAO's total meat supply figure for 1993 is 56,176 t. Assuming that this already *includes* an estimate for bushmeat in the category 'other meat' (20,275 t), Steel's estimate would correspond to 31.9 per cent of total meat supply. If bushmeat were *additional*, the meat-supply share would be 24.2 per cent.

57 Most urban people in Gabon live near ports, where imported meat is relatively cheap, while transported bushmeat is more expensive. The aggregate value figures given in Steel (1994) suggest an average urban bushmeat price of 2.71 US$/kg, compared to 1.91 US$/kg in rural areas.

58 Chief Mboula Thaopile, Nioungou village, cited in Adams and McShane (1996: 207).

59 This paragraph draws on a second interview with Mr John Bickerton, Operations Manager of Amerada Hess, Owendo. It was carried out two years after the first interview, on 28 June 2002.

60 The time-lag between exploration and production is short enough for the companies to be able to base their investment decisions on current, rather than expected future prices.

61 One reason for offering better conditions is that Gabon faces increasing regional competition from countries that have far more promising oil deposits, such as Equatorial Guinea and Angola.

62 Future priorities were defined in the Law of Agricultural Orientation (*Loi d'Orientation Agricole*) of 1998, where agricultural production is planned to increase by 45 per cent until the year 2010, mainly for reasons of food security, import substitution and export promotion and to raise rural living standards (Marchés Tropicaux 1998: 17).

63 See Wunder (2003: section 8) for a detailed discussion of future development scenarios.

5 Venezuela

Venezuela is another highly specialised oil exporter. Was oil just as important a forest-protecting factor as in Gabon? As we will see in this chapter, many similar mechanisms were at work. Forests expanded as croplands receded. However, cattle-ranching had a contrary land-use impact, much helped by protectionist trade policies. Still, this was insufficient to reverse the overall trend: the secular rise of oil significantly contained deforestation and forest degradation in Venezuela.

Deforestation in Venezuela

Vegetation history

Venezuela has three main regions with forests: the Andes and Coastal Cordilleras, the Orinoco *llanos*, and the Guyana–Amazonas basin (Harcourt and Sayer 1996) (see Map 5.1).[1] About half of Venezuela's land area is forested today, but historically that share has been larger.[2] The only non-forested areas were mountain peaks, natural high-altitude steppe in the Andes (*páramo*) and part of the central lowland savannahs (*llanos*). Thus, the *llanos* comprise both natural and man-made grasslands, including those derived from pre-Columbian human impacts. A large proportion of the *llanos* and *páramos* have previously been forested; current vegetation cover constitutes a 'plagio-climax' or 'fire climax' created by repeated human intervention.[3] Hence, areas now cleared for agriculture and cattle-ranching have been derived partly from grasslands but mainly from forests, especially from deciduous forests in the Orinoco *llanos* (Gill 1931: 204).

Considerable uncertainty persists about the size of pre-Columbian settlements and populations. Probably about 1 million indigenous people lived in the territory now known as Venezuela. This figure was much reduced after the conquest, mainly through the spread of European diseases (Denevan 1976: 291). Indigenous cultures had concentrated along two north–south axes, producing maize in the west and manioc in the east.[4] Archaeological findings show that large parts of the *llanos* have been occupied and cultivated, through both migratory slash-and-burn and more permanent settlement systems.[5] The large numerical decline of the indigenous population, though less dramatic than in the Central Andes (see Chapter 7 on Ecuador), must also have reduced areas under cultivation, allowing them to revert back to forest.

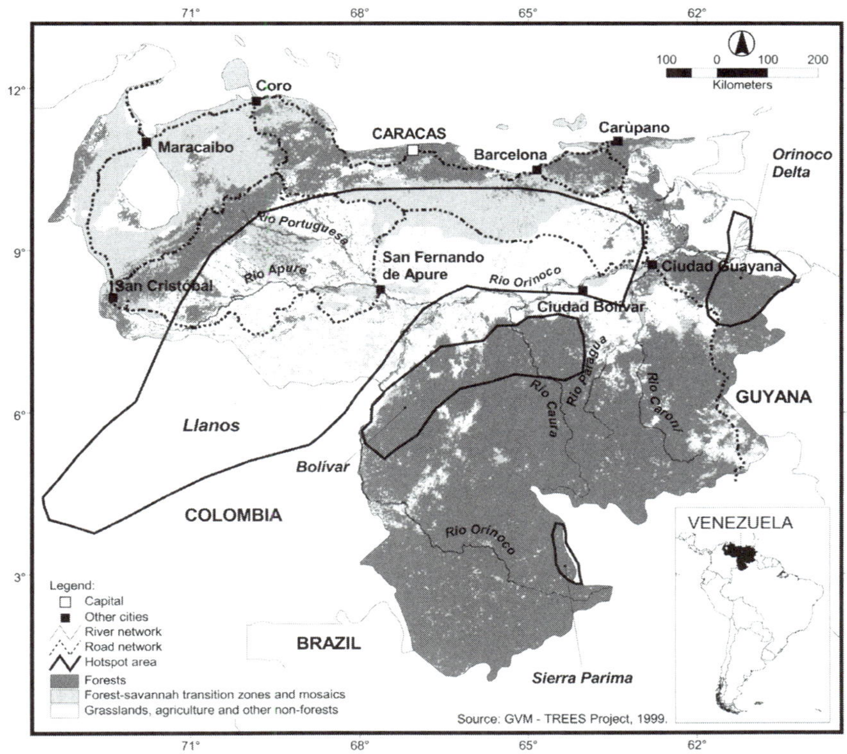

Map 5.1 Venezuela.

Whereas indigenous settlement had focused on the interior, the trade orientation of the colonial economy provided incentives to occupy the northern coast, producing a shift in the regional focus of deforestation. The diversification of colonial agriculture (maize, tobacco, cocoa, wheat, cotton, sugar, cattle) gained momentum until the mid-seventeenth century. Cattle pastures and, especially, cocoa plantations increased (Vivas 1993). Timber exports (e.g. *guayacán*, brazilwood) and other forms of forest extraction were also begun in the colonial period, but they were restricted to the highly selective exploitation of easily accessible high-value species.[6] On the whole, therefore, forest cover in the interior is likely to have increased during the early colonial period, whereas some coastal forests gave way to cultivation. Net forest cover in Venezuela probably increased after the conquest as a result of population decline. It was only between 1850 and 1890 that population growth in Venezuela took off, rising from 700,000 to around 2 million people.

From 1830 onwards, export plantations of cocoa (the traditional colonial crop) and coffee (the prime exportable in the republican period, especially after 1870) expanded rapidly. Cocoa plantations were concentrated in the central valleys, but the different climatic and soil requirements for coffee led to the gradual colonisation of hillsides in coastal and

Andean areas (Carvallo and Rios de Hernández 1984: 19–22, 57–9). Tobacco, cotton, cattle-skins and sugarcane were additional exportables, supplemented by various slash-and-burn cultivated subsistence crops (*conucos*). Valuable timber (cedar, mahogany) was exported on a small scale from the ports of Maracaibo, La Guaira and Puerto Cabello (Carrero 1995: 9). Brito (1996: 461–2) estimates that by the first decade of the twentieth century, 14 per cent of private lands were dedicated to agricultural production, of which plantations made up about a quarter. In policy terms, the Venezuelan state responded to the greater demand for agricultural land *inter alia* by passing successive land-colonisation laws between 1821 and 1936. These aimed at the dismantling of indigenous communal land-tenure systems that had been partly protected during Spanish rule. Colonisation facilitated the expansion of the agricultural frontier and of production by designating indigenous-controlled areas 'idle lands' (*tierras baldías*) and thus opening them up for squatter occupation (Arvelo-Jiménez 1984: 105–8). At this stage, therefore, Venezuelan land-use patterns did not differ significantly from those in other Andean countries, specialising in cash-crop exports.

Oil discoveries and production from the Lake Maracaibo region profoundly altered the structure of the Venezuelan economy in the 1920s. From 12 million *bolívares* in 1921, the value of oil exports rose to 247 million *bolívares* in 1926 and 634 million in 1930. In 1920, export orientation had almost exclusively been based on agriculture and cattle, but by 1930 the situation had been completely reversed. A geographically diversified agro-export base, nationally financed, had largely been replaced by an oil-export economy with the growing participation of foreign capital and an increasing North American trade-orientation (Brito 1996: 461–77).

There is a general perception that the 1920s marked the first Dutch Disease period in Venezuela. Oil exports and real currency appreciation crowded out agricultural tradables, due to a loss of competitiveness accompanied by an exodus of rural labour to the *campos petroleros*, the main cities, and road and other construction work (Hausmann 1990: 3–4). In political terms, the oil-led economy weakened the influential rural landowning elite in favour of a politically articulated urban middle class (Karl 1986). Coffee, the mainstay of the agro-export economy, was already in crisis before the rise of oil, due to an expanding productivity gap *vis-à-vis* its main world market competitors, Brazil and Colombia (Aranda 1984: 82–6). The real appreciation of the *bolívar* then sealed the miserable fate of export agriculture throughout the 1940s and 1950s, in spite of incipient subsidies, road-building and other measures designed to favour the rural sectors. The rise of oil rents favoured the state-led development of urban sectors. The corresponding neglect of agriculture in national development reduced the pressures on forests from the 1920s onwards. As Veillon shows (1976: 97), forest area expanded during 1920–50, especially in the agriculturally important Orinoco *llanos*. This was due to the abandonment of agriculture and the accelerated rural–urban migration caused by oil incomes (Veillon 1997: 59/85).

Deforestation only returned in the 1950s, following the aggressive expansion of the national road network. But conversion pressures differed greatly according to region. Around 95 per cent of Venezuela's population live north of the Orinoco. As transport connections improved and urban markets grew, the north was subject to growing pressure from logging and agricultural conversion. The Orinoco *llanos* also have soils with good potential for cultivation. Their evergreen and semi-deciduous forest (in the west) and dry deciduous forests (in the east) were thus increasingly cleared, due to pressures from both agriculture (corn, sorghum, groundnuts, etc.) and cattle-ranching (Rojas 1993; Harcourt and Sayer 1996: 312; Veillon 1997). The Venezuelan Andes contain peaks of 4,000–5,000 masl; the Coastal Cordillera is somewhat lower (2,000–3,000 masl). A large variety of forests has developed, from the dry and sub-montane coastal forests to the montane and cloud forests in the Andes (Plonczak 1997: 55). Here, forest loss follows a common land-use cycle of selective logging, agricultural conversion and extensive cattle-ranching (Rojas 1993; Harcourt and Sayer 1996: 312). Forest areas close to roads, growing urban centres and the hinterland of the Maracaibo oil region were the most severely affected.

On the other hand, the land south of the Orinoco has been too far from urban markets, only sparsely inhabited by indigenous groups, and with its forests largely untouched by modern development. The Guayana region still contains about 80 per cent of Venezuelan forests, mainly tropical rain and moist deciduous forests (FAO 1993: table 8c). Only with the crisis in the oil-based economy over the past decade has there been a renewed interest in developing southern Venezuela on a larger scale. Bolívar state in particular is experiencing pressures from logging and mining (Miranda 1998).

Current forest loss

Forest cover in Venezuela is estimated at between 44 and 49 million ha, or 49–54 per cent of land area (see Table 5.1).[7] The size of forest loss over the last two decades has been much disputed; estimates vary by a factor of four for comparable time-periods. Official, conservative estimates have been around 150,000–250,000 ha, while the FAO's estimates have been 500,000–600,000 ha. Based on the following discussion of numerous forest-cover and land-use data, I will argue that annual deforestation over the last couple of decades has probably been in the intermediate range of 250,000–400,000 ha. Comparing that span to the forest-stock range from above yields a yearly forest-loss interval of 0.5–0.9 per cent, low to moderate compared to elsewhere in the tropics.

O. Carrero's 1977 vegetation map (scale 1 : 250,000) from the Ministry of Environment and Renewable Natural Resources (MARNR 1982), based on satellite images and aerial photos, is the main tool used in the past assessments of both the FAO (1993, 1997a) and the IUCN's Conservation Atlas for the Americas (Harcourt and Sayer 1996). This put forest area (including gallery forests, mangroves and plantations) at 56,985,121 ha, which is 63.2 per cent of land area (MARNR 1982: table III-I). The FAO-FRA's more recent figures in Table 5.1 (1990: 46,512,000 ha; 1995: 43,995,000 ha)[8] are model-based extrapolations; apart from the map by Huber and Alarcón (1988), two decades passed without a direct assessment.[9] In the meantime, some observers used their own forest definitions and often arbitrary sources of information, such as the FAO's *Production Yearbook* (FAO 1996), MARNR *et al.* (1996) or Matute (1984). But the FRA 1990 figures were widely accepted and used by others, both in Venezuela (e.g. Centeno 1990, 1997) and internationally (e.g. WRI 1992, 1994).

Analysing NOAA-AVHRR images from 1991 to 95, the TREES estimate by Mayaux *et al.* (1998), of 49,250,000 ha evergreen and semi-deciduous forests with at least 70 per cent tree-crown cover, is a significant upward revision of the FAO-FRA's extrapolations. There are two complementary explanations for the discrepancy. First, the FAO had

Table 5.1 Venezuela: forest cover and deforestation estimates

Author	*Forest cover* (in ha)	*Year*	*Annual deforestation* (in ha)	*Relative decline* (%)	*Period*	*Source type*	*Coverage notes*
FAO (1997a) FRA	46,512,000 43,995,000	1990 1995	503,000	1.1	1990–5	Model estimate	Natural forests
FAO (1993) FRA	45,691,000	1990	599,000	1.2	1980–90	Model estimate	Natural forests
FAO (1998)	53,946,000	1977	—	—	—	Satellite, air photos	All forests
Mayaux *et al.* (1998)	49,250,000	1991–95	—	—	—	NOAA-AVHHR satellite	Evergreen and semi-decidous, <70 %
WRI (1992)	48,086,400	1980	245,000 —	0.7 —	1980–5 —	'FAO and other sources'	Forests and other wooded areas
WRI (1994)	51,681,000 45,690,000	1980 1990	245,000 599,000 953,200	0.5 1.2 1.9	1980–5 1985–90 1986–90	FAO etc. FAO[a]	All forests
FAO (1996) Production Yearbook	33,365,000 31,915,000 30,465,000 30,000,000	1979 1984 1989 1994	290,000 290,000 93,000	0.9 0.9 0.3	1979–84 1984–9 1989–94	Forestry agency reporting	Prod. forests + other categ.[b]
Harcourt and Sayer (1996)/ MARNR (1982)	54,268,200	1982	—	—	—	Aerial Photos 1:50,000/ 100,000	Closed broadleaf forests[c]
Myers (1994)	35,000,000	1989	150,000	0.4[a]	1980–8	Unknown	Closed forests
MARNR/MEM/ UNEP/US-CSP (1996)	58,000,000	1980	517,090	0.9	1980–90	Various	All forests, north of Orinoco
Matute(1984)/ MARNR	27,064,145[a]	1978	219,063[a]	0.8[a]	1963–79	Reported deforest. permits	'High and medium vegetation'
Catalán (1993)/ MARNR*	8,189,268 5,248,942	1975 1988	216,000	2.7	1975–88	Satellite and aerial photos[f]	Maracaibo and western plains[d]
Veillon (1976)*	3,962,500 2,673,500	1950 1975	51,560	1.3	1950–75	Satellite and aerial photos[e]	Only western plains
MARNR (1996)*	48,226,621 44,831,817	1982 1995	261,139	0.5	1982–95	Satellite and aerial photos	13 states, complementary to[d]

Notes

a Own calculations using figures indicated in the specific source.

b Production forests + other wooded land + intended reforestation − recreation forests.

c Excludes dry and dry deciduous forests.

d Ten out of twenty-three states with assumed high deforestation, representing 16 per cent of all forest area registered in 1975.

e Scale predominantly 1 : 500,000.

f Scale predominantly 1 : 250,000.

* Regional estimates.

overestimated deforestation at 599,000 ha/yr (1.2 per cent) for 1980–90 and 503,000 ha (1.1 per cent) annually for 1990–5. Second, the more consolidated method employed by TREES (AVHRR images at 1.1 km resolution and Landsat area correction) detected forest fragments in the transition zones that were not counted in 1977. This is particularly important in Venezuela, because savannahs and open forests occupy large areas and have also been subject to marked conversion pressures.

For the other sources, several implicit assumptions are problematic. The FAO itself classified the underlying map from 1977 as solid information, but the extrapolation of that point estimate as of 'low reliability' (FAO 1998). The World Resources Institute adopted the FAO-FRA figures for 1981–90 (e.g. WRI 1994: 307), although previously a lower estimate of 245,000 ha had been published for 1980–5 (WRI 1992); the two are bound to be inconsistent. The FAO *Production Yearbook* (e.g. FAO 1996) based its deforestation estimates on official agencies' land-use reports, but sources and forest definitions are inadequate (see Chapter 1). Forest size (30–33.4 million ha) is much lower, and yearly forest loss is reported to drop to an unlikely 93,000 ha for 1989–94.

Official deforestation estimates by MARNR have generally been much lower than the FAO's, but Table 5.1 also points to marked inconsistencies in the methodology, coverage and results of the Ministry's own studies.[10] For instance, Matute (1984) refers only to 'legal' deforestation, that is, annual deforestation permits granted by the Ministry of Agriculture and Breeding (MAB), which suggest a figure of an average of 219,063 ha of forest loss for 1963–79. Even as a lower boundary, the validity of this is dubious. At the other end, a joint study (MARNR *et al.* 1996) on Venezuelan carbon emissions supports the high-end estimates, with a deforestation figure for the 1980s of 517,090 ha /yr; this even excludes areas south of the Orinoco. Nevertheless, the methodology in this study is not very transparent and seems to overestimate deforestation.[11] Bevilacqua *et al.* (2001: 12–16; 62–3) conclude that it is simply impossible to compare official forest figures over time, due to a series of problems: non-transparent or inconsistent methods, unclear reference dates for estimates and even simple computational errors in the published statistics.

Regional-coverage studies (marked with an asterix in Table 5.1) can give some clues. Veillon (1976) studied the western *llanos*, an area with particularly high agricultural conversion pressures, and found a 1.3 per cent of annual forest loss for 1950–75. For the subsequent 1975–88 period, the MARNR study by Catalán (1993) found twice that high a rate (2.7 per cent) for the *llanos*, indicating an accelerated deforestation after 1975. The analysis also points to a high geographic concentration of pressures: 216,000 ha were lost annually in a zone that represents only 16 per cent of national forest area registered in 1975. This lends some credibility to high-end national deforestation figures for the 1980s, but the coarse resolution probably also overestimated forest clearing.[12] The annex volume to the 'Environmental Status for Venezuela' (MARNR 1996) reproduced the earlier forest-loss figures from Catalán (1993), and supplements them with deforestation estimates for the thirteen additional federal states at 261,139 ha (ibid.: table 2.7-B). But the study periods coincide only partially (Catalán 1993: 1975–88; MARNR 1996: 1982–95), and a closer inspection also reveals serious inconsistencies.[13] In other words, these studies generally indicate an acceleration of deforestation after 1975, but the data are poor and there are many caveats.

In which regions of Venezuela do we currently find strong deforestation pressures? The expert consultation in 1997 (Achard *et al.* 1998) identified four 'hot-spots' (see Map 5.1). First, forest remnants in the Orinoco *llanos* continue to be logged, often followed by agricultural conversion, especially in the east. This borders the second area, the Orinoco delta region, where mangroves are being converted for charcoal in the north, timber and oil concessions are expanding, and gold-mining is being undertaken in the southern part. Third, route 19 to Puerto Ayacucho (Bolívar state) is promoting the agricultural conversion of humid and montane forest. Finally, the Sierra Parima area (Amazonas state), near the Brazilian border, faces little agricultural conversion, but strong degradation pressure from small-scale gold-miners entering from Brazil (*garimpeiros*).

How have deforestation rates changed over time, and do these changes relate systematically to cycles and trends in the oil economy? What seems unambiguous is that, since the start of the oil era in the 1920s, Venezuela has retained a larger amount of forest cover, especially south of the Orinoco, than would have been the case had it not become a quasi mono-exporter of oil. Compared to Gabon, this structural shift just happened earlier. Agriculture and other land-using sectors have experienced a long-term decline, and the drive towards opening up remote forest frontiers has been less pronounced than in neighbouring countries. As will be argued below, the long-term structural-change argument linking high oil exports to low deforestation is clearly confirmed for Venezuela.

It is much less clear how deforestation relates to the short-term boom-and-bust cycles of oil exports: when average deforestation estimates differ by a factor of four, a further decomposition into the relevant oil-cycle sub-periods would seem completely speculative. In the light of these limitations, additional clues may be gained from agricultural survey data on the net expansion of cultivated area (see Figure 5.1). How good a substitute is this in calculating forest loss? First, a large proportion of cultivated area has historically been drawn from non-forest areas, especially pastures established in natural grasslands, but during recent decades most area expansion has been drawn from forests. Second, only actually used or harvested areas are measured in the statistics. By excluding fallow areas, the degree of underlying forest conversion is underestimated. Centeno (1990: 1) calculates that 21.8 million ha of land were under some sort of cultivation in 1988. An additional 12.1 million ha (56 per cent) of fallow and abandoned areas were cleared but without any agricultural use that year.[14] Third, land-use data also have their limitations, as definitions differ between sources. In particular, in 1978 cultivated pastures occupied 76 per cent of cultivated area, but after that year pastures were not surveyed regularly and had to be estimated in Figure 5.1 using cattle-herd statistics. Pastures were estimated for 1978–96 by regression, extrapolating the relationship observed during 1945–77 between the stock of cattle and the size of cultivated and natural pastures.[15] Overall, one may therefore expect agricultural area expansion to provide at least a reasonable lower boundary for the area that has been deforested.

As can be seen, expansion is highly variable from year to year. Pastures expanded in most years and dominated total change, but cropped areas also contracted for certain periods. Specifically during oil booms, it seems that pastures expanded and cropped areas were reduced (see discussion in the section on 'The competitiveness of land-using sectors'). The annual average for pastures and crop expansion combined is 206,000 ha for the entire

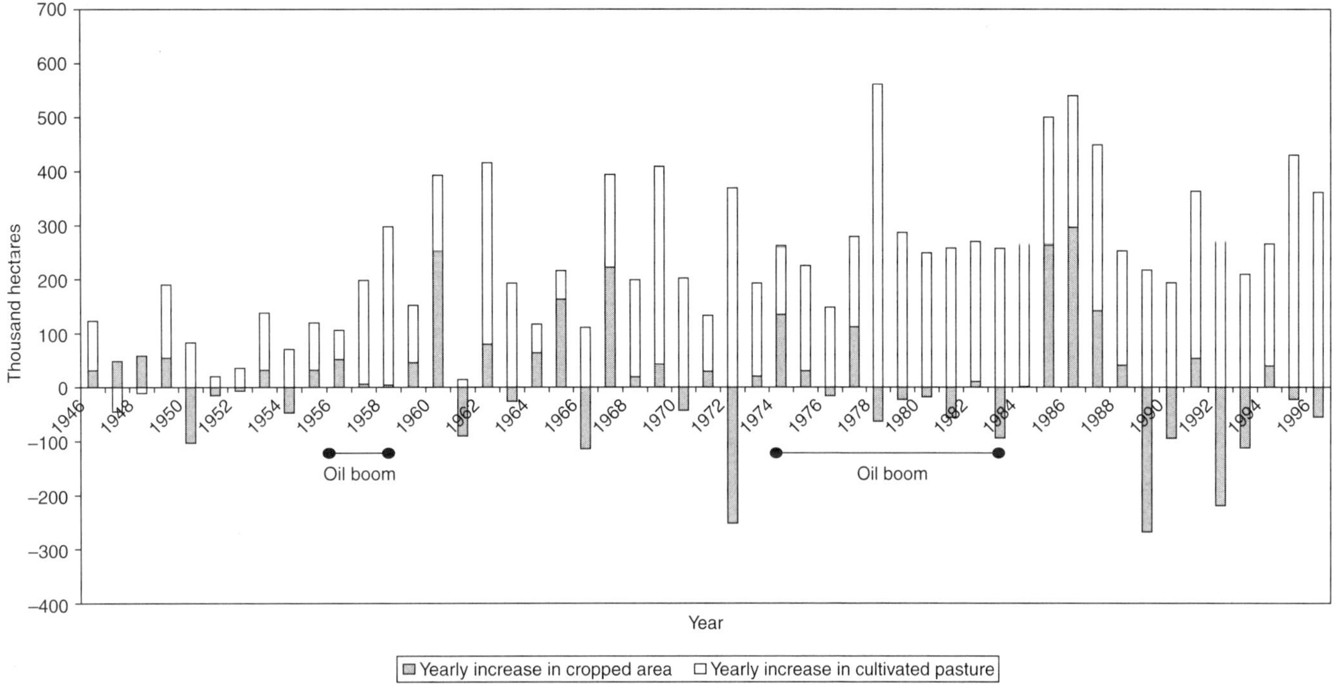

Figure 5.1 Venezuela: yearly change in cultivated area, 1946–96.

Sources: Banco Central (1977); MAB (various years) Anuario Estadístico Agropecuario; Unpublished MAB data.

Note
Pasture expansion after 1977 estimated from change in cattle herds.

period, but with fluctuations and a general acceleration over time. To compare with oil-boom periods, it would be best to look at period averages:

1946–55	Moderate oil revenues	66,000 ha yearly area expansion
1956–58	Oil boom	201,000 ha yearly area expansion
1959–73	Moderate oil revenues	199,000 ha yearly area expansion
1974–83	Oil boom	252,000 ha yearly area expansion
1984–95	Moderate oil revenues	273,000 ha yearly area expansion

From a simple comparison of the five sub-periods, no clear relationship between oil booms and low area expansion is apparent. Actually, during the two boom periods, area expansion seems to have accelerated, remaining high afterwards.[16] The fact that there is no simplistic relationship between oil-boom sub-periods and reduced levels of conversion calls for a more detailed scrutiny of sectors, namely agriculture and logging (see section on 'The competitiveness of land-using sectors'), sector-relevant budgets and policies (section on 'Windfall impacts on government spending') and, in the following, the forest impact of oil and mining proper.

The effect of mineral production on forests

Oil

With proven reserves of 73 billion barrels, Venezuela has the largest oil deposits in the Western Hemisphere (EIU 1999b: 18). Oil activities have not had a direct impact on the forests, because until now the resource has come almost exclusively from non-forested areas. The two main oil-production zones, in and near Lake Maracaibo and the eastern *llanos* (especially Anzoátegui state), have vegetation dominated by savannah and scrub land, supplemented by a smaller proportion of deciduous forests. Furthermore, oil production has been land-intensive (i.e. concentrated on a very reduced surface space), meaning that the impact on vegetation has been limited.

From the 1920s onwards, higher royalties and taxation gradually reduced foreign companies' high oil rents. This culminated in the nationalisation of the oil industry in the 1970s, as a result of which exploration and production activities became rather stagnant for two decades. The recent strategy of renewed openness (*apertura petrolera*) has implied a wave of concessions for exploration and production. Future threats to forests from oil may occur in the Orinoco delta (especially to mangroves) and the middle Orinoco region. Here, the extraction of sizeable heavy crude deposits occurs in areas that, unlike the traditional ones, are largely forested (Harcourt and Sayer 1996: 313). The environmental risks and potential impacts on the indigenous Waroa people (RAN 2002) resemble the notorious case of devastating oil drilling in Nigeria's mangroves (see Chapter 9). Yet, at present it seems unlikely that oil production will become a major driver of Venezuelan deforestation. Activities in the Orinoco delta are still concentrated on exploration rather than production.[17] Second, the adoption of new, more environmentally friendly production technologies reduces deforestation impacts (A. Paolilla, personal communication).

In addition to the direct effects, one may consider the indirect impact of oil production in 'opening up' new areas for settlement and alternative uses, in particular through road construction. Oil roads have been mentioned as a cause of forest conversion.[18] However, these impacts mostly date back to the 1940s and 1950s, and do not play a major role today. Oil development also attracts labour and promotes settlement. The growth of the city of Maracaibo led to rapid forest conversion in the surrounding Zulia state, especially for cattle-ranching. At the beginning of the oil era, Gill (1931) had still been able to refer to 'the great timber region around Lake Maracaibo' (ibid.: 44) and he described the southern end of the lake as 'a region which, because of its remoteness and the presence of hostile Indians, has been little explored. Its forests are continuous' (ibid.: 207). This picture changed radically with the rise of Maracaibo as 'the capital of oil'. Borcherdt (1988: 21) describes forest conversion during the 1960s, at the southern end of the lake and in the Andean foothills, for sugarcane, plantains and pastures devoted to milk production. Pastures became the main destiny of the region's forests in the following decades, as Venturini (1969) and Catalán (1993) note. MARNR (1996: 13) estimates that the forest area of Zulia state was reduced from 3,949,197 ha in 1982 to only 2,244,565 ha in 1995. With a yearly deforestation of 172,659 ha (4.4 per cent) for the period, this actually makes Zulia the main locus of recent deforestation in Venezuela. But except for this indirect, regional effect, oil-related activities are likely to have had a very limited forest impact, in the past as in the present.

Other mining

Unlike oil-drilling, mining gold and diamonds is carried out from alluvial surface deposits in the Guyana shield, on extensive areas that are typically covered by forests. This implies a potentially much larger forest impact from gold (mostly in greenstone belts that are distributed all over Bolívar state) and diamonds (concentrated near the Brazilian border; see Miranda 1998: 15). Gold-mining has played a role for at least a century, but significant expansion occurred only during 1980–8. During the 1990s production has been stagnant, *inter alia* because declining world-market gold prices have made costly production activities in remote areas less attractive (Rodríguez 1994: 24–33). A politically sensitive example of the conflict between forest conservation and mining is the Imataca forest reserve, which is estimated to contain a huge deposit of 11.8 million ounces of gold (Miranda 1998: 15). Under Presidential Decree 1850, dated 28 May 1997, and the new Management Plan, 1.4 million ha (38 per cent of the Reserve's area) have been allocated for current or future mining concessions (Centeno 1997b).

There are numerous direct environmental impacts from small-scale mining. In Venezuela, the most frequently mentioned are mercury pollution, watershed sedimentation, fishery decline, garbage disposal and wildlife decline due to habitat loss and subsistence hunting by miners (Miranda 1998). It is primarily the use of high-pressure water and suction dredges (hydraulic mining) that actually destroys forest cover, both trees and lower vegetation, which washes away the topsoil and seed banks necessary for subsequent forest regeneration. Additional damage results from road construction and soil compactation caused by heavy machinery (Miranda 1998: 18–24). Yet, it was estimated that mining until 1985 had only affected 257,289 ha, which corresponds to 0.28 per cent of Venezuela's land area (Bisbal 1988).

Mining techniques and their derived forest impacts vary with the scale of operation (Müller 1997). Small-scale, artisan gold-mining in and near the rivers produce low impacts;

although the number of miners is large, tree-cover removal is between 0–25 per cent, and should thus rather be regarded as forest degradation (ibid.: 47). Medium-scale, semi-mechanised surface mining of placer deposits is probably most destructive, due to the extensive use of hydraulic mining: tree-cover removal is 75–100 per cent. Industrial mining is carried out mostly by foreign companies using capital-intensive techniques. In August 2001, three industrial mines were operating, two of them open-pit mines. Several others were planned, but uncertainties remain about their future impacts (M. Miranda, personal e-communication, 8 August 2001).

Only scattered attempts have been made to quantify deforestation from gold- and diamond-mining. A study by Barreto and Pérez-Puelles (1991) in a 1.84 million ha study area of the upper Caroní watershed found that about 16,000 people employed directly or indirectly in mining had deforested about 2,000 ha, corresponding to 0.12 per cent of the area's forest.[19] This indicates that forest-clearing is very localised and that degradation of the forested watershed through sedimentation and mercury pollution is more important (Bevilacqua *et al.* 2001: 56). It is thought that between 60,000 and 80,000 people are directly employed in mining in Venezuela (Rodriguez 1994: 32). Estimates of small and medium-scale producers range from 30,000 (Miranda 1998: 16) to 40,000 (Müller 1997: 49) and 60,000 (Harcourt and Sayer 1996: 315). If, as an upper boundary, we assume that each directly employed miner deforests on average half a hectare per year (net of forest regrowth), using Müller's intermediate estimate of 40,000 miners, then 20,000 ha /yr would be lost, corresponding to 5–8 per cent of the likely annual forest loss in Venezuela (see section on 'Deforestation in Venezuela').

The macroeconomic impact of the mineral boom

In 1930, oil already accounted for 83 per cent of Venezuela's exports (Gelb with Bourguignon 1988: 290), and in the 1930s the country temporarily became the world's largest oil exporter. Oil remained largely an enclave with few linkages to the rest of the economy. Oil revenues basically became a rent accruing from abroad, with the petro-state as the prime rentier (Karl 1995). Thus it was in the 1930s that the term 'to sow the oil' (*sembrar el petróleo*) was born. This meant developing strategies to distribute a growing, publicly accruing rent to different sectors and competing interest groups in a balanced way, in order to share welfare gains and sustain growth in the non-oil economy (Heimberger 1994: 74–8). Historically, several specific distributive instruments have been used to reach these objectives. Among these are:[20]

- real currency appreciation (increasing general import capacity)
- high public expenditure (both current consumption and investment)
- low taxation (low tax rates; a large and tolerated degree of tax evasion)
- economy-wide subsidies for production, consumption, credit, etc.
- high income transfers.

The long-term availability and reliance on oil rents was a national privilege, both economically and in creating political stability and broad support for democracy. *Inter alia*, it facilitated the intra-elite *Pacto de Punto Fijo* in coming to a basic consensus involving the two main political parties, the AD (Social Democrats) and COPEI (Christian Democrats). However, oil also produced structural features in the Venezuelan political economy that are less desirable from a productive point of view. Among these were several that we are already familiar with from the Gabon case: the increase in rent-seeking and spread of corruption, a high level of state interventionism and centralisation of power, an unclear separation of the public and private sectors, and a lack of incentives to correct inefficiencies in the structure of production. This all amounts to what has been called 'a more fundamental opposition between production and rent appropriation' (Coronil 1997: 287). As Karl has expressed it:

> Oil underwrote pacts that rested on the capacity to grant extensive state favors, contracts, and infrastructure to entrepreneurs while charging the lowest taxes on the continent, permitting some of the highest profits and supporting the highest wages, price controls and food subsidies in Latin America. In short, the petro-state made it possible to have a democracy with very few losers.
>
> (Karl 1995: 33)

Obviously, high dependence on oil also implied vulnerability *vis-à-vis* fluctuations in oil prices. In 1956, Venezuela experienced its first boom-and-bust cycle. This exposed patterns that would be repeated later: a short-lived boom caused by the Suez crisis increased fiscal revenues, and a public-spending spree occurred shortly afterwards. When the boom ended, fiscal cutbacks were asymmetrical and slow, making it necessary to adopt an adjustment package that included a cumulative devaluation of 35 per cent until 1964 (Hausmann 1990: 5).

From 1960 to 1972, real oil exports remained stagnant, and even showed a fall in per-capita terms (García *et al.* 1997 and Figure 5.2). Compared to this, the size and impact of the first large oil boom in 1973–6 was huge. The value of petroleum exports increased fourfold, from US$2.9 billion in 1972 to US$4.3 billion in 1973 and US$10.5 billion in 1974 (IMF 1996). During 1972–8, non-oil GDP rose at a yearly rate of 8.4 per cent, private consumption 12 per cent and gross investment 15 per cent (Gelb with Bourguignon 1988: 289). The boom was perceived by most actors as a permanent rather than a temporary price hike (Zambrano 1995: 59). The administration of Carlos Andrés Pérez (1974–9) was initially inclined to save most boom inflows abroad through the Venezuelan Investment Fund (FIV), until domestic investment opportunities gradually developed (Hausmann 1990: 5). However, after 1976 internal pressures to increase public spending could no longer be resisted, and the public sector even became a net foreign borrower (Gelb with Bourguignon 1988: 297).

It was believed that by 'sowing' oil wealth in strategic sectors, Venezuela could industrialise rapidly and create a 'Great Venezuela' that would enter the circle of developed nations.[21] Funding was made available from the mid-1970s for ambitious industrialisation projects (large investments in iron ore, steel, aluminium and electricity). Pérez' electoral promises of social expenditure and increasing public employment and wages suddenly became affordable. Oil wealth also fuelled a nationalistic drive towards more aggressive policies *vis-à-vis* multinational companies, but an over-assessment of the country's own

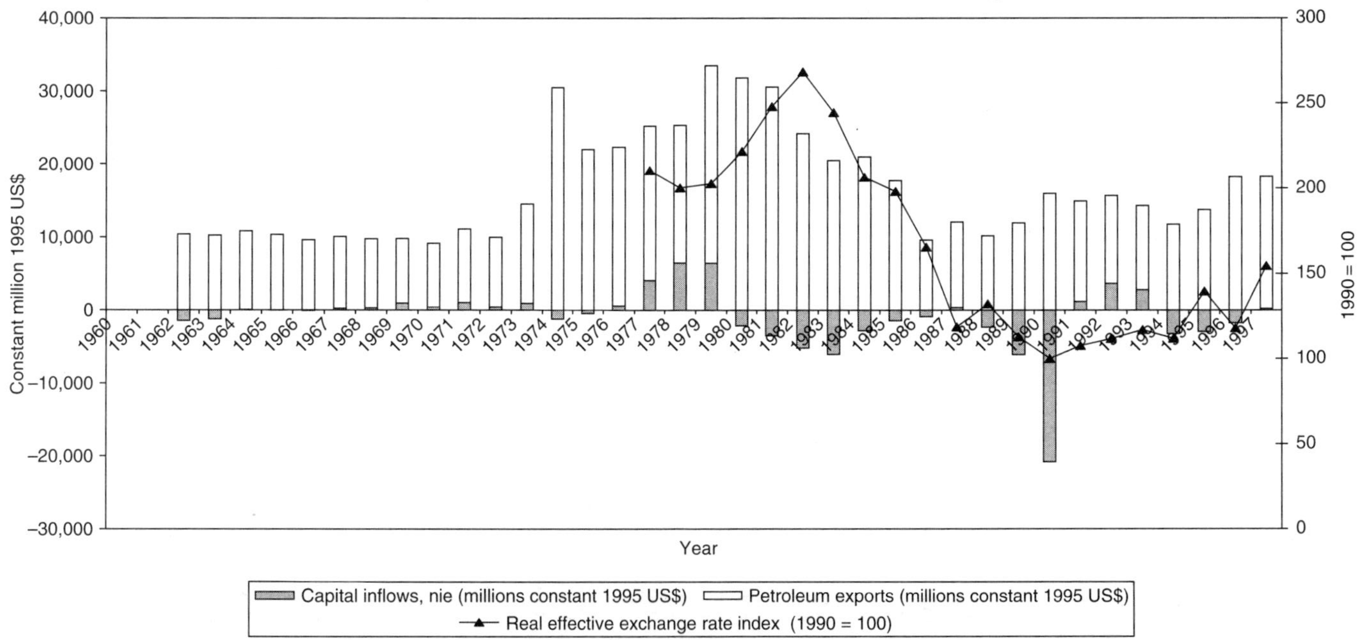

Figure 5.2 Venezuela: capital inflows, petroleum exports and real effective exchange rate, 1962–98.

Sources: IMF (1990, 1992, 1999) World Bank (1999a).

Notes

1 Real effective exchange rate, 1977–8: 1990 weights; 1979: 1985 weights; 1980–97: 1990 weights.

2 Capital inflows, not including exceptional financing, 1962–91: other capital, total, nie, 1992–8; financial account, nie.

3 Petroleum exports, 1962–98.

capacity and bargaining power led to considerable resources being wasted.[22] The use of the first oil windfall (1974–8) was thus characterised by traditional distributive mechanisms (see above), but also by new features, which included:

- large capital-intensive parastatals engaged in non-oil production
- growing public wages and employment
- subsidies (energy, food, etc.) and transfers to the private sector
- infrastructure investments (extension of Guri dam, Caracas metro, etc.).

Venezuela's second oil-price boom (1979–82) was about equal in size to its first. However, rather than accelerating economic growth, it caused stagflation. The combination of two oil booms, excessive foreign borrowing, distortions caused by increasing state intervention and overwhelmingly inefficient investments in 'white elephants'[23] implied that the country was unable to produce a supply response that could match overheated demand. GDP growth turned negative in 1980 (−2 per cent), and inflation rose from 2.8 per cent in 1972 to 12.4 per cent in 1979 and 21.5 per cent in 1980 (García *et al.* 1997: 16). Figure 5.2 summarises the role of foreign exchange inflows, capital inflows and relative prices (RP) in this picture.

Real oil receipts peaked in 1974 and again 1979–81, but then started to decline markedly until 1986, before reaching a higher and more stable level throughout the 1990s. Foreign borrowing occurred mostly during 1977–9, but was not as important as in other countries in this book. There were financial capital outflows during the 1980s, which became more accentuated in 1990–1. Accumulated capital flight from Venezuela is estimated at US$60–90 billion, or two to three times the country's foreign debt (Coronil 1997: 382). The RER appreciated significantly in the 1970s, but then took a dramatic 60 per cent slide throughout the 1980s. After a full decade of nominal exchange-rate stability (1973–83), the *bolívar* was devalued sharply from US$4.30 to US$9.03 (1984) and US$15.34 (1986).[24]

During the 1980s, the economic policy responses of the administrations of Herrera Campins (1979–84) and Lusinchi (1984–9) were generally inadequate. Hausmann (1990) notes the unexpected nature and unfortunate timing of the oil-price bust in 1986. The government was committed to fiscal expansion, and thus chose to delay politically costly economic adjustment. Foreign borrowing became a cushion for public accounts, thus avoiding cuts (Karl 1995). The shift from unified to multiple exchange rates proved to be a new source of rent-seeking, with actors competing for the largest quota of the cheapest available rate (ibid.: 45).

When Pérez was elected for a second term in 1989, the disastrous state of the economy finally led him to adopt a structural adjustment programme (SAP), following the suggestions of the IMF. The SAP involved cuts in public employment and wage freezes, the elimination of price and interest regulations and of a range of subsidies, and the unification of multiple exchange rates. Foreign exchange controls were eased and tariffs reduced, and financial assistance from the World Bank and the IMF was agreed upon (CENDES 1995: 2). However, fiscal reform was delayed (Zambrano 1995: 64), and the discrepancy between Pérez' populist programme and the proposed neo-liberal adjustment package, known as his 'big turnaround' (*el gran viraje*), proved to be politically unsustainable in a country that had not experienced such measures in the postwar period. The three subsequent years saw severe riots and violent military repression, two aborted military coups and the fall of Pérez due to corruption charges (Smith and McCoy 1995: 6). The SAP was eventually abandoned in 1993, but the broad social consensus of the postwar period had disappeared.

After a severe 8.6 per cent fall in GDP in 1989, high oil prices brought a short-lived expansion, with an average growth of 7.5 per cent during 1990–2 (García *et al.* 1997: 17). But in 1993–4, this was abruptly put to an end by a severe banking crisis. The government chose to bail out thirteen failed banks and their depositors at a cost of US$8 billion, corresponding to no less than 14 per cent of GDP (Smith and McCoy 1995: 9). The fact that this huge transfer was implemented, favouring mostly the well-to-do, illustrates the power of vested interests. It also shows the inability of the Venezuelan state to adjust in a timely, disciplined, socially balanced manner to the emerging economic crisis.

The Caldera administration (1994–9), elected on a populist base outside the two dominant parties, returned initially to a traditional interventionist economic programme, but was forced by economic hardship to sign a new deal with the IMF in 1996. As oil continues to account for about three-quarters of all exports, the country's fate in the late 1990s was closely linked to the marked fluctuations in the oil price, with moderately low prices in 1994–5, rising prices in 1996–7, record low prices in 1998–9, and a price boom in 1999–2001 under the new Chávez administration. RERs have fluctuated with oil revenues. The introduction of a crawling-peg band *vis-à-vis* the US$ in July 1996 was intended to stabilise the real value of the *bolívar*, but there has been a significant real appreciation of the currency in the last few years (EIU 1999b: 40).

On the whole, policies accompanying the two oil bonanzas of the 1970s and early 1980s were 'far from optimal' (Gelb with Bourguignon 1988: 289), and the adjustments to the busts of the 1980s and 1990s were delayed until their costs were almost unbearable. Real GDP per capita, which had grown steadily during 1950–77, was not only lower by 1996 than before the first boom, it had even fallen back to its 1963 level. Like Gabon, Venezuela also reached a development stage where even an oil bonanza could not fully make the economic crisis vanish. Already by the end of the 1970s, the huge waste of financial resources and the increasing corruption had become apparent. Instead of being called 'black gold', oil was now increasingly referred to as 'the devil's excrement'.[25] A root cause of this calamitous outcome was the public sector's poor performance as a guardian of oil windfalls. The caveats of fiscal policy are central in this regard. Basically no non-oil tax base had been created, and an almost uninterrupted fiscal deficit was maintained for two decades. There was an ever-growing public debt; debt service and state investment commitments crowded out essential social expenditure, and pro-cyclical spending tended to reinforce external fluctuations instead of stabilising them (García *et al.* 1997). The deep Venezuelan post-1979 crisis is thus not a technical matter of policy fixes: its roots are located in the political economy sphere.

The competitiveness of land-using sectors

Agriculture

Most analysts agree that agriculture and cattle-ranching are the principal causes of forest-cover loss in Venezuela (Borcherdt 1985; Catalán 1993; Rojas 1993; Gómez 1995; Infante

1995; Centeno 1997a). Centeno (1997a: 3) estimates that, during the 1980s, about three-quarters of forest loss could be attributed to agricultural conversion. This pressure is particularly pronounced in the forest reserves north of the Orinoco, where population growth, market access and soil richness have increased the opportunity costs of forest conservation, both historically and today, and have led to widespread conversion (Rojas 1993: 20–1). Ample case-study evidence supports the hypothesis that crops and cattle are the overwhelming causes of deforestation.[26] However, in Venezuela agricultural expansion has not followed a deterministic logic of impoverished peasants striving for survival by extending slash-and-burn agriculture. Rather, the availability of public forested land for private appropriation ('homesteading') in areas gradually made accessible by more and better roads has enabled private entrepreneurs to push forward land colonisation, especially for cattle-ranching. Subsequent use of converted land is often extremely wasteful (Centeno 1990: 1; Rojas 1993: 30).[27] The logic of converting forest is well captured by a quote from Haase for Sucre state: 'The producers define forests … mainly as non-productive areas that can be made productive by deforestation and transformation into agricultural areas' (Haase 1997: 19, my translation from the Spanish).

How have different agricultural sectors fared during the oil booms and busts, and to what extent did Dutch Disease constraints on their development protect forests? The trade-policy section below will show that agriculture as a whole was a semi-traded sector; given an export value of 3 per cent, agricultural exports (coffee, rice, etc.) are minuscule, and import-competing sub-sectors were not fully and continuously exposed to international trade. However, as the present section will demonstrate, even this level of exposure was sufficient for agriculture to be hit by the Dutch Disease, although production trends and land demand differed considerably between the crop and pasture sub-sectors: crops were subject to a decline, while cattle-ranching continued to expand.

Returning to the land-use data given in the section on 'Deforestation in Venezuela', national cultivated area expanded 431 per cent from 1945 to 1996 (Figure 5.3), from 2,397,000 ha (1,011,000 ha crops; 1,386,000 ha pasture) to 12,717,000 ha (1,654,000 ha crops; 11,063,000 ha pasture).[28] Hence, cropped areas grew by 64 per cent (strongest in cereals and fruits), but pastureland by no less than 994 per cent. Pastureland accounted for no less than 94 per cent of the rise in cultivated area. In other words, Venezuelan deforestation in the postwar period mostly took the form of the conversion of forests to extensive pastureland, often passing through an intermediate crop stage of 'nutrient mining', that is, of capitalising on soil fertility after burning (Centeno 1995: 5).

An appreciated currency, combined with trade protection and a range of subsidies, provided a strong incentive to import machinery and inputs, favouring the rise of a large-scale, capital-intensive agro-industrial sub-sector that developed few linkages with the small-scale *campesino* economy (Borcherdt 1985). One difference from the other cases in this book is that the crowding out of agriculture started very early. From 1937 to 1958, the share of imported foodstuffs already grew from 9.8 to 45 per cent (Esteves, cited in Borcherdt 1985: 86). Historically, sharply rising rural wages in particular drove up the costs of production, and the rural exodus increased because of well-remunerated jobs in the urban non-traded sectors. This impact was only partially dampened by Colombian immigrants working in Venezuelan agriculture.

With the crisis from the mid-1980s onwards, things have changed somewhat. A fall in the RER and in real rural wages worked in favour of agriculture. Cropped area expanded

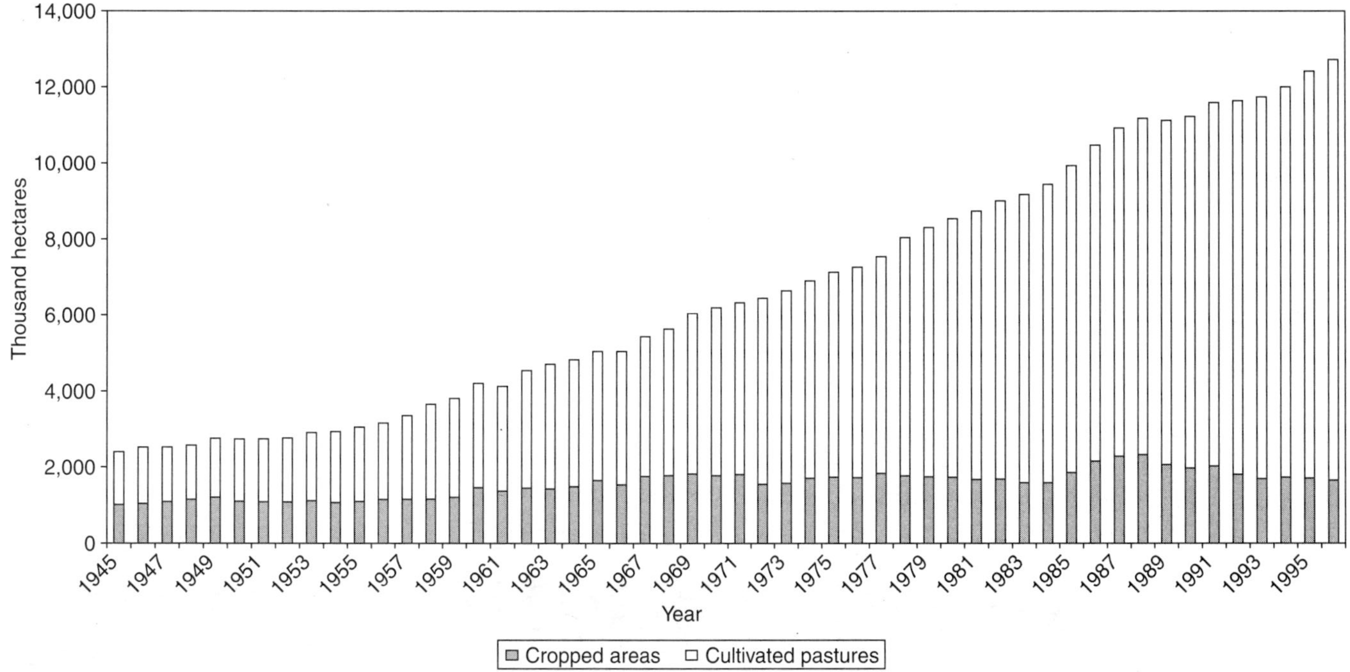

Figure 5.3 Venezuela: cultivated area, disaggregated into crops and pasture, 1945–96.

Sources: Banco Central (1977); MAB (various years) Anuario Estadístico Agropecuario; Unpublished MAB data.

Note
Pasture expansion after 1977 estimated from change in cattle herds.

from 1,591,000 ha in 1984 to 2,023,000 ha in 1991 (27 per cent). As shown in Figure 5.4, real agricultural value-added grew remarkably in the 1960s, but then stagnated during the oil boom. Post-boom production increased in value remarkably again (1985–9), coinciding with the real currency depreciation, followed by some fluctuations in the 1990s. In 1982, MAB and MARNR jointly targeted an increase in cultivated areas from 24.2 million ha in 1980 to 38.7 million ha in 2000. This 59.1 per cent increase would have corresponded to an agricultural surface expansion of 14.5 million ha (794,444 ha/yr), of which 13.5 million ha were for pasture and 1 million ha for food crops. The target was not achieved at all, but it is remarkable that it received the support of international agencies such as UNEP and CEPAL, although it was clear that most of this expansion would have to be achieved by deforestation. MARNR's report from 1982 concludes quite frankly: 'It is evident that we must sacrifice part of our natural vegetation to put the soil to productive use'.[29]

Forestry

At the start of the 1930s, Gill (1931) praised the 'great, untapped reservoir of forest wealth' in Venezuela (ibid.: 207), and predicted that '[i]t seems inevitable that the lumber industry [in Venezuela and Colombia] will become one of the important sources of national revenue as soon as world demand seeks out tropical species' (ibid.: 44). However, Venezuelan forestry never fulfilled these optimistic predictions. As for agriculture, growing oil incomes, a strong *bolívar*, high production costs and a lack of competitiveness played a major role in the lagging development of forestry, encouraging growing imports of a wide range of wood products, in spite of the country's own rich natural resources. This was particularly pronounced during the oil boom. From 1975 to 1982, national policies actively discouraged wood production from deforested areas. Coupled with RER appreciation, by 1978 almost half of the consumption of processed wood products was imported. Wood exports never took off. Only 1,000 m^3 of roundwood were exported in 1996, demonstrating the sector's almost exclusive home market orientation (Centeno 1995: 15). With the economic crisis since the mid-1980s, significant import substitution has occurred, but in the 1990s the value of wood imports still exceeded that of exports for every single year, the export value of all forest products being a meagre US$101 million in 1999 (FAO 2000b).

In other words, RER appreciation has both hampered the long-term development of the sector and created fluctuations during certain periods. How much forest has been affected by timber harvesting? Unfortunately, the size of allocated concessions, which rose remarkably in the 1990s, is a poor indicator. The majority of concessions are still unused, especially south of the Orinoco; they constitute a claim on future exploitation rather than a present production activity. It is thus best to stick to timber production figures. Figure 5.4 above shows the development of wood production over the past four decades as recorded in FAO (2000b). The main categories in 1999 were saw- and veneer-logs, paper and paperboard and sawnwood, but the ranking between them has changed notably over time, so data in Figure 5.4 have been aggregated to their roundwood equivalent. This aggregate shows a typical Dutch Disease pattern: there was a notable production expansion in the pre-boom period (1960–73), a slow-down and fall in production during the oil boom (1974–83), a sustained rise after the boom (1984–92), but a less favourable development in the rest of the 1990s. The latter is mainly due to a decline in the category of saw- and veneer-logs.

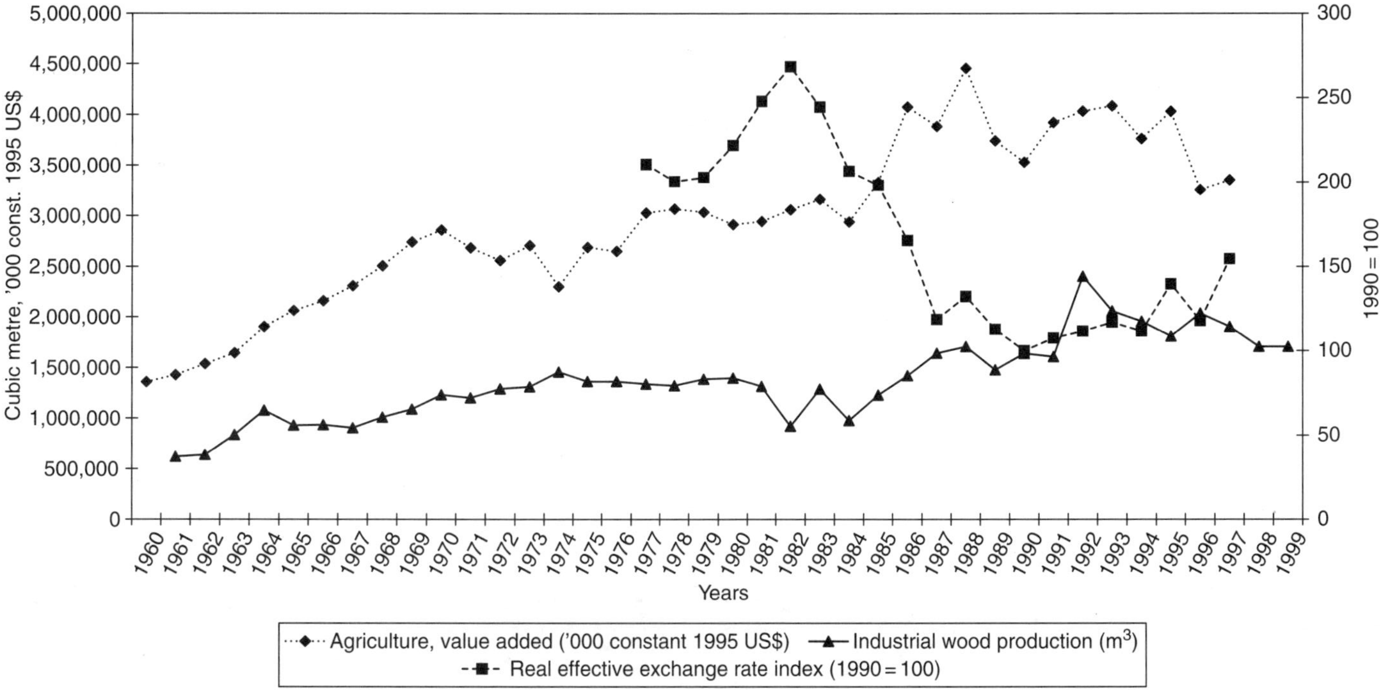

Figure 5.4 Venezuela: industrial wood production, agricultural value-added and real effective exchange rate, 1960–99.

Sources: FAOSTAT (2001, at http://apps.fao.org), World Bank (1999a). For RER, see Figure 5.1.

Note

1 Industrial wood production in roundwood equivalent (Industrial roundwood: 1; Plywood: 2.3; Sawnwood: 1.82; Veneer sheet: 1.9).

Where does the wood come from? In the 1990s, some production started to come from plantations, 90 per cent of which is Caribbean pine. There are two sources of wood production from natural forests: first, private concessions granted in public Forest Reserves (and other production forests with management plans); and second, privately owned forests, including annual deforestation permits, a source that declined greatly in the 1990s. Most of the rise in production from 1982 to 1995 came from concessions. In 1983, forest concessions accounted for 17 per cent of total supply (MARNR 1995: 29); by 1995 the share had risen to 63 per cent (MARNR 1996: 11).

In 1998, the shift towards long-term concessions was accompanied by a restructuring in which the Venezuelan Forest Service (SEFORVEN) was replaced by the Forest Department under MARNR, which was renamed Ministerio del Ambiente y de los Recursos Naturales (MARN). The move also underlined the importance of industrial forestry in policy (MARNR 1996: 12). Nevertheless, critical observers note that the dramatic increase in forest concessions (from about 500,000 ha in 1986 to 2,700,000 in 1992) has not been accompanied by more sustainable practices on the ground. In principle, concessionaires need to present management plans every five years and cutting plans every year (including the replanting of valuable species and the setting aside of conservation areas); rotation periods of 20–40 years are assumed to ensure sustainable yields (Miranda 1998: 12–13). In practice, however, logging firms find ways to accelerate harvesting and avoid replanting, thanks especially to the Forest Department's limited presence in the field.

For many years, forest concessions (typically 80,000–160,000 ha in size) produced extremely low royalties. For the main species, they had been set at a fixed *bolívar* value corresponding in 1982 to US$16/m^3, but in the following decades the real value was totally undermined by high inflation rates, so that Venezuelan logging fees were among the lowest in the tropics (Centeno 1995: 11–13; Miranda 1998: 34–5). In 1999, laws were changed to raise public revenues to about US$12/m^3 and to keep that value indexed to inflation. Thus, the development was very similar to what occurred in Gabon (Chapter 4): only when oil revenues declined and signs of economic crisis appeared did the government start to worry about seriously capturing timber rents.

The impact of logging operations in forest reserves north of the Orinoco has been both to degrade forests and to provide access for deforestation. The reserves to the north represent homogeneous forests with high densities of commercial timber and good agricultural soils that are also close to markets. Here, timber extraction thus had both direct degradation and strong indirect impacts that facilitated the progressive agricultural conversion of forest reserves such as those at Turén, Caparo and Ticoporo (Rojas 1993; Gutiérrez 1996; Veillon 1997: 40–53).[30] Agriculture and especially cattle-ranching were the driving forces, with repeated forest burning for grazing, leading to lasting changes in vegetation cover (Veillon 1997). Even within the former SEFORVEN, one finds the view expressed that the remaining forest reserves are still 'unrealistically large' and will necessarily be converted, though this should be 'planned' instead of 'spontaneous' (Gómez 1995).

South of the Orinoco, in the rapidly expanding forest concessions of Bolívar state, the density of commercial species is low, markets are distant, and harvesting is highly selective.[31] Silvicultural techniques and sawmill-processing minimise investments and are highly rudimentary; up to 30–40 per cent of forest cover can be damaged during extraction, while up to half the wood may be wasted during sawmill-processing (Centeno 1995: 8–13;

Miranda 1998: 14). Recent research shows that endemic biodiversity (e.g. bats and bird species) in logged-over forests is greatly reduced, and that regeneration of commercial tree species may be deficient (M. Miranda, personal communication, June 2001). However, these direct impacts should be seen as degradation rather than deforestation. With some exceptions, notably the conversion in the Imataca reserve, the indirect impacts of opening up areas for squatter settlement are still limited in scale by the predominance of low soil fertility and remoteness from the centres of economic development.

Trade policy impacts

It is doubtful whether any commodity in Venezuela can genuinely be called 'tradable'. State interventions in the economy have been far-reaching, with shifting degrees of import protection and oil rents subsidising various non-oil sectors on a discontinuous basis, according to shifting political regimes, alliances and favours. This makes it very difficult to judge *ex ante* what products and sectors are hurt how much and for how long by declining competitiveness.

Trade protection and subsidies to agriculture, and partly to the forestry sector, began quite early in the twentieth century, as a reaction to the 1929 global economic crisis (E. Cabrera, personal communication). From 1937 to 1958, the value of agricultural exports (in million *bolívares*, at 1957 prices) fell from 254.8 to 194.5, while production for internal food consumption rose from 802.5 to 1327.5 (Aranda 1984: 165–7). This shift towards an inward-looking agricultural sector was made possible by both market and policy factors. Quantitative restrictions and tariffs on food imports, floor prices, guaranteed sale volumes and a range of government subsidies (for credits, infrastructure, transport, research and development and technical assistance) helped sustain high agricultural growth rates (see next section).

During the oil boom, agriculture received higher subsidies that were financed directly out of oil revenues. Trade protection thus tended to cushion the decline of agriculture, which would otherwise have been even more pronounced. On the other hand, some products were subjected to trade liberalisation to help control inflation. Meat import restrictions are a good example of 'stop–go' policies that also had important implications for land use. In 1971–3, a restrictive regime held meat imports at insignificant levels (800–1,000 t). During 1974–8, trade liberalisation helped raise imports by a factor of eighty to 83,071 t. They remained high until the end of the boom, but were then reduced drastically in 1984–6, to 5119, 2086 and 93 t, respectively. Subsequently imports have remained low, except for short-lived periods of liberalisation in 1987–8 and 1992–3.

This example also illustrates that the first policy response to the economic crisis of the 1980s was an increase in protectionism and interventionism. The Lusinchi administration (1984–9) introduced a combination of preferential exchange rates, import and export restrictions, consumer price controls, higher producer prices and subsidies. In 1988, the second government of Carlos Andrés Pérez liberalised agricultural trade and removed most food subsidies as part of the SAP, but price controls for cereals, flour, sugar, meat, eggs, fruit and vegetables were later reintroduced. In 1994, when the Caldera administration took office, import controls rose again, though state monopolies in the export of coffee and cocoa were abandoned (Josling 1997: 19–22). The new Chávez administration reintroduced the protection of maize, palm oil, sugar, fishing and forestry, in the hope of expanding import substitution (EIU 1999b: 29).

Recent trade policies have thus been contradictory, over time and across products, ranging from half-hearted liberalisation efforts to undisguised protectionism. Consequently, it is also difficult to determine the net production and deforestation impact of trade policies at the sector level. Even today, after renewed protectionism, 64 per cent of all food was imported in 1998 (EIU 1999b: 28). Import substitution has therefore not been very successful. Yet protectionist policies made most agricultural sectors semi-traded, that is, partly protected from the long-term loss of competitiveness. There might also be a suspicion that the shift from traditional export crops (coffee, cocoa) to some protected home-market commodities (grains, cattle-ranching) may have caused additional deforestation, as these protected sectors are more land-extensive than traditional export sectors. On the whole, trade protectionism thus cushioned Dutch Disease effects at the margin and accelerated deforestation, especially in the form of clearing land for cattle-ranching.

A quantitative view

Agriculture and forestry have both been 'semi-traded' in Venezuela, that is, partly protected by trade policy. Nevertheless, this has not been enough to protect them from Dutch Disease impacts. Real currency appreciation has been a major impediment to the expansion of semi-traded sectors using land, which has also helped significantly to preserve a large amount of forest cover, though with the major exception of cattle-ranching. Only recently, with the severe crisis from the end of the 1980s and the sequence of mini-booms and -busts in the late 1990s, have there been any signs of expansion in forestry and agriculture. The intention of Table 5.2 is to test our key hypotheses over the last two decades (1977–97) through regression analysis. The equation in row 1 confirms, as expected, that the RER is positively correlated with oil export revenues (at 1 per cent significance level); the coefficient for capital inflows also has the expected positive sign, but is not significant. The latter probably reflects the fact that foreign borrowing only had a sizeable impact on the economy during limited sub-periods. Still, the two-variable equation explains almost two-thirds of the variation in the RER ($R^2 = 63$ per cent).

How did RP affect production quantities? The equation in row 2 shows that, in spite of the semi-traded status of agriculture, the growth of agricultural value-added was highly dependent on a depreciated RER, with a negative coefficient that is significant at the 1 per cent level ($R^2 = 43$ per cent). This means that trade protection and other interventions were insufficient to 'cure' Venezuelan agriculture from Dutch Disease. Did this also affect the speed of agricultural area expansion? Regressions 3 and 4 clearly answer this question in the affirmative. As most of the expanded areas during this period had to be taken from forests, they also confirm the core mechanism: oil-led decline in competitiveness was a powerful protector of Venezuelan forests.

The equation in row 3 investigates cropped area, the minor component in total cultivated area (13 per cent). Again, RER appreciation had a dampening effect on crop expansion, which was highly significant (1 per cent level). As a control variable, non-agricultural GDP was included here to test the extent to which area expansion depended on increasing urban demand. The negative coefficient (significant at 1 per cent level) reveals that this

Table 5.2 Venezuela: relating oil wealth to relative prices and traded sector production. Regression results, 1977–97

Dependent / independent	*Petroleum exports (constant million 1995 US$)*	*Capital inflows (constant million 1995 US$)*	*Real effective exchange rate (1990 = 100)*	*Non-agricultural GDP (constant million 1995 US$)*	R^2	*F-value*	*Years*
1 Real effective exchange rate (1990 = 100)							
Coefficient	0.006496949	0.000886184			0.63	15.17	1977–97
T-value	5.44283829***	0.649275691					
2 Agricultural value added (Constant 1995US$)							
Coefficient			−6.662787846		0.57	25.04	1977–97
T-value			−5.0040795***				
3 Cropped area ('000 ha)							
Coefficient			−4.24835	−0.02629255	0.57	11.36	1977–96
T-value			−4.74487***	−3.67858***			
4 Cultivated pasture ('000 ha)							
Coefficient			−13.1352	0.100768882	0.74	24.23	1977–96
T-value			−2.70763**	2.60209**			
5 Industrial wood production (m^3)							
Coefficient			−5291.357801		0.60	28.66	1977–97
T-value			−5.3540370***				
6 Industrial wood production (m^3)							
Coefficient			−2459.34847	30.88838214	0.83	43.85	1977–97
T-value			−2.7974927**	4.912242***			

Notes

** Parameter *T*-value significant at the 5 per cent level.

*** Parameter *T*-value significant at the 1 per cent level.

was *not* the case. There was a clear negative interaction between crop expansion and urban growth, which indicates the dominant impact of third urban-related variables (urban labour migration, urban investments, etc.) that competed with crop production.[32] More than half of the variation in cropped area is explained (R^2 = 57 per cent).

The equation in row 4 looks at the same equation for cultivated pasture, which generates half of agricultural production value but makes up no less than 87 per cent of total cultivated area and 94 per cent of its expansion since 1945: pasture expansion out-competed crop expansion by a factor of seventeen. Somewhat surprisingly, the RER also had a negative impact on pasture expansion. However, the coefficient is less significant (5 per cent level) than in the equation in row 3. Its size is about three times that of the corresponding coefficient in the equation in row 3: a one percentage-point real-currency depreciation triggers, *ceteris paribus*, a 13,125 ha increase in pasture and a 4,284 ha increase in cropped area. Yet, note that crop expansion is *relatively more sensitive* than pasture size to changing competitiveness. An interesting result is that pasture growth is positively correlated with non-agricultural GDP (significant at 5 per cent level), that is, just opposite the sign for cropped area in the equation in row 3. This reflects the fact that, unlike crop production, cattle expansion was closely linked to growth in urban demand for meat and dairy products. Oil bonanzas and booming urban sectors would thus clearly stimulate a rising demand for pastureland (see section on 'Structural changes in income and demand'). Almost three-quarters of the variation in pasture area is explained by the equation in row 4 (R^2 = 74 per cent).

Another semi-traded sector with land-use implications is timber. The equation in row 5 confirms that, over the last two decades, the total quantities of industrial wood produced in Venezuela (aggregated in roundwood equivalents) were also clearly sensitive to RER changes. The coefficient is significant at the 1 per cent level and the equation explains the bulk of variation in wood production (R^2 = 60 per cent). However, introducing non-agricultural GDP in equation in row 6 makes this new variable significant at the 1 per cent level, and explanatory power is much improved (R^2 = 83 per cent). Simultaneously, the numerical size of the competitiveness parameter is now cut to less than half, and its significance is reduced to a 5 per cent level. This shows that, as for cattle-ranching but unlike agricultural crops, semi-sheltered timber production was positively linked to growth in urban income during the oil bonanza, especially during the boom in urban construction that greatly increased the demand for wood products.

On the whole, declining competitiveness from oil revenues played a key role in retarding the expansion of cropped area; it also cushioned logging pressures, but it did little to restrict cattle-ranching, which was mainly stimulated by the rise in urban incomes. At present, new pressures on natural resources can be observed in response to the economic crisis. State policies try to attract mining and other investment from larger companies, mainly for the large forested areas south of the Orinoco (Aicher *et al*. 1998: 4–5). A recent National Development Plan (1996–9) had proposed an expansion of timber concessions to almost 12 million ha, but was scrapped by the Chávez government. Agricultural sectors like rice and cocoa, as well as paper and pulp industries, were recently pointed out as areas with an explicit comparative advantage for exports, suitable for future expansion (Enright and Saavedra 1995: 15–16). The quest for diversification, agricultural expansion and alternatives to over-reliance on oil is thus still an incipient feature, but it has probably come to stay. If oil prices continue to be volatile, Venezuela may gradually come to resemble its

Latin American neighbours more in future decades, with their greater emphasis on land-extensive, natural resource-dependent sectors.

Windfall impacts on government spending

Agriculture

In spite of being hit by relative price deterioration, as shown in the previous section agriculture actually received an important allocation of windfall revenues during the oil boom. An intricate system of producer and consumer subsidies ensured profitability in production, in spite of food-price controls. The banks were forced to reserve a fixed share of credit volume for agriculture (Borcherdt 1985: 89). In addition, new financial institutions were created for the agricultural sector. Funding was provided to support the two main export sectors (coffee and cocoa), but more important was a series of new, fast-disbursing institutions for the entire agricultural sector. For instance, the *Fondo de Desarrollo Agropecuario* was created in 1974, and spent 2,167 million *bolívares* (US$503 million) in only one and a half years. As shown in Table 5.3, this was supplemented by a steadily growing volume of subsidised credits from other institutions during 1973–5, with a total of 5,780 million *bolívares* (US$1,344 million) (Aranda 1984: 255–64). Most credits were absorbed by capital-intensive, entrepreneurial agriculture, whereas small-scale *campesinos* received less institutional support (Borcherdt 1985). During the 1980s, a large share of agricultural credit continued to go into capital-intensive establishments, especially for irrigation projects and mechanisation. With the 1989 adjustment programme, the MAB budget, dominated by expenditure on state-owned agro-industrial enterprises, was cut by 46 per cent (Blarcom *et al.* 1993).

A fundamental question is to what extent windfall funding allocated to agricultural agencies actually functioned as an effective incentive for increased production. As in Gabon, much funding was absorbed by 'white elephants' and rent-seekers, and thus did not act as land-use incentives. First, some resources were deviated: 'corruption, in its myriad forms, had grown to match the dimensions of the state's expanded income' (Coronil 1997: 357). Second, a large share of spending went into inefficient state enterprises. For instance, Blarcom *et al.* (1993: 9) noted that 51 per cent of the MAB's investment budget was devoted to the construction of the FANATRACTO tractor factory in Ciudad Bolívar. Due to disagreements with multilateral co-investors, it did not produce a single tractor before it was finally abandoned (Coronil 1997: ch. 7). Thus 'the state could break commitments,

Table 5.3 Venezuela: yearly disbursement of agricultural credit (million of current *bolívares*) during the first oil boom (1973–5)

Institution / Year	*1973*	*1974*	*1975*	*Total*
Instituto de Crédito Agrícola y Pecuario	366.8	581.4	950.8	1899.0
Banco de Desarrollo Agropecuario	368.5	511.1	1287.1	2166.7
Fondo de Crédito Agropecuario	—	64.8	1649.4	1714.2
Total	735.3	1157.3	3887.3	5779.9

Source: Aranda (1984: 264).

violate agreements, and waste investments and ... its decisions could find social acceptance' (ibid.: 320). The trade policy and subsidy nexus was probably more important in expanding (or maintaining) cultivated area. Under the Lusinchi administration (1984–9), state subsidies actually reached a size that equalled one-third of total agricultural production value (1987 figure; Blarcom *et al.* 1993: 75).

Forestry and conservation

In terms of forest conservation efforts, the 1970s marked a decade of significant improvement. Prior to 1970, only eight national parks had been created, but the Venezuelan park system was already considered among the most advanced in tropical countries (Amend 1990: 35–8). In the 1970s, eighteen new parks were added, but none were created between 1978 and 1987 (Gabaldón 1995: table 1). There was thus a clear correlation over time between oil wealth and public-sector conservation efforts. Some parts of the policy could not have been implemented without boom revenues. For instance, compensating people for being removed from protected areas was relatively costly, and could not be sustained in the 1980s (Miranda 1998: 31).

However, INPARQUES and SEFORVEN, the conservation and forestry agencies, never experienced budgetary increases in the 1970s comparable to those for agriculture. Also, many NGO representatives and other observers question MARN's budgetary priorities and capacity to implement policy. For instance, Romero (1994: 122) claimed that only four INPARQUES rangers were employed in the largest national park, Canaima (3,000,000 ha), while MARNR at that time had 11,000 employees. MARNR as a whole suffered major cutbacks in the 1980s and 1990s (CENDES 1995: 11). But the cuts also had positive side-effects, since they forced MARNR increasingly to decentralise activities and seek greater cooperation with NGOs (Reed 1996: 212–13). INPARQUES maintained its limited budget, and its technical expertise was generally regarded as good (Amend 1990). With a current base allocation of US$5 million, its options in fulfilling a mandate covering 15 per cent of national territory (the size of the protected areas) obviously depends also on external funding (M. Gabaldón, personal communication, Caracas, November 1998). Also, INPARQUES dedicates only part of its resources to forest conservation. Urban recreational parks alone consumed 61 per cent of the agency's budget in 1991 (Miranda 1998: 29).

Obviously, oil money did not do the conservation trick alone. Many achievements in the 1970s were helped by a growing environmental awareness under the Pérez administration (1974–9). This is confirmed by the fact that during Pérez' second term (1989–93), another expansion of the park system took place, in spite of the extreme fiscal austerity. Pérez managed to hold back the Conquest of the South (CODESUR) plan, which had previously been launched by the Christian Democrats (Arvelo-Jiménez 1984: 120). A large number of park management plans were drawn up, and even politically unpopular decisions were made in order to protect particularly fragile environments.[33] In comparison, during the recent Caldera administration (1994–9), nine different presidents were nominated to INPARQUES – an extreme degree of discontinuity inducing many experienced professionals to leave the institute (F. Gabaldón, personal communication, Caracas, November 1998). In other words, political cycles are important complements to oil cycles in understanding the shifting success of the state in achieving forest conservation.

Roads and other infrastructure

In keeping with claims made in the international deforestation debate, in Venezuela too, road-building through virgin areas is a major, if not *the* major cause of deforestation (Harcourt and Sayer 1996: 315, 70–1). However, this process is now historical rather than contemporary. Most of the opening-up of forest areas through road-building occurred in the 1940s and especially the 1950s (Plonczak 1997: 55). Oil wealth enabled the government to extend the road network much earlier than Venezuela's Andean neighbours. The provision of other public infrastructure was also important for rural areas, such as bridge construction (E. Cabrera, personal communication), electrification and, in particular, malaria control programmes that provided the key to settlement in humid tropical forest areas (Venturini 1969; Borcherdt 1985: 85).[34]

Venezuelan roads directly occupied an area of 62,523 ha in 1985 (Bisbal 1988: 229), which is only 0.07 per cent of land area. But, as elsewhere, roads indirectly provided access for settlement and extraction, thus increasing pressures on forests. This is confirmed by various historical case studies, such as the construction of the Pan-American highway (1953–5) and subsequent loss of forest south of Lake Maracaibo (Venturini 1969), or the Troncal 05 in 1964 and the following deforestation of the *llanos altos* of Barinas state (Rojas 1993: 34–5; Gutiérrez 1996: 54). A current example is deforestation around route 19 to Puerto Ayacucho in Bolívar state. Agriculture is not always the original motive for road-building: it may be oil (Rojas 1993: 59) or timber extraction (Lux 1997: 31–2), followed by a wave of migration that takes advantage of road access to establish *conucos* (Borcherdt 1988: 36). In the western *llanos*, road-building changed timber-extraction from the selective harvesting of high-value species to a broader extraction of multiple species, as reduced transport costs now made this extraction economically feasible (Rojas 1993: 17).

Particularly instructive is the case of the colonisation of the Andean foothills, south of Lake Maracaibo (Venturini 1969). Prior to the construction of the Pan-American Highway, the study area of 3,000 km^2 was almost exclusively virgin forest. Work was completed in 1953–5, and from 1953 an intensive wave of occupation from the densely populated highlands and from Colombia began. From 1959 to 1967, migration to the area slowed down and economic growth stagnated. Forest interventions were initially led by wood extraction, then 2–3 years of cropping, followed by conversion to pastureland. Pioneer squatters were increasingly bought out by large cattle-ranchers. By 1967, fifteen years after road construction began, pastureland had become the single most important vegetation cover in the study region.

Although roads north of the Orinoco had strongest impacts historically, road construction also seemed to have expanded following the second oil boom, but stagnated in the late 1980s and 1990s. Data are incomplete, but spending on new construction (including major improvements) soared to US$301.1 million in 1983, only to be cut back drastically to US$53.1 million in 1984 and US$95.2 million in 1985 (IRF 1988: 200). That delayed several projects, but by 1986 total road length was 100,571 km, up 61 per cent from 62,440 km in 1983 (ibid.: 19). The frequency of vehicles on this road network was, at 44.6 km in 1987, impressively high – higher than in France, the US or Sweden. Cheap petrol prices (see below), good roads and a large car park made people and goods extremely mobile in Venezuela. Yet the economic crisis produced severe stagnation. Compared to the peak 1986 figure, road length actually declined by 6 per cent to 94,923 km in 1994, and then expanded marginally to 96,155 km in 1997 (IRF 2000: 18). Still, the country remains privileged compared to the rest of Latin America, with the highest percentage (60 per cent) of surfaced highways (EIU 1999b: 19).

In principle, energy subsidies can have a comparable effect in lowering transport costs to promote the development of marginal forest-rich areas. In Venezuela, huge subsidies have kept domestic energy prices artificially low. During the first oil boom (1974–6), implicit transfers to domestic energy consumers amounted to 6.1 billion *bolívares*, or about 8 per cent of the total windfall (Bourguignon 1980: table 18.1). In the 1980s, prices did not keep up with rising inflation rates, so that their real value was eroded. In 1989, the overdue rise in petrol prices was the single most precarious element in the IMF-led adjustment package, causing massive riots, with hundreds of deaths. However, even in the mid-1990s, Centeno (1995: 2) could still observe laconically that 'gasoline is sold at less than 5 cents of a dollar per litre, much cheaper than bottled water'. In March 1996, President Caldera had to declare a 600 per cent increase in petrol prices as part of the new IMF austerity package, but the price of petrol in Venezuela is still very low compared to that in other Latin American countries. I know of no study that analyses the spatial production impacts of Venezuelan energy subsidies. But for primary commodities with a small value/weight ratio (e.g. grains or low-value timber), transport is an essential cost element. Consequently, huge energy subsidies must have increased the mobility of these commodities significantly, thus favouring agricultural production and conversion in marginal areas with long marketing distances.

Finally, two caveats on the alleged pro-agricultural and deforestation-promoting impacts of Venezuelan roads are necessary. Historically, new roads were sometimes actually opposed by local agricultural interests for fear that they would increase rural labour mobility, encouraging a sequence of urban commuting, rural–urban migration, rural labour scarcity and higher wages, and therefore rising costs for landowners (Aranda 1984: 85). As in Gabon, roads thus helped to 'empty out' the countryside, reinforcing the urbanisation trend described in the section on 'Structural changes in income and demand'. Second, roads not only transport nationally produced goods but also imported ones. In 1995, 151 million t of cargo were transported by freight-haulage trucks that handled 98 per cent of all arriving goods (EIU 1999b: 19). Generally, roads can facilitate the movement of competing goods *into* rural forested areas, and the exit of production factors *out of* these areas. One would expect this forest-protecting counter-effect to dominate during oil bonanzas with a highly appreciated exchange rate, while the flows may be reversed during downturns when prices are more favourable to local agricultural production.

Directed settlement

Efforts in respect of directed settlement have been scattered in Venezuela, and at present they do not constitute a major cause of deforestation. Historically, the first attempt to direct human settlement in Venezuela was made by the missions in the Amazon and Guyana frontier regions, attempting to concentrate a traditionally scattered indigenous population into colonisation clusters (Arvelo-Jiménez 1984). At the end of the 1920s, the Venezuelan state sponsored agricultural colonisation through the directed settlement of European immigrants. But the development of oil and the decline of export agriculture marked a severe setback to these efforts. In 1938, with the founding of the Technical Institute for

Immigration and Colonisation (ITIC), the state tried to revive directed colonisation as a means of modernising agriculture and occupying apparently empty spaces. However, most of these efforts were either abandoned, or did not reach the targets that had been set initially (Rojas 1993: 15).

The Conquest of the South (CODESUR) plan mentioned earlier and its successor of the 1980s, PREDESSUR, represented plans for semi-directed settlement – both were generally unsuccessful. The aim was to secure claims to a region that has frequently been encroached upon by Brazilian gold-miners (*garimpeiros*) in the south and Colombian farmers in the west. This has been a source of preoccupation among the military, who would like an increased Venezuelan presence in this zone. Buschbacher (1987) describes a small government project to develop 40 ha of pastures in San Carlos near the Brazilian frontier, but he also describes how oil wealth in the 1970s made it extremely difficult to implement geopolitical projects, thus explaining their lack of success:

> It was difficult to entice Venezuelan citizens to live in remote border areas such as near San Carlos. In the Amazon Territory of Venezuela, the standard of living was relatively low. Venezuelans living in the Amazon Territory often emigrated to find steady employment and a higher standard of living in cities to the north. As Venezuelans left the region around San Carlos, Colombians and Brazilians migrated across the border and established their own sites of shifting cultivation within Venezuela.
>
> (Buschbacher 1987: 47)

In other words, in the 1970s oil was a prime explanation of why directed settlement *did not succeed*. At present, large parts of the south remain protected areas, designated for conservation and indigenous use. However, there is a continuous political struggle between 'developers' on one hand, and conservationists and the Indian rights movement on the other (Harcourt and Sayer 1996: 70–1). Although no large-scale government plans have been implemented, the economic crisis has given more pressures to inhabit marginal zones and more political initiative to the developers.

Truly directed settlement has thus not been successful in Venezuela. But land colonisation has not been 'spontaneous', individualistic and atomistic, either. Occupation of forest areas is often collective, organised by either peasant leaders or private entrepreneurs (Rojas 1993: 54–60). State support is also vital. Squatter occupation by poor farmers is tolerated, encouraged or even paid for by local politicians, as part of the Venezuelan land-reform process, and in direct antagonism to the forest-reserve system operated north of the Orinoco (Miranda 1998). Often, land occupation occurs in two steps. First-generation squatters act as social and physical bulldozers (and reap transitory land-use benefits), while large landowners often take over in the second phase and gain permanent land rights (Venturini 1969; Rojas 1993: 54–60).

'Homesteading' rules, where illegal land occupation and deforestation qualify the squatter for title to land (Centeno 1990: 22), are common in Latin America (see also Chapter 7 on Ecuador and Chapter 9 on Mexico). But what is exceptional in Venezuela is that, even after a short period of land occupation, squatters obtain rights to compensation in cases of expulsion (*cancelar las bienhechurías*). While land occupation in other countries is a risky venture that may lead to *uncompensated* expulsion, Venezuelan practice means that a squatter cannot lose, thus providing a clear incentive for occupation. It is thus hardly surprising that forced resettlement with compensation cannot stop the process of illegal occupation (Rojas 1993: 55); indeed, 'this fact creates a precedent, and in one way or the other it encourages subsequent occupations in other areas' (Gutiérrez 1996: 60, my translation from the Spanish). Compensation rules have also fluctuated with the degree of oil wealth; more was paid in the wealthy 1970s than in the 1980s and 1990s in compensation for invasions. 'Keeping everybody pleased' to avoid social tensions became a major goal for the petro-state, even at the cost of creating perverse incentives. Although the size of the population engaged in small-scale, land-extensive farming and ranching is uncertain,[35] this more indirect settlement support has clearly accelerated forest loss in Venezuela.

Structural changes in income and demand

Poverty alleviation

The sustained economic crisis, with falling per-capita incomes, has also increased poverty in a country that, thanks to its oil wealth, has been less poverty-stricken than most of Latin America. Table 5.4 shows that overall income inequality (measured by the Gini index) was reduced significantly from 1975 to 1992, in particular during the oil boom (1975–82). However, while the incidence of poverty stayed the same during the boom, with 33–33.5 per cent of the population being classified as 'moderately poor' nationally, this proportion rose rapidly and continuously after the boom, to 70.5 per cent in 1995. This indicates that the sequence of boom and bust has an impoverishing effect. The absolute number of poor tripled from 1982 to 1995; most of this rise occurred in urban areas, in particular in the early 1980s. This is related to the shrinkage of the urban middle classes and the growing informal sector (Gómez and Malavé 1998). Rural poverty, which is likely to be associated with land-extensive agriculture (*conucos*), more land invasions, etc., increased from 2,417,211 persons in 1975 to 4,222,229 in 1995, that is, by 1,805,018 (74.7 per cent), only about half of which can be attributed to rural population growth.

There have been many allegations but few facts concerning how changing poverty affects the environment, including forest conditions.[36] There is a strong popular belief that 'more poverty causes more deforestation',[37] but no specific analyses have been made to back up this claim. Variables like foreign debt, structural adjustment and peasant impoverishment are all being integrated into the poverty argument (Harcourt and Sayer 1996: 318; Centeno 1997a; Reed and Sheng 1998: A14), but the links are never made explicit. Based on the analysis in the present chapter, I would expect rises in both urban and rural poverty to have an *ambiguous* rather than unequivocal effect on Venezuelan forests. Higher rural poverty lowers the opportunity costs of labour, making it more attractive to practice *conuco* farming and the forest-degrading extraction of low-value products. But it also lowers the means available for rural investment, particularly in cattle-ranching. Higher urban poverty lowers incentives for urban migration, thus increasing rural population pressures. But, as indicated by the results in Table 5.2 (equation in row 4), it would also lower urban demand for meat and dairy products, thus reducing pasture expansion (see next section). 'Cattle' is thus an important factor in making the impact of poverty on forests a dubious matter.

Historically, there is some evidence that impoverished rural producers from the densely populated Andes colonised the lowlands through a combination of 'push' migration and

Table 5.4 Venezuela: poverty* and inequality 1975–95

Indicator / Year	*1975*	*1982*	*1988*	*1990*	*1992*	*1995*
Total population	12,522,721	15,687,032	18,516,716	19,544,889	20,638,453	22,200,840
Urban population	8,540,496	11,451,533	13,913,564	14,858,320	15,865,109	17,353,195
Rural population	3,982,225	4,235,499	4,603,152	4,686,569	4,773,344	4,847,645
Share of urban population in total population	68.2	73.0	75.1	76.0	76.8	78.1
National level, percentage of poor	33.0	33.5	53.5	64.2	61.5	70.5
National level, absolute number of poor	4,132,498	5,255,156	9,906,443	12,547,819	12,692,649	15,651,592
National level, Gini coefficient	50.1	44.0	44.9	43.8	42.4	46.5
Urban level, percentage of poor	24.5	26.6	48.4	59.8	57.0	66.5
Urban level, absolute number of poor	2,092,422	3,046,108	6,734,165	8,885,275	9,043,112	11,539,875
Urban level, Gini coefficient	47.1	42.2	43.8	42.9	42.4	45.8
Rural level, percentage of poor	60.7	59.9	78.6	84.8	82.3	87.1
Rural level, absolute number of poor	2,417,211	2,537,064	3,618,077	3,974,211	3,928,462	4,222,299
Rural level, Gini coefficient	50.2	42.1	44.9	40.8	40.9	42.8
Rural poor as share of total poor	58.5	48.3	36.5	31.7	31.0	27.0

Sources: Population data from Baptista (1997). Poverty and inequality: unpublished data from Professor Matias Riutort (IIES – UCAB).

Note

* Poverty is defined as 'moderate poverty' in relation to the number of people receiving an income below the value of basic food consumption and other necessities.

deforestation. For instance, Venturini (1969: 33) showed that 67 per cent of migrants from the Andes to the Lake Maracaibo region had been landless in their place of origin, 57 per cent had lived in houses with a dirt floor, and 58 per cent had had no access to medical services. However, as described in the previous section, push motives were often mingled with pull-led entrepreneurial interests to appropriate land. Much poverty alleviation was associated with rural–urban migration, taking advantage of better remuneration in the urban sectors (see next section). Although Venezuela therefore achieved relatively low levels of absolute poverty, income distribution was highly unequal, even by Latin American standards. The bias towards capital-intensive sectors meant that, even prior to the oil boom, functional income distribution favoured capital: the share of labour in national income plummeted from 77.2 per cent in 1960 to 42.0 per cent in 1973 (Aranda 1984: 213–20). Policy biases also played a role, such as a marked under-investment in human capital: the share of GDP taken by education fell continuously from 7.4 per cent in 1983 to 3.8 per cent in 1998; only 22 per cent of spending goes on the primary education system, in spite of it having 75 per cent of the students (EIU 1999b: 17).

Poverty has increasingly become an urban problem and, as one poverty researcher expresses it, the environmental pressures that arise from increasing poverty are predominantly urban, with no clear relation to forests (L. P. España, personal communication, Caracas, December 1998). About 2 million extra rural poor have probably increased pressures on forests over the last two decades, but it may well be that the much more pronounced rise in urban poverty actually mitigated forest pressures through a slowdown in meat and dairy demand. The relation between poverty changes and forests thus remains ambiguous, and it is difficult to determine the relative strength of contrary effects.

Rural–urban migration

Together with the Southern Cone countries, Venezuela is the most urbanised nation in Latin America. Definitions of urbanisation vary with one's sources, but the latest figures are between 85.4 per cent (for 1995; the Venezuelan statistical office OCEI 1997: 14), 86.4 per cent (for 1997; World Bank 1999a) and 92.8 per cent (for 1998; EIU 1999b: 14). Baptista (1997) applies a less inclusive definition of 'urban' (see Table 5.4 above), producing a present proportion of urbanisation of 78.2 per cent (in 1995), but his historical figures show that the share in 1936 was only 26.5 per cent. This indicates clearly that extraordinary urbanisation as a long-term structural phenomenon is linked to emerging oil wealth. Also, the speed of urbanisation has been more rapid during boom periods. Baptista's figures for 1975–82 show an increase from 68.2 to 73 per cent in just seven years, with less yearly progress in the following thirteen years (1982–95; up to 78.2 per cent). The annual data published by the World Bank (1999a), using different definitions, tell an even clearer story: urbanisation progressed rapidly from 74.1 per cent in 1973 to 83.6 per cent in 1982 (rising 1.05 percentage points/yr), compared to a modest rise of up to 86.4 per cent in 1997 (a rise of 0.19 percentage points/yr).

So, no matter what definition one uses, it is clear that urbanisation is linked to the urban spending of oil rents. From around 1930, oil stimulated the growth of the urban NT sectors (public employment, private services, construction) and, from 1950 onwards, manufacturing, a protected, 'quasi non-traded' sector (Karl 1986). Up to 1950, urban growth was heavily concentrated in Caracas and in the oil production centres, but it then gradually spread to other urban areas (Salazar-Carillo 1976: 194). Rural population growth largely took the form of a residual difference between fertility-determined growth and the out-migration rate, which was determined by labour absorption in the urban sectors. When the petro-state entered into crisis in the mid-1980s, so did the urban NT sectors that were nourished by oil rents. This did not, as in Cameroon (see Chapter 7), cause outright return migration to the countryside, but at least the drift towards urbanisation decreased significantly.

One sometimes finds the view that urbanisation in itself is a large *direct* deforestation source, through the physical space that residential areas and urban infrastructure occupy. For instance, it has been found that squatter settlements in Caracas expand at an extraordinary rate that consumes about 1 ha/day, which also constitutes a source of forest loss (CENDES 1995: 7). Notwithstanding the importance this may have for local land use, its relative size is minuscule. The alleged 365 ha of annual forest loss make up only 0.09–0.15 per cent of total annual deforestation in Venezuela with respect to the range of 250,000–400,000 ha in the section on 'Deforestation in Venezuela'.

On aggregate, Venezuela's urban bias has clearly reduced deforestation. Some of the population that stayed behind in rural areas since the 1980s are likely to have engaged in land-use activities that caused direct pressures on forests. Admittedly, some recent urban growth south of the Orinoco (the Ciudad Bolívar-Upata-Ciudad Guayana complex) was not 'forest-saving', due to the fact that it was related directly to growth in forest resource-extracting and access-providing sectors like mining and timber (A. Mansutti and M. Miranda, personal e-communication, 13 June and 8 August 2001, respectively). Urban population growth also caused indirect pressure, mainly through the demand for cattle and dairy products (see next section) or timber for construction. Nevertheless, the aggregate long-term impact of Venezuelan urbanisation on forest loss is bound to have been counteractive, as it was associated with a move of people from resource-dependent rural activities to urban services and value-added production that required very little land.

The structure of consumption

Degree of urbanisation, combined with changes in personal income, alters not only the level, but also the structure of aggregate demand: urban and richer people spend their money on different things than a poor and rural population. Venezuela is no exception to this pattern. Borcherdt (1985: 97) observes that the consumption of animal products (cattle, chicken and pigmeat, dairy products) increased particularly towards the end of the 1970s, the time of peak oil revenues. On the other hand, Gutiérrez (1992) identifies a reversal during the crisis-hit 1980s. With falling per-capita incomes, the food budget share of households rose, but within this amount the share spent on animal calories deteriorated, while sources of vegetable calories from cheaper foodstuffs (cereals, rice and flour) increased.

Lacking time-series for commodity consumption, cross-sectional household data can give useful insights into income-led consumption changes through a comparison of the income elasticities of high- and low-income groups. Table 5.5 gives income elasticities based on 1988 household-consumption data from the Caracas Metropolitan Area. Both the

Table 5.5 Venezuela: income elasticities of urban household consumption (selected commodities, Caracas Metropolitan Area, 1988)

Item	*II Quartile elasticity*	*IV Quartile elasticity*
Cereals and derivatives	0.29	0.15
Rice	0.29	−0.05
Maize flour	0.18	−0.34
Oranges	1.85	0.03
Beef	0.79	0.34
Milk and derivatives	0.71	0.27
White cheese	0.78	−0.02
Fish and seafood	0.55	0.75
Foodstuffs, total	0.64	0.34

Source: CORDIPLAN figures from Padrón and Ledezma (1991), cited in Gutiérrez (1992: 72).

second income quartile (e.g. a blue-collar worker household) and the fourth quartile (a middle-class household) are shown. The former responds to a 1 per cent increase in income by raising the value of his food basket by 0.64 per cent. For the latter, the *ex ante* richer group, the figure is only 0.34 per cent. A disaggregation of food items shows that absolute consumption of staple crops rises only a little for the blue-collar household, while it actually declines in the high-income group. The latter only spend significantly more on fish and seafood, and to a certain extent on beef and dairy products. Fruit, beef and dairy products are also clearly the top priority of the blue-collar household in terms of increased spending on food.

What does this mean for deforestation? The period of high economic growth from 1940 to 1980 implied a significant transition of low- to middle-income households in Venezuela, more or less like transferring consumption weight from the first to the second column in Table 5.5. This has produced a pronounced shift in aggregate food demand, which is also influenced by shifting nutritional habits from a traditionally rural to an increasingly urban population (Rojas 1993: 28). Some observations from Table 5.5 go hand in hand with structural changes in cultivated area. For instance, fruit has a high income-elasticity, and also sextupled cultivated area from 1945 to 1996. Over the same period, cereals, other grains and tubers doubled cultivated area, a low level of expansion over half a century, which coincided with the low income-elasticity of staple crops. But most important in land-use terms was the expansion of cattle pasture, which coincided with the high income-elasticity for beef and dairy products. As the cattle sector remained semi-sheltered from import competition, domestic production expanded rapidly; the stock of cattle grew at an annual average rate of 2.2 per cent from 1945 to 1996. Per-capita meat-consumption peaked by the end of the oil boom (1983), then faced a marked decline during the crisis of the 1980s and a fluctuating level during the 1990s (FAOSTAT 2001b). Simultaneously, cattle production showed no sign of land-saving technological change over the same period. On the contrary, our regression results in the section on 'Deforestation in Venezuela' seem to imply 'diminishing ranching returns to land', that is, a *falling* average carrying-capacity of pastures. Hence, rising demand for cattle-derived products and higher domestic livestock production translated into a huge expansion in pastureland (see Figure 5.3), which over time increasingly occurred at the cost of forests.

Synthesis and conclusion

Among the countries studied in this book, Venezuela is clearly the most oil-rich, and the one that has been blessed with oil for the longest period. Consequently, it is a particularly interesting case for the evaluation of long-term structural change. From the 1920s onwards, oil rents have dominated the national economy and provided the basis for high income-growth. This translated into expanding urban NT sectors, the long-term real appreciation of the exchange rate, and labour being massively drawn out of rural areas and into the cities, especially during the oil-booms in the 1920s, 1950s and 1970s. Correspondingly, both the traditional agro-export sectors (cocoa, coffee) and import-competing agriculture lost competitiveness and declined radically. Other primary land-using sectors (mining, forestry) lacked the incentives to become really developed.

As in Gabon, all this translated into the *de facto* conservation of forests. But cattle-ranching was an obstructing factor in Venezuela. North of the Orinoco, especially in the *llanos* region, the Maracaibo upland and the Andean foothills, a combination of road-building, logging and conversion to pastures actually lead to extensive forest-clearing. Yet, economic development was concentrated in the north, while the frontier forests south of the Orinoco (Guyana and Amazonas states) were left intact. In Venezuela, oil itself is mainly found in non-forested areas, and its exploitation caused only indirect forest-loss impacts. Long-term deforestation was thus modest – about half of Venezuela's land area remains forested. Current deforestation estimates are highly contradictory, varying by a factor of four, but likely to be in the range of 250,000–400,000 h (0.5–0.9 per cent). This is lower than the FAO's 1990 FRA figure (599,000 ha), but higher than both the official estimates of 150,000–250,000 ha and than the average clearing since the 1920s.

The long-term structuralist impact of oil wealth is thus clearly confirmed: the inflow of rents has clearly reduced deforestation over the past eight decades. What about the medium-term link? Did periods of oil wealth coincide with enhanced forest conservation and, conversely, oil busts with increased forest degradation and deforestation? This was only partially the case. Lower oil prices and bad macroeconomic management triggered a deep economic crisis in the 1980s and 1990s, and real per-capita incomes fell back to their mid-1960 level. This eventually caused both strategic policy changes and relative price shifts that increased the emphasis on land-using sectors, such as agriculture, logging and mining. Two large devaluations were implemented, in 1983 and 1989. New forest frontiers in southern Venezuela are now being opened up for development. But, looking at the postwar period as whole, there was no clear evidence that forest-clearing actually declined during oil booms when compared to the pre-boom periods.

The most coherent way of testing medium-term linkages is the quantitative analysis carried out in Table 5.2. The regression results strongly confirmed the impact of fluctuating oil-export revenues on RP, and the corresponding link between price competitiveness and production in the forestry and agricultural sectors. In spite of the myriad of government interventions to protect these sectors, the Dutch-Disease effect was too strong for them to remain effectively sheltered. Cropped area basically followed the expected oil-cycle pattern, being negatively correlated with both the RER and urban income growth. But trade restrictions and changing demand patterns altered the result for other semi-traded sectors. Pastures, the more extensive agricultural land use, *expanded* significantly with growth in urban incomes. The same applied to forestry, another inward-looking home-market sector that was greatly dependent on urban construction. Although both sectors were sensitive to changes in competitiveness, they were much more dependent on rising domestic demand.

For cattle-ranching, the positive interplay with urban growth resulted from a combination of factors. On the demand side, a rapidly growing population moved to the cities, earned higher incomes, and developed a maturing appetite for meat and dairy products. On the supply side, large amounts of half-heartedly protected 'forest reserves' were readily available for colonisation and the reinvestment of petro-dollars in cattle, supported by the state's perverse incentives in the form of compensation for squatters. Large-scale cattle-ranchers took advantage of this to privatise and convert abundant forest land, often using poor squatters as a social justification. Thus, although total cultivated land quintupled after the Second World War from 2.4 million ha in 1945 to 12.7 million ha in 1996, 94 per cent of this net increase was due to a single land use: pastures. Add to this picture the technologically backward characteristics of stagnant cattle carrying-capacity, and it becomes clear why extensive pastures became the biomass-poor end-use of most converted forests.

What was the relative weight of different factors in Venezuela's long-term relationship between oil wealth and forest cover? Table 5.6 provides an overview. Two strong, interrelated oil-wealth effects protected forests. First, there was a marked long-term decline in the competitiveness of land-using sectors (2). This caused mobile factors of production to move out of these sectors permanently. Second, the expanding NT sectors were all urban, leading to an extraordinary degree of urbanisation, even judged by the standard of other oil countries (1). Taken together, these two dominant effects imply that a relatively large proportion of Venezuelan forests was not converted to alternative uses.

The majority of 'subordinate' impacts worked in favour of deforestation, but they were not sufficient to reverse the overall outcome of low forest loss. The wealth-induced shift towards higher consumption of meat and dairy products (3) was the strongest of these contrary impacts, triggering a remarkable expansion of pastureland. Road construction (4) provided the necessary access to forested areas, especially in the 1950s. Together with transport subsidies, this probably represented a medium-sized impact over the whole period. The same is true of agricultural protection (5), which faced important fluctuations over time and products, but was successful in sheltering cattle-ranching from the Dutch Disease (5).

The impact of oil wealth on government budgets was variable. Additional funding for agricultural expansion (6) was much higher than that allocated to forest conservation and management (7), but inefficiencies and waste markedly affected the quality of spending. Also, much of the funding for agriculture went into capital-intensive projects with little forest impact. In terms of conservation, even an under-funded national park system has made rather efficient use of scarce resources, achieving results that compare favourably with those in the rest of the continent. The aggregate budgetary effect was probably to promote higher forest loss, but not in any significant manner.

Not all the potential explanatory factors were found to be equally important. Rising poverty has frequently been used to explain Venezuelan deforestation since the onset of the

Table 5.6 Venezuela: oil wealth and deforestation – an overview of long-term impacts

Economic and productive impacts			*Links to deforestation*		*Deforestation impact*	
No.	*Type*	*Intensity*	*Type*	*Strength*	*Type*	*Intensity*
1	High urban labour absorption (services, industry)	Very strong ●●●	Low rural population pressures on forested land	Close ➔	Less forest conversion	Very strong ᐯᐯ
2	Long-term loss of primary sectors' competitiveness	Strong ●●	Hampers development of crops, forestry and mining	Close ➔	Less forest conversion (works jointly with impact 1)	Strong ᐯᐯ
3	Higher urban incomes shift food demand	Strong ●●	Expands particularly cattle-ranching	Close ➔	Forest conversion to pasture	Strong ᐱᐱ
4	New road construction (and energy subsidies)	Medium, for sub-periods ●	Opening up frontiers for commercialisation	Medium ⇨	Conversion in the 1940/50s; weaker today	Medium ᐱ
5	Agricultural protectionism and subsidies	Medium ●	Shelters the agricultural sector from effect 2	Medium ⇨	Enlarges cropped and ranched area	Medium ᐱ
6	High budgets of development agencies	Medium, for sub-periods ●	Support for colonisation and compensation of squatters	Medium ⇨	Frontier expansion and more speculation	Medium ᐱ
7	High budgets of forestry/park agencies	Weak ❍	Improving forest and protected area management	Medium ⇨	Less encroachment and degradation	Weak
8	Non-sustainable forestry and non-oil mining	Weak, but growing ❍	Intervention into primary forests	Medium ⇨	Degradation and access provision for settlement	Weak Δ
9	Others: oil sector site-impacts, directed settlement	Negligible ○	Intervention into primary forests	Weak →	Little effect as yet: weak, historical	Negligible ○
10	Poverty-reduction	Medium ●	Raises costs of deforestation (2); reduces cattle demand (3)	Variable ?	Ambiguous	Variable ?

Note
1 and 2 area – reduces deforestation; 3–6 and 8–9 area – increases deforestation; 10 area – ambiguous.

crisis in the 1980s (10), but the impacts of increasing poverty on forests are ambiguous. Oil exploration and production do not affect forest cover seriously at present, though they may represent threats to forests in the future (9). Directed settlement has not played any role in recent Venezuelan history (9). The development of primary non-oil sectors such as forestry and mining (8) is a recent phenomenon whose long-term impact has been weak. Selective logging has played an active role in opening up forest reserves north of the Orinoco; today it is one of the main factors causing forest degradation in the south.

Another is gold- and diamond-mining, where the technologies used by small- and medium-scale miners are so detrimental to forests that they can actually lead to deforestation. The amount of forest loss due to mining is not currently known, but it is small compared to that from the conversion of forests to agricultural use.

A strategic policy decision for Venezuela today is whether to continue diversification by making the non-oil traded sectors more competitive, including the land-using sectors of mining, agriculture and logging. Although the diversification-*cum*-devaluation argument has gained supporters with the fluctuations in oil prices, there is still strong resistance in Venezuela to this scenario. Instead, it is often argued that the main problem is the internal (fiscal) imbalance, and that devaluation has been an erratic element in previous adjustment packages (Zambrano 1995). Much depends on the development of world-market energy prices, which will ultimately determine the likelihood of an enhanced diversification strategy being adopted.

What may be particularly worrisome is the prospect of the intensified exploitation of renewable natural resources in a country that is accustomed to the 'easy life' of resource-mining, rent-seeking and land abundance. Current experiences point to the danger that the stewardship of these resources may come to resemble the extractive logic of the petro-society. An even greater danger lies in intensified import-substitution efforts designed to increase the protection of technologically stagnant sectors, like the timber and cattle sectors, from import competition. The only comparable advantage of these protected home-market sectors then becomes their privileged access to unrestricted natural resources, such as idle land or timber, which they prove to consume in great quantities. In order not to waste its natural resource potential in a myopic effort to compensate for lower oil revenues, Venezuela will need to develop its institutions and at the same time discourage development scenarios that build on a perception of limitless resource abundance.

Notes

1 Plonczak (1997: 55–6) divides the first category into five sub-regions: Coastal Cordillera, Margarita Islands, Falcón, Maracaibo and the Andes.

2 Bryant *et al.* (1997: 20) estimate that Venezuela at present has lost 41 per cent of its *frontier* forest. Myers (1994: 30) estimates the original area of *tropical moist forests* alone at 42 million ha (46 per cent of land area); to this must be added dry, submontane, montane forests and mangroves.

3 Pittier (1948: 175–6) maintains the hypothesis that most of the savannas of the *llanos* were originally covered by forest and calls them 'post-sylvan meadows' (*praderas post-selváticas*). Veillon (1989: 57–60) also sees dry forests as the true climax vegetation of most of the *sábanas*, and wet sub-alpine forest as that of the *páramo* grasslands between 3,000 and 4,000 masl.

4 This is called the 'H theory' of pre-Columbian settlement in Venezuela (Strauss 1993).

5 As Grau states (1997: 139): 'Discoverers and conquerors did not arrive in empty, depopulated spaces dominated by virgin nature. On the contrary, they came to landscapes that had been inhabited and intervened in by different indigenous ethnic groups for thousands of years, with multiple intensive and extensive landscape transformations, in both continental and insular territory' (my translation from the Spanish).

6 Plonczak (1997: 56) refers to colonial period logging activities in *los llanos*, Gill (1931: 44) in the Maracaibo and Guyana regions.

7 Venezuelan land area (90,135,883 ha) here excludes lakes and dams (1,520,617 ha), as well as the territory disputed with Guyana (*zona de reclamación*).

8 1995 figures includes 600,000 ha of plantations of *pinus caribea* (90 per cent) and eucalyptus (10 per cent).

9 Because of a more exclusive forest definition than MARNR (1982), the FAO used a lower 1977 figure (53,946,000 ha) for its model extrapolation.

10 In addition to the more formal studies covered in Table 5.1, 'loose' Ministry estimates without specified sources exist, for example, Yañez (1995: 24): current yearly deforestation of 240,000 ha (1.3 per cent); or MARNR (1995: 29): 0.94 per cent forest loss between 1960 and 1985.

11 The study seems to draw on the high regional estimates of Catalán (1993) and to extrapolate other equally high estimates from the north-east. The total figure may thus represent an overestimation of current deforestation. Catalán reports a yearly regional loss figure of 217,000 ha, which does not seem to be used in the carbon emission study. The deforested area for different forest types does not seem to add up to the suggested total (ibid.: 45, table IV-1).

12 Both the Veillon (1976) and Catalán (1993) studies are based on map scales of 1 : 500,000 and 1 : 250,000, which implies low precision. The less detailed the map scale, the more likely forest fragments are to be overlooked, resulting in an underestimate of forest cover (see Chapter 1). Veillon (1976: 98) acknowledges this problem and suggests that his estimates be adjusted upwards by 10 per cent; however, judging from the Chapter 6 on Ecuador, this correction may be too low.

13 Juxtaposing the Catalán (1993: 12) forest cover figures for 1988 with the 1995 estimates in MARN (1996: 8), the latter are in fact *higher* than the former for seven out of the ten states. Net reforestation during 1988 to 1995 is not likely for any of these states, so Catalán's forest-cover estimates were based on a much less inclusive definition of forest cover than in MARN (1996).

14 Unfortunately, Centeno provides no source for his estimate.

15 A simple linear regression was estimated for 1945–77, using cattle stock (number of heads) as the independent (X) and all pastures (cultivated and natural, in ha) as the dependent variable (Y). Natural pastures (11,506 ha, according to figures published in the mid-1970s) made up 92 per cent of all pastures back in 1945, and even in 1996 still accounted for about half. The estimated equation, $Y = 7{,}660{,}000 + 0.98\ X$, indicates a falling 'carrying capacity' of head of cattle per ha of cultivated pasture, yielding a R^2 of 78 per cent. Residuals are auto-correlated, so estimation should be improved for more sophisticated purposes, but it probably gives a reasonable prediction of pastures for 1978–96.

16 Rojas (1993: 16–17, 28–9) points out that agricultural colonisation and deforestation, especially around Lake Maracaibo and in the western *llanos*, were stronger during the 1950s, including the mini-boom of 1956–8, than in the 1960s.

17 A. Peñate and C. Rodner, personal communications, Caracas, December 1998, respectively.

18 For instance, Rojas (1993: 59) for Unit IV of the Ticoporo Forest Reserve.

19 Ninety per cent of the study area is forest-covered (ibid.: 1). I thank Marta Miranda of WRI for sending me the original study, and for clarifying an error in an earlier quotation of this study, where it had been claimed mistakenly that as much as 60 per cent of the forest had been cleared (Miranda 1998: 19).

20 Main categories used by Prof. C. Domingo: 'Naturaleza y perspectivas de la economía venezolana', seminar at ULA-Mérida, Faculty of Economics, 3 December 1998.

21 As President Pérez stated: 'We have nothing to fear tomorrow, when the sources of our oil become exhausted, because Venezuelan industry and agriculture will be the solid and firm basis on which the well-being of Venezuelans will rest' (*El Universal*, 23 September 1978, cited in Coronil 1997: 288).

22 Coronil (1997: ch. 6) convincingly describes the case of the automotive industry, where old plans to develop a fully integrated national vehicle industry were now pushed forward, with the unrealistic requirement to reach 90 per cent local value added by 1985. This ran counter to the worldwide strategy among multinationals to globalise production and, combined with vested but conflicting interests within the government, led to the 'motor wars' of the late 1970s.

23 Gelb with Bourguignon (1988: 308) reports an incremental capital–output ratio of 5.6 for the public companies that received a large share of public investments during the first boom.

24 The latter two rates apply to non-petroleum exports (Baptista 1997: tables 8, 9).

25 Term used by Juan Pablo Pérez Alfonzo, one of the founders of OPEC.

26 For instance, Catalán (1993) describes this gradual process for the eastern *llanos* and the forests south of Lake Maracaibo during 1975–88. López *et al.* (1996) denounce conversion in the Caparo Forest Reserve during 1987–94, Gutiérrez (1996) in the Ticoporo Forest Reserve since the 1970s. Venturini (1969) explains forest loss in the north-west Andean foothills in the 1950s and 1960s, while Rojas (1993) does the same for both Ticoporo and Caparo in historical perspective.

27 Also, post-harvest wastage is huge. An MAB study found in 1995 that about 40 per cent of small-scale food production is lost before final consumption, due to wastage in transport and storage (J. C. Centeno, personal communication, Mérida, December 1998).

28 The Agricultural Census should in principle be a more reliable source than the survey data, but it has been carried out too infrequently and inconsistently. Originally a census was done every ten years, as in 1950 (agriculture 1,426,200 ha; pastures 1,659,600 ha) and 1961 (A: 1,669, 400 ha; P: 2,761,000), but the last was in 1971 (A: 1,732,700 ha; P: 4,104,000 ha – see Rojas 1993: 29).

29 Figures and quotation draw on Centeno (1995: 6).

30 Kammesheidt *et al.* (1995) show for the Caparo Forest Reserve that selective harvesting has had a long-term *degradation* impact (very low density of re-growth of commercial species).

31 2–10 trees/ha (Miranda 1998: 9); the average is 5–6 trees/ha (J. Rojas, personal communication, November 1998). For instance, the 3.6 million ha Imataca Forest Reserve in Bolívar state was opened up for mining and logging concessions in 1997. Half of its territory has been allocated (Centeno 1997a).

32 For the sake of symmetry, I originally also included non-agricultural GDP in the estimate of the equation in row 2, but it did not come out as significant. This probably reflects the fact that agricultural value-added could not be disaggregated into crops and cattle-ranching.

33 For instance, about 2,000 illegally built cabins and holiday flats in the Morrocoy region were physically removed (Amend 1990: 38).

34 Venturini (1969: 13–14) vividly describes the indirect protection from colonisation that malaria historically provided in the lowland area south of Lake Maracaibo. 'The insalubrity of the region was of such a degree that, for the inhabitants of the mountains, descent to the zone south of the Lake implied almost certain death' (my translation from the Spanish).

35 One guess is 2 million people, including both Venezuelan and Colombian immigrant families (J. C. Centeno, personal communication, Merida, December 1998).

36 For instance, Reed (1996: 214) writes: '*It is possible* that increasing agricultural encroachment and illegal fishing, hunting, removal of fauna, and logging are taking place in the national parks as a result of the growth of poverty' (emphasis added).

37 See O. Nuñez (cited in Infante 1995: 32): 'Deforestación y pobreza: um binomio indisoluble', *El Universal*, 2 March 1995, Caracas.

6 Cameroon

Cameroon's oil boom differed markedly from those of Gabon and Venezuela in being much smaller in size and shorter in duration. Thus, agriculture also remained much more important for the national economy – before, during and after the boom. In particular, the shifting cultivation of food crops for a growing population was a stronger vehicle for land-use change than in the two previous country cases. Yet, forest loss and degradation from oil wealth also decreased in this context, though as an effect that was superimposed on pre-existing trends. Shifting oil wealth contributed significantly to a boom-and-bust cycle that was stronger than in any other country of this book, allowing us to test for asymmetries in the adjustment to wealth changes.

Deforestation in Cameroon

Vegetation history

Geographers often refer to Cameroon as a 'mini-Africa', due to the high diversity in vegetation and ecosystems represented within one country. Situated on the boundary between central and western Africa, its climate varies from humid–tropical along the Atlantic coast and in the south to semi-arid in the north. Yearly precipitation ranges from 4,000 mm in Douala (the country's main port, in Littoral Province) to less than 600 mm in the Lake Chad region (MINEFI 1998b: 4). Elevation goes from sea level to high plateau and to mountain peaks of around 4,000 masl. The overall ethnic composition is highly heterogeneous, and twenty-four major African languages are spoken in the country (CIA 1999b: 3).

Cameroon's forests are highly diverse, but concentrated in the southern part (see Map 6.1). Main categories range from evergreen Guineo-Congolian (south) to semi-deciduous forests (centre), coastal lowland forest and mangroves (west), as well as sub-montane and afro-montane forests (northwest) (Gartlan 1989: 8–12). In addition, there are open woodland formations in the transition zone to the savannahs in the non-forested north. Depending on the definition of forest, the historical 'original' forest cover has been estimated at between 22 million ha (Myers 1994) and 37.4 million ha (WRI data, cited in Bikié *et al.* 2000a: 41). As argued below, current forest-cover estimate, using a broad forest definition (Laporte *et al.* 1998), is 23.9 million ha (1992–3; 51 per cent of land area), of which 17.3 million ha are closed forests (37.1 per cent of land area). On these grounds, the assertion that 'perhaps 50–60 per cent of the forest [in Cameroon] has been cleared' (WCMC 1996: 1)

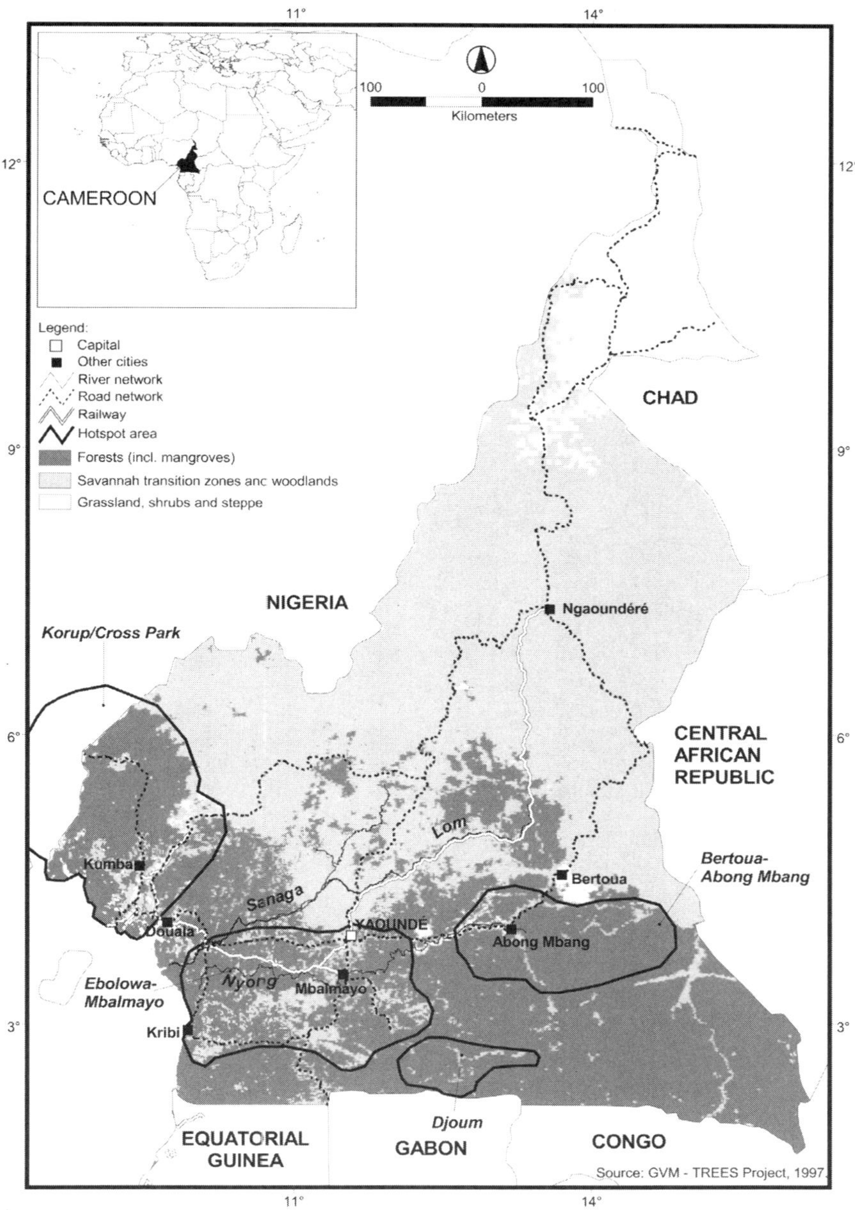

Map 6.1 Cameroon.

is exaggerated: the majority of Cameroon's forests have been *affected* by diverse human impacts, but until now they have survived conversion pressures.[1] Of the countries discussed in this book, the distinction between temporary and permanent forest-clearing is particularly important for Cameroon, due to the widespread use of shifting cultivation. It was Bantu tribes who most extensively colonised forest areas, mainly from the nineteenth century onwards (Diaw 1997: 7–9).

Most forests in Cameroon are located in what will be referred to in the following as the Humid Forest Zone (HFZ) in the southern and western part of the country, which is divided into five provinces: East, South, Littoral, Centre and South-West. The discussion of land-use impacts will therefore be focused on the HFZ. By far most of closed forest area is lowland rainforest, but other types are important in qualitative terms. For instance, the montane forests (>1 per cent of land area) are more threatened,[2] and deserve special attention from a biodiversity viewpoint.[3] Extensive mangroves near Douala (Littoral Province) and the Nigerian border (Southwest Province) fulfil important protection functions *vis-à-vis* fishery resources (ibid.: 112).

The expansion of food-crop production for subsistence uses has always been an important source of land-use change in Cameroon. Tubers and plantain dominate in the HFZ, while cereals (e.g. millet and sorghum) abound in the north (Adamaoua, North and extreme North Provinces). But in colonial times, from the end of the nineteenth century, German rule strongly promoted cash-crop plantations, continued later by the English and French: cocoa, coffee, oil palm, tea, tobacco and rubber were introduced to the south; cotton and peanuts to the north. Raising livestock, especially cattle, was concentrated in the north, outside the forest zone. Some of the best soils in trade-accessible areas were chosen to create plantations: the coastal region, especially fertile areas around Mount Cameroon (palm oil, tea, rubber and cocoa), and the western plateau (coffee and tea). Investments in roads, railways and the export port of Douala opened up new areas for settlement and trade. Direct clearing for plantations reduced forest cover in departments such as Fako, La Mémé and Ndian. The indirect effects of increased settlement and food-crop cultivation subsequently affected forests in the entire coastal zone (Neba 1987: 163–5).

The impact of agricultural expansion on forests thus depends greatly on the product and area in question. For instance, increased demand for cotton or livestock products would be likely to lead to pressures on savannahs rather than forest areas, because production mainly originates in non-forested areas in the north. However, higher production of plantains, tubers, cocoa, tobacco, sugarcane and *robusta* coffee is likely to lead to forest loss, as these products come mostly from the HFZ. Case-study results indicate that cropped-area expansion in the land-extensive production systems of the HFZ is achieved mostly at the expense of forests.[4] Expansion of HFZ cropped area can thus be used as a proxy indicator for deforestation.[5] As will be argued throughout this chapter, a sequence of two decisions made by HFZ rural households had a fundamental and cyclically shifting impact on forest cover:

1. how many resources to allocate to farming versus off-farm (principally urban) activities
2. within farming systems, how many resources to allocate to land-intensive cash crops (coffee, cocoa) versus land-extensive food crops (plantain, tubers, etc.).

Hence, the larger the weight of off-farm activities (inducing urban migration) and the lower the weight of food crops within farming, the lower the pressure to convert forests to agriculture.

Current forest loss

Size estimates of forest cover in Cameroon depend greatly on the classification of forest transition zones and open woodlands. Table 6.1 gives an overview. The FAO-FRA for 1990 (FAO 1993) calculates 1990 forest cover at 20,244,000 ha, but forest and woodland cover, the latter including open forest areas and transition zones, at twice that (35,905,000 ha). The FRA builds on forest-area measurement from two surveys dating from 1975 and 1987 respectively, and changes over time are extrapolated from the FORIS model. The FAO's own assessment of these results is that the Cameroon survey data are of low to medium reliability.

The best forest-stock data are those derived from the TREES satellite imagery (NOAA-AVHRR) of 1992–3. A recent Global Forest Watch study by WRI (Bikié *et al.* 2000a) directly measures forest size from the TREES Central Africa map, yielding 17,915,200 ha of closed forest and 22,794,300 ha in total, including 'degraded' forests. Laporte *et al.* (1995, 1998) and Mayaux *et al.* (1998) both use correction procedures, comparing the NOAA-AVHRR to Landsat images to eliminate method-specific biases. Their closed-forest estimates are marginally lower (17.3 million ha).

All other sources suffer from the same weaknesses that we found for the other case countries described thus far. The WRI *Yearbook* data are based mostly on FAO sources, the FAO *Production Yearbook* repeatedly publishes static forest-cover figures from national agency reports, and the IUCN *Forest Conservation Atlas* uses an old coarse-resolution map that neglects forest fragments and thus comes out with a very low estimate (15,533,000 ha). Consequently, the TREES survey provides the best approximations of forest cover, while Mayaux's closed-cover estimate based on the TREES data (17,378,000 ha) is perhaps based on the most convincing method.

Deforestation has not been well documented for Cameroon, and is subject to the same variability in definitions. For instance, Myers's high annual figure of 230,000 ha (1.4 per cent) is due to his inclusion of forest degradation through logging. As mentioned, the model-driven FAO-FRA data show a yearly forest loss in the 1980s of 129,000 ha, with a marginal reduction to 120,300 ha in the first half of the 1990s. But, as I will argue below, forest loss actually seems to have accelerated since 1986. The only study that attempts a satellite imagery-based comparison is Laporte *et al.* (1995). Closed forest area in her NOAA-AVHRR 1989 map is 2,351,300 ha lower than in the FAO map of 1975, which yields an average yearly net loss of 167,950 ha of closed forest over this fourteen-year period. With 0.9 per cent per year, this would represent a higher estimate than the FAO's figures (0.6 per cent). Yet, only 33,179 ha (20 per cent) actually represented net forest loss, while 80 per cent figured as a net increase in 'degraded forest', a category that is dominated by shifting cultivation, that is by a mosaic of crops and fallows or regenerating forest areas. This calculation shows the importance of shifting cultivation for deforestation, but also how difficult it is to derive clear conclusions from the comparison of different types of satellite images with variable forest classifications (see Chapter 1). At least the

Table 6.1 Cameroon: forest cover and deforestation estimates

Author	*Forest Cover* (in ha)	*Year*	*Annual deforest.* (in ha)	*Relative decline* (%)	*Period*	*Source type*	*Coverage notes*
FAO (1997) FRA	20,244,000 19,598,000	1990 1995	129,000	0.6	1990–5	Model estimate	Total forests, >10% tree cover
FAO (1993) FRA	20,373,000 35,905,000	1990 1990	120,300	0.6	1980–90	Model estimate	> 10% tree cover Incl. woodlands
TREES – Mayaux *et al.* (1998)	17,378,000	1991–5	—	—	—	NOAA-AVHRR satellite (corrected)	Evergreen and semi-deciduous forests, > 70% tree cover
Bikié *et al.* (2000a), using TREES data	17,915,200 22,794,300	1992–3 1992–3	—	—	—	NOAA-AVHRR satellite images	Closed forest Total forests
Laporte *et al.* (1998)	17,385,000 23,862,300	1992–3 1992–3				NOAA-AVHRR satellite	Closed forest Total forests
Laporte *et al.* (1995)	16,808,700	1989–90	167,950 192,333	0.9 1.1	1975–89 1989–92	(corrected)	Closed forests Loss of closed forest
WRI (1998)	21,573,000 20,244,000 19,598,000	1980 1990 1995	132,900 129,200	0.6 0.6	1980–90 1990–5	FAO and ITTO data	All forests (including plantations)
FAO (1996) Production Yearbook	35,900,000 35,900,000 35,900,000 35,900,000	1979 1984 1989 1994	0 0 0	0 0 0	1979–84 1984–9 1989–94	Forestry agency reporting	Forests and woodlands[b]
Sayer *et al.* (1992)	15,533,000[c]	1985	—	—	—	Map ICIV, Toulouse, (Letouzey)	Rain forests[c]
Myers (1994)	22,000,000 16,400,000	'original' 1989	230,000	1.4	1989	Unknown	Tropical moist forests
Mertens and Lambin (2000)[*,a]	101,393 97,675 93,929	1973 1986 1991	286 749	0.3 0.8	1973–86 1986–91	Landsat and SPOT images	>30% tree cover Bertoua area, East Province
Mertens *et al.* (2000)[*,a]	— —	— —	144 594 631	— — —	1973–86 1986–91 1991–6	Landsat and SPOT images	All forests, Ndélélé area, East Province

Notes

a Figures recalculated based on authors' estimates.

b Production forests + other wooded land + intended reforestation ÷ recreation forests.

c According to the source, forest cover is overestimated because forest fragmentation is underestimated.

* Regional estimates.

closed-forest estimates for 1989–90 in Laporte *et al.* (1995) and for 1992–3 in Laporte *et al.* (1998) should be comparable, as both are derived from NOAA-AVHRR imagery. Apparently, there was an average loss of 192,333 (1.1 per cent) ha over these three years in the early 1990s – a high figure, compared to the previous period (see above).

Is this evidence of an accelerating trend in deforestation over time? An answer to that question would require at least three comparable point estimates for national forest cover, which no study has produced hitherto on the national level. Yet a number of recent case studies at a sub-national level (the last three rows, marked with '*' in Table 6.1) fulfil this requirement. The study by Mertens *et al.* (2000) found for the HFZ area of Ndélélé in East Province that yearly deforestation levels in perivillage forests quadrupled in the 1986–91 period (594 ha) compared to 1973–86 (144 ha), rising further in 1991–6 (631 ha). Similarly, for the Bertoua area (also East Province), Mertens and Lambin (2000) found that deforestation almost tripled, from 0.3 per cent in 1973–86 to 0.8 per cent in 1986–91. Sunderlin *et al.* (2000) confirm a similar, though less dramatic trend in a periurban area near the capital, Yaoundé. Deforestation doubled during 1987–95, compared to the previous period (1973–88). Finally, survey data from the HFZ among 648 households from fifty-four villages in three HFZ provinces (South, Centre and East) suggest that clearing of forests increased in the 1990s. The share of households that reported an increase in the area of plantain, coffee and other crops was larger for 1993–7 than for 1985–93 (Bikié *et al.* 2000b).

As we would expect, deforestation rates were thus much lower during 1979–86 than after the onset of the economic crisis in 1986. Deforestation in the 1990s may also have been significantly larger than the FAO-FRA figures imply. This picture is also consistent with the detailed analysis of agricultural land uses (see the section on 'The Competitiveness of agriculture and forestry'). Where does this accelerated deforestation occur? Cameroon currently holds four of the nineteen deforestation 'hot spots' in West and Central Africa that have been identified by a global expert group (Achard *et al.* 1998: 47). All four are located in the HFZ (see Map 6.1): the Korup National Park (Southwest and Littoral Provinces), Mbalmayo–Ebolowa (Centre and South Provinces), Djoum (South) and Bertoua–Abong Mbang (East Province). In all of them, agricultural conversion is the main source of pressure, especially following recent road construction (see section on 'Windfall impacts on government spending'). The expansion in logging since the currency devaluation in 1994 has also been instrumental in providing access for conversion, especially in the East province. The rest of the chapter will deal with potential explanations for the shifts in trends over time in the speed of forest loss. A first question is whether the oil sector itself contributed to pressures on forests.

The effect of mineral production on forests

Although Cameroon's forests are increasingly being exploited and converted, it is not the oil and mining sectors that one should look to for an explanation. Currently, the deforestation impact of Cameroonian oil and mineral extraction is basically zero, although that situation may ultimately change. In principle, Cameroon is a mineral-rich country,[6] but its mining sectors have remained underdeveloped, due to insufficient exploration, infrastructure and finance (Neba 1987: 110). The oil industry remains vital for the economy, being responsible for about half the country's export value. But production and exploration are overwhelmingly offshore, with no impacts on forests. For instance, the most important field today is the Elf-operated Kole field (60,000 barrels/day), a marine field in the Rio del Rey basin, near the Nigerian border.

Discoveries of commercially extractable oil were made in Cameroon from 1972 onwards in the Rio del Rey, but 1978 was the first year with significant production levels (Jua 1993: 131). Additional exploration further south was initiated in the 1980s (Douala basin) and 1990s (Kribi-Campo basin). While this helped to identify mainly natural gas deposits, national oil production and reserves have been declining. From a peak of 158,000 barrels/day in 1985, daily production is now around 100,000 barrels/day, a quarter of which is consumed domestically. This still makes the country Sub-Saharan Africa's fifth largest oil producer. Currently known reserves (400 million barrels) correspond to about eleven years of production at present levels (EIA 2000). In other words, Cameroon's oil boom started late, was less intense than in Gabon, and the prospects are that it will be rather short-lived.

A future oil project with major forest implications is the US$3.5 billion underground oil pipeline, which will connect landlocked oilfields in southern Chad to a port in Kribi on the southern coast of Cameroon. The project, financed by the World Bank, has been subject to lengthy international controversies for its environmental and social risks (Bikié *et al.* 2000a: box 4), but a consortium headed by Exxon Mobil recently announced that construction has commenced (Mbendi 2000). One of several concerns is that road construction will open up pristine forest areas for conversion, which will also threaten indigenous forest-dwelling populations such as the Bakola pygmies (Amigransa *et al.* 1997: 28; Mbendi 2000). Cameroon's future economic (direct and indirect) benefits from the project have been estimated at US$900 million. But a critical point for our study is that, whatever the forest impacts, the project itself is not related to *Cameroon's* but to Chad's oil wealth. Cameroon will be a service provider of transport facilities, but impacts on its forests caused by this development cannot be attributed to its own oil production. This means that the pipeline should rather be seen as a foreign-exchange diversification source to *reduce* (domestic) oil dependence. This makes the pipeline more comparable to the sectors analysed in the section on 'The competitiveness of agriculture and forestry'.

On the whole, Cameroon's own oil production does not impact on forests, whether through direct or indirect (access-providing) effects. But things may change in the future. One source mentions that mangrove forests have been affected by oil activities, but without specifying the type or range of impacts (Amigransa *et al.* 1997: 25). Indeed, new sizeable deposits have been identified on the Bakassi peninsula, which is largely covered by mangroves. But this area is currently disputed with Nigeria, so production has not yet begun. Also, the new Block 1 and Block 6 that together cover 2,033 square kilometres are located in an onshore, forested area between Douala and Kribi, but production has hitherto not been commercially viable (EIA 2000). It can thus be concluded that the forest impacts of Cameroon's oil and mining industries remain negligible, at least at present. Let us now turn to the derived economy-wide impacts of oil, starting with a description of the main macroeconomic trends in the country since independence.

The macroeconomic impact of the oil boom

None of the countries analysed in this book has faced such pronounced economic cycles as Cameroon. Since independence, four distinct sub-periods can be distinguished: the pre-oil boom period (1960–79), the oil-boom years (1979–86), the crisis and structural adjustment period (1986–94), and the post-devaluation period with gradual economic recovery (from 1994).

The pre-boom period (1960–79)

The period from independence to 1979 was characterised by a stable and balanced growth of the economy at around 4 per cent/annum (about 1 per cent per capita). Ahidjo, Cameroon's first president, pursued a development path known as 'planned liberalism', which was based on five-year development plans. There were no strong discontinuities or marked shifts in emphasis between the budgets of the first three plans. Resources were devoted to industrialisation, infrastructure and social sectors. In particular, the Ahidjo government did not fall into the traps of biasing policy strongly towards urban areas or pursuing 'forced' industrialisation. The Second Development Plan was known as the 'Peasant Plan', due to its emphasis on smallholders and on reducing disparities between rural and urban incomes (Ndongko and Vivekananda 1989: 233–4). Agricultural exports like coffee and cocoa were 'leading sectors' in the economy, but a rather diversified base of cash crops developed, including rubber, tea, cotton and tobacco. In 1970, Cameroon's population was 6.6 million and its GDP export share was 19.6 per cent (World Bank 1999a). The country had many features of a diversified, small, open economy.

In the last half of the 1970s, Cameroon's two main export commodities, coffee and cocoa, experienced strong world-market price-hikes (IMF 1990: 266–7). The expansionary demand effect from booming revenues did cause some RER appreciation, but most of the windfall was 'siphoned off' by passing only a small share of the price rise on to farmers. The boom thus strengthened foreign exchange reserves, but did not cause an upsurge in incomes or inflation. Furthermore, unlike other booming African economies, the country did not succumb to the temptation to borrow against future revenues and become heavily indebted. In other words, Cameroon was a sub-Saharan showcase of prudent economic policies for export-oriented growth.

The oil-boom period (1979–86)

When oil revenues started to grow from 1978 onwards, this prudence was still evident, as expressed in a contemporary political slogan: 'Before Oil We Had Agriculture and After Oil We Will Still Have Agriculture' (DeLancey 1989: 121). In a speech made in March 1981, President Ahidjo put it even more strongly: 'We will be guilty and condemned if we submit to the mirage of oil and neglect the development of our agricultural riches. Some other countries have done that and are regretting it' (cited in Jua 1993: 138–40). However, some attitudes would change after the shift to the Paul Biya administration in 1982. Figure 6.1 summarises oil exports and financial capital inflows over the last two decades. Oil exports contributed to rising GDP growth rates from 1978 onwards, but 1978–9 was still

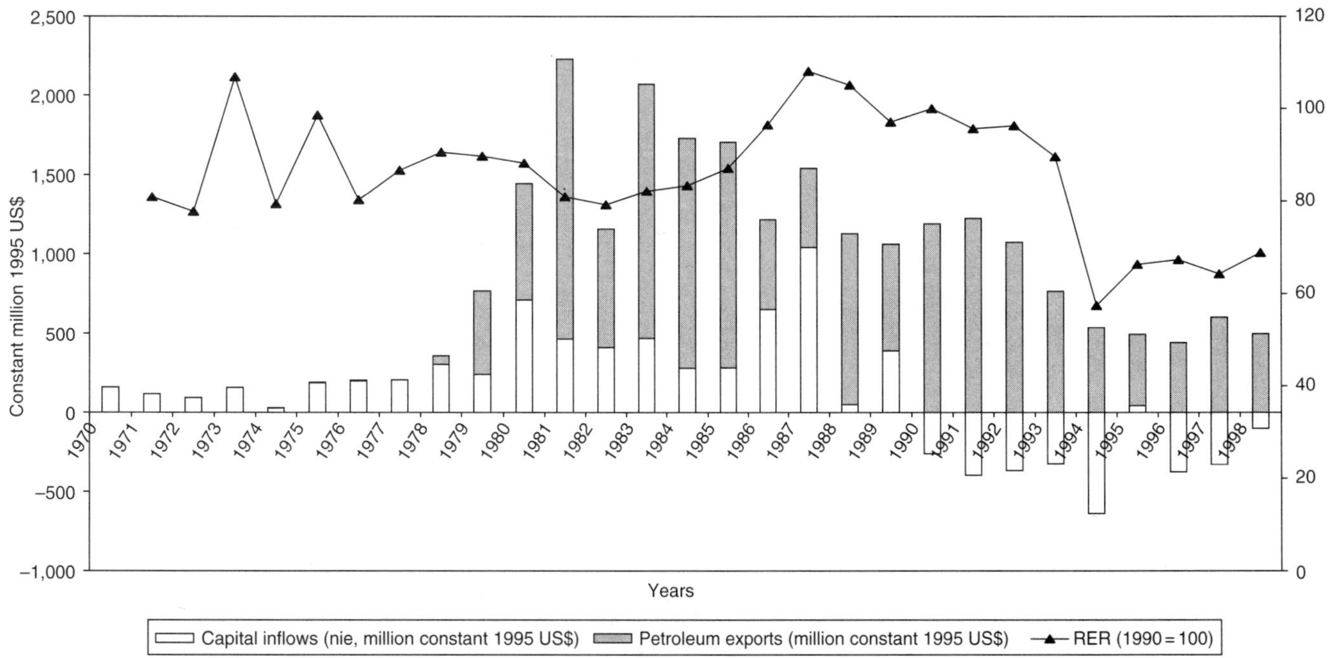

Figure 6.1 Cameroon: capital inflows, petroleum exports and RER, 1970–98.

Sources: Capital inflows: 1970–87: IMF (1990) Other short-term capital + other long-term capital; 1988–95: IMF (1999a) Financial account, nie; 1996–8: IMF (2000) Long-term capital + short-term capital. Petroleum exports: 1970–83: World Bank (1999a) Fuel exports; 1984–9: Blandford *et al*. (1994) Petroleum exports; 1990–3: Marches Tropicaux (1994) Fuel exports; 1994: EIU (1995) Petroleum exports; 1995–8: IMF (1999a) Petroleum exports. RER: 1971–6: Amin (1996) non-tradable prices divided by import prices; 1977–9: IMF (1990) RER; 1980–98: World Bank (1999a) RER.

dominated by the separate boom of the traditional export sectors, cocoa and coffee.[7] Thus Figure 6.1 indicates that, in terms of foreign-exchange inflows, the oil boom started in 1979–80.[8]

A full time series of oil-export revenues is essential for analysing Dutch Disease adjustments of production and land use to the oil boom, but for Cameroon the quantification of oil revenues is an intricate research topic in its own right. Tough negotiations with the oil companies yielded a rising public share of oil revenues (Jua 1993: 135–6), but not all of these publicly accruing revenues were counted in the official trade statistics, instead being channelled directly into accounts held abroad. Apparently, the aim of this policy of 'oil secrecy' was to avoid the domestic spending pressures that had occurred in Gabon and especially Nigeria, and thus to administer the boom in a cautious manner (DeLancey 1989: 140). Another interpretation is that, because repatriation from extra-budgetary CHB accounts (*comptes hors budget*) was at the discretion of the President, he had personal control over half of all public investment expenditure (Jua 1993: 139).[9]

CHBs saved abroad made up three-quarters of all oil revenues, according to Devarajan and de Melo (1987: 451); another source estimates that about half of oil revenues were expatriated (Blandford *et al.* 1994: 139–40). Probably the Ahidjo administration delayed spending, which then accelerated during Biya's boom years; by 1986 apparently only 14 per cent of oil revenues were still being held abroad (World Bank report, cited in Jua 1993: 137). Oil secrecy is also reflected in diverging estimates of the state's oil receipts.[10] Even standard international statistics, like the World Bank's World Development Indicators, are published with significant 'holes', that is, years for which oil-export data are unavailable (World Bank 1999a). The time series in Figure 6.1 links six different but comparable statistical sources, focusing on *repatriated* oil revenues.[11] Except for 1982, oil revenues exceeded US$1 billion during all boom years (IMF 1990; Blandford *et al.* 1994), but in real 1995 US$, the trend was slightly regressive. High revenues mainly resulted from high prices at constant production levels: between 1979 and 1985, the price per barrel of Cameroon's crude stayed continuously above US$27 (MINEFI 1998b: 20). Oil revenues grew steadily in importance, from 1 per cent of GDP in 1978 to a maximum of 20 per cent in 1985 (Blandford *et al.* 1994: 134). Furthermore, Figure 6.1 shows that the real inflow of foreign capital was continuously higher than in the 1970s, thus amplifying the impacts of the oil boom.

What was the oil money spent on? There were five main channels of absorption, basically in the following quantitative order:

- higher public employment, wages and fringe benefits
- the financing of parastatal companies
- infrastructural projects
- consumer and producer subsidies
- urban 'prestige' projects.

First, in contrast to fiscal prudence in 1980–1, public employment rose sharply from 105,907 in 1981 to 176,068 in 1987–8, and in only three years (1981–4) the total wage and salary bill grew by 117 per cent (Jua 1993: 141). Also, generous fringe benefits (free housing, cars, etc.) were granted to public-sector employees. It is no coincidence that this growth occurred after the shift in government. Many people from the southern Beti group were installed in public positions and key state-controlled businesses, for example, issuing import–export licenses (Konings 1996: 250–2). A class of *nouveaux riches* emerged or, in the colourful words of Jua (1993: 155), of '*va-nu-pied subitement devenus millionnaires*'.[12] In addition to side benefits accruing to a 'parasitic class' (ibid.: 146), there was also a rise in outright corruption and diversion of funds (Konings 1996: 252). A quite common example was the creation of fictitious firms for the delivery of non-existent or grossly overvalued supplies to the public sector. From fiscal prudence, Cameroon drifted more and more towards a 'full-belly policy' (*la politique du ventre*; Jua 1993: 147–8).

A second emphasis was on the more than 150 parastatal companies, many of which were in the industrial and agro-industrial sectors. There were two pathways of absorption: investment costs, and subsidies to cover recurrent deficits. Operation subsidies to parastatals became an accelerating burden on public coffers: in 1984, they made up 150 billion CFA francs (FCFA) (US$343.28 million), which represented 18 per cent of total public expenditure for that year (Jua 1993: 150). The land-use impacts of parastatal agro-industries will be discussed in the section on 'Windfall impacts on government spending'. But we can already say that much of the money for parastatals was wasted, as planned activities were not completed and resources were diverted into non-productive employment and personal consumption.

A third and generally more productive pathway was investment in infrastructure, such as roads, rail, energy and ports. This item will be dealt with in the section on 'Windfall impacts on government spending'. The fourth was subsidies to producers (such as imported fertilisers) and consumers (such as rice), which will be discussed in the section on 'The competitiveness of agriculture and forestry'. Finally, a lot of money was spent on urban prestige-type investments, frequently with dubious socio-economic returns. This is in particular true for Yaoundé, which received an oversized international airport, office skyscrapers, international hotels and a modern television network. Even outside the capital, the size and location of new airports is an example of spending that was justified more by political cachet than potential demand. Many of these contracts were awarded to foreign companies, especially French firms, thus limiting the backward linkages of spending (Jua 1993: 141–2).

What did these oil-windfall spending-priorities mean for forests? The planning of new roads was directly linked to timber-harvesting and indirectly to agricultural conversion, so this priority had a direct impact on forests. Also, free-housing benefits and higher urban incomes triggered a construction boom that led to increased timber demand (for both factors, see the section on 'Windfall impacts on government spending'). Agro-industrial plantations expanded into new areas that were taken from forested land. But most spending created urban benefits and had no direct impact on forests. Indeed, the stimulus to urban areas indirectly transferred labour out of agriculture and into the cities, which reduced forest-clearing (see the section on 'Structural changes in income and demand').

Economic crisis and structural adjustment (1986–94)

By the mid-1980s, Cameroon was 'considered to be one of the most successful of African economies' (DeLancey 1989: 143). But over the next decade, the country fell into an unprecedented economic crisis. During the next seven years, GDP fell by 30 per cent and real per-capita income was halved (Tchoungui *et al.* 1995: 41). How can a 'success story' suddenly turn into a sustained fiasco? One element was the combination of high trade dependency and 'bad luck': Cameroon's major commodity-export revenues declined simultaneously in the mid-1980s. Crude oil prices were halved to US$13.8 per barrel in 1986 (MINEFI 1998b: 20), taking oil revenues with them, which fell to about US$0.4 billion (IMF 1990; Blandford *et al.* 1994; Figure 6.1). Oil exports were denominated in US$, but the dollar depreciated by 40 per cent against the franc between 1985 and 1988, thus reducing the value of oil in local currency, and therefore state revenues, even further. Cocoa and coffee prices also declined after the mini-boom in 1986. The net result was that Cameroon's terms of trade deteriorated dramatically by 65 per cent between 1985–6 and 1986–7 (Ndoye and Kaimowitz 2000: 17). To make things worse, oil production also started to decline steadily from the peak of 9.16 million t in 1985 to 7.2 million t in 1991 and 5.2 million t in 1995–6 (EIU 1995: 24; EIU 1997: 29).

Within the CFA monetary union, the FCFA was collectively tied to the French franc at a fixed rate of 1 : 50. Cameroon thus lacked an important economic policy tool with which it could react to deteriorating terms of trade. It could not decide to devalue on its own, and several CFA member countries successfully resisted devaluation. An overvalued currency thus made it difficult to restore competitiveness to the non-oil traded sectors, especially cash-crop exports.

Thus trade openness, 'bad luck' and the inflexible exchange-rate regime bear a large part of the blame for the crisis. But oil wealth itself had also reduced the economy's ability to adjust to external shocks. Fiscal policy was a key part of this. Even in 1986–7, when it seemed clear that oil prices would remain low, the government was not able to scale back its expenditure, which actually rocketed by 33 per cent, before finally being cut in 1987–8 (Blandford *et al.* 1994: 154). Figure 6.1 also shows that a first response to the foreign-exchange crisis in 1986 and 1987 was a massive increase in foreign borrowing, which deferred adjustment. In 1988, a presidential decree finally endorsed cutbacks in public salaries, but implementation was incomplete (Jua 1993: 151). A severe banking crisis emerged from imprudent and corrupt practices, and from the sudden shortage of liquidity.

This was thus the background for the adoption of a SAP in 1988–9, in collaboration with the IMF and the World Bank. As a first measure, half of the 150 parastatal companies were examined by a joint commission of experts, which proposed that 15 of them be liquidated, 12 privatised, 4 merged and 38 rehabilitated. Implementation of the privatisation plan was, however, extremely slow. Another cardinal aspect of the SAP was the unsatisfactory state of the fiscal balance. The government was not able to consolidate a non-oil tax base. Also, public employment had started to grow again (4.9 per cent from 1988 to 1990), while public wages had been frozen but not cut (Blandford *et al.* 1994: 157–9). Finally, the SAP required producer 'price-stabilisation' schemes for export cash crops – really an ill-disguised taxation scheme – to be reformed (see the section on 'The competitiveness of agriculture and forestry').

Cameroon maintained a strong position in the SAP negotiations with the IMF, and the first two conditions (fiscal balance, privatisation) were only implemented very reluctantly (Konings 1996). Agricultural trade was indeed deregulated, but the net impact of structural adjustment on export crops was negative prior to the 1994 devaluation: cocoa, cotton, and *robusta* and *arabica* coffee production all fared worse compared to the boom period (Amin 1995: 21–2). As shown in a case study of cocoa farmers in East Province, the sudden scaling back of public services (transport, subsidised inputs, marketing systems, replanting subsidies) hit producers hard (Tchoungui 1995: 80–9). Similar factors affected coffee-growers in West and Littoral Provinces. Maintenance of plantations was neglected, and productivity per hectare plummeted (Gockowski 1994). Farmers left cash-crop plantations semi-abandoned, but many instead cleared new areas for food crops – partly for themselves, but also to satisfy higher national food-crop demand because the capacity to import foodstuffs plummeted. In fact, local or urban sale of the so-called 'food crops' often became a vital cash generator for rural households. The net effect of that process was a large increase in forest-clearing, as indicated in the case studies in Table 6.1 (section on 'Deforestation in Cameroon').

Devaluation and slow recovery (from 1994)

Most of the countries in the CFA zone faced external trade deficits, and in 1994 the decision was finally taken to devalue the FCFA collectively by 50 per cent. This was a turning point towards a cautious economic revival for Cameroon. Real GDP picked up from negative rates to a growth rate of 3.3 per cent in 1994–5 and around 5 per cent in the following three years, or about 2 per cent in per-capita terms (EIU 1999c: 18, 32). All land-using sectors expanded production, but at different speeds. Timber exports grew most rapidly (see next section). Bananas benefited from the favourable EU banana-import regime. Food-crop output expanded moderately. Food imports (such as cereals and sugar), which had expanded greatly during the oil boom, now became too expensive and were increasingly replaced by domestically produced commodities (Ndoye and Kaimowitz 2000). Perennial exports from the HFZ (coffee, cocoa) rose, but much of this was achieved from pre-established plots that were under-utilised prior to 1994. New planting of perennials lagged behind because of the long-run decapitalisation of the sector, the disruptive impacts of market liberalisation, and perhaps also a certain lack of confidence based on the fear that surpluses might be taxed away. Farmers became generally risk-averse and diversified into food crops, fruit trees, non-timber forest products, etc. (O. Ndoye, Yaoundé, personal communication, 5 July 2001.) As yet, there has been little new deforestation for cocoa and coffee.

On the whole, the collapse in urban activities, growing unemployment and poverty after 1987 drove many people back to the countryside, many falling back by default on slash-and-burn agriculture. Structural adjustment prior to 1994 was half-hearted, and it did not provide answers to the severe economic crisis. The 1994 devaluation initiated an incipient economic revival. This underscores the crucial impact of competitiveness, which will be analysed in the following section.

The competitiveness of agriculture and forestry

The RER mechanism

As one should expect from the Dutch Disease theory, Cameroon experienced strong currency appreciation related to the combined oil and foreign-borrowing boom (1982–7), and strong real depreciation during the crisis (1987–94). Yet price effects occurred with a lag, compared to oil and borrowing inflows (Devarajan and de Melo 1987: 451), as can also be seen in Figure 6.1. There are four possible explanations for this. First, booming foreign-exchange inflows from the traditionally leading export sectors, coffee and cocoa, had already prior to the oil boom made the RER appreciate. Second, repatriation of oil revenues from CHBs held abroad was stronger afterwards, accelerating demand and real appreciation. Third, widespread price controls hindered a rise in 'official' prices, but from 1980 to 1988 non-controlled prices rose by 50 per cent more than official prices (Roubaud 1994: 67), so that RERs using 'official' rates of inflation are an underestimate. Finally, data quality for the RER is compromised for 1971–6, as different indices had to be linked for the full-time series in Figure 6.1.[13]

Up to the 1994 devaluation, competitiveness under a fixed exchange-rate regime could be improved only by reducing the inflation differential *vis-à-vis* Cameroon's trading partners or by using protectionist trade policies, such as export subsidies and import tariffs, for a 'quasi-devaluation' (Devarajan and de Melo 1987). Figure 6.1 shows that the economic crisis caused some real depreciation after 1987, but domestic inflation adjusted only slowly, held up by 'sticky' prices, extensive foreign borrowing (1986–7) and the sustained fiscal deficit. After 1990, there were increasing expectations that the FCFA would be devalued.[14] This encouraged an increasing net outflow of financial capital (EIU 1999c: 18; Figure 6.1). Adjustment under fixed exchange rates was definitely not a success story in Cameroon.

Failure to make timely fiscal adjustments contributed heavily to overvalued exchange rates, which impeded the revival of non-oil exports. A modelling exercise by Kalulumia (1997) concludes that the combination of terms-of-trade shock and sustained fiscal imbalance may in itself provide a full explanation for the prolonged economic depression in Cameroon. Another macroeconomic model (Benjamin 1996) illustrates that agricultural interests, in particular poor farmers, suffered most from the sustained overvaluation of Cameroon's currency. Let us examine this more closely.

Agriculture

Which crops had a significant land-use impact on forest cover?[15] To what extent was land demand for these crops affected by oil-induced changes in competitiveness? These two questions will be answered in this sub-section.

Starting with forest impacts, it is important to distinguish between different agricultural systems in the HFZ. First, medium- and large-scale plantations produce oil palm, sugar, tea, bananas, etc. At the height of the boom, in 1984, this 'modern sector' occupied 124,000 ha or less than 5 per cent of the land area in the HFZ. Second, small farmers used 458,000 ha for tree crops, basically cocoa and *robusta* coffee. These perennials are grown in a land-extensive manner in Cameroon compared to other countries,[16] but usually the same plot is used for about forty years. Third, 324,000 ha of small farmers' plots were under a variety of food crops, and 285,000 ha of this area was under shifting cultivation. One to two years of production are, on average, followed by 5–10 years of fallow. Clearing practices also differ according to two field types (*esep* versus *afub awondo*), which again depends on the (sequence of) planted crops.[17]

What was the structure of HFZ land use in the 1984 Agricultural Census year? Ndoye and Kaimowitz (2000: table 6.4) estimate that each of the 284,300 ha of food crops under shifting cultivation required 5–10 years of fallow, so total fallow area was 1.4–2.8 million ha (5.2–10.4 per cent of HFZ land area). Large plantations (123,900 ha) and small-scale tree (516,100 ha) and food crops (284,300 ha) added up to a total cultivated area of 924,300 ha. Total area *affected* by agriculture (i.e. crops plus fallows) is thus 2.3–3.8 million ha, or between 9 and 14 per cent of the HFZ land area.

What did this mean for deforestation? Fallow areas regenerate over time and become young secondary forests, so not all fallows are non-forested areas by our definition of 10 per cent tree cover at a minimum height of 5 m. As in the Gabon chapter, let us assume conservatively that non-forest fallows equal twice the area under food crops. This means that deforested area for agriculture equalled cultivated area (924,300 ha) plus non-forest fallows (284,300 × 2 = 568,600 ha), that is, a total of 1,492,900 ha, or 5.5 per cent of the HFZ land area. These assumptions illustrate that shifting food-crop systems have a crucial multiplier impact on the size of deforestation: within two years, 1 ha shifted from tree crops into food crops creates a 2 ha increase in non-forest fallows, and thus a similar net incremental deforestation. Structural changes promoting food crops within farming systems in the HFZ thus accelerate forest loss. Thiele and Wiebelt (1994: 163) claimed that shifting cultivation makes up as much as 95 per cent of all deforestation in Cameroon.

The second question is economic: how sensitive are different crops to changes in competitiveness? In Cameroon, only export cash crops (like coffee and cocoa) are genuine 'tradables', in the sense of being consistently exposed to foreign competition. Plantation cash crops for export are also tradables, but those for domestic use have partly been protected. Food-crops are difficult to classify. Some are tradables, for example, in urban and periurban areas, where they compete with substitutes from abroad or from other parts of the country, or when surplus production is exported to Gabon or Nigeria. In isolated rural areas, food crops are non-tradables that are largely detached from the market economy.

Following an oil boom with an appreciating RER, one would expect cash crops like coffee and cocoa to suffer, and 'semi-traded' food and plantation crops to show mixed results. But in Cameroon, the performance of cash crops was compromised in a slightly different way. First and foremost, spending pressures and real appreciation came with a time lag and thus hit the cash-crop sector hardest from the onset of the crisis until devaluation (1987–94). Second, international cocoa and coffee prices were highly favourable in 1983–6.[18] Third, in 1987–8 there was the political will to keep domestic cash-crop producer-prices high (see next section). The combined effect was that cash crops fared reasonably well during the boom itself, but declined markedly with the economic crisis.

Tentative estimates for HFZ land-use trends in sub-sectors and for sub-periods are summarised in Table 6.2 (adapted from Ndoye and Kaimowitz 2000: table 7). Prior to the oil boom (1960–79) Cameroon was predominantly rural, and land-use expansion was driven by the subsistence demands of a growing population for (non-traded) food crops and a steady expansion of (traded) cash crops. The large-scale plantation sector grew slowly.

During the oil boom (1979–86), parastatal and other plantations expanded more rapidly, but the additional area occupied remained restricted. Among smallholder tree crops, coffee production increased, which was mostly due to growth in productivity per hectare associated with subsidised inputs (see below). Conversely, cocoa area expanded but production increased very slowly, due to growing disease problems and an ageing stock of

Table 6.2 Cameroon: shifting land-use trends in the humid forest zone after independence

Sector	*I. Pre oil boom (1960–79)*	*II. Oil boom (1979–86)*	*III. Crisis and fixed exchange rate (1986–94)*	*IV. Devaluation and slow recovery (1994–)*
1 *Cocoa*				
• Area planted (ha/yr)	9,200 (1964–71); 7,600 (1972–80)	12,000 (1980–84)	• Moderate production decline • Stagnant area size	Moderate rise
2 *Robusta coffee*				
• Area planted (ha/yr)	3,000 (1972–80)	3,200 (1981–4)	• Strong production decline • Extensification	Moderate rise
3 *Food crops*[a]	• Low commercialisation • Area proportional to rural population size • Long fallow	• Slow growth • Low urban commercialisation • Shorter fallow	• Strong growth • Extensive crops	• Strong growth • Extensive growth
4 *Plantation crops*[b]				
• Area planted (ha/yr)	1,934 (1951–80)	6,400 (1987–7)	Stagnant	Stagnant
Timber				
• Industrial roundwood production volume (m^3)	1,178,529 (1961–77)	2,365,000 (1978–85)	2,906,000 (1986–93)	3,373,400 (1994–8)
• Industrial roundwood exports volume (m^3)	395,294 (1961–77)	595,625 (1978–85)	586,750 (1986–93)	1,356,800 (1994–8)

Sources: Timber data from Bikié *et al.* (2000a: 43); agricultural data from Ndoye and Kaimowitz (2000).

Notes

a Mainly plantain, cassava, yams, cocoyams, groundnuts, *ngon* melons and maize.

b Mainly rubber, oil palm, sugar cane, banana and tea.

cocoa trees. HFZ food-crop production was 'delinked' from population growth because of rapid rural–urban migration, which considerably changed consumption patterns (see the section on 'Structural changes in income and demand'). No reliable statistics are available for food crops, but World Bank estimates from 1988 (cited in Ndoye and Kaimowitz 2000: 29) indicate that production and cultivated area stagnated, especially in remote areas. Urbanisation, demand growth and shifting consumption patterns increased the significance of foodstuffs produced outside the HFZ. Imports of cereals and sugar increased, as did the consumption of plantation crops from the northern provinces, such as rice (Ndoye and Kaimowitz 2000). New urban employment options raised the opportunity costs of labour, and migration out of rural areas decreased the labour available for food-crop production. On the whole, agricultural land use in southern Cameroon is likely to have increased only very slowly, and forest loss was thus also limited.

In the post-boom crisis period (1986–94) plantations stagnated, among other things due to the financial difficulties of parastatal companies. HFZ cash crops declined drastically, due to the simultaneous impacts of a collapse in international prices, a steadily overvalued exchange rate, and the drastic removal of support systems and subsidies (see next section). Falling incomes and a scarcity of foreign exchange made consumers return to 'inferior' food crops. Rural areas again had to absorb local population growth, and in some cases also impoverished migrants returning from urban areas. Thus there was a massive return to HFZ food crops. As explained for the HFZ case studies in the section on 'Deforestation in Cameroon' this fuelled a large rise in forest-clearing. First, coffee and cocoa trees were seldom replaced and often left semi-abandoned while growers waited for prices to improve, but cleared new areas for food crops in the meantime. Consequently, cash-crop production declined much more than the area occupied (e.g. Tchoungui *et al.* 1995: 82–7). Second, as mentioned above food crops required non-forest fallows of at least double the size of cultivated areas. This is especially pronounced for a crop like plantains, which require high soil fertility, which is obtained from burning. The net impact of the marked shift from cash to food crops was thus a significant rise in land demand, predominantly at the expense of forests.

The 1994 devaluation seems to have been generally beneficial for the agricultural sector. It seems that both food and cash crops in the HFZ expanded, but cash crops have done better. From 1994–5 to 1997–8, production of cocoa and in particular *robusta* coffee grew significantly. For comparison, tuber and plantain production grew marginally. Plantation crops like tea and rubber stagnated, bananas expanded rapidly, and palm oil declined (EIU 1999c: 33; MINEFI 1998a: 89–91). The ability of HFZ cash-crop sectors (coffee, cocoa) to respond to devaluation has until now been limited by the decapitalisation of the sector, ageing tree stocks, the lack of new credit and deteriorating product quality. A revival may also concentrate on fewer producers and more specialised regions. For instance, the recent rise in cocoa production has been largely confined to the southwest, where productivity is double that of the south and centre. At the same time, about 500 cocoa exporters in 1990 have now been reduced to around 50, 10 of which control 98 per cent of export volumes (Marchés Tropicaux 1999: 22–5). In other words, HFZ cash crops may eventually come back, but not necessarily in their widespread pre-crisis role as the small HFZ farmer's general cash alternative to food crops.

There is thus a shifting balance between cash and food crops, first during the oil boom (from food to cash crops), then with the onset of the crisis (from cash to food crops), with perhaps a more balanced growth after devaluation. This shifting balance had major implications for land use. But what does the aggregate, national picture for agricultural production look like? Agriculture's share of GDP fluctuated around 30 per cent in the 1970s, but then dropped during the oil boom to a low of 20.6 per cent in 1985. With the crisis, it experienced a moderate revival to 27.3 per cent in 1993, which in the face of falling per-capita GDP meant that it did not decline as much as the urban sectors. Following devaluation, agriculture's GDP share again expanded rapidly to 42.4 per cent (World Bank 1999a; IMF 2000). Agriculture was thus definitely hit by the Dutch Disease, deforestation being limited during the oil boom, but accelerating with the economic crisis. RER movements played a key role in this process. However, forest impacts were not directly proportional to the decline of agriculture. This depended on whether the affected crops came from forested areas and, if so, how the balance between food and cash crops was affected.

Forestry

With an industrial roundwood export volume of 1.28 million m^3 (1998), Cameroon is among the world's top five exporters of tropical logs (Bikié *et al.* 2000a: 13). Forestry has increased its importance in the economy, currently producing 7 per cent of GDP (EIU 1999c: 20). Direct forestry employment is 33,000. Standing timber represents a significant 'natural capital' for Cameroon. In 1995, commercial timber stocks were estimated at 310 million m^3, with an approximate value of US$70 billion.[19] Around 80 per cent of the forest area is potentially exploitable, but access has been the main restriction in the past. Three species have traditionally accounted for more than half of all production: *ayous (Triplochiton scleroxylon), sapelli (Entandrophragma cylindricum)* and *azobe (Lophira alata)* (Marchés Tropicaux 1994: 761). Licences are given for both large-scale concessions and small-scale exploitation (*vente de coupe*). On the downside, timber extraction has traditionally been a capital-intensive enclave dominated by foreign interests, with very unequally distributed revenues (Essama-Nssah and Gockowski 2000: 11; Bikié *et al.* 2000a) that only offers low-skill employment: 92 per cent are low-paid workers (Eba'a Atyi 1998).

Important structural changes have occurred in the sector over the last decade. Since the 1994 devaluation, timber exports have risen rapidly. As in Gabon, Asia replaced Europe as the main export market in 1997 (Bikié *et al.* 2000a: 14). Asian timber firms are also operating large concessions in Cameroon, often through national intermediaries, as in the case of Rimbunan Hijau (Debroux and Karsenty 1997). The serious Asian crisis in 1997–8 reduced the price and value of Cameroon's log exports, but the decline seems temporary. Until 1987, that is, throughout the oil-boom years, logging in Cameroon remained largely a European business. But the economic crisis also stimulated domestic capital to invest in alternative export sectors. The rise of new domestic logging enterprises has been remarkable in the 1990s; they now make up a majority of authorised firms, although some of them are indirectly controlled by or closely collaborating with foreign interests (Eba'a Atyi 1998: 1; Bikié *et al.* 2000a: 8).

The upswing in timber-harvesting was originally not accompanied by a rise in domestic primary processing capacity: most timber was always exported in its raw state (Eba'a Atyi 1998: 10). But in 1999, Cameroon announced the gradual introduction of a species-dependent log-export ban, a policy that has been controversial with the World Bank and main bilateral trading partners. Many loopholes have reduced the efficiency of the ban, but the industry's response has still been a major increase in domestic processing capacity. Log export volumes halved between 1997 and 1999, from about 2 to only 1 million m^3 (ATIBT 2000: 35). Correspondingly, the export volume of processed wood (sawnwood and veneer) doubled between fiscal years 1996–7 and 1998–9, from 314,134 m^3 to 610,591 m^3 (MINEFI 2000).

However, data on timber production and exports in Cameroon are generally not very reliable. The official estimate of 1998–9 production (exports and domestic consumption) is 3.2 million m^3 (MINEFI 2000: 21), but this underestimates the growth in illegal cutting and informal-sector production. The latter include significant unauthorised log exports overland to Nigeria and Chad and through the port of Douala. In the main production zone in East Province, more than half of the logs may be harvested illegally (Karsenty 2000: 28). The same source estimates yearly production rather at 4.5–5.0 million m^3 (ibid.: 3). For comparison, the FAO (2000b) statistics used below are based on yearly questionnaires to official sources. They show certain underestimation biases, especially *vis-à-vis* exports, but their advantage is that they go back to 1961. Industrial roundwood has historically dominated wood production and exports, but with the recent rise in domestic processing, sawnwood and veneer panels should also be considered in any long-term comparison of production levels.

With these caveats in mind, Figure 6.2 shows how the FAO (2000b) aggregated production and export volumes of industrial roundwood, sawnwood and veneer panels (in roundwood equivalents)[20] over the last four decades, compared to RER movements (available only for 1971–98). A steady and substantial growth in production from 1970 to 1986 was followed by stagnation in the crisis years of 1987–93 and a sharp rise in 1994–6. The pair-wise correlation between production and changing competitiveness appears ambiguous. In 1977–8 and 1982–6, rising timber production went along with a higher RER index, which would seem counterintuitive, as the sector lost competitiveness. The opposite, expected pattern is found from 1994 onwards, when a devaluated currency was accompanied by higher timber production.

The explanation seems to be that the 1970–86 expansion was driven by a growing domestic market – a trend that has surprisingly been ignored in the recent booming literature on logging in Cameroon. High-value species are mainly exported but large quantities of wood are used for domestic construction and furniture. As already mentioned, the FAO's figures may overestimate domestic production, though alternative estimates are not much lower.[21] Some domestic processing is rudimentary, with large levels of wastage of wood, especially in the informal sector (Carret and Clément 1993: 172–3). Home-market timber has remained highly protected from competing imports, and is thus a 'quasi non-tradable'.[22] The drivers of home-market expansion have generally been under-researched, but population growth, urbanisation and growing national income are the most likely candidates (see below). During the oil bonanza, a major impetus was urban construction.

Domestic demand weakened with the crisis, and after 1993 growth was export-led, with cost competitiveness being a key parameter (see regression results below). Timber harvesting is perhaps the single activity that has benefited most from Cameroon's sequence of crisis, structural adjustment and devaluation (Tchouingui *et al.* 1995). Whereas cash-crop exports suffered greatly from the collapse of support systems, the 'pure' extraction

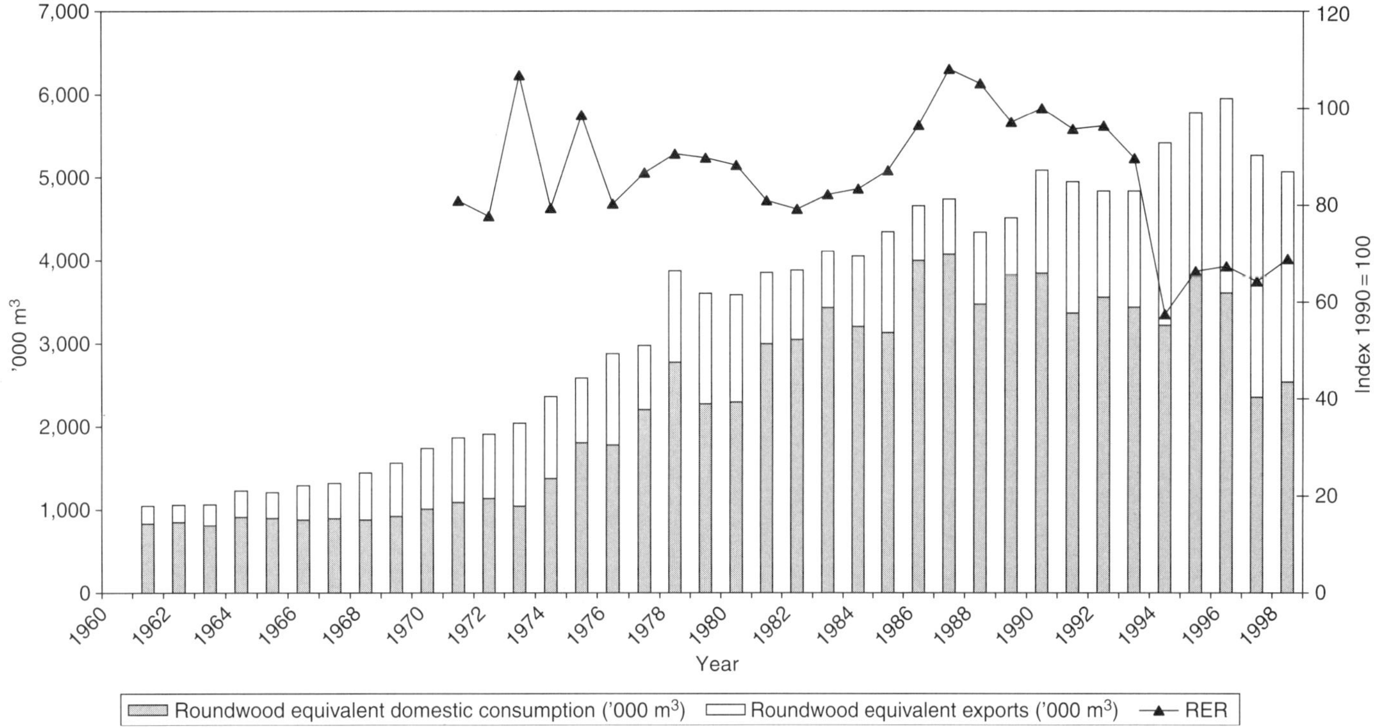

Figure 6.2 Cameroon: industrial wood production, exports and competitiveness,[1] 1961–98. Industrial roundwood, sawnwood and veneer panels (in roundwood equivalents).

Sources: FAO (2000b), MINEFI (2000).

Note

1 Real trade-weighted exchange rate – see Figure 6.1 for sources and definitions.

of timber rents mostly depends on favourable prices, investments in roads, and cheap transport and labour costs.

Has this expansion in log exports in the aftermath of devaluation been accompanied by a similar expansion in concession areas? Apparently, area expansion was inferior to the rise in production. A comparison of four maps published in Bikié *et al.* (2000a: 59) of concessions in 1959, 1971, 1995 and 1999 show that the bulk of forest areas had already been allocated to concession holders between 1971 and 1995; basically no further changes occurred in 1995–9. In terms of areas actually opened up for logging, the main expansion occurred prior to the 1994 devaluation (Eba'a Atyi 1998: 19). The quest for access to and control over Cameroon's valuable timber resources was thus decided before the RER depreciation. It probably responded to other factors, such as the exhaustion of stocks in other producer countries and the expansion of roads in Cameroon (see the section on 'Windfall impacts on government spending').

Was the production increase accompanied by a proportional rise in logged-over areas? Apparently not: the number of species exploited and the quantity of wood extracted per land unit both rose, in order to satisfy higher and shifting demand for a greater variety of secondary species. Previously, loggers tended to 'skim' (high-grade) the forest for high-value species only (Toornstra *et al.* 1994: 5), but over the past decade the number of species exploited in Cameroon increased from 40–50 to more than 70. This relates to the broader character of Asian demand. One example is the less-valued *fraké* (*Terminalia superba*) species, which was not harvested from remote areas due to high transport costs. After devaluation it became profitable, and was temporarily the second-most exploited species. A higher per-hectare rate of extraction consolidated the profitability of a logging sector that operates with high fixed costs per unit of intervened area (Eba'a Atyi 1998: 19).

To what extent are the new players and the increase in per-hectare harvesting causing an acceleration in the deforestation or degradation of Cameroon's forests? Although the Asian firms tend to 'mine' their concessions without forest-management plans, their environmental impacts are probably not very different from those created by earlier producers (Debroux and Karsenty 1997). The more diversified exploitation of species has increased ecological impacts somewhat. Still, it is estimated that barely 5–10 per cent of the pre-existing forest canopy is being opened and, on average, only 7 m^3 are being harvested (Essama-Nssah and Gockowski 2000: 5). As a joint research report drawn up by Tropenbos and CIFOR states, the rapid opening-up of large areas for logging is worrying, but 'it is difficult to conclude that the current level of timber harvesting per hectare is a threat to the sustainability of the forest' (Eba'a Atyi 1998: 2). Selective harvesting has reduced the frequency of species such as *moabi* in the Dja forest and *padouk*, both in East Province (Delvingt *et al.* 2000: 33). But indirect impacts are more important. As Brown and Ekoko show (2001), this includes improved hunting access stimulating the bushmeat trade (section on 'Structural changes in income and demand'), demand from timber firms favouring the generation of local incomes in terms of food sales, room rents and compensations, better access for in- and return-migration (section on 'Structural changes in income and demand') and, in particular, agricultural conversion facilitated by logging roads (section on 'Windfall impacts on government spending').

Trade policy impacts

Historically, Cameroon's trade policy has been characterised by a traditional emphasis on home-market protection, import-substituting industrialisation and an anti-export bias (Milner 1990). The level of protectionism in Cameroon has been higher than in the average developing country (Thiele and Wiebelt 1994: 172). When the Central African Economic and Customs Union (UDEAC) was transformed in 1994 into the Central African Economic and Monetary Community (CEMAC), trade policy also changed. Simultaneous liberalisation within the CEMAC implied large potential trade benefits for Cameroon, in particular for its agricultural exports (Bakoup and Tarr 1998). Liberalisation has been an additional factor favouring export growth after 1994.

The 'hidden' trade policy of agricultural pricing has been the most important tool affecting agricultural competitiveness. The Office National de Commercialisation des Produits de Base (ONCPB) had a key role in setting agricultural prices throughout the marketing chain. For food crops, until the mid-1990s the government attempted to control key prices for rice, flour, edible oils, sugar and bread in the Yaoundé and Douala markets, but this was not very effective (Amin 1996: 21). For cash crops, ONCPB's mandate was to ensure fair trade margins and reasonable consumer prices and to 'stabilise' producer prices. In practice, however, the ONCPB siphoned off export revenues during price booms, such as those in coffee and cocoa in the late 1970s, without compensating growers during busts. For the period 1970–85, the ONCPB retained 48 per cent of coffee and 46 per cent of cocoa revenues, very little of which was ploughed back into the cash-crop sector (Amin 1996: 21). The phenomenon of 'price stabilisation' was thus *de facto* heavy taxation that in the long run drained the cash-crop sector of resources. Stabilisation funds were diverted to other sectors, as well as being used for bureaucratic expansion within ONCPB; the scheme effectively promoted rent-seeking (Blandford *et al.* 1994: 149–53). To some extent, high taxes on exported cash crops were deliberately intended to favour producers going over to food crops, in an attempt to promote national food security (Hoogeveen and van Soest 1993: 19).

How did pricing policies vary between sub-periods for the two most important cash crops in the forest zone, cocoa and *robusta* coffee? External prices were favourable in 1983–6, and the government resisted the temptation to tax away the whole windfall, as had happened during the previous boom of 1977–9: it realised that cash-crop producers were already being punished by the oil-led RER appreciation. A genuine stabilisation effort was even made during 1987–9, when external prices were almost halved and the producer price was raised to about 80 per cent of the external price. But when international prices dropped even further in 1990–1, ONCPB became practically insolvent. In the meantime, its work force had grown to 2,800 employees and, as part of the SAP, it was wound up and replaced in 1991 by the Office National du Café et du Cacao (ONCC), which had a staff of only 157 (Gockowski 1994: 15–16). ONCC was to set producer prices closer to export prices, as indicated by the ratio of internal to external prices.[23]

In terms of spatial impacts, it is interesting that liberalisation also eliminated the old system of pan-territorial pricing. ONCPB had paid approximately the same 'just' price to coffee and cocoa producers everywhere in the HFZ (C. Diaw, personal communication, Bogor, 30 November 2000). Obviously other factors, such as low transport frequency and high middlemen margins, worked against farmers in remote areas. But pan-territorial pricing was a partial element that worked in favour of remote cash-crop farmers in being practically a transport subsidy. Also, the government's provision of marketing services and technical assistance even in remote areas was helpful to frontier farmers, as became clear when the system was liberalised and services were cut.

One trade policy that worked in favour of cash crops during the oil boom was the subsidisation of imported fertilisers, insecticides and fungicides. By the mid-1980s, 60 per cent of imported fertilisers were being subsidised at 67 per cent of their full cost. However, many producers remained rationed in their access to subsidised fertiliser, and a parallel market developed. The combination of high subsidies and rationing was an ideal recipe for bureaucratic distribution, rent-seeking and inefficient use. Farmers usually received and applied fertilisers too late, in October instead of April. Cheap fertiliser was often used for less appropriate crops: in some areas 90 per cent of coffee fertiliser was used for food crops. The size of margins and rents became clear after subsidies were cut by 75 per cent as part of the SAP. This only led to a 30 per cent rise in the purchase price, implying that the lion's share of subsidies had previously been absorbed by margins and profits (Blandford *et al.* 1994: 150–3, 160–1).

In other words, input subsidies did benefit cash-crop (and some food-crop) producers. They probably also reduced deforestation and forest degradation marginally by making some cropping systems more land-intensive. There is tentative evidence that fallow length has been reduced since the 1970s,[24] and that coffee productivity per hectare rose during the oil boom, thanks especially to fertilisers. But in productive terms, probably a large proportion of the subsidies was diverted or wasted.

For the last three decades as whole, it is correct to say that agricultural smallholders suffered from Cameroon's trade policy (Benjamin *et al.* 1989). Industries and agro-industries were heavily protected and thus expanded during the oil boom in a sheltered market, just like the construction and service sectors. This reinforced urban biases and the trend toward urbanisation (see the section on 'Structural changes in income and demand'). The policy bias against cash crops was somewhat reduced during the boom. High subsidies for the urban consumption of rice produced in the non-forested north greatly hurt competing HFZ food crops during the oil boom, which further accelerated their decline and alleviated forest-conversion pressures (D. Kaimowitz, personal communication, 18 February 2001). Consequently, the aggregate impact of trade policies was to favour urban over rural sectors. Within the rural sector, the trend during the oil boom was to favour the more land-intensive cash crops, at the cost of land-extensive food crops. At the margin of other trends, trade policies thus probably curbed the demand for converted land, helping to contain deforestation.

A quantitative view

In this section, we have identified an intricate pattern of land-use adjustment to changing competitiveness across primary land-using sectors. An aggregated, quantitative approach can give us a supplementary test of how important the 'core' Dutch Disease shift from oil boom to declining competitiveness and to falling demand for agricultural land and timber production was in Cameroon. Table 6.3 provides regression results for the time-series that were pulled together for this study.

Table 6.3 Cameroon: relating oil wealth to relative prices and traded sector production: regression results, 1971–98

Dependent/independent	*Petroleum exports (constant million 1995 US$)*	*Capital inflows (nie, million constant 1995 US$)*	*RER (1990 = 100)*	*Non-agricultural GDP (constant million 1995 US$)*	R^2 (%)	*F-value*	*Years*
1 RER (1990 = 100)							
Coefficient	0.0019	0.014			17.6	2.667	1971–98
T-value	0.44750	2.25497**					
2 RER (1990 = 100)							
Coefficient	0.05	0.029			92.1	58.123	1986–98
T-value	8.89754***	8.47349***					
3 Agricultural value added (% of GDP)							
Coefficient			−0.33147		50.6	26.622	1971–98
T-value			−5.15966***				
4 Agricultural value added (% of GDP)							
Coefficient			−0.42933		89.9	97.649	1986–98
T-value			−9.88175***				
5 Industrial wood production ('000 cubic metre)							
Coefficient			−21.53439		5.9	1.642	1971–98
T-value			−1.28154				
6 Industrial wood production ('000 cubic metre)							
Coefficient			−21.16518		64.1	19.647	1986–98
T-value			−4.43248***				
7 Industrial wood export ('000 cubic metre)							
Coefficient			−30.37277		42.6	19.325	1971–98
T-value			−4.39605***				
8 Industrial wood production ('000 cubic metre)							
Coefficient			−36.65755	0.56596	70.7	30.137	1971–98
T-value			−3.74777***	7.43024***			

Notes

** Parameter *T*-value significant at the 5 per cent level.

*** Parameter *T*-value significant at the 1 per cent level.

Regression 1 shows the impact of oil revenues and capital inflows on real currency appreciation during the full 1971–98 period. Both parameters have the expected positive sign: higher oil-export revenues and financial capital inflows reduce price competitiveness, but only the latter parameter is significant (at the 5 per cent level). The two-variable model explains less than one-fifth of actual variation (R^2 = 17.6 per cent). We attribute this deficient model fit to the factors listed previously: coffee- and cocoa-boom impacts in the 1970s, the uncertain level of oil-revenue repatriation, an imperfect RER index for 1971–6, and the distortions of official inflation indexes caused by price controls.

We know that all these biases became less important after 1986. Regression 2 thus examines the same model for the post-boom period only (1986–98). The differences are astonishing. Almost all the RER variation is now explained by the model (R^2 = 92.1 per cent), and both its explanatory variables are highly significant (at the 1 per cent level). During the post-boom periods of economic crisis, devaluation and subsequent slow recovery, the changes in RP are thus well explained by changes in oil revenues and capital inflows.[25] The Dutch Disease RP mechanism was indeed at work in Cameroon.

How did RP affect agriculture's share of GDP? Regression 3 shows that real appreciation did have the expected negative impact on the position of agriculture in the economy: the negative parameter is significant at the 1 per cent level, and about half the variation in the agricultural share of GDP is explained. In regression 4, we try out the same model for the 1986–98 period only. Again, model fit is greatly improved (R^2 = 90 per cent). The transition from RP to agricultural production is therefore explained better after the onset of the crisis. Agriculture was the true loser from Cameroon's Dutch Disease. However, as we have seen, forest impacts depended not only on production values, but also greatly on the balance between specific HFZ cash and food crops. Food crops came back massively with the economic crisis, and this significantly drove up deforestation.

The second production variable of interest is timber output. Regression 5 explains aggregate industrial wood production (see the definition in Figure 6.2) for 1971–98 by RER movements. The results are very poor. The parameter has the expected negative but insignificant sign, and the R^2 of 5.9 per cent is extremely low. Again, by applying the same model to 1986–98 in regression 6, the results improve markedly, with parameter significance and an R^2 of 64.1 per cent.

Does this mean that the causal link between RP and timber production was tremendously poor prior to the crisis? There are major discontinuities between the Cameroonian economy both before and after 1986, but the picture is probably even more divided by markets than by time-periods. Up to 1986, much industrial roundwood production was taken by a growing home market undergoing a construction boom that was effectively sheltered from foreign competition. Thus total production was only marginally hampered by relative price variations. That changed radically after 1986, when the home market started to collapse and export markets increased rapidly in importance, especially after the 1994 devaluation.

Some proof for this interpretation is provided in regressions 7 and 8. In regression 7, wood exports alone are shown to be well explained by relative price changes for the entire period (1971–98). The parameter is highly significant and the model's R^2 is 42.6 per cent. But can we also explain total production over the whole period? Regression 8 re-examines model 5 – total industrial wood production regressed on the RER – but in this version non-agricultural GDP (in fixed 1995 prices) is added as a proxy indicator for demand on a sheltered home-market. The results change remarkably. R^2 was 6 per cent in regression 5, but now rises to 71 per cent, and both parameters have the expected signs, at 1 per cent significance level. This underscores the point made above about the importance of domestic wood demand. At the margin of this expansion, the RER thus had the expected competitiveness effect during the whole period, in particular for the export component. Below we shall examine how budgetary priorities affected the adjustment of agriculture, forestry and nature conservation to the sequence of boom and bust.

Windfall impacts on government spending

Agriculture and forestry

At the beginning of the oil boom, the government's stated intention was to use a substantial share of oil revenues to provide new financing for agriculture. Indeed, this happened: CHB extra-budgetary funding allocated at presidential discretion to the Ministry of Agriculture jumped from CFAF 300 million in 1981–2 (US$1 million) to CFAF 11.82 billion in 1984–5 (US$26.67 million) (Jua 1993: 150). Fertiliser subsidies became the most expensive item in the Ministry's budget: in 1987, the total cost was CFAF 6 billion (US$200 million) (Blandford *et al.* 1994). Subsidies to the SODECAO leaped from CFAF 200 million in 1983–4 (US$0.49 million) to 10.5 billion CFAF in 1984–5 (US$25.67 million) (Jua 1993: 150). In other words, a lot of oil money went into a number of agricultural budgets.

In terms of achieving agricultural development, the problem was not the size of funding, but the direction and quality of spending. Some of the money fell into the hands of a 'kleptocracy': for instance, in 1988 one-third of the stationary budget of the Ministry of Agriculture went on stationary that was never delivered (Jua 1993: 148). Bureaucracies administering the new funding rose rapidly. A large portion also went into parastatals, but this sector never came to contribute more than 10 per cent of the value of agricultural output (ibid.: 150). The most successful agro-industries were in the North. Production of crops like cotton and rice were stimulated, though at a high cost (Blandford *et al.* 1994: 149). As can be seen in Table 6.2, plantation investments and area planted in the HFZ by the Cameroonian Development Corporation (CDC) and other parastatals also benefited from the oil boom, but the increase in annual area was about 7,000 ha, or 5–10 per cent of estimated total deforestation. On the other hand, small-scale producers received only 5 per cent of the total volume of credit during 1977/8–1984/5. The small-scale sector remained confined to informal credit, the *tontine* system, where interest rates were much higher (ibid.: 146). Tight monetary policies during the boom aggravated farmers' interest-rate costs (Manning 1991).

There were also some incentive schemes for small-scale cash-crop farmers to expand their land-use activities. For instance, planting subsidies for coffee (*robusta* and *arabica*) were offered in the 1980s. But these were applied in established agricultural areas and never reached remote producers at the forest margins, where most deforestation occurs (Essama-Nssah and Gockowski 2000: 10). Together with input subsidies and a less discriminatory cash-crop pricing policy, the government was able to redirect some of the oil money into cash-crop agriculture. But the deforestation impact was very limited. On the one hand, agricultural production rose little, due to an over-emphasis on large-scale investments, bureaucratic expansion, corruption and resource waste. The type of production that benefited most – large-scale plantations and cash-crops in established agricultural areas – was rather capital- and land-intensive, and thus had modest expansionary land-use and forest-conversion impacts.

A second question in this sub-section is how public forestry administration was affected by oil wealth. Did it benefit from higher budget allocations, which translated into greater efficiency in regulating forest management and protecting the resource on the ground? As mentioned above, domestic timber consumption grew rapidly during the oil boom. The Forest Law of 1981 proved to be increasingly inept at providing a management framework to cope with the growing pressures on forest resources. Oil wealth meant that policy-makers had neither the incentive nor the need to raise taxes on forests. In particular, the low level of area taxes and concession prices promoted land-extensive logging operations with inefficient resource use (Brunner and Ekoko 2000: 63). By the end of the 1980s, stumpage and area taxes were heavily undervalued, and the discretionary allocation of concessions contained an important element of rent distribution to vested interests.

When the crisis hit Cameroon, this was the background for the World Bank's efforts to use structural adjustment conditionalities to push for reforms of Cameroon's forest policy. The reform elements in the 1993 draft proposal for a new forestry law, prepared jointly by the Bank and the Ministry of Environment and Forests (MINEF), were:[26]

1. to introduce a competitive auctioning system for the allocation of concessions;
2. to increase tax rates, especially area taxes, to improve resource efficiency and raise revenues;
3. to make forest management plans obligatory for all larger concessions; and
4. to give local communities the rights to forest management and greater logging compensations.

The proposed reforms contained a balanced mix of fiscal, equity and efficiency objectives, including environmental considerations. The decline in oil wealth and the severity of the economic crisis gave the Bank a strong bargaining position in the structural adjustment process. Nevertheless, in a year-long process of negotiation it proved impossible to implement more than a fraction of the proposal. Forestry taxes were actually raised, and are now about double the size of taxes in neighbouring Gabon (Bikié *et al.* 2000a: 27–8). But the auctioning system was only partially implemented, due to the failure to overcome strong vested interests in both the public and private sectors. As a result, illegal logging soared (Brunner and Ekoko 2000; Karsenty 2000).

Oil wealth thus delayed legal, and in particular fiscal steps to regulate the growing timber sector. The oil bust raised fiscal measures and awareness of the need for regulation, but the capacity to implement these measures was weakened.[27] One asymmetry here is that rent-seeking and corrupt practices that were nourished during the oil boom proved highly resistant after the economic breakdown. Furthermore, fiscal cuts during structural adjustment also reduced the MINEF's ability to enforce regulations on the ground (Bikié *et al.* 2000a: box 4). Thus contradictory forces affected the link between oil wealth and efficient forestry regulation.

Finally, did domestic funding for protected forest areas benefit from oil richness and, vice versa, did it suffer fiscal austerity from the economic crisis? There were some fluctuations in that respect, but protected-area funding was overwhelmingly external and thus detached from the oil cycles. Most of Cameroon's protected areas were created in colonial times, especially between 1932 and 1950 (Sournia 1998). With a broad area coverage, by the mid-1970s Cameroon was considered one of the most advanced countries in francophone Africa (Depierre and Ole 1976: 5). However, conservation was never a spending priority for Cameroon, not before, during or after the oil boom. Culverwell (1998), cited in Wilkie and Carpenter (1998), estimates government spending on protected areas at US$143,325 for 1996. That was 0.01 per cent of the state budget, down from 0.04 per cent in 1990. But the 1996 figure represented only 3 per cent of total annual protected-area spending: on average, foreign donors had spent US$4.6 million/yr since 1993. Furthermore, external donations rocketed after the crisis started in 1986, especially for the newly created Korup National Park (ibid.). In other words, Cameroon left basically all the financing of protected areas to foreign donors. Their overwhelming financial predominance throughout the period meant that the small oil-related budget cuts had little if any impact on conservation.

Roads and other infrastructure

For countries that specialise in primary commodities, transport is essential to the development of commercial links, both between regions and for external trade. In Cameroon, access to transport continues to be a bottleneck for primary sectors such as mining, cash-crop agriculture and forestry. The total road network was 34,200 km in 1995, 12.5 per cent of which was surfaced, and road density was 0.07 km/km^2. This is very low by international standards, but comparable to that of Central African neighbours (IRF 2000: 10). Road density is much higher in the south, but distances between towns are greater in the northern provinces. The rudimentary state of Cameroon's physical infrastructure by the end of the 1970s is best described by the fact that the two main cities, the capital Yaoundé and the port of Douala, were connected neither by an all-weather road nor by a functioning railroad. Most roads are dry-weather with irregular access during the rainy season(s). Maintenance of existing roads is a serious problem, so heavy risks and costs may be involved in travelling on poorly constructed and badly maintained roads in marginal areas (S. Hauser, personal communication, Yaoundé, 19 May 2000). Furthermore, traffic density is limited by the very low number of vehicles. The frequency of four-wheeled vehicles per km of road in 1987 was 2.3, compared to 3.7 for Bolivia, 5.1 for Botswana and 24.4 for South Africa (IRF 1988: 28).

Infrequent, costly and risky transport on bad roads in Cameroon seriously limits the viability of commercial production in remote areas. Furthermore, the low transport frequency favours middlemen monopolies or oligopolies with high profit margins, thus lowering the incomes of small-scale producers and worsening income distribution (Benjamin 1996: 1007–8). Isolated rural communities thus tend to favour more and better roads, in particular because it allows them to sell surplus agricultural production.

Building durable roads is often one of a village's requirements in negotiating with logging companies to enter forested areas (Carrière 1999; Dounias 1999). Improved road conditions permit a wider range of economic activities and also a larger degree of land clearing (Brown and Ekoko 2001). Road *improvement* also enhances deforestation, though typically less than the construction of new roads.[28]

The oil-boom years brought a notable expansion in national transport networks, which became one of the government's spending priorities. The budgetary expansion was strongest in the fiscal years 1981–2 and 1982–3, based on CHB oil funds and foreign loans (Ministry of Equipment 1984). Road expenditure (including loans) rose sharply from US$178 million in 1983 to US$294 million in 1986, before falling back to US$211 million in 1987 after the start of the economic crisis.[29] This was an area where results on the ground were actually achieved. Between 1983 and 1986, the length of main and national roads increased by 17 per cent and of secondary roads by 13 per cent (IRF 1988: 10). In particular, the Douala–Yaoundé road was asphalted, and the old German-built railway was rehabilitated in the early 1980s. At the same time, boom money also led to a significant increase in the number of vehicles from 1983 to 1986, by 53 per cent for private cars, 76 per cent for buses and 35 per cent for goods vehicles (ibid.: 84). A highly ambitious plan to (re)construct 10,564 km of roads was drawn up (DeLancey 1989: 121), but after the crisis only a fraction was actually built (Jua 1993: 141).

What impact on forests had this oil wealth-induced expansion of roads, infrastructure and transport? Again, in a country as diverse as Cameroon, it is necessary to distinguish impacts according to geographical area: some road-building, for example, in the remote and non-forested north, obviously had no detectable impact on forests. However, from the 1970s onwards logging interests pressed for more road-building in the HFZ (DeLancey 1989: 116–20). Unlike Gabon, Cameroon lacks a large river on which extracted logs can be evacuated to ports and is therefore more dependent on roads and rail. The link between roads and timber was intensified during the oil boom, which allowed new logging frontiers in East Province to be opened up, this being a main factor in the expanding timber sector. Thus to the extent that increased timber extraction directly or indirectly caused forest degradation (see the discussion in the section on 'The competitiveness of agriculture and forestry'), road-building was highly instrumental in that process.

Table 6.4 documents boom-led road expansion into forested areas, based on contemporary ministerial data on the length of constructed and upgraded asphalt roads. Road projects are ordered according to their year of completion, whether they are in the HFZ or not,[30] and whether they were restorations or new constructions. Prior to independence, about the same length of roads was constructed (681 km) as during the next one-and-a-half decades (628 km). During 1975–9, that is, just prior to the oil boom, construction accelerated to 1,069 km, but this was further expanded to 1,437 km during the boom years of 1980–4. There were also shifts in geographical distribution. Prior to the boom, there had been a clear emphasis on construction in the north. The annual HFZ construction rate was only 18.6 km (44 per cent of the period total) for 1960–74 and 58.8 km (28 per cent) for 1975–9. This was reversed during the oil boom, when the HFZ became the favoured region for road projects (151.2 km/yr; 53 per cent of the period total). Finally, within the regional total, the emphasis of the HFZ road programme during 1975–9 had clearly been on rehabilitation and upgrading existing roads (92.5 per cent of total

Table 6.4 Cameroon: construction and upgrading of asphalted roads, 1960–84[a]

		Km	Km/yr	*Percentage*
I	*Before 1960*	681		
	HFZ[b] (km)	681		100
	Construction (%)		100	
	Upgrading (%)		—	
	Other regions (km)	—		
	Construction (%)		—	
	Upgrading (%)		—	
II	*1960–1974*	628		
	HFZ[b] (km)	279		100
	Annual Humid Forest Zone (km/yr)		18.6	
	Construction (%)			92.1
	Upgrading (%)			7.9
	Other regions (km)	349		100
	Annual other regions (km/yr)		23.2	
	Construction (%)		100	
	Upgrading (%)		—	
III	*1975–1979*	1,069		
	HFZ[b] (km)	294		100
	Annual Humid Forest Zone (km/yr)		58.8	
	Construction (%)			7.5
	Upgrading (%)			92.5
	Other regions (km)	775		100
	Annual other regions (km/yr)		155	
	Construction (%)			100
	Upgrading (%)		—	
IV	*1980–1984*	1,437		
	HFZ[b] (km)	756		100
	Annual Humid Forest Zone (km/yr)		151.2	
	Construction (%)			63.1
	Upgrading (%)			36.9
	Other regions (km)	681		100
	Annual other regions (km/yr)		136.2	
	Construction (%)			44.9
	Upgrading (%)			55.1

Source: Ministry of Equipment.

Notes

a Classification refers to year of project completion.

b Humid Forest Zone: East, South, Littoral, Centre and Southwest Provinces.

length), but during the boom years this shifted abruptly to new construction (63.1 per cent). Thus new road construction in the HFZ in 1980–4 was 95 km/yr, compared to barely 16 km/yr for 1960–74 and 5 km/yr for 1975–9. Normally, new road construction triggers larger land-use impacts than upgrading (see Chapter 3), so a nineteen-fold increase in road construction over the previous period is a momentous change.

To what extent did this road expansion of roads in the HFZ actually have a negative impact on the forests? Indirect impacts are by far the most important, and these differ in space according to the strategic importance and type of economic opportunities that such roads permit. The asphalting of the Douala–Yaoundé road was important because it connected central Cameroon to the main port, and the extension of the road into East Province was vital for timber extraction. For instance, it is reported that the eastward stretch between Yaoundé and Ayos had serious impacts on the population of elephants and large primates, as well as the indigenous Baka forest-dwellers (Essama-Nssah and Gockowski 2000: 10). Roads and logging together permitted over-hunting and forest degradation. Of course this is also true for dirt and gravel roads, which are not counted in Table 6.4. In areas with reasonable market access, for example, near Yaoundé, new or improved roads also increased deforestation proper. In a spatial regression model using sample points at 5 km intervals in the entire HFZ, it was found that the average forest-cover proportion increases monotonously by 0.7 percentage points for each kilometre of distance from roads (Mamingi *et al.* 1996). So, the closer a forest area is to a road, the more likely it is to be cleared. The likelihood is twice as large as that for the corresponding distance to railways. Road distance also proves to be more significant in determining forest loss in Cameroon than in Zaire, mainly because commercial agriculture is more important (ibid.).

A number of factors qualify the general deforestation impact. Roads seem to gain in importance in the second (the post-logging) phase of colonisation, when basic production structures have already been developed (Mertens and Lambin 2000: 26). Also, an important constraint on colonisation in areas opened up by roads is the strength of customary systems of land rights, which provide impediments to the movement of different ethnic groups into other groups' territories. Finally, the immediate effect of road-building in some remote areas of the HFZ has also sometimes increased out-migration, caused by increased mobility and contact to urban areas (Franqueville 1984). This shows that roads and lower transport costs can have ambiguous impacts: at least in the short run, they guarantee neither rural development nor increased deforestation.

In other words, oil money did have a highly expansionary effect on road budgets. In addition, new roads into forested areas of the HFZ had a disproportionately high share among road projects, compared to road upgrading. Many projects were motivated by timber interests; indeed, roads were the *sine qua non* for an increase in logging. Road upgrading, and in particular new road construction, had the expected forest-degradation impacts (poaching in particular), and in many places they also accelerated deforestation. With the economic crisis and the fiscal cuts under structural adjustment, road-building stagnated totally, and poor maintenance caused the road network to decline, which especially hurt agriculture (Amin 1996: 22). However, the sheer existence of new roads proved to be an asymmetry that affected forests well into the crisis period.

Directed settlement

Another policy influence on forests might be the use of oil money to finance public (re)settlement programmes that change the occupation of rural space. As we saw, resettlement was an important issue for rural land use in neighbouring Gabon. In Cameroon, historically the colonial powers also resettled forest-based populations to near-road areas by force, in order to increase political control and integration into cash-crop markets, and to mobilise labour for work in the plantations (Franqueville 1984: 436; 1987: 554–60). One of the objectives of these colonial interventions was the 'sedentarization' of the semi-nomadic (mostly Bantu) tribes that inhabited most of the HFZ (Diaw 1997: 3).

Since independence, there have been some resettlement programmes affecting the HFZ, such as in the Yaoundé area (1964–8) or the Yabassi-Bafang initiative to resettle Bamiléké people from West Province into Littoral Province (1970). However, their net impact on forest cover is not clear, and none was implemented during the oil-boom period. Outside the HFZ, resettlement programmes did coincide in part with the oil bonanza, such as the Baygone Plain Programme in 1982 (Bamiléké, Foumban area, West Province), and the Northeast Benoue (1973–85) and Southeast Benoue (1978–86) projects in the savannah zone of the Far North Province (O. Ndoye, personal communication, 29 June 2001). While some of these initiatives led to more bushland conversion (e.g. in the northern Mandara mountains – see Campbell and Riddell 1982), none of them had an impact on forests.

Post-independence resettlement efforts were much less prominent than in Gabon. An intuitive explanation is that trade in agricultural products, for both export and emerging urban markets, developed much more vigorously in Cameroon than in Gabon. This means that market incentives gradually induced rural people to live closer to roads and urban areas anyway, without the necessity of forced government intervention (Franqueville 1987: 539–43). It has been shown that road density is significantly correlated with population density in the HFZ, and that population density is correlated with the proportion of deforested areas (van Soest 1995). On the other hand, the overall feasibility of farmer resettlement from the densely populated northern regions into the rural HFZ has been restricted by ethnic and cultural barriers, as well as highly articulated systems of land rights in the in the HFZ (Toornstra 1994: 6; Diaw 1997).

Structural changes in income and demand

Poverty alleviation

Cameroon's marked fluctuations in national income, with rapid growth (1975–85) followed by a similarly rapid fall (1986–94), triggered significant changes in well-being, employment and consumption patterns across different population groups. The derived changes also affected agricultural land use and the harvesting of forest products. This is particularly true, as poverty in Cameroon has a strong regional and spatial dimension. The incidence of poverty is generally much higher in the rural savannah, plateau and forest regions than in Douala, Yaoundé and other towns (MINEFI 1998a: 61). In the 1996 household survey, income per adult in the rural HFZ was actually the lowest in Cameroon (CFAF 165,600; US$324). The highest was in the HFZ cities, Douala (CFAF 552,500; US$1,080) and Yaoundé (CFAF 442,000; US$864) (UNDP 1999d: 35).

The sequence of coffee, cocoa and oil booms allowed Cameroon's per-capita GDP (in fixed 1987 dollars) to rise by an astonishing 61 per cent, from US$735 in 1975 to US$1,183 in 1985 (UNDP 1999c: 2). Although oil revenues were far from equally distributed, increased urban demand and employment and their derived effects are likely to have reduced poverty markedly. In 1984, 40 per cent of households and 46 per cent of the population were still below a poverty line defined mainly in terms of the cost of a food-consumption basket (World Bank 1999b: 196; UNDP 1999d: 15–19). In the subsequent decade of economic crisis, real per-capita income fell back to its 1975 level (US$756 in 1997; UNDP 1999c: 2). 'Poverty, which was a marginal and limited phenomenon in the early 1980s, has spread all over the country and is affecting all levels of society' (UNDP 1999d: 11). From the poverty lines used in the same source, the increase looks less dramatic: 38.4 per cent of households and 50.5 per cent of population were living below the poverty line in 1996, representing only marginal changes over 1984 of 40 per cent and 46 per cent, respectively (ibid.: 18). Alternatively, IMF estimates show that the incidence of poverty rose from 40 per cent in 1980–5 to 51 per cent in 1997 (IMF 1999c: table 4). Non-income poverty criteria show a mixture of trends over the last one-and-a-half decades: housing conditions, literacy and some health indicators have improved, while malnutrition seems to have worsened (ECAM 1997: 46; IMF 1999c: table 4; UNDP 1999d: 20).

What are the main dimensions of poverty in Cameroon? Large household size, low educational levels and a lack of formal-sector employment all strongly promote higher poverty across the sample in the latest household survey (ECAM 1997: 45). Also, growing unemployment has transformed the urban economy: 'The spread of urban unemployment due to the recession of the 1985–95 decade was the major cause of poverty, especially in urban areas' (UNDP 1999d: 50). However, not all unemployed people are poor people: actually the incidence of unemployment rises with educational level and is higher among the non-poor than the poor (ibid.: 52–3). This underlines the fact that unemployment alone is a poor poverty indicator, although it doubtless contributed to the rise in urban poverty.

How does this dynamic pattern of poverty relate to deforestation? Obviously, the disparity between rural and urban parts of the HFZ was highly conducive to rural–urban migration, which has been a main route of poverty-alleviation as well as causing significant land-use impacts (see next section). But rural immiseration has also induced rural–rural migration and forest colonisation. One specific example has been forest conversion for *robusta* coffee in the Moungo Division of Littoral Province by rural migrants from West Province, who are responding to high population growth and increasing poverty in the sending highland areas (Essema-Nssah and Gockowski 2000: 9). Similarly, people from the plateau region hit by drought in 1982–3 started to move to the high slopes of the forested volcanoes in Western Cameroon and to clear the forests for crops and pastures (Tsalefac 1994).

Higher poverty and lower labour remuneration not only triggers land colonisation, it can also lead to the over-exploitation of marginal open-access forest products. These can provide livelihood options for poor producers and cheap supplies for urban consumers, but may also cause the degradation of forest resources. Ndoye *et al.* (1998) indeed found that the crisis promoted over-exploitation of medicinal plants by impoverished producers, following a large increase in the prices of imported drugs after devaluation. This link also applies to a wider range of non-timber forest products (NTFPs): in fifty-four HFZ villages in South, Centre and East Provinces, there was a significant increase between 1985 and 1997 in the number of households collecting NTFPs (Bikié *et al.* 2000b). The poverty argument is sometimes extended to other forest products, like bushmeat, and taken to the point that forest conservation was effectively made impossible by economic crisis, currency devaluation and natural-resource pressures (Tandjeu 1998).

However, it would be dangerous to generalise. For bushmeat, Bennett and Robinson (2000), using Central and Western African studies, rather conclude that fewer income-poor hunters increase the use of '[f]irearms, cartridges, batteries, outboard motors, motor vehicles and fuel', which all raise harvesting levels and cause defaunation. There is also no empirical support for the claim that bushmeat is an 'inferior' product preferred by crisis-hit poor consumers (see the section on 'Structural changes in income and demand'). For other forest products, like firewood, the crisis-induced substitution of hydrocarbons has indeed increased urban consumption and periurban harvesting (Essama-Nssah and Gockowski 2000: 7), although this probably caused little additional forest loss, beyond what was being cleared for food crops as the primary motivation.

In other words, the oil boom probably reduced poverty in both rural and urban areas. This trend was certainly reversed during the economic crisis period. By then, poor producers and consumers were turning much more to the 'free' (over-)harvesting of cheap forest products, although effects may differ across products. Some impoverished urban migrants returned to the countryside to take up low-remuneration food-crop activities. Rural–rural 'push' migration into forested areas occurred in response to natural disasters or unsustainable population densities. On the other hand, some dimensions of poverty also reduced forest-clearing. For instance, sustained poverty and credit constraints probably prevented cocoa- and coffee-producers from extending their cultivated areas in response to devaluation and favourable world-market prices. On the whole, it seems that in Cameroon rising poverty overall increased forest pressures. The most important links between poverty and land use were created through rural–urban migration, which will be analysed in the following section.

Rural–urban migration

In a country where smallholder agriculture is the decisive factor in the extent of forest-clearing, the size and distribution of the agricultural labour force will have important influences on land use. Among Central African countries, Cameroon has always been one of the more rural, but its rate of urbanisation (urban over total population) has risen considerably over the last half-century, from 10 to around 50 per cent (Moriconi-Ebrard 1993, cited in Wolff n.d.: 5). How much has this long-term trend towards greater urbanisation been influenced by oil wealth?

Again, the vigour of our analysis is limited by an acute lack of statistics. Cameroon's last population census was in 1987 and its last agricultural census in 1984. National statistics like MINEFI (1998a: 22–7) do publish more recent population figures, which would seem to indicate a linear continuation of Cameroon's rapid urbanisation: 47.2 per cent in 1997, up from 37.8 per cent in 1987 and 28.5 per cent in 1976. Amazingly, the figures have made their way directly into standard international statistics like World Bank (1999a) and FAO (2000b), but they are simply extrapolations of previous trends and are not based on new empirical findings. This makes them completely inadequate for our purpose, namely the analysis of changing migration patterns between sub-periods. For the period after 1987, we thus have to look at sub-national data to seek the answers.

How pronounced was rural–urban migration within the HFZ prior to the oil boom? In 1974–5 Franqueville (1987) carried out a demographic and land-use survey of 2,479 households in 38 villages in Centre and South Provinces, including both the periurban upland of Yaoundé and more remote forest areas. The area covers one-third of the HFZ land area and can claim some representativeness for the HFZ as a whole. Franqueville found a strong urban drift during this period in regions dominated by coffee and cocoa. Prices for these products were sluggish until 1977, taxes were heavy, and import-substitution policies favoured the urban sectors.

The oil boom opened up further urban employment options in construction, parastatals and public and private services. This reinforced the pre-boom pattern of urbanisation, especially in the capital, Yaoundé: between 1976 and 1987, its population more than doubled from 318,700 to 649,000. Growth in the largest city, Douala, was 77 per cent; medium-sized towns grew by 92 per cent and small ones by 89 per cent (MINEFI 1998a: 27). Average annual urban growth was 5 per cent, and about half of this was due to migration. Obviously, urban migrants did not only come from the HFZ. But rural population growth in the five HFZ provinces was lower (0.9 per cent) than in other rural areas (1.4 per cent). The rate of urbanisation jumped from 27.8 per cent in 1976 to 37.5 per cent in 1987. Also, that rate was much higher in the HFZ than in the highland and savannah provinces (World Bank 1999a; Essama-Nssah and Gockowski 2000: 8).

The bulk of the surplus of rural population was thus absorbed into the cities, where most of the oil money was spent. This significantly reduced the pressure to clear the forests for new fields in the HFZ. Yet, in terms of households' ability to cultivate and clear land, not only do population figures matter, but also age structure. Agricultural census data indicate that in 1984, by the end of the boom, the average age of cocoa-growers was 52, while that of all farmers was 47 years (Blandford *et al.* 1994: 149). Many young people left for education and more rewarding employment in the cities. Those people who stayed behind in the villages had less capacity to cultivate the land or clear forests.

What was the impact of the economic crisis and, subsequently, of the devaluation, on migration flows between the rural HFZ and urban areas? No census data are available, so the best source is a demographic and land-use survey undertaken by Sunderlin and Pokam (2002) in the South and Centre Provinces, repeating the survey questions posed by Franqueville (1987) in the same thirty-eight villages two decades earlier. Annual population growth had been 0.75 per cent between 1976 and 1987, but went as high as 4.6 per cent in the post-boom period (1987–97).[31] Growth was largest in the areas closest to Yaoundé. Most of the increment was due to natural population growth, reflecting that, in this first phase of the crisis, fewer people migrated to the crisis-hit urban economy. However, as the crisis continued (after 1992), there was a net return migration from urban areas back to the countryside. As various observers have noted, migrants returned to rural areas both inside and outside the HFZ.[32] This was closely associated with the massive increase in the area of food crops. Men became increasingly involved in cultivation, which earlier had been done exclusively by women, probably also reflecting a larger degree of commercialisation. Because food crops require shifting cultivation with fallows, households also tended to clear more forests.

Overall, at 2.6–2.9 per cent over the last three decades (IMF 1999c: table 4), Cameroon's annual population growth is high, even by sub-Saharan standards, and has undoubtedly had notable long-term impacts on land use and forests. But, as both Sunderlin and Pokam (2002) and Ndoye and Kaimowitz (2000) point out, the link between population and deforestation was an indirect one, which was also significantly interrupted by the decade of oil wealth. Rural–urban migration proved to be an important intermediate variable, which was led by the dramatic shifts in the macroeconomic environment. Later, the economic crisis reversed this trend, severely reducing urban labour absorption, which triggered a substantial return to land-extensive food-crop cultivation and forest-clearing.

The structure of consumption

During the oil boom, rising per-capita income and rapid urbanisation significantly changed food-consumption patterns. While sources analysing consumption changes over time are scarce, recent household consumption surveys provide a cross-sectional perspective on income dynamics. The poor generally use a higher share of income on food, especially cereals and fish, and less on starchy food and meat (ECAM 1997). Higher income increases in particular the consumption of dairy products and meat (UNDP 1999c). Which of these boom-and-bust changes in income and speed of urbanisation affected consumption in a way that triggered land-use changes in the HFZ?

Higher demand for meat and dairy products tends to stimulate cattle-ranching. This also happened in Cameroon, but unlike the Latin American cases in this book, this did not lead to massive forest-clearing. One reason is that domestic supplies were exposed to the competition of cheap imported meat. The FAO's per-capita livestock-production index rose less than 1 per cent annually during the oil boom (1979–86). At the same time, meat imports exploded by a factor 15, from 1,556 t in 1979 to 23,047 t in 1986, followed by a sharp cutback in the economic crisis to 5,738 t in 1988 (FAO 2000b). Second, any minor stimulus to national production did not affect the forest zone: cattle-ranching is strongly rooted in the savannah zone of the north but remains negligible in the HFZ, for both veterinary and cultural reasons.[33] Thus until now cattle have not caused deforestation in the humid forest zone.[34]

A supplementary reason for this is that the forest itself also supplies meat to the HFZ. As in Gabon, bushmeat consumption is widespread, and harvesting is occurring at unsustainable rates. As in other parts of Central Africa, defaunation is perhaps the most important challenge to sustainable forest management (Delvingt *et al.* 2000; Bikié *et al.* 2000a: 17). The most common bushmeat species are artiodactyls, mainly duikers (*Cephalophinae*) and bushpigs (*Potamochoerus porcus*),[35] but site-specific pressures on primates (e.g. gorillas and chimpanzees) are significant. It has been suggested that bushmeat provides cheap meat to poor urban consumers, and that as a consequence consumption rises as a direct response to economic crisis (Tandjeu 1998). However, there is no evidence that bushmeat is cheaper than other meat, and consumption levels seem to be fairly equal across income classes (Essama-Nssah and Gockowski 2000: 5): it is definitely not a 'poor man's product'. Two Cameroonian field studies[36] even suggest the opposite, namely that wild meat is preferred to domestic livestock, and therefore is closer to being a luxury good. Consumption levels of bushmeat may actually have risen during the oil boom, partly because road-building facilitated increased supply of it.

A final factor to consider in this section is the role of shifting staple crops. It is difficult to divorce structural shifts in preferences from competitiveness effects (see the section on

'The competitiveness of agriculture and forestry'). The indexed value (1986 = 100) of food imports (excluding fish) more than doubled with the appreciating RER in the 1980s, from 54 in 1981 to 126 in 1990. As expected, imports dropped sharply after the 50 per cent devaluation in 1994, to index 38 in 1997 (FAO 2000b). During the oil boom, grains such as rice and maize had entered solidly into the diet of urban dwellers in the south. Rice came both from the north and from Asia. Rice imports multiplied from 11,039 t in 1981 to 74,871 t in 1987 and 90,287 t in 1990, before being halved in 1991, including because of increasing foreign exchange shortages (FAO 2000b). Rice reduced pressure on HFZ forest cover, as it reduced the demand for HFZ food crops. Conversely, a comparison of the 1993 and 1996 household-consumption surveys shows that devaluation had an immediate substitution effect. Consumption of domestic food crops like banana, plantains and igname all increased at the expense of imported rice (ECAM 1997). However, Asian currency devaluations in 1997 reversed this trend again, with record rice imports in 1997–8 (FAO 2000b). Essentially, price effects seem to have had a primacy, but rice in particular is a staple that has come to stay even in the face of short-run shifts in RP. This will help to reduce future pressures on forest conversion in the HFZ.

Changing consumption structures induced by the oil boom-and-bust pattern thus did have an impact on forest conditions. Some factors, like bushmeat over-exploitation, were driven more by the supply side, while other trends, like rising bovine meat demand, did not affect the HFZ but only the north. The most notable change was probably in staple crops from starchy tubers and plantains to rice, which reinforced the decline of HFZ food crops during the boom and alleviated forest-conversion pressures, followed by partial reversal in the 1990s.

Synthesis and conclusion

With about 17 million ha of closed forest (37.1 per cent of land area) and 24 million ha of total forest cover (1992–3; 51 per cent of land area), Cameroon is still a forest-rich country. The bulk of this forest is situated in the five southern provinces that together make up the HFZ. What sorts of pressures exist on the forests in the country? Cameroon is an important timber exporter, and logging has recently expanded rapidly into the eastern part of the HFZ. But timber extraction remains fairly selective. Its main degradation impact is defaunation from the expanding bushmeat trade. The FAO has estimated deforestation rates at 0.6 per cent over the last two decades, but the rates may be underestimated for the 1990s. There have been important fluctuations between sub-periods: little forest conversion took place during the oil-boom years (1979–86), while forest loss accelerated with the economic crisis (1987–93) and the post-devaluation period (from 1994).

The direct motives for deforestation in Cameroon are almost exclusively agricultural: oil, mining and other sectors play negligible roles. Expanding cultivation in the north draws mainly on savannah areas, and in the south on forests. Three sub-sectors may be distinguished within agriculture in the south of the HFZ. Shifting food-crop cultivation (plantain, tubers, etc.) is the most land-extensive type, and a predominant determinant of demand for land and forest conversion pressures. At the farm level, it competes with HFZ smallholder's cash crops, basically cocoa and coffee. Third, there are large-scale cash-crop plantations (banana, tea, tobacco, etc.) in coastal areas, but they form a minor proportion.

In slightly simplified terms, Cameroonian net forest loss is determined by the allocation of factors of production between three different sectors: urban (causing little conversion but some degradation pressures), HFZ cash crops (some deforestation, but rather stationary land use) and HFZ food crops (land-extensive systems with large deforestation impacts). The most significant farm-level decisions for forest conversion are thus, first, whether or not to pursue employment options in urban areas, which reduces that household's capacity to cultivate land or clear forest in their place of origin. The second choice for rural households is between land-extensive food crops and more land-intensive cash crops.

The oil boom of 1979–85 changed the fate of a small open economy that had successfully been relying on exports of cash crops. Suddenly blessed with oil wealth, the intended policy was to sterilise any Dutch Disease impacts by saving much of the windfall in extra-budgetary accounts (CHB) abroad in order to avoid inflationary disruptions to agricultural development. But after some years of booming oil revenues, spending pressures became too high. Public salary expenditure, subsidies to inefficient parastatals, and urban prestige and infrastructure projects began to absorb not only the resources generated by the oil sector, but also those that were repatriated from CHBs and some foreign borrowing. The RER, which had been stable, started to appreciate, and urban biases were increased still further. In spite of all initial preventive measures, Cameroon had caught the Dutch Disease.

When a dramatic terms-of-trade shock hit the economy in the mid-1980s, Cameroon was unable to respond adequately. The fixed exchange-rate regime made it impossible to devalue and restore competitiveness. A rent-seeking class and a variety of urban interests were able to resist fiscal adjustment, thus nourishing sustained currency overvaluation and increased foreign borrowing. This asymmetry in adjustment obviously made matters worse. The crisis caused a collapse in the formal urban economy, an increase in the informal sector and a labour return 'by default' to subsistence-oriented food-crop agriculture. Since the sharp 50 per cent devaluation in 1994, the economy has been recovering gradually. Cash-crop agriculture has slowly come back, including the export crops in the HFZ, cocoa and coffee.

How did this sequence of severe macroeconomic fluctuations affect the forests? Table 6.5 classifies ten effects of the oil boom of 1979–86. Macroeconomic impacts are linked to land-use changes and forest impacts. The primary emphasis is on deforestation, but degradation pressures are also outlined. A quick overview shows why deforestation decreased during the oil boom: seven factors curbing forest loss (areas 1–4, 7, 9 and 10) outweighed three factors favouring deforestation (areas 5, 6 and 8), and the curbing factors were on average stronger (last column).

Most of the boom spending accrued in urban areas, so rural–urban migration (factor 1) from the HFZ to Yaoundé, Douala and smaller towns was strong, and rural population growth was reduced significantly. This was probably the most powerful mechanism in reducing the rural labour available for both food- and cash-crop cultivation, and thus curbing forest-clearing. It was reinforced by a shift in urban food demand towards staple crops from outside the HFZ, notably rice (4). These trends were also linked to (urban) poverty alleviation (3), associated with higher remuneration and growing employment in the urban sectors. Boom-induced poverty alleviation, urbanisation and changing demand patterns thus deferred deforestation considerably.

Table 6.5 Cameroon: oil wealth and deforestation – an overview of impacts

Economic and productive impacts			*Links to deforestation*		*Deforestation impact*	
No.	*Type*	*Intensity*	*Type*	*Strength*	*Type*	*Intensity*
1	Rural–urban migration	Strong ●●	Modest growth in rural farming population	Close ➔	Less conversion and degradation pressures	Strong ▼▼▼
2	Loss of traded-sector competitiveness	Strong ●●	Cash (and food) crops stagnate; imports rise	Medium ⇨	Less forest conversion	Medium ▼
3	Reduced absolute poverty	Strong ●●	Higher labour opportunity cost (mostly) reduces forest pressures	Medium ⇨	Less forest conversion	Medium ▼
4	Higher urban income shifts food demand	Medium ●	Grains rise, substituting HFZ food crops	Medium ⇨	Less forest conversion	Medium ▼
5	New road and rail construction	Strong, lagged ●●	Opening up for logging, bush-meat, conversion	Medium ⇨	Degradation strong, conversion medium	Medium ⮝
6	Higher budgets of agricultural agencies	Medium ●	Capital- and land-intensive agro-industries	Weak →	More forest conversion	Weak ↑
7	Agricultural trade policy	Medium ●	Taxation slightly reduced, high input subsidies	Weak →	Reduced land-extensiveness of agriculture	Weak ↓
8	Oil and mining production	Close to zero	On-site direct and indirect forest loss	Weak →	Point impacts in coastal area	Close to zero
9	Higher budgets of forestry/park agencies	Close to zero	Weak domestic policy priority	Medium ⇨	Improved forest-law enforcement?	Close to zero
10	Directed (re)settlement	Zero – no relation to oil	Process finished before oil boom	Medium ⇨	Villages concentrated near roads	Zero

Notes
1–4, 7, 9 and 10 areas – effect reduces deforestation; 5, 6 and 8 areas – effect augments deforestation.

The Dutch Disease 'core mechanism' of declining competitiveness (2) was also clearly at work, though not entirely parallel to the oil boom. Real currency appreciation occurred in 1982–7, but overvaluation remained a problem for the traded sectors until the 1994 devaluation. Among the HFZ products, this certainly hurt timber and agricultural exports (cocoa and *robusta* coffee), that is, the 'pure' tradables. Plantation cash crops for domestic consumption enjoyed greater trade protection. Food crops were traded in part, that is, as exports to Nigeria or Gabon or goods competing in urban markets with imported or national substitutes, while, for example, the share grown for household consumption in remote areas was *de facto* non-traded, and thus not hit by relative price effects. Finally, while timber exports declined in the 1980s, this was more than compensated for by a booming, sheltered home market, which benefited from the urban construction boom. So, while competitiveness indeed hindered forest conversion to agricultural purposes and forest degradation from timber harvesting, the effect differed greatly across products.

At the margin of these two main forest-protecting clusters (urbanisation and Dutch Disease), government policies had varying effects on forests. Trade policy (7) had traditionally discriminated against agriculture. During the oil boom, administered cash-crop prices were raised and imported inputs (fertilisers, pesticides, etc.) were heavily subsidised, favouring the land-intensive HFZ cash crops. In turn, the land-extensive HFZ food crops were hit by subsidies to competing rice consumption. Thus, without being a major factor, oil-boom trade policies *ceteris paribus* reduced forest loss. New road construction into forested areas (5) expanded heavily, which was strongly linked to logging firms gaining access to new extraction areas. This increased forest degradation, probably more through over-hunting and other indirect effects than through the highly selective logging operations proper. However, forest-conversion from roads built by oil money was not as dramatic as in countries like Ecuador or Venezuela. The reasons for this were both economic (e.g. underdeveloped markets, absence of cattle in forested areas, out-migration effects) and non-economic (e.g. immigration being restricted by customary land rights). Finally, the massive funding made available for agricultural development (6) resulted in relatively little deforestation or degradation. One reason for this was that not many resources worked all the way through bureaucratic and rent-seeking filters to the point of actually raising production. Those that did, like agro-industrial plantations along the coast, were land-intensive and resulted in fairly modest area expansions of not more than 10 per cent of the national FAO forest-loss figures.

It may also be interesting to examine the factors that did *not* prove to be important in Cameroon for the overall forest outcome. The oil and mining sector itself (8) caused negligible forest impacts because most production has been offshore, although that may change in the future. The increase in the budgets of forestry and conservation agencies (9), as well as corresponding cuts during the crisis, also had little effect. Conservation was never a priority for government spending. Financing was almost entirely left to the donor community, and external funds for Cameroon's protected area system actually seem to have expanded after the onset of the crisis in 1986. Forestry regulation was extremely weak before the oil boom, and the legal framework only advanced under external pressures linked to crisis-induced structural adjustment. Directed settlement (factor 10) existed historically, but forest impacts were ambiguous and did not affect the HFZ during the oil boom.

Did the economic crisis merely imply a simple reversal of trends in the oil boom? The crisis response was greatly divided between the period prior to (1987–93) and following devaluation (from 1994), but on the whole many oil-boom impacts were in fact directly reversed. As in the approaches used by Sunderlin *et al.* (2000) and in Table 6.5 (factor references in parentheses), one can list three main crisis-related clusters of factors that all stimulated forest-clearing:

- rural farming population and their food-subsistence demands rose rapidly, triggered by natural population growth plus return migration from an impoverished urban economy (1, 3);
- consumption shifted back from imported foodstuffs and previously subsidised Northern rice to the 'inferior' but now more competitive and affordable HFZ food crops (2, 4, 7);
- production shifted from the crisis-hit land-intensive HFZ tree crops (especially prior to 1994) to land-extensive HFZ crops (plantains, tubers, etc.), due to cuts in producer prices, input subsidies, and an overvalued FCFA exchange rate (2, 7).

In addition, logging also expanded considerably after the 1994 devaluation (Eba'a Atyi 1998), which may have accelerated both forest degradation and access for more forest conversion. Clearly, other factors from Table 6.5 were also at work, but were less important. In particular, less money was available for new road-building (5), although the expansion from the mid-1980s still had an impact in the early phase of the crisis. Agricultural development (6) and the state's forestry and conservation budgets (9) all suffered, but none of these was imperative.

Some asymmetries between boom and bust were also present. In general, government spending and inflation proved much more difficult to cut than to expand. This caused an important overvaluation asymmetry in the adjustment of the RER, which hurt the traded sectors. In part, overspending was related to the persistence of both rent-seeking pressures and corruption. The roots of corruption in Cameroon may be diverse, but many authors point to the fact that oil rents greatly increased it. In the 1999 *Corruption Perception Index* published by the Berlin-based NGO Transparency International, Cameroon occupies a sad leading position among a sample of ninety-nine countries, implying that a number of business people, risk analysts, etc. have classified the country as having the most corrupt practices (Transparency International 1999). Once corrupt practices have become common, it seems that they easily spread to all other sectors where economic rents can be gained. The timber sector has been an obvious candidate in that respect, suffering a large increase in illegal logging, tax evasion and clandestine exports. The effect of more recent anti-corruption initiatives (IDA and IMF 2000: 16) remains uncertain.

Another asymmetry concerns the HFZ cash-crop sectors, cocoa and coffee. During the oil boom, favourable international prices and internal pricing policies compensated producers for losses from real appreciation. When external prices declined, domestic prices were initially kept high. But when the former collapsed even further at the beginning of the 1990s, the producers' situation became catastrophic. Domestic prices were heavily cut, input subsidies eliminated and the abrupt liberalisation of the marketing system disrupted the level of services available to producers. At the same time, competitiveness remained low until the 1994 devaluation. The *timing* of this policy response, especially the cuts in subsidies and services mandated by structural adjustment, thus proved fatal. The sequence of events forced farmers back into food crops. However, they did not fell their cocoa and coffee trees to plant plantains and tubers, and cocoa and coffee plots were seldom allowed to revert back to forests but were often left aside 'for better times', while forest was cleared to make way for food crops.

This seems to point to three more general lessons. First, in an environment of extreme volatility in terms of both external prices and government policies, risk-averse farmers will tend to keep a larger area under (extensive) cultivation as a safety measure should prices or policies change. In Cameroon, sharp fluctuations and uncertainties thus caused farmers to maintain their tree-crop orchards even though they were not harvesting them, while clearing more forest for expanding food crops.[37] Second, in the early 1990s the bad timing of rapidly introduced liberalisation and fiscal cuts for the cash-crop sector proved to be a 'lose–lose' policy, since both producers and the forest suffered. This is a message one should keep in mind in designing SAPs in the future. Third, prior to the boom, deforestation had been closely linked with (rural) population growth. The oil boom 'de-linked' the two variables by providing a growth in urban employment, but the urban crisis restored the former link between population and deforestation, as a rapidly growing rural population became directly dependent on forest nutrients for slash-and-burn agriculture. This shows that one should not play down the long-run deforestation impact of population growth over shifting economic cycles, but also that, in the medium term, oil wealth was a powerful intermediary factor between population and deforestation. On aggregate, Cameroon's boom-and-bust cycle also confirmed the 'core hypothesis' of this book in *relative* terms: forest area did not grow in absolute terms during the oil boom (1979–86), but deforestation rates were strikingly lower than after the onset of the economic crisis in 1986.

Notes

1 For a more thorough discussion of the quantitative estimates, see Table 6.1.
2 Threats here are more from conversion than degradation, due to the predominance of fertile volcanic soils with good agricultural potential (Toornstra *et al.* 1994: 3).
3 It is estimated that there are 9,000 plant species in Cameroon, of which 156 are endemic to the country; on Mount Cameroon alone, 45 endemics are found (Sayer *et al.* 1992: 113).
4 However, fallow length is also often reduced when land becomes scarce, which may reduce the amount of forest clearing per unit of crop output (see discussion below).
5 For the Ndélélé area (East Province), Mertens *et al.* (2000) found a correlation coefficient of 88.4 per cent between, on the one hand, remotely sensed deforestation between 1991 and 1996 and, on the other, household survey information on increases in cultivated area.
6 Mineral occurrences include bauxite, gemstones, gold (current production around 1000 kg/yr), iron ore, nickel, phosphate, tin and zinc (US Geological Survey 2000).
7 The combined export share of cocoa and coffee was 65 per cent in both 1978 and 1979 and 49 per cent in 1980; the share of oil jumped from 3 per cent in 1978 to 24 per cent in 1979 and 31 per cent in 1980 (IMF 1990: 267).
8 Most Cameroonian statistics are calculated on the basis of fiscal years (June to July). We follow the common practice of converting this to calendar years by attributing the values to the last year in the split notation: for example, 1980 refers to 1979–80 (Blandford *et al.* 1994: 131). This implies, for example, that the world-market oil-price hike during 1979 is attributed to the year 1980. To calculate currency conversions for split-year figures, we use the average of the two end-of-year exchange rates.
9 Secrecy over oil revenues was not limited to the oil boom. Greater transparency on oil revenues has been a recurrent issue of discussion with the Bretton-Woods institutions. Only in 1997 did the national oil company start submitting annual international audits, and recently virtually all oil revenues seem to appear in the budget (IDA and IMF 2000: 16).
10 For instance, Jua (1993: 135) estimates gross oil revenues (repatriated and non-repatriated) for 1982 at US$1,229 million. This is more than double the figure of US$483 million based on trade statistics that is reported in the World Development Indicators (World Bank 1999a), or the US$669 million estimate calculated on the basis of the share of oil in exports in Blandford *et al.* (1994: 138).
11 Oil revenues that the government held abroad are not interesting from a domestic absorption viewpoint, since they do not affect RP and production until they are repatriated.
12 In English, 'the barefooted suddenly becoming millionaires'.
13 For 1977–98, different trade-weighted RER indices were linked from various World Bank and IMF sources. Trade-weights may differ between the indices, but distortions should be minor. However, for 1971–7 no trade-weighted exchange rate data were available. Instead, three relative price series from Amin (1996) were inspected: non-tradables (services and construction) over import prices, non-tradables over non-oil export prices, and domestic consumer prices over food prices. I calculated correlation coefficients for each of these with the real trade-weighted exchange rate for the overlapping period (1978–84). The index of non-tradables over import prices had the closest fit (a Pearson coefficient of 0.75). On this basis, relative changes in this index were used to extrapolate back the real trade-weighted index for 1971–7.

14 Although difficult to measure, one opinion is that the RER prior to 1994 was overvalued by 77 per cent (Amin 1996: 48).

15 The analysis in this sub-section draws heavily on Ndoye and Kaimowitz (2000).

16 Amin (1995: 22) quotes a report from the Ministry of Agriculture for an average cocoa yield of 248 kg/ha, compared to 600 kg/ha in the Ivory Coast and over 1,000 kg/ha in Malaysia; similarly, *robusta* coffee yields 170–200 kg/ha in Cameroon, but about 1,000 kg/ha in Latin America.

17 *Esep* fields are mainly used for plantain, cassava and *ngon* melons (*Cucumeropsis mannii*); they typically require forest clearing, which is done by men. *Afub awondo* are used for groundnuts, cassava, maize, cocoyams and yams; they are mostly managed by women on previous *esep* fields and short-fallow plots. In other words, establishing *esep* fields tends to require deforestation, while *afub awondo* does not (Diaw 1997: 16–20; Ndoye and Kaimowitz 2000: 9).

18 The average kg price of cocoa jumped from FCFA 577 during 1980–2 to 929 for 1983–6; for *robusta* coffee, the price went from 702 in 1980–2 to 1,161 in 1983–6 (MINEFI 1998b: 20).

19 Figures from the forestry agency ONADEF, cited in Essama-Nssah and Gockowski (2000: 10).

20 The conversion factors to roundwood equivalents are 3.0 for sawnwood and 2.2 for veneer panels, as suggested by MINEFI (2000: Section 'Les impacts économiques', p. 3).

21 Karsenty (2000: 27) believes that wood currently processed for the domestic market is in the range of 500,000–800,000 m^3. Carret and Clément (1993: 172) estimated domestic sawnwood production alone at about 300,000 m^3 in 1985, which would correspond to around 900,000 m^3 in roundwood equivalents.

22 Milner (1990: 145) classifies different sectors according to their effective protection rates. He finds that forestry products and wood and paper products are among the three sectors with the highest ranges of effective protection on domestic markets, together with leather and rubber, and textiles.

23 Producer prices' share of FOB export prices developed as follows for cocoa: 1960–9: 48 per cent; 1970–9: 42 per cent; 1980–6: 56 per cent; 1987–9: 87 per cent 1990–6: 55 per cent. For *robusta* coffee, the trend was parallel: 1960–9: 68 per cent; 1970–9: 37 per cent; 1980–6: 46 per cent; 1987–9: 79 per cent 1990–6: 49 per cent. The shares are calculated from Ndoye and Kaimowitz (2000: table 5) and MINEFI (1998b: 20–1).

24 See Ndoye and Kaimowitz (2000: table 3) for an overview of regional studies. Higher population density seems to be the main driver for fallow reduction. Increased fertiliser use may curb nutrient mining and diminish the forest area *affected* by agriculture. However, with the assumptions made in this chapter that fallows of three years and above already count as secondary forests, only drastic fallow reductions would be able to reduce deforestation.

25 To optimise the sub-period division, we also tried regressions for 1976–98 and 1982–98 respectively, but the results were disappointing (R^2 = 22.0 and 32.1 per cent, respectively), indicating that 1986–98 is the most appropriate period for applying the model.

26 The sources used for this section are Brunner and Ekoko (2000), Essama-Nssah and Gockowski (2000), and Bikié *et al.* (2000a).

27 For instance, for the East and Centre Provinces, Bikié *et al.* (2000a: 32) find that the number of logging violation reports declined dramatically from more than 100 during 1986–9 to 0–20 during 1995–9. However, this does not seem to be due to fewer violations, but rather to a breakdown in the legal enforcement systems: the percentage of violation cases followed through the judicial system declined rapidly, from 40–60 per cent in 1988–91 to almost zero in 1992–3.

28 Own observation and E. Siriak (Mengomo), personal communication, on field trip to southern Cameroon (Yaoundé, Mbalmayo–Ebolowa–Mengomo). On the unpaved stretch from Ebolowa to Mengomo, much less forest is cleared than along the Mbalmayo-Ebolowa paved stretch. Personal transport costs from Mengomo to Ebolowa are normally moderate (FCFA 1,000; US$1.66), but get extremely high during the two rainy seasons (FCFA 3,000; US$4.98) when the risk of cars becoming stuck rises dramatically.

29 These figures are from IFR (1988: 192) and have been converted from SDR to US$.

30 Roads that are only partly in the HFZ were also counted as HFZ roads.

31 One village in the metropolitan area, Ngoulemakong, had a disproportionate influence on the results, as it basically became a suburb of Yaoundé. Removing it from the survey, annual

population growth in the remaining thirty-seven villages becomes 2.4 per cent, still much higher than during the oil boom.

32 Pokam and Sunderlin (1999), Gubry *et al*. (1996), Hoogeveen and van Soest (1993), Toornstra *et al*. (1994: 10), Ndoye and Kaimowitz (2000) and Essama-Nssah and Gockowski (2000: 8).

33 The main obstacle to cattle-ranching is the tsetse fly, which it is technically feasible but too expensive to eradicate locally. Also, the more egalitarian culture of the south is less receptive to cattle-ranching as a source of capital-accumulation than is the case for some of the northern tribes (S. Hauser, personal communication, Yaoundé, 19 May 2000). On the other hand, it has been shown hypothetically that introducing cattle into the HFZ has the potential to multiply forest conversion (Toornstra *et al*. 1994: 6).

34 However, cattle have caused some forest clearing outside the HFZ, such as around Mount Oku and in the Bamboutou mountains in western Cameroon (Tsalefac 1994).

35 In a summary produced by Wilkie and Carpenter (1998), the volume share of artiodactyls in three study sites of Cameroon was 85–88 per cent.

36 King (1994) and McRae (1997), cited in Bennett and Robinson (2000: Section 'Income of hunters and consumers').

37 See Sunderlin *et al*. (2001) for a similar argument, comparing Cameroon to Indonesia.

7 Ecuador

In contrast to the previous three countries, in Ecuador oil wealth did not help to protect forests. As in Venezuela, almost all the net deforestation was due to pastureland, led by higher demand for cattle products from higher urban incomes and rapid population growth. But the policy package accompanying the oil boom was decisive in turning Ecuador into a deviating case. Mostly led by military governments, Ecuador opted for a strong modernisation strategy focused on new roads and high transport subsidies, partly financed by high foreign borrowing. When in the 1980s oil revenues, interest rates and relative prices turned around and foreign debt had to be serviced at high cost, the presence of roads greatly facilitated further land extensification, providing an asymmetry in the oil wealth-deforestation link.

Deforestation in Ecuador

Vegetation history

It is estimated that 90–4 per cent of Ecuador's land area was originally covered by a wide variety of forests (Cabarle *et al.* 1989). Except for snow-capped volcanoes, *páramo* (highland steppe), swamps and natural grasslands, forest was the overwhelmingly dominant natural vegetation. Today forest cover has been reduced to about half the land area (see section on 'Current forest loss'). The country's three main regions are shown on Map 7.1. The Andean highlands (Sierra), the Amazon lowlands in the east (Oriente) and the western lowlands towards the Pacific coast (Costa) have been subject to rather different land-use histories. In the Sierra, the Inter-Andean Valley with rich agricultural soils of volcanic origin had already been widely deforested when the Spanish arrived. Archaeological research shows that large cultivated areas in the Sierra reverted back to forests in the colonial period, as warfare and exotic European diseases caused a dramatic decline to just one sixth of the pre-Columbian indigenous population.[1]

Deforestation in the modern period from the late nineteenth century to the 1970s has been primarily due to forest loss in the Costa. This was previously a sparsely populated region, dominated by the Chocó moist forest in the north (Esmeraldas Province) and mostly dry forest in the south. The effects, both direct and indirect, of two agro-export bonanzas encouraged colonisation of the coastal lowlands. The first was the cocoa boom from the turn of the century to 1930. The second and most significant was the banana

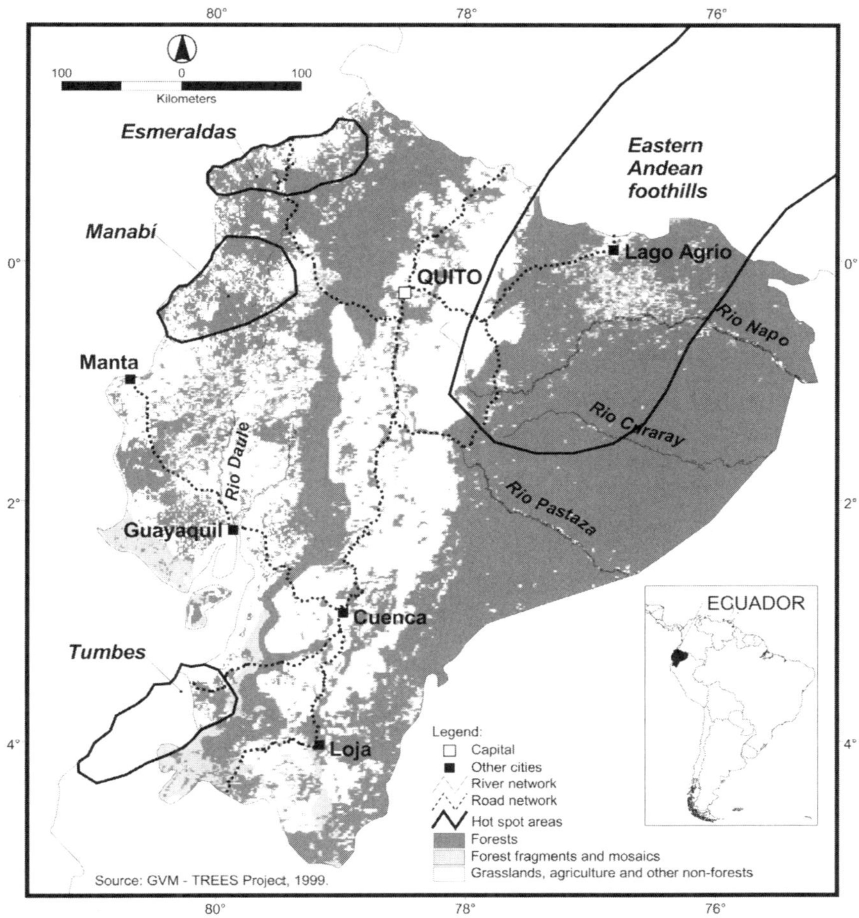

Map 7.1 Ecuador.

boom, which experienced a particularly strong expansion in cultivated area from 1950 to 1965 (Bromley 1981; Wunder 2001). Cattle-ranching and cash crops, such as coffee, soybeans and African oil palm, also played a role in forest cover loss, mainly in the post-Second World War period (Benalcázar 1989). In particular, the Guayas river basin possesses one of the best settings for agriculture along the Pacific coast of South America, with fertile soils, abundant water and the absence of hurricanes. Demand for land for plantations provided the initial impetus to the opening up of new forest frontiers. But the subsequent incremental clearing process between agro-export booms, including of forest remnants, was sustained by a number of explanatory factors (Wunder 2000: ch. 4). Among these were infrastructural improvements (railways, roads), the growth in income, demand and inter-regional trade, and, especially in the post-Second World War period, high population growth.[2] This discontinuous historical pattern of spatial occupation and shifting land uses seems to support Rudel's hypothesis (1993) that deforestation in frontier forests (large, previously untouched forest tracts) came in waves of 'big push', while population growth promoted the gradual clearing of fragmented forests in already established agricultural areas.

Whereas deforestation was spread over two millennia in the Sierra, in the Costa forest loss has occurred in the last century. Currently, however, deforestation is concentrated in the Oriente, Ecuador's oil-rich Amazon region. This area was isolated until the 1950s, with colonisation only in the Andean foothills, along four penetration roads that served urban markets with sugarcane and fruit (Uquillas 1984). Both lowland regions (Oriente and Costa) hosted tropical diseases, such as malaria and yellow fever, which were major factors in protecting lowland forests from colonisation (Bromley 1981). Oil exploration began in the 1940s, but production did not take off until in the 1970s. The direct impact of oil on forests will be dealt with in the section on 'The effect of oil production on forests'.

Current forest loss

As shown in Table 7.1, estimates of current forest stock range from 11.1 to 15.4 million ha. The latter, higher figure, an estimate based on the 1994–5 forest cover map from CLIRSEN (the Ecuadorian Centre for the Integrated Survey of Natural Resources through Remote Sensing), appears to be the most reliable. The FAO's FRA (FAO 1993, 1997a) reports a forest area of 11,962,000 ha for 1990. The *Conservation Atlas for the Americas* (Harcourt and Sayer 1996) gives a forest cover estimate of 14,237,000 ha (no yearly change data), but only for broadleaf forests (the *Atlas* excludes dry deciduous forests), based on maps collected between 1977 and 1987. This supports CLIRSEN's upward revision of the FAO's forest stock figures. CLIRSEN uses satellite imagery on a more detailed scale, detecting forest fragments that previously had been overlooked. This increased forest estimates for the Sierra in particular. In spite of these contradictory assessments, the basic message is that there is more forest left in Ecuador than was previously believed.[3]

Accordingly, estimates of yearly deforestation differ widely (Table 7.1). Based on a comparison with agricultural census data, I shall argue below that average annual deforestation from the mid-1970s to the early 1990s was in the range of 180,000 ha to 244,000 ha, before probably declining somewhat over time. This corresponds to relative ranges of 1.2–1.6 per cent (using the 15.4 million ha stock estimate) or 1.6–1.8 per cent (using the FAO 11.1 million ha estimate). According to the FRA 1990, annual deforestation for 1980–90 was 238,000 ha. Out of this, 142,300 ha (59.8 per cent) were tropical rainforests, 34,100 ha (14.3 per cent) deciduous forests and 61,700 ha (25.9 per cent) hill and montane forests. This amounts to an annual deforestation rate of 1.8 per cent, the FRA's highest rate in South America except for Paraguay (2.7 per cent). For Ecuador, the FRA data were based on two surveys, made in 1963 and 1987 respectively (FAO 1997b). The estimate for 1990–5 projected a slow-down in annual forest loss to 189,000 ha (1.6 per cent), but this is a *model-based* result closely linked to slower population growth. Finally, the FRA for 2000 (not shown in Table 7.1) records an annual forest-loss figure as low as 137,000 ha (see Chapter 1).

Alternative sources give both higher and lower annual figures than the FRA. Even for comparable periods, estimates differ considerably: INEFAN (1995) reports 0.8 per cent

Table 7.1 Ecuador: forest-cover and deforestation estimates

Author	*Forest cover* (in ha)	*Year*	*Annual deforest.* (in ha)	*Relative decline*[a] (%)	*Period*	*Source type*	*Coverage notes*
FAO (1997a)	11,137,000	1995	189,000	1.6	1990–5	Model estimate	All forests
FAO (1993)	11,962,000	1990	238,000	1.8	1980–90	Model estimate	All forests
FAO (1997b) Sánchez and Toro	12,483,000	1987	—	—	—	Satellite image	All forests
WRI (1992)	14,773,000	1980	340,000	2.3	1980–5	Unknown	All forests
WRI (1994)	11,962,000 (FAO)	1990	340,000 136,000[b]	2.3 1.2[b]	1980–5 1985–90	Unknown FAO	All forests
FAO (1996)	15,600,000	1994	−6,666[a]	−0.1[a]	1979–94	Agency reporting	Prod. forests + other categ.[c]
Harcourt and Sayer (1996)	14,237,000	1987	—	—	—	Maps from 1977–87	Closed broadleaf forests
SUFOREN (1991) (*)	8,070,000	1984	120,000	1.3[b]	1965–84	Inventories	Only seven provinces[d]
ITTO and INEFAN (1994b)	15,642,000 12,405,000 11,437,000	1962 1985 1988	140,739 330,000	1.0[b] 2.7[b]	1962–85 1985–8	Various	All forests
INEFAN (1995)	11,578,000[e]	1995[e]	106,000	0.8[e]	1965–95[e]	Unknown	All forests
Amelung and Diehl (1992)[f]	—	—	306,000	—	1980–8	Model estimate	All forests
Cabarle *et al.* (1989)[g]	26,200,000 17,500,000 7,270,000	'Original' 1958 1988	341,000	2.9[b]	1958–88	Unknown	All forests
Sierra (1996) (*)	546,180 489,628[b] 337,401[b]	'Original' 1983 1993	15,223	2.0	1983–93	Remote sensing	Only NW region
Wunder 2000 (CLIRSEN)	15,417,465	1994/95	—	—	—	Satellite and survey	All forests

Notes

a Negative figure indicates net reforestation, rounded up to one decimal place.

b Own calculations using figures indicated in the specific source.

c Production forests + other wooded land + intended reforestation ÷ recreation forests.

d Esmeraldas, Pichincha, Morona-Santiago, Napo, Pastaza, Sucumbios, Zamora-Chinchipe.

e Calculated from figures indicated; years are not stated explicitly.

f Cited in Sierra (1996: table 1).

g Data reproduced in Harcourt and Sayer (1996: 265).

* Regional estimates.

forest loss for 1965–95; a WRI report by Cabarle *et al.* (1989) gives 2.9 per cent for 1958–88. The static figures in the FAO's *Production Yearbook* are generally inadequate.[4] Amelung and Diehl (1992) report 306,000 ha in the 1980s. The WRI (WRI 1992) reproduces an estimate of 340,000 ha (2.3 per cent) for 1981–5, but this does not seem compatible with the WRI reported data for the whole decade (WRI 1994: 307).[5] National estimates by the Ecuadorian forestry agency (SUFOREN, later INEFAN and MoE) are generally much lower, typically around 1 per cent/yr. For instance, the SUFOREN (1991) estimate of only 120,000 ha yearly forest loss for 1964–84 draws on land-use data for only seven out of the total, by then, of thirteen provinces (Oriente and northern Costa).

Faced with this contradictory evidence, a supplementary approach is to look at agricultural census and survey data from the Ministry of Agriculture (MAG), published by the National Institute of Statistics and Censuses (INEC) (see Table 7.2). These are based on the last agricultural census, held in 1974, supplemented by sample-based annual survey data – (see SICA 2001 for a description of the methodology used). Most striking is the dramatic increase in pastures, which almost tripled from 1972–3 to 1988–9; total cropland

Table 7.2 Ecuador: agropastoral land-use trends, 1972–3 to 1988–9 (in thousand ha)

Land use	*1972/73*[a]	*1984/85*[a]	*1988/89*[b]	*Change 1972/73 to 1984/85*	*Change 1972/73 to 1988/89*
Land planted to major highland crops[c]	503	249	325	−254	−178
Land planted to major tropical crops[d]	1090	1364	1393	274	303
• in Costa	1060	1304	1258	244	198
• in Oriente	30	60	135	30	105
Total cropland	1593	1613	1718	20	125
Pasture	2241	4406	6021	2165	3780
• 10 Sierrra provinces	1024	1917	2349	893	1325
• 5 Costa provinces	833	2005	2792	1172	1959
• 5 Oriente provinces	384	484	880	100	496
Net change in total land use	—	—	—	2185	3905
Yearly change	—	—	—	182	244

Sources: Southgate and Whitaker (1992: 19) and own calculations.

Notes

a MAG annual surveys.

b INEC annual surveys – methodology not fully comparable to that of MAG's surveys.

c Major highland crops: barley, legumes, potatoes, soft corn, wheat, temperate vegetables and fruits.

d Major tropical crops: bananas, cocoa, cassava, coffee, rice, plantains, soybeans, cotton, sugarcane, hard corn, oil palm, manila hemp, peanuts, castor oil, lowland fruit and vegetables.

increased only slightly. During 1982–7, growth of cropland and pasture combined was higher than in every other Latin American country except for Surinam (Southgate and Whitaker 1994: 46). However, there were important regional differences. Highland crops (maize, grains, etc.) faced a decline; there was specialisation on cattle-ranching in the Sierra and on cash crops in the Costa. The last two columns in Table 7.2 show that total agro-pastoral land use expanded by 2,185,000 ha from 1972–3 to 1984–5 and by 3,905,000 ha from 1972–3 to 1988–9.[6] This corresponds to an annual net increase of 182,083 and 244,063 ha, respectively. This provides us with a basic check on deforestation figures: annual deforestation was probably in the range of the FAO-FRA predicted figures (238,000 ha during 1980–90).

Unfortunately, agricultural land-use data ceased to give us a solid reality check in the 1990s. The growing economic crisis and fiscal cuts meant that the annual survey was no longer being carried out, so that the data published in SICA (2001) are desk-based extra-polations of a partial set of agricultural production figures. They indicate an outright *decline* in the overall agropastoral land area after 1994.[7] This is indeed inconsistent with what happened in the two principal areas of current deforestation, the northern Oriente and the northern Costa (see below). However, it may well be true that there has been a *slow-down* in Ecuadorian forest loss in the 1990s, maybe even to a greater extent than registered by FAO-FRA. Unpublished MAG figures on pasture – by far the most expansionary land use in Ecuador[8] – indicate a doubling in 1971–82 (7.2 per cent/yr), reduced growth during 1983–7 (2.1 per cent/yr) and a stagnation for 1988–93 (0.4 per cent/yr). Only within the next couple of years will nation-wide data become available that consolidate our knowledge on recent Ecuadorian land-use changes.[9]

In the light of inconsistent figures, assessing changes in forest loss between sub-periods must remain a tentative exercise. In synthesis, Tables 7.1 and 7.2 indicate that there was probably a peak in deforestation from the mid-1970s to the late 1980s, then a slow-down during the 1990s. This is supported by some regional deforestation studies. Schmidt,[10] for example, estimated in 1990 that deforestation in the Amazon region – a 'hot spot' for forest clearing – had by then declined to about 60,000 ha/yr. The same pattern is found in case studies in the southern Oriente, where, forest-clearing experienced a spurt in the 1975–80 period.[11] As the FAO-FRA model stipulates, this may be related to marginally slower population growth, but as will be shown in what follows, macroeconomic factors and new infrastructure actually played a more vital role.

Where does current clearing mainly occur? The deforestation 'hot-spots' identified by Achard *et al.* (1998) have been marked in Map 7.1 above. Esmeraldas Province on the northern coast is the prime logging area in Ecuador; access roads for logging are gradually being used by squatters to convert forests to agriculture. Further south, clearing in northern Manabí Province is for both plantations and smallholders cultivating a variety of crops. In both hot spots mangroves have also been cleared for the establishment of shrimp ponds, which is also affecting the small coastal border area up to Tumbes (Peru). The currently most important hot spot is the eastern flanks of the Andes with its transition zone towards the Amazon lowlands. In its northern part, especially between the Colombian border and the Napo river, deforestation has been led by oil roads (see next section); in the south other roads have been constructed, with conversion for coffee, fruit, oil palm and particularly cattle. In-migrating smallholders dominate the area.

With respect to our basic oil-cycle hypothesis, there is no indication that forest loss, whether in absolute or relative terms, would have been lower during the oil boom of the 1970s, compared to the periods before or after. On the contrary, compared to the period before forest loss accelerated in the 1970s, and it did not go into reverse during the crisis of the 1980s. The 'core hypothesis', namely that oil wealth leads to a reduction in forest loss and degradation, would thus not seem to apply to Ecuador. In what follows, the challenge will be to explain why and how Ecuador deviated from the expected pattern. The direct deforestation impact of the oil sector in the highly forested Oriente is the first candidate that might contribute to this explanation.

The effect of oil production on forests

Direct oil impacts

Oil exploration started on a small scale in Ecuador in 1911 on the Santa Elena Pensinsula (on the Pacific coast). In the Amazon, Shell started exploration just before the Second World War, and efforts were intensified when sizeable discoveries were made in the area just north of the Colombian border (Amigransa *et al.* 1997: 63; Hiraoka and Yamamoto 1980). The country emerged as a new oil exporter in 1972, when the share of oil in total exports jumped rapidly to 18.4 per cent. Oil became the dominant export commodity; its export share has remained beyond 20 per cent ever since, and even reached a peak of 74 per cent in 1983 (World Bank 1999a). Accordingly, oil's share of GDP over the past three decades has been in the 7–20 per cent range, and its share of the state budget has been 30–60 per cent. With currently known reserves, Ecuador will remain a significant oil exporter for at least three more decades (Ruiz 2000: 65; Ajamil and Carillo 1999).

More than 99 per cent of Ecuadorian oil comes from the Amazon (northern part), that is, from the region that retains the most continuous tropical forest cover. Ecuador has also been among the most controversial cases of oil production in the tropics, in terms of the public outcry over the environmental and social impacts of oil companies' practices. This culminated in several US-based lawsuits by indigenous groups in the Amazon against Texaco, the predominant company involved until 1990 (Ruiz 2000: 66). Today, almost 80 per cent of oil is produced by the state-owned company PETROECUADOR. In a pioneering book, Kimerling (1991) brought the careless production methods of both Texaco and PETROECUADOR to the forefront of international attention. Since 1995, however, under increasing pressure from NGOs and indigenous groups, oil technologies have moved a long way towards reducing environmental impacts, especially in the case of foreign companies that need to be increasingly concerned about their environmental images in their home countries.

Let us first examine more closely direct oil-related deforestation. It is estimated that under the rudimentary practices before 1995, each exploration platform triggered the complete clear-felling of 2–5 ha; an additional 15 ha were heavily degraded by extracting construction timber, etc. (Kimerling 1991: 56; Southgate and Whitaker 1994: 79–89). One platform thus affects up to 20 ha. Kimerling (1991: 47) reported that 392 wells were in production in 1988; Ruiz (2000: 65) claims that a total of 622 wells were opened between 1970 and 1990. Using the latter figure as a basis for extrapolation, continued expansion of exploration and production in the 1990s would lead us to expect that, in total, about 1,000 wells have been opened at the time of writing (2002). This corresponds to a deforestation figure of 2,000–5,000 ha and a degradation figure of 15,000 ha, which seem quite consistent with what has been reported from detailed case studies. One authority estimates 'affected forests' in the main oil-production zone of Napo–Sucumbíos at 5,485 ha, that is, less than 2 per cent of the study area.[12] Another study finds that, in a 200,000 ha concession area, an accumulated total of 1,046 ha (0.52 per cent) had been fully cleared. These numbers probably all include a fair share of transitory forest loss, due to the limited canopy opening created by some of the clearing.[13]

A second direct impact comes from the construction of transport infrastructure for the oil sector. This includes a 500 km pipeline built to transport oil across the Andes to the Pacific (*Oleoducto Trans-Ecuatoriano*) and an extensive road network built into the Amazon. Beyond the few pre-existing access roads into the region (e.g. Quito–Baeza), most of the roads in the northern Amazon were somehow related to oil, while those in the southern part were not. Data are not available classifying the purpose of roads built. A conservative estimate by USAID (cited in Kimerling 1991: 75) states that by the mid-1980s more than 500 km of roads had been built by the oil industry into forest areas. For an updated and perhaps more comprehensive guesstimate, we might assume that 40 per cent of the current Amazonian road network (i.e. most of the northern part) was built for some oil-related purpose, including secondary roads, access roads to oil workers' camps, services, etc. We can then try to quantify the total impact by using road-network data from Ruiz (2000: 73). In 1996, primary roads in the Oriente totalled 980 km – including, according to our estimate, 40 per cent (392 km) created by or for the oil sector – secondary roads 534 km (oil-related: 214 km) and tertiary roads 1,483 km (oil-related: 593 km). If, furthermore, we assume an average clearing width of 20, 15 and 10 m for each of the three categories of road, the aggregate direct forest-loss impact of road building would be 1,579 ha.[14]

In deforestation terms, direct oil impacts are thus certainly very limited. Even using fairly non-conservative assumptions, our estimates add up to no more than 3,000–6,500 ha, which corresponds to between 0.04 and 0.09 per cent of the forest left in the Ecuadorian Amazon.[15] The area affected by oil-motivated timber extraction may be three times as high, but it is still a very small figure. In addition, there is recent evidence that direct oil-related forest loss declined in 1986–97 compared to 1971–85 (Torres 2000), probably because of the adoption of less rudimentary techniques.

However, the oil sector's degradation impacts other than forest-clearing are much more pronounced, and have affected larger areas. Most of these relate to production rather than exploration (Southgate and Whitaker 1994: 80). Pollution impacts were serious, and not restricted to the Oriente. The Trans-Ecuadorian pipeline to the Pacific has been affected by a series of accidents, including the 1987 earthquake, which together may have spilled 16.8 million gallons (Kimerling 1991: 69). By 1990, the daily on-site discharge of production wastewater (toxic brine) and chemicals seems to have been as much as 4.3 million gallons, affecting both river and groundwater quality.[16] Adding to this other impacts, such as deposits of toxic drilling mud, the overflow from rudimentary waste pits, the use of dynamite for exploration and the burning of natural gas, oil production seriously affected fish and wildlife populations in the northern Amazon. It therefore also had disastrous nutritional and health impacts on indigenous forest people, including direct effects (skin, respiratory and other diseases), reduced availability of natural resources, and detrimental social impacts (Kimerling 1991: 77–84; Amigransa *et al.* 1997: 75–6).

Indirect oil impacts

For deforestation, far more important than direct site impacts are the indirect ones, most of which relate to land colonisation around new roads. It is generally recognised that oil roads decisively 'opened up' new agricultural frontiers in the northern Amazon region (Sierra 2000: 6): about 60 per cent of the Oriente's economically active population work in agriculture (Southgate *et al.* 1991: 1146). Prominent examples of roads being followed by intense colonisation are the Texaco-built Lago Agrio road in the 1970s (see Map 7.1) and the Tarapoa–Tipishca road in the 1980s. A large number of workers came into the region to work for the oil companies or to build roads, and the vast majority of them intended to stay on to start a farm. While clearing was modest initially, mainly for establishing coffee plots, the gradual move into cattle-ranching has accelerated the process over time (T. Rudel, personal e-communication, 15 May 2002).

How much land is cleared around a road? A high-end estimate is that the range is 2–12 km, depending *inter alia* on market closeness and soil potentials, which would mean that for every new km of road, somewhere between 400 and 2,400 ha of forest are cleared (Kimerling 1991: 75). For the sake of illustration, if we use the road-network figures from above and assume an average indirect forest-loss impact of 2,400 ha/km of primary, 1,000 ha of secondary and 400 ha of tertiary road, the aggregate effect would add up to a staggering 1,178,520 ha.[17] This is more than the accumulated clearing in the whole of northeastern Ecuador.[18] The estimate is inflated because of Kimerling's simplistic assumption that the entire area near a road is cleared. Post-road on-farm clearing processes may take decades, as has been shown by several studies using data gathered by remote sensing. For one particular site (Joya de las Sachas), Sierra (2000) found that 9 per cent had been cleared in 1977, rising to 26 per cent in 1986 and 46 per cent in 1996. Torres (2000: 29–30) found, for 78,000 ha of individual farmland in Napo-Sucumbíos Province, that only one-third of the land had been cleared (most of it for pasture); the rest remained forested because of farmers' resource constraints in 'developing' their farms. He also found that, in spite of the above-mentioned decline in direct oil-led deforestation from 1971–85 to 1986–97, indirect deforestation apparently tripled in the second period, which was closely related to an upsurge in in-migration to northeastern Ecuador (Sierra and Brown 1994). For the whole Napo region, Sierra (2000: 5) confirms that the deforested share of land increased from 7.25 per cent in 1986 to 13.4 per cent in 1996. These examples show that indirect impacts clearly predominate over direct ones, but that they can take a considerable time, and that they are heavily influenced by a series of factors that are unrelated to the oil sector.

Oil roads thus act as promoters and spatial determinants of deforestation; even in advance of their actual construction, they promoted speculative land claims along the planned routes (Pichón 1997: 71). In the first wave of post-road construction interventions, access is used for logging operations in some cases. Then agricultural squatters follow, gradually clearing the land using slash-and-mulch methods[19] and using it mostly for commercial crops and extensive cattle-ranching. Poor soil quality and the rapid exhaustion of nutrients through transitory farming techniques may induce smallholders to degrade land quickly and clear new plots (Thapa *et al.* 1996: 1321). Finally, road access has also made it feasible for industrial entrepreneurs to establish some large-scale African palm-oil plantations. Besides road construction, other oil-sector 'pull-factors' inducing deforestation include social infrastructure (medical posts, schools, shops, etc.) and occasional off-farm employment opportunities. These factors all helped to consolidate settlement at the forested frontier.

For the squatter, the process of securing land titles to an occupied plot can take many years, and his success in 'homesteading' has been contingent on his continuous clearing of land. For many years, the Ecuadorian land-tenure authorities required evidence that the land had been 'worked' before they would provide formal land titles, which in its turn is a requirement for obtaining credit. Even in the region's protected areas, forest-clearing has been a way of legalising land occupation by *fait accompli*. As a park ranger in the Cuyabeno Wildlife Reserve expressed it: 'It's against the law for the settlers to come into the reserve, but now it's not really against the law anymore because its already done' (cited in Kimerling 1991: xvii). In a statistical analysis of deforestation in the northern Ecuadorian Amazon, it has been shown that prolonged tenure insecurity is a significant, independent factor in accelerating deforestation (Southgate *et al.* 1991). The same has been confirmed in case studies from the southern Amazon (Rudel 1993). Tenure rules have thus clearly accelerated frontier deforestation in Ecuador.

In principle, assessing the true impact of oil is related to the difficult counterfactual questions of where and how much clearing would have occurred without oil production. It seems obvious that some roads would have been built anyhow, and that some migrants might have gone to clear land elsewhere, for example, in the southern Amazon. The point is illustrated by the remote-sensing findings of Sierra (2000): in the adjacent Peruvian section of the Napo area, no deforestation occurred, but clearing in the Colombian part was actually *higher* than in northeastern Ecuador, partly because of coca production. But many figures speak in favour of oil's leading role in the colonisation 'take-off'. There was little migration into the Oriente before the 1970s, but during 1974–90 population grew at 6.7 per cent a year, by far the highest rate in the country. Between 1974 and 1982, 92,700 people moved to the region (Southgate *et al.* 1991: 1147; Thapa *et al.* 1996: 1321).

Comparing the four different Amazon provinces, deforestation data for 1965–84 from SUFOREN (1991) (see caveats above) and land area figures from SICA (2001) confirm that the 'oil province' (Napo–Sucumbíos) actually had the highest forest loss among the provinces (770,000 ha; 14.3 per cent of land area). This figure is much higher than regional forest loss in neighbouring Pastaza Province (250,000 ha; 8.6 per cent). But as a share of land area, it is actually less than in the southern Oriente, in Morona–Santiago (550,000 ha; 22.4 per cent) and Zamora–Chinchipe (400,000 ha; 17.4 per cent of land area). This clearly indicates that oil was an important but not the fundamental reason for Amazon deforestation. Non-oil factors, such as absorption of a highland population surplus and the southern Oriente's progressive economic integration with the southern Sierra provinces, were at least as powerful as oil development in triggering regional forest loss.

To sum up, the expansion of oil production in the northern Amazon region caused both forest-loss and degradation impacts. Direct clear-felling by the oil companies was very limited, but pollution impacts were serious, and the indirect impacts, in the form of oil roads making virgin forest areas accessible and attractive for agricultural settlement, were substantial. How much deforestation would have occurred if the boom had entailed pure spending effects, that is, if it had been a financial transfer alone, without the need to undertake clearings in the Oriente? In the rest of this chapter, the question of these derived oil-wealth effects on forests will be discussed.

The macroeconomic impact of the oil boom

As a new oil exporter, the size and timing of Ecuador's bonanza was determined by both price and quantity. Oil production rose only gradually after 1973, due to unresolved conflicts between foreign oil companies and the Ecuadorian state over royalties and concessions. In terms of foreign-exchange inflows, Ecuador thus did not take full advantage of the 1973–4 price hike (Gelb and Marshall-Silva 1988: 176). In addition, being a new exporter with difficult access to crude extraction and transport, a non-trivial share of oil receipts during the early years had to be allocated to investments in the oil sector proper. Oil revenues were greatest following the 1979–80 price boom (see Figure 7.1). In the 1982–6 period, Ecuadorian crude exports doubled in quantity, thus partially compensating for the falls in oil prices, first in 1983, and much sharper in 1986 (IMF 1991: 342–3). Thus, the real value of crude petroleum exports rose to a significant level in 1974, remained fairly high in the 1970s, but was highest in absolute terms in 1979–85.

However, the Ecuadorian Dutch Disease was not only a story about oil exports. Like many oil exporters, the country used its new creditworthiness and easy access to international capital markets to undertake external borrowing. From 1977 to 1981, Ecuador indulged in a short but intense period of borrowing. Foreign capital inflows, which had been insignificant prior to the oil boom, rose to significant levels in 1976–82. Long-term loans jumped from US$159.8 million in 1976 to US$633.2 million in 1977, actually more than petroleum export-revenues for that year (US$478.2 million). In 1981, long-term loans soared to US$1,275 million. Yet, with the Mexican crisis in 1982, the figure abruptly turned negative. External debt had risen tenfold in only six years to US$7,705 million (1982). With the sudden rise in US real interest rates, long-term interest payments went up to US$764.7 million, causing net capital outflows throughout the 1980s and beginning of the 1990s (Figure 7.1 and World Bank 1992: 234–5). Net capital in- and outflows were thus closely associated with the oil boom.

How should we distinguish between boom and bust periods for Ecuador? From the combined picture of the two inflows, I will refer in the following to 1974–82 as 'the oil-boom period', though with stronger economic impacts in 1979–82 than in 1974–8, because foreign borrowing contributed more to foreign-exchange abundance. Conversely, in 1983–6 oil revenues remained high due to rising oil production, but some of these revenues were consumed by net capital outflows. An overheated economy with an accelerating rate of inflation had already created internal limits to continued economic growth under the Hurtado administration (1982–4). Still, external conditions in 1983–5 were much more favourable than in 1986–93, as clearly can be observed from Figure 7.1.

The development of relative prices over three decades (1970–98) generally supports this classification of sub-periods. The nominal exchange rate was held constant for an entire

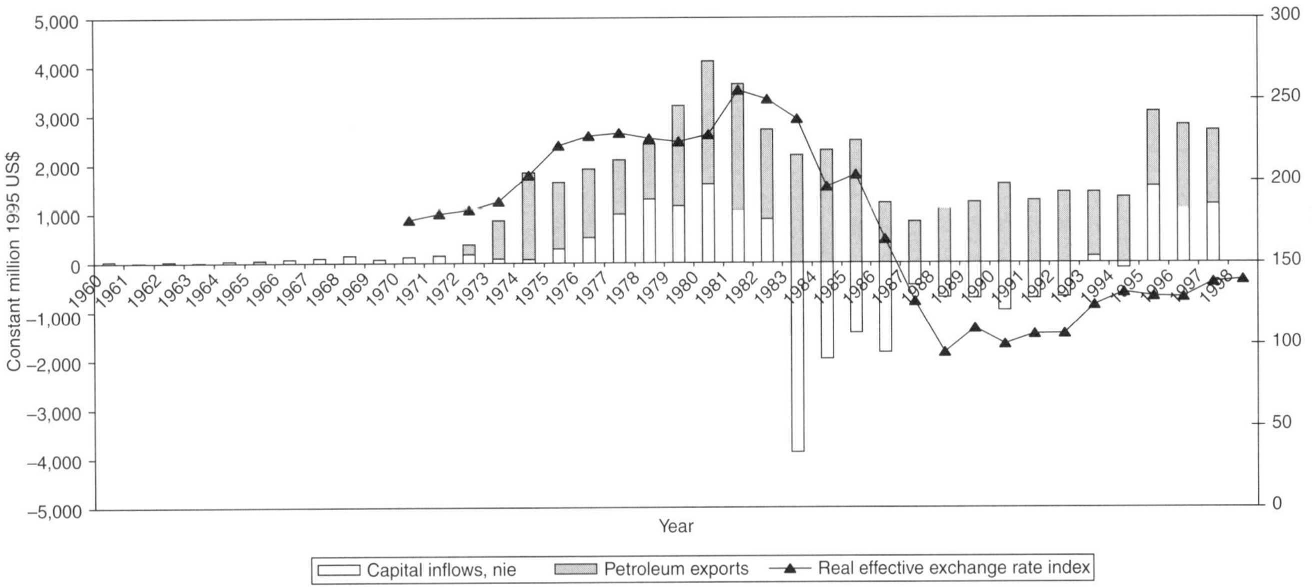

Figure 7.1 Ecuador: capital inflows, petroleum exports and real effective exchange rate, 1960–98.

Sources: IMF (1990, 1992, 1999); Larrea (1992); World Bank (1999).

Notes

1 RER, 1970–6: 1970 weights; 1977–9: 1985 weights; 1980–97: 1990 weights; 1998: 1995 weights.
2 Capital inflows, 1960–91: Other capital nie; 1992–7: Financial account nie.
3 Petroleum exports, 1962–89: Crude petroleum exports; 1990–7: Fuel exports.

decade, from 1971 to 1981, at 25 *sucre* to the dollar (IMF 1991: 340–1). This allowed oil wealth to have a full impact on relative prices. Yearly inflation, measured by the consumer price index (CPI), rose gradually from 7.5 per cent in 1973 to 21.7 per cent in 1981 (World Bank 1992: 232–3). As these rates continuously exceeded the corresponding US$ inflation rates, the real trade-weighted exchange rate appreciated significantly during the 1970s, reaching its peak in 1981 (Figure 7.1). Downward adjustment after 1981 was initially moderate, but then the *sucre's* real value was halved between 1986 and 1989, to a level that was much lower than before the oil boom.

What was the growth impact of these sizeable foreign exchange inflows? Real appreciation favoured imports of consumer goods and, in particular, of imported machinery and equipment for industry, an effect that equalled between 3.8 and 5.3 per cent of Ecuadorian GDP in 1975–81 (Gelb and Marshall-Silva 1988: 182). The macroeconomic impacts of oil income were remarkable. With real growth rates averaging 9 per cent during the 1970s, Ecuador jumped from the status of 'low-income country' to that of 'middle-income country', a wealth effect that was only marginally reversed by the economic crisis of the 1980s. Post-boom growth rates were 0.9 per cent in 1982–4, 3.2 per cent in 1984–8, 4.5 per cent in 1988–92 and 2–3 per cent throughout the 1990s (Araujo 1998: 105). The impact of this boom on the economy differed fundamentally from the agro-export booms (coffee, cocoa, bananas) that Ecuador had experienced previously, in the sense that boom revenues accrued to the public sector, not to private business (Abril-Ojeda 1991). The principal recipient sectors of oil incomes were the following:

- growing public employment and wages
- subsidies for import-substituting industrialisation
- improved education and health
- infrastructure development (roads, seaports, airports, etc.)
- subsidies for transport and energy consumption.

What was the weight between these expanding spending categories? According to CEA (1998 – cited in Whitaker and Greene 1990: 26), 79 per cent of the additional boom revenues was used for higher public employment and salaries. This would indicate the marked urban bias of the boom. However, such a figure is highly sensitive to the assumption about counterfactual scenarios. More important, much of the additional foreign borrowing was used for infrastructure and other investments. We will return to these spending categories in the section on 'Windfall impacts on government spending', but for the military governments that ruled Ecuador during most of the boom, the goal of increased national integration was key, especially between the two main regions, Sierra and Costa. Cross-regional infrastructure and rural-sector investments were thus much more important than in the other country cases in this book (except for Indonesia).

In 1982, most of these spending items had to be cut back significantly, even more so after the abrupt fall in the oil price in 1986, followed by the 1987 earthquake that damaged the Trans-Ecuadorian oil-pipeline. In sectoral growth terms, agriculture resumed its leading position after 1984. The toughest load of adjustment came under the government of Febres-Cordero (1984–8), who took steps to restore internal and external balance. He also applied orthodox policies that clearly favoured the coastal agri-business interests he represented, while severely hurting interest groups in the Sierra (de Janvry *et al.* 1991; Mosley 1991). Perhaps the most important of these policies was to restore the competitiveness of the non-oil traded sectors through currency devaluations. This led to a sharp fall in the RER index from a peak of 255 in 1981 to a low of 95 in 1988, just seven years later (see Figure 7.1).

The 1980s and 1990s both represented periods of 'muddling through' for Ecuador, with shifts between orthodox stabilisation policies under Febres-Cordero and Durán Ballen (1992–6) to populist periods of more expansionary fiscal policies, such as part of the Hurtado (1982–4), Borja (1988–92) and notably the Bucaram governments (1996–7) (Araujo 1998). The latter lasted only half a year, but it initiated one of the country's deepest political and economic crises, which carried over into the interim government of Alarcón and the collapsing administration of Mahuad, and eventually led to the dollarisation of the economy in 2000.

The competitiveness of agriculture and forestry

Agriculture and shrimp farming

After this short macroeconomic overview, let us turn to the analysis of those specific sectors that eventually influenced forests. Did real currency appreciation hamper agriculture and other primary non-oil sectors? If so, did a crisis in agriculture and other primary production also diminish the pressures to convert forests? And how did adjustment policies influence the outcome? Let us start by examining agriculture and shrimp farming.

Unlike other countries, such as Indonesia, Ecuador did not use currency devaluation as a tool to shelter agriculture from the Dutch Disease. Real currency appreciation was not only tolerated, but actively used as a means of distributing the windfall gains to protected producers, also favouring agricultural mechanisation with imported machinery. Trade policy was used to shelter part of agriculture from declining competitiveness (see next section). Growing national income and purchasing power thus increased both prices and production in sheltered and semi-sheltered sectors. The expansionary monetary policy of the 1970s, with controlled nominal (and negative real) interest rates, made subsidised credits from the semi-public National Development Bank (Banco Nacional de Fomento, BNF) available mainly to coastal agriculture and Sierra cattle-ranching. No less than eighty-five different preferential interest rates were offered to agriculture (Mosley 1991: 419). The combination of higher capital intensity and unfavourable relative prices caused an outright decline in agricultural employment, from 873,000 in 1974 (48 per cent of the work force) to 773,000 in 1982 (35 per cent) (Larrea 1992: 274–5). This in turn accelerated rural–urban migration (see section on 'Structural changes in income and demand').

Development trends were highly uneven across agricultural sub-sectors, the main differences being between the 'purely traded' export crops and the 'semi-traded' home market sectors. The prime candidates for decline – Ecuador's traditional export crops, bananas, coffee and cocoa – coincidentally also faced favourable economic climates. As in Cameroon, coffee and cocoa benefited from the beverage-price boom in the second half of the 1970s. Banana exports stagnated during the boom period, and only began a period of sustained growth from 1984 onwards, but more favourable world-market banana prices stimulated Ecuadorian production and improved earnings from the mid-1980s onwards. But this expansionary process did not cause any significant deforestation, because the new Cavendish banana variety with mechanised technology was heavily land-saving (Wunder 2001).

For the main exportables, sector-specific trends thus mixed with Dutch Disease effects. For home-market crops, important structural changes also occurred, with the substitution of rice for highland staples and the increasing highland specialisation in cattle-ranching, mainly dairy farming. Furthermore, many domestic staple crops remained partially trade-protected (see below). Figure 7.2 shows that aggregate agricultural production, which had grown fairly rapidly during the pre-boom years (1969–73), first expanded at a lower rate (1973–7) and then declined in absolute terms (1978–82). Agro-exports clearly stagnated during the oil boom (2.6 per cent annually in quantum terms) compared to their post-boom growth (7 per cent in 1982–9; 11.4 per cent in 1983–8 in value terms), which was among the highest in Latin America.[20]

Agriculture's decline relative to the booming economy as a whole was even more pronounced. Its share of GDP was halved, from 24 per cent in 1970 to 12 per cent in 1982, before recuperating partially to 15 per cent in 1987 (World Bank 1999a). Its per-capita output grew only 0.1 per cent annually in 1972–82, compared to 6.9 per cent in manufacturing, 5.8 per cent in the tertiary sectors, and 4.1 per cent for the GDP as a whole (Larrea 1992: 209). This is the picture we would expect from the Dutch Disease: relative prices turn against the non-booming traded sector, and factors of production move to the (mostly urban) NT sectors. But semi-protected sub-sectors were sheltered from import competition and took advantage of the rapid rise in domestic demand. New infrastructure (roads, irrigation), tariff exemptions, cheap imported inputs and credit subsidies (at negative real interest rates) favoured particularly the rise of capital-intensive agriculture, including livestock. For the sector as a whole, the policy package probably cushioned the decline in agriculture, compared to what relative prices alone would have done to the sector.

Another land-using primary sector with deforestation impacts has been Ecuador's shrimp exports, which became a miracle export sector in the 1980s and 1990s.[21] Shrimp exports started to grow rapidly from 1981 to 1988, and again in 1991, making Ecuador the largest shrimp producer in the Western Hemisphere in less than a decade. Production is carried out in ponds built along the Pacific coast. Producers adopt land-extensive, low-cost, low-yield production methods, and the sector's dramatic growth in the 1980s has had a marked impact on ecosystem degradation and mangrove deforestation. From 203,700 ha in 1969, mangroves declined to 182,100 ha in 1984 and 175,100 ha in 1987.[22] Obviously, the accelerating real currency depreciation after 1982 helped this new sector to penetrate export markets and accumulate capital. In other words, had the Ecuadorian *sucre* remained heavily overvalued throughout the 1980s, both production growth and mangrove deforestation would have been less spectacular.

Forestry

Forestry is another potential T-sector victim of the Dutch Disease, the contraction of which might have eased deforestation and degradation pressures. Timber production

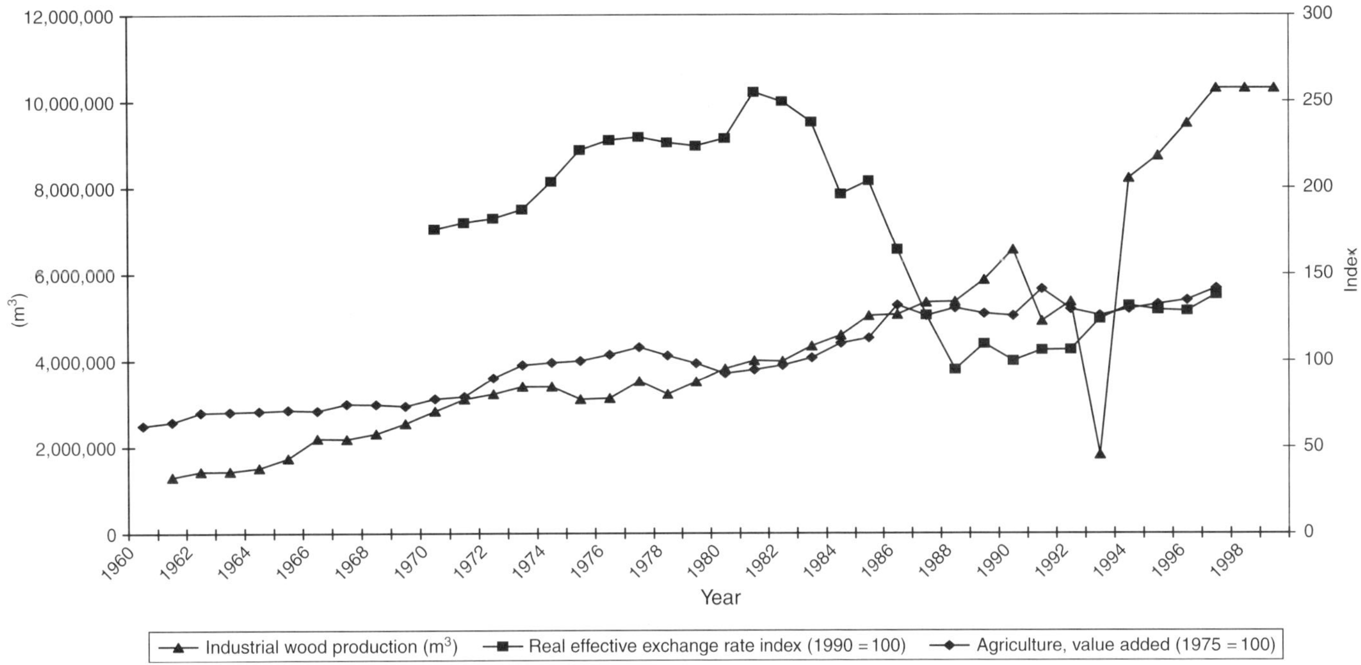

Figure 7.2 Ecuador: industrial wood production, agricultural value-added and real effective exchange rate, 1960–98.

Sources: FAO (2001b, at http://apps.fao.org), World Bank (1999a). For RER, see Figure 7.1.

Note
Industrial wood production in roundwood equivalent (Industrial roundwood: 1; Plywood: 2.3; Sawnwood: 1.82; Veneer sheet: 1.9).

characteristics in Ecuador are region-specific. Historically, Sierra natural forests provided significant supplies, but these have been gradually exhausted; today, only a few valuable species are still being extracted (from the Andean flanks), while ordinary species, destined for the construction sector, are harvested from high-altitude areas (Wunder 1997). Exotic plantations of pine and eucalyptus are expanding gradually in the highlands. Some timber comes from the Oriente, but mainly as a by-product of frontier expansion, with highly selective extraction: only the most valuable species can compensate for the high costs of transport. Two-thirds of sawnwood and almost all roundwood consumption in Ecuador comes from natural forests (ITTO and INEFAN 1994a), and about 70 per cent of this is from the biodiverse Chocó forest on the northern Costa (Esmeraldas Province).

It is only in northwestern Ecuador that deforestation can be claimed to be led by logging, for sawnwood and especially veneer production. For the period 1983–92, Sierra (2001) here compares satellite imagery for land-use change in a 600,000 ha study area with official timber transport data (scaled up by an alleged 60 per cent because of under-reporting). Using regression analysis, he finds that veneer consumption explains about 70 per cent of what he calls 'deforestation', namely the 86,108 ha that were degraded to less than 30 per cent tree-crown cover during the ten years under analysis. This 'forest degradation' (in our and the FAO's terminology) implies the removal of on average seventeen trees per ha, with significant changes in forest composition and structure (ibid.: 330–3). However, in other cases logging was more selective, removing a maximum of five trees/ha (Sierra and Stallings 1998: 141). Case studies show that remote forest communities only converted about 20 per cent of the logged-over area to agricultural uses – in other words, logging interventions ran far ahead of the needs for agricultural conversion. Unlike the other cases in this book, timber production at the forest level is dominated by small-scale producers (Sierra and Stallings 1998). On the other hand, veneer firms are the main providers of new roads, and even finance the maintenance of state roads, which stimulates in-migration into areas with good access. To use the terminology of Rudel (1993), timber firms are 'lead actors' in opening up the area to migrant squatters, but also in facilitating high-grade forest logging by local, indigenous dwellers.

In other words, logging has had severe degradation impacts and indirect deforestation effects in northwestern Ecuador. But to what extent were these interventions affected by Dutch Disease changes in competitiveness? The Ecuadorian timber industry is entirely focused on the domestic market, especially for furniture and construction (Sierra and Stallings 1998). This pattern has been promoted by partial trade protection and other policies favouring cheap timber supplies from natural forest. There are low investments, pronounced processing inefficiency (with about 40 per cent wastage in roundwood production) and oligopolistic output markets with high profits: for sawnwood, margins average 130 per cent (MoE 1999: 10; Comafors and IPS 2001). For northwestern Ecuador, the quantities produced have thus fluctuated with domestic demand, as well as with site-specific supply conditions. Strong El Niño climatic obstacles jeopardised production in the early 1980s, while a sustained expansion in 1986–91 coincided with new road-building in the area (Sierra and Stallings 1998: figure 3). In spatial terms, river and road access explain more than 90 per cent of the location of logging sites (ibid.: 152).

Timber exports have accounted for less than 2 per cent of industrial production, and are dominated by balsa wood and eucalyptus, both of which come from plantations (ITTO and

INEFAN 1994b). The trade balance of the Ecuadorian forestry sector remains negative, due to the country's large imports of pulp and paper.[23] Extraction from natural forests is thus almost entirely for the home market, not for export. Even so, annual roundwood production reached 5.95 million m^3 in 1998, a spectacular 41 per cent rise on 1994. Recently, roundwood exports rose to 123,850 m^3 (US$23 million) and plywood to 31,420 m^3 (US$37 million) (MoE 1999), but these figures remain small compared to the domestic market.

Changing price-competitiveness is thus unlikely to be the decisive driving force for Ecuadorian forestry. Sectoral growth rates behaved contrary to what a Dutch Disease leads us to expect: forestry grew 9.6 per cent in 1965–81, but only 2.4 per cent in 1982–9 (Southgate and Whitaker 1994: 18). Figure 7.2 above shows how the production of industrial wood (in RWE) evolved over the past four decades, according to FAO (2001b). These figures reflect the Ecuadorian forestry agencies' annual reports to the FAO and, as revealed by interviews with different forestry officials, for several years their size is dubious. Nevertheless, it seems clear that long-run growth was basically pro-cyclical to aggregate demand, while RER appreciation had a secondary role (see regression results below).

Trade policy impacts

Ecuador is a country with a significant tradition of import-substitution policies, and trade policies also influenced the adjustment to oil wealth. First, manufacturing and sub-sectors of agriculture enjoyed a high level of trade protection that turned them into 'quasi non-tradables';[24] the 'deepened' import-substitution strategy created a modern capital- and import-intensive industrial enclave (Larrea 1992: 237–42). In indirect terms, this urban bias hurt agriculture by helping to turn relative prices against that sector, simultaneously drawing labour from the countryside to urban areas (see section on 'Rural–urban migration'). But meat is an important example of a 'quasi non-traded' rural commodity that had huge land-use impacts on pasture expansion (see section on 'Structural changes in income and demand').

Second, trade policy was used as an instrument in stop–go policies to regulate aggregate demand and control inflation. Many agricultural goods became 'semi-traded' in the sense of not being fully or continuously exposed to the competition of importables. Here, trade policy had highly differential impacts. Some domestically produced crops were protected by food-import prohibitions, but their prices were not fully sheltered because of close imported substitutes. This applied, for example, to rice and vegetable oils. Much of agriculture was also favoured by subsidies, public purchases to support minimum wholesale prices, etc., on either a temporary or a permanent basis (Mosley 1991: 419). This was especially the case for capital- or import-intensive 'modern' agriculture (e.g. oil palm, large cattle- and rice-producers).

At the other extreme, trade policy deliberately 'sacrificed' highland wheat – a product that had been highly protected but could not compete with coastal and foreign grains once roads were improved. On top, wheat imports were massively subsidised to hold bread prices down (at a cost averaging 0.5 per cent of GDP in 1972–82), so that imports covered 92 per cent of wheat demand by the late 1980s (Larrea 1992: 254–6). Staple crops other than rice (potatoes, tubers, grains) also fared less well: they remained widely protected from direct import competition, but substitution effects with respect to subsidised wheat imports held their prices down. Trade policy was thus partially liberalised, but with widely differing impacts on different agricultural sub-sectors.

During the post-boom adjustment period of the 1980s, trade policy remained inconsistent over time and across products. This also reflected significant disagreement between Ecuadorian governments and the Bretton-Woods institutions. However, since 1989 Ecuador has reduced tariff rates significantly, in part triggered by the requirements of harmonisation within the Andean Pact. Tariffs were reduced to about 5 per cent for grains, 10–15 per cent for vegetable oils and derivatives, and 20 per cent for milk powder and wheat flour (Josling 1997: 14). Consequently, 'semi-traded' agricultural goods became much more exposed to international competition in the 1990s.

For forestry, its inward-oriented structure has been shaped decisively by past trade policies. Ecuador implemented a roundwood export ban until 1989. One result was that the mainly small-scale primary producers did not have access to exports, which were restricted to heavily concentrated processing sectors, such as plywood and veneers. Second, the roundwood and sawnwood prices paid to forest producers became heavily depressed, thus promoting excessive harvesting rates and large-scale wastage of wood (Southgate and Whitaker 1994). Ever since the ban was lifted, industry structure and resource-extraction patterns have adjusted very little (Sierra 2001: 334); distortions have had a lasting impact on an industry that has remained a typical home-market sector.

A quantitative view

This description indicates that forest- and land-using sectors were to some extent influenced by the oil boom and bust, but sector-specific trends, trade policy and shifting domestic demand all seemed to play their part in the adjustment of the economy. The only way to obtain more clarity on the general impact of competitiveness is to take an aggregate view. In Table 7.3, the main sector hypotheses are tested by regression analysis at the sector level, using data from three decades (1970–97).

Regression 1 shows that the oil exports and capital inflows (both inflation-corrected) influenced the RER in an upward (appreciating) direction, as we would expect. However, while oil exports are significant at the 10 per cent level, capital inflows are not, and low F-value and R^2 (14.5 per cent) indicate a relatively poor fit. Much of this is due to the model structure and the low impact of oil on the economy in the early 1970s, when bananas, cocoa and coffee were actually the leading sectors for foreign-exchange generation. Running the same regression for the last twenty years only (1977–97) leads to results that are statistically significant (not shown in Table 7.3).[25] The second question is how these relative price changes affected production patterns. In spite of the trade-protected status of some sub-sectors, we can see from regression 2 that agriculture's aggregate share in GDP was heavily influenced by the RER (coefficient significant at the 1 per cent level). Relative price changes can explain more than half the fluctuations in the sector's economic participation ($R^2 = 56.3$ per cent). Oil wealth thus definitely had an impact on Ecuadorian agriculture.

Turning to forestry, as a 'purely traded' sector, wood-export quantities (regression 3) should be highly sensitive to changing competitiveness. The problem is that wood exports

Table 7.3 Ecuador: relating oil wealth to relative prices and traded sector production. Regression results, 1970–97

Dependent / independent	*Petroleum exports (constant million 1995 US$)*	*Capital inflows (constant million 1995 US$)*	*RER (1990 = 100)*	*Non-agricultural GDP (constant 1995 US$)*	R^2	*F-value*	*Years*
1 RER (1990 = 100)							
Coefficient	0.026306596	0.008158676			0.145	2.126	1970–97
T-value	1.860500226*	1.050389569					
2 Agricultural value added (constant 1995US$)							
Coefficient			−4377948.151		0.563	33.554	1970–97
T-value			−5.79259483***				
3 Industrial wood exports (m^3)							
Coefficient			−650.129869		0.065	1.811	1970–97
T-value			−1.345818476				
4 Industrial wood production (m^3)							
Coefficient			−21850.94186		0.286	10.436	1970–97
T-value			−3.2305424***				
5 Industrial wood production (m^3)							
Coefficient			−8148.141446	0.000378071	0.578	17.142	1970–97
T-value			−1.305269919	4.159818***			
6 Industrial wood production (m^3)							
Coefficient			−10313.48559	0.000421146	0.776	41.569	1970–92 and 1994–7
T-value			−2.302043103**	6.430549***			

Notes

* Parameter *T*-value significant at the 10 per cent level.
** Parameter *T*-value significant at the 5 per cent level.
*** Parameter *T*-value significant at the 1 per cent level.

were totally insignificant until the last half of the 1990s (less than 20,000 m^3) and concentrated on a few sectors and plantation species with their own trends (FAO 2001b). As a result, the relative price coefficient has the expected negative sign, but is insignificant. The situation is different if we look at total industrial wood production, thus including the domestic market (regressions 4, 5 and 6). Regression 4 shows that the RER explains about one-quarter of the variation in production over the three decades (negative coefficient, significant at the 1 per cent level). However, we noted above the overwhelming importance of the home market and the partial protection of the timber sector. Regression 5 thus includes non-agricultural GDP as a proxy indicator of the mainly urban demand for furniture and construction goods. This explanatory variable has the expected positive sign, is highly significant, and considerably improves the equation's explanatory power (R^2 more than doubles), but it leaves the RER insignificant. This indicates that the RER had picked up some of the home-market effect in regression 4. Finally, we observed that all wood-production regression results were generally influenced by an outlier (1993), due to erroneous agency reporting to the FAO for that year.[26] Deleting that outlier in a new regression 6 yielded a much more satisfactory model performance, with both variables significant (at the 5 and 1 per cent levels) and explaining more than three-quarters of the variation in wood production.

Thus industrial wood production was affected by declining competitiveness, but rising domestic demand was its main driving force. Agricultural production directly stagnated during the boom, as a direct result of declining competitiveness; it became more mechanised and capital-intensive, and was subject to significant structural change in terms of crops and regions. However, we still need to make the link with land use: how did these production trends affect land demand and forests?

Unfortunately, area statistics are much more scattered. Higher timber production obviously increased forest degradation, especially in the timber-supplying forests of northwest Ecuador. Degradation here was therefore stimulated by high urban demand, but held somewhat back by competitiveness factors. However, increasing demand in the 1990s was also partially satisfied by plantation forests. In other words, the rise in degraded forest area was probably less than proportionate to the rise in timber production.

Turning to crop- and pastureland, data by R. Vos (cited in Larrea 1992: 255) indicate that total cropped area actually declined by a marginal 0.9 per cent from 1972 (1,563,200 ha) to 1982 (1,549,000 ha), followed by the aforementioned drastic expansion during 1982–7. With 2.5 per cent population growth in the 1970s, the decline in area is in itself a considerable 'achievement' of oil wealth. Tremendous structural changes between regions and products occurred; basic foodstuff areas lost one-third (because cereals declined 275,000 ha), while agribusiness crops expanded correspondingly. These technology and product changes can trigger forest loss in their own right, as new areas are converted, while abandoned ones are not allowed to revert back to forest cover (Wunder 2001a). Among the expanding coastal crops, coffee, cocoa and partially rice were planted on new lands; soybeans, hard corn, plantains and oil palm were planted mostly on land previously cultivated in bananas and cotton. However, the overwhelming land-use change during this period was a 60 per cent expansion of pastureland, from 3,346,100 ha in 1972 to 5,968,700 ha in 1982. Most of this was in the Sierra, with 40 per cent coming from abandoned cropland and 60 per cent from area expansion – mostly deforestation in the

Andean foothills (Whitaker and Alzamora 1990: 136–41). So, although the Dutch Disease actually hit agriculture as a whole, this particularly land-extensive sub-sector was allowed to expand, with a tremendous impact on forests. Ecuadorian deforestation during this period was thus basically a 'cowboy story', which we will return to in the section on 'Structural changes in income and demand'.

Windfall impacts on government spending

Agriculture and forestry

As we have seen, oil wealth caused relative prices and production trends to go against agriculture, with the notable exception of cattle ranching and a number of cash crops. But what about the budget allocations made available for these sectors? Did they grow with oil wealth and, if so, did that affect land use significantly? Certainly, agriculture did receive additional funding in a direct manner. The Central Bank channelled oil revenues through the Financial Funds Mechanism to selected sectors, including agriculture. The state-owned BNF and the Centre for the Economic Recovery of Azuay, Cañar and Morona-Santiago (CREA) are two prominent examples; others are agencies with a direct mandate in settlement and land colonisation (see section on 'Structural changes in income and demand').

The agenda of these institutions was in direct contradiction with the objectives of forest conservation, and the additional money they received did promote land extensification to a certain extent. In the Sierra, BNF generally earmarked subsidised credit almost exclusively for cattle-ranching, the most extensive type of land use and the 'end use' of most converted forestland. In the Costa, BNF broadly assisted the expansion of crops, especially rice, which, as we have seen, was drawn partly from newly colonised lands (Ramos and Robison 1990). CREA, a regional development institution for southern Ecuador, saw its main task as linking the highland industrial area of Cuenca to the southern Amazon region by promoting infrastructure and agricultural trade between the two regions, which gave a major incentive to forest-clearing in that region (see section on 'The effect of oil production of forests'). Rudel (1993: 56–7) describes how, in this sense, the oil bonanza created a highly exceptional situation, given that government agencies normally tended to be short of funding to assist the *colonos*, while now there were abundant resources available to support the ongoing process of land colonisation.

However, not all agricultural funding achieved its alleged objectives. Although inefficiencies in government spending on agriculture were less pronounced than for the three previous countries in this book, there was much deviation of subsidised agricultural credit. As de la Torre notes (cited in Larrea 1992: 245), 'credit diversion may have been significant ... the higher profitability of non-agricultural credit ... created strong incentives for diverting a part of the credit given ostensibly to agriculture towards other sectors'. As the 14.7 per cent of farms that are larger than 20 ha obtained 70 per cent of the credit (often at negative real interest rates), a large share of this funding became simply a rent-seeking target for large, influential landholders (Larrea 1992: 245). Part of it ended up financing the purchase of land or of consumer goods, such as 'cars, boats and vacations abroad' (Ramos and Robison 1990: 239). Finally, some of the public funding that actually was invested in agriculture produced meagre returns. The best example is the state irrigation and water projects. They accounted in the 1970s and 1980s for, on average, 40 per cent of public-sector spending on agriculture, but effectively irrigated only 6 per cent of cropland by 1985 and produced only 6 per cent of agricultural GDP (Whitaker and Alzamora 1990a: 176–7).

In forestry, additional funding for forest and protected-area management was made available during the bonanza, which could potentially have helped to better protect forest resources. One of the resulting achievements was the planning and creation of the majority of Ecuador's national parks during the 1970s, culminating in the establishment of the Ecuadorian Protected Area System in 1981. By 1995, the system included eighteen protected areas, corresponding to an impressive 11 per cent of national land area (Figueroa 1995: 223). Despite this positive example, there are two reasons for questioning the general effectiveness of increased public-sector funding in curbing Ecuadorian deforestation and forest degradation. First, the resources flowing into forest conservation and management could simply not match the funding made available for agricultural development agencies, as just described. Second, bureaucratic and centralised institutional structures made it difficult for the forestry agency to use additional funds to achieve the desired results in the field, especially in new areas such as nature conservation. On several occasions, political changes have led to existing forestry agencies being closed and new institutions being set up. In some cases, these institutional discontinuities have been coupled with local irregularities in the enforcement of Ecuadorian forest laws.[27] Until recently, the forestry agency was widely perceived as the 'prolonged arm' of large timber firms. In fact, despite the genuine efforts of individual employees of public agencies, it can be suggested that NGOs, enabled by international funding, have been the most proactive and consistent agents of forest conservation in Ecuador over the past two decades.

Roads and transport subsidies

The previous section showed that the oil boom's budgetary impacts on strengthened forest and protected-area management was overshadowed by the greater boom-induced financial injections to development institutions that directly counteracted forest conservation objectives. In spite of 'leakages' and inefficiencies in spending budgets acting both for and 'against' forest conservation, it is clear that, on balance, the criteria for the allocation of soaring government budgets reinforced deforestation.

This conclusion becomes even more evident if we include road-building in the analysis. The extension of the poor pre-boom road infrastructure was a *sine qua non* for the implementation of a strategy of national integration, increased factor and goods mobility, and specialisation according to regional comparative advantage. This is illustrated by the overwhelming priority given to roads in budgetary terms. In 1974, no less than 48.4 per cent of public investment was channelled into road construction, a share that only gradually declined to 18.2 per cent in 1981 (Gelb and Marshall-Silva 1988: 184). Improved infrastructure and access also helped rural areas to take advantage of rapidly growing urban markets, shift from subsistence to commercial practices (see the section on 'Structural changes in income and demand'), and increase off-farm employment (Commander and Peek 1986). This led to a variety of ways of interaction with the urban economy, including temporal migration (see Waters 1997, and the section on 'Structural changes in income and demand').

The massive subsidies to domestic energy consumption, amounting to an astonishing 7.3 per cent of GDP in 1980 and 9 per cent in 1986,[28] worked in the same direction. By lowering transport costs drastically, rural mobility and market access for agricultural products were increased, creating a favourable climate for the integration of remote areas into a regional economy. In the 1960s, only about 20 per cent of peasant production was marketed; this share rose to over 60 per cent in the 1970s (Larrea 1992: 350–1). Energy subsidies were a boom-spending category that was built into the expectations of the electorate even after the boom, and thus for more than a decade proved politically too costly to eliminate.

In Ecuador, the crucial role of transport and roads in promoting land-use change had already been exemplified by the completion of the road from Quito to Santo Domingo in 1964, which was a benchmark in the colonisation of the coastal lowlands and the western flanks of the Andes.[29] The western flanks were also those most heavily affected by new, oil-financed road-building, due to their strategic location between the two economic poles, the Costa and the Sierra. In the section on 'The effect of oil production on forests', we commented on the deforestation impact of roads for oil production in the northern Oriente. Even in the southern Amazon provinces, where there was no oil, roads were vital for colonisation. In his analysis of Morona–Santiago Province, Rudel (1993) demonstrates how the first wave of *colonos* arrived in a given area in anticipation of planned road construction. The main factor determining the success or failure of the settlement effort was whether the road was actually constructed or not. Similar speculative land-occupation patterns that focused on road construction can be observed elsewhere in Ecuador.[30]

On the whole, the ambitious road construction programme of the 1970s probably had an immediate, large impact on frontier clearing (in the first 3–5 years). However, as indicated in the section on 'The effect of oil production on forests', the limited labour and capital available to most farmers meant that on-farm clearing on a typical 40–50 ha frontier plot would be a decade-long process.[31] In other words, while some of the deforestation impact of roads became apparent immediately, a good part of it only occurred in the 1980s, when relative prices had turned in favour of agriculture (section on 'The macroeconomic impact of the oil boom'). Thus what seems to be a slowdown in deforestation in the 1990s may be linked more to the fact that few new road projects were initiated during the crisis of the 1980s, while the lagged impact of the old oil-boom roads had died out. In 1983–6, the total length of the road network only expanded marginally from 35,662 to 36,187 km; a significant new expansion was only achieved in the early 1990s, to 43,200 km in 1994 (IRF 1994, 2000).

Directed settlement

Historically, there has been little tradition of directed settlement programmes in Ecuador. This led one observer to conclude that 'colonisation has been more an accidental by-product of government policies than a direct result of them' (Bromley 1981:15). This verdict seems exaggerated, given the fact that many road projects served the explicit purpose of promoting or consolidating *spontaneous* frontier expansion, thus relieving population and land-reform pressures in the Sierra. In legislative terms, the promotion of spontaneous colonisation has long roots,[32] as has the support of successive national colonisation institutes (INC, IERAC, INDA) or regional development agencies (e.g. CREA or INCRAE; see the section on 'Windfall impacts on government spending'). While the mandate of these institutions was to promote agricultural development, rather than to occupy specific regions or to redistribute population, they did in fact have these latter effects, and oil wealth also helped them to some extent to achieve more of their goals.

The Ecuadorian Institute for Agrarian Reform and Colonisation (IERAC) followed in the 1970s and 1980s the traditional concept of 'bringing people without land to land without people', with continuous on-farm forest-clearing as a prerequisite for granting land titles and, at the local level, engaging in dubious land transactions. In 1974–5 alone the oil revenues directly allocated to IERAC caused a tripling of its budget, thus underlining the link to oil wealth. Although historical experiences of 'semi-official' colonisation projects also exist in the Costa,[33] most current efforts are concentrated in the Oriente: Harcourt and Sayer (1996: 69) refer to a total of seven official development schemes in the Amazon region. However, these state-directed projects (or rather, attempts to control and rationalise existing spontaneous settlements) have a poor success rate (Uquillas 1984). For instance, Hiraoka and Yamamoto (1980) describe the 3,400 ha IERAC-financed Shushifindi project, plagued by a number of problems including its cooperative organisation that individual settlers were unaccustomed to.

After the 1941 war with Peru, Ecuador lost the bulk of the Oriente. As a response, the government established or reinforced military outposts in the border area and combined it with crop-cultivation schemes. Yet, in most cases these settlements were too isolated to cause much deforestation (Hiraoka and Yamamoto 1990: 425–6). The most significant motivation for directed settlements has thus been geopolitical. But there is no evidence that these minor projects were systematically correlated with the oil boom: rather, they are determined by the shifting strength and urgency of military interests over time. For instance, in the aftermath of the last armed conflict with Peru in 1995, government programmes intensified Ecuadorian settlement in the disputed *Cordillera del Condor* border area in the South.

Structural changes in income and demand

Poverty alleviation

Did poverty decline during the oil boom, and correspondingly rise during the following economic crisis period? And how were changes in poverty and human welfare related to the forest? First, an oil-led per-capita GDP growth in a country the size of Ecuador should make us expect poverty to be reduced, even though the growing capital intensity of the economy and reduced rural employment may pull in the opposite direction. Indeed, income poverty was reduced somewhat, although the data are extremely scarce. World Bank estimates (cited in Tabatabai 1996: 59–60) suggest that in 1975, 65 per cent of the rural population lived below a poverty line of US$183/cap/yr, but by 1980 the same population share was below a significantly higher income line, US$320/cap/yr. For the urban population, 40 per cent were below a corresponding line of US$269/cap/yr, but rising five years later to US$470/cap/yr. Later data (for 1987) are not directly comparable, but there may have been an increase in income poverty.

Perhaps more importantly, targeted government investments of oil revenues in the social sectors triggered fairly remarkable improvements in general living standards, though admittedly from a very low pre-boom level. From 1974 to 1982, housing standards improved greatly, with a share of households with 'adequate dwellings' jumping from 38 to 64 per cent (in rural areas, from 28 to 58 per cent); the electrification of rural areas tripled. Illiteracy rates were almost halved; mean years of schooling improved from 3.6 to 4.7. Life expectancy improved from 56.8 (1965–75) to 64.3 (1980–5), but a basic needs factor such as the incidence of child malnutrition remained around 50 per cent (Larrea 1992: 386–94). In other words, there were significant improvements, though many indicators remained well below Latin American averages.

How did this poverty reduction affect forests? There are little data to illuminate this relationship. But one factor is that labour became more expensive as a result of oil wealth. Notably, the more than doubling of minimum rural wages (in fixed 1975 *sucres* per month), from 854 in 1973 to 2,115 in 1980, is likely to have reduced labour-intensive deforestation and forest degradation (firewood gathering, unsustainable NTFP collection, etc.) by significantly increasing the opportunity costs for labour. Similarly, the opposite must have happened in the 1980s, when the same rate gradually fell back to 1,156 *sucres* a month. Yet improved rural infrastructure and living standards also indicate that there were some incentives for the poor to remain in their rural occupations.

Most frontier deforestation in Ecuador may be opportunity- rather than poverty-driven. It is seldom the poorest, landless, deprived peasant who conquers the frontier, but rather the entrepreneurial lower middle-classes seeking to improve their livelihoods, and having the necessary minimum capital outlay to finance 'land improvements'. This image of the typical frontier farmer has been confirmed for a broad range of areas of colonisation in Ecuador.[34] Pull factors, such as favourable market prices, (anticipated) road construction and the resulting generation of land rents, subsidised credits, etc., prove to be more important deforestation incentives than poverty-led push factors. This characteristic was already noted by Casagrande *et al.* (1964: 292) for the early colonisation of the western Andean flanks: 'While the requirements, in terms of capital, for establishing a homestead and acquiring title to the land are not great, they may be substantial enough to make the move economically impossible for the really destitute *Sierran*'. The oil boom may have helped many to meet exactly these requirements. This refers in particular to the accumulated savings and availability of cheap rural credit, to road-building and to the increased support of development institutions. In combination, this mixture of incentives favoured frontier expansion, in particular capital-demanding, land-extensive, small-scale cattle-ranching. The improvement in economic capacity may particularly help to explain sustained expansion of pastureland, the most important factor in creating forest loss (see the section on 'Structural changes in income and demand').

These examples show that poverty alleviation had multiple and ambiguous impacts on forests. The effects depended on what type of 'poor people' were affected, what was their specific interaction with natural resources, and how relative price effects compared to investment-constraint effects. In addition, an important demand-side factor was that the reduction in urban poverty caused a higher demand for meat and dairy products, stimulating cattle-ranching significantly (see section on 'The structure of consumption'). A best guess is that a reduction in poverty also reduced low-remunerative forest degradation, while actually *increasing* cattle-led deforestation.

Rural–urban migration

It is notable that the sectors with the largest growth rates under the oil boom, the NT sectors (construction, public and private services) and the quasi NT sectors (industry), are all largely concentrated in urban areas. To move factors of production to urban areas, incentives had to change: for instance, the ratio of rural to urban minimum wages declined drastically from 88 per cent in 1970 to around 70 per cent in 1972–82, before gradually rising back to 80 per cent in 1983–9 (Larrea 1992: 329). As a result, the oil boom accelerated migration to urban areas (Commander and Peek 1986). Urbanisation rose from 40.6 per cent in 1972 to 48.7 per cent in 1982. This continued throughout the 1980s, reaching 55.1 per cent in 1990 and 60.3 per cent in 1997, but a statistical reclassification actually overestimates urbanisation in the 1980s.[35] On the other hand, agriculture both stagnated and became more mechanised, which reduced agricultural employment between 1974 and 1982 by 90,000 jobs, this being a main explanation for the rural exodus. This means that boom-led urban prosperity and employment options most probably reduced the number of people seeking to degrade and convert forest areas, whether it be forest frontiers or on-farm forest fragments.

But migration data alone underestimate the growing overall importance of urban employment opportunities created by oil wealth. Rural–urban commuting in the 1970s integrated many of those staying behind in rural areas into the urban economy, thus reducing their dependence on agricultural activities. Commander and Peek (1986) showed that the smallest farms (0–5 ha) received more than 50 per cent of their household income from off-farm sources, for example, from wage-labour in urban construction. Higher labour mobility was generally supported by road construction and huge energy and transport subsidies (see the section on 'The competitiveness of agriculture and forestry'). These policies became a means of achieving a 'trickle-down' effect from rich to poor and from urban to rural areas. Their impact on forests seems to have been variable. On the one hand, it allowed agricultural goods from the hinterland to be marketed, thus increasing forest pressures (see the section on 'Windfall impacts on government spending'). On the other hand, cheap transport also allowed people to interact more closely with the urban economy. In Pichincha Province, in the peripheral area around Quito, cheaper transport connections allowed for flexible patterns of daily, temporal and cyclical migration of people seeking urban informal jobs but maintaining farming on a part-time basis. In a survey from 1985, this 'quasi-urbanisation' meant that 10 per cent of the economically active population of Quito resided outside the city (Waters 1997).

In regional terms, the boom also partly shifted economic power from the agribusiness centre and main coastal export port of Guayaquil to the capital, Quito, in the Sierra: 40 per cent of all industrial investments were made in Quito. As boom revenues accrued to the public sector, rent-seeking and rent-allocation naturally had their starting points in the capital. But small towns also grew quite rapidly during the boom, favoured by state investments in the social sectors. There were important differences in demographic trends across types of rural area. Larrea (1992) compares 1974 and 1982 rural population census data at cantonal level, using cluster analysis to distinguish types of rural areas. He finds that pure export agriculture areas, mainly banana-producing ones, experienced heavy reductions in population, due to labour-saving technologies, while rural populations in areas of periurban capitalist agriculture grew because of increased urban demand. What is interesting for our purposes is that rural frontier areas of new settlement, mainly in the Amazon (Larrea 1992: 382, 609), generally *increased* their workforce in absolute terms: the economically active rural population in 'colonisation areas' grew from 41,265 in 1974 to 53,580 in 1982.[36] The observation is in line with the land-use changes from Table 7.2 above, showing an accelerated expansion in the Amazon only after 1984. This shows that urbanisation could not halt expansion at the Amazon frontier but, in spite of the access-providing effects of oil roads (see the section on 'The effect of oil production on forests'), it was much slower than the subsequent expansion of the 1980s.

The structure of consumption

Income growth not only increases the demand for non-tradable goods disproportionally, there are other long-run shifts in consumption too, substituting 'inferior' for 'luxury' goods. Specifically, the agricultural sector is likely to face shifting production incentives for different types of foodstuffs. For a country like Ecuador that experienced a 50 per cent rise in per-capita income, of which 40 per cent proved to be lasting throughout the next decade, we would expect such demand shifts to be prominent.

Foodstuffs as a whole are likely to decline in importance when incomes grow. This 'Engel effect' was indeed confirmed by Southgate and Whitaker (1994: 16) for Ecuador: the income elasticity for foodstuffs was 0.55 for 1965–81, that is, for a 1 per cent rise in income, food demand only grew by 0.55 per cent. But there was also a marked redistribution within the food consumption basket, as shown by the following percentage production changes between 1970 and 1980. Traditional staple crops like soft maize (−70.7 per cent), barley (−70 per cent), potatoes (−40 per cent) and cassava (−14 per cent) lost a great deal of ground to rice (+297 per cent) and imported wheat. On the other hand, demand increased production for 'luxury' foodstuffs, such as beef (+67 per cent), other meat (+91 per cent), milk (+43 per cent), vegetables like tomatoes (+74 per cent), African palm oil (+1,061 per cent, from a tiny base) and fruits like oranges (+248 per cent) (de la Torre, cited in Larrea 1992: 257–8).

As we have seen in the section on 'The competitiveness of agriculture and forestry', the net land-use effect of these drastic structural changes on total cropped area was roughly neutral. In land-use terms, the most important factor was the high income elasticity for livestock products (Southgate and Whitaker 1994: 16): a richer urban population buys more dairy products and has more barbecues. Production in the livestock sector grew by an impressive 4.6 per cent yearly during 1965–81, while growth rates declined to 1.9 per cent for the post-boom period of 1981–9 (Southgate and Whitaker 1994: 18). The sector thus followed a cycle that was positively correlated with the oil boom and expanding domestic demand, rather than being hurt by the Dutch Disease. A basic reason for this is that it largely remained a trade-protected sector. Unlike the other countries analysed in this book, such as Venezuela and Gabon, where meat imports soared with growing oil wealth, Ecuador's meat imports never passed 100 million t during the whole 1973–82 period.

Only in the 1990s was trade opened up to such an extent that meat imports increased to the 2,000–3,000 million t range (FAO 2001b). For dairy products trends were variable, but most of them remained protected; for milk powder, the tariff rate only in the 1990s came down to 20 per cent (Josling 1997: 14).

The shift in consumption towards livestock products, combined with protectionist trade policies, had a startling impact on land use. From 1972 to 1982, pastureland increased by more than 2.5 million ha, an area the size of the US State of Maryland. This was a 60 per cent increase in one decade, which meant that pastures in 1982 came to make up three-quarters of all land under agro-pastoral use (Larrea 1992: 259). As mentioned, most of this increase was derived from forested land. For areas close to the two big cities, Guayaquil and Quito, and regions that had been integrated into growing urban markets by newly constructed roads, this link was very direct. The example of the northeastern part of Pichincha Province, in the vicinity of Quito, with dramatic levels of forest conversion being almost exclusively destined for pasture, is highly illustrative of this.[37]

Notably, cattle expansion was achieved by sheer *extensification*: the livestock-to-land ratio declined markedly, from 1.1 cattle head/ha in 1972 to 0.7 in 1982 (Larrea 1992: 259). This reflects the expansion of ranching into still more marginal areas.[38] Probably some of what is counted as pastureland was degraded land, which now is used only marginally. Another part is probably land acquired by urban investors for speculation purposes; stocking it with a few head of cattle helps sustain their tenure claim by a visible 'productive' land use. In frontier areas, smallholders' highly land-extensive ranches are often plagued by disease problems, and they lack technical expertise and assistance. Nevertheless, as a general observation, it is remarkable that the booming livestock sector developed no land-saving technological progress over this period; rather, the opposite occurred. Obviously, this also reflects the fact that land did not develop a sufficient scarcity value, which is rooted in the package of pro-extensification policies (road-building, tenure rules, cattle credits, etc.) that favoured accelerated deforestation.

Synthesis and conclusion

In Ecuador, the 1974–82 oil boom did not reduce deforestation or forest degradation. On the contrary, forest loss increased from low pre-boom levels and continued to be high throughout the crisis of the 1980s, before probably levelling off somewhat in the 1990s. Correspondingly, non-sustainable logging experienced a remarkable growth during the bonanza years. This happened in spite of the fact that, as a new oil exporter, Ecuador received a new, semi-permanent transfer of wealth, increased its foreign-exchange inflows in the late 1970s by borrowing heavily in international capital markets, and had an economy that grew significantly during the boom.

What made Ecuador deviate from the Dutch Disease core hypothesis that oil wealth relieves pressures on forests? One possible explanation is that the oil itself came from the heavily forested northern Amazon region, where it caused large-scale forest loss. Our assessment showed that this is only partially true. Although there were severe forest-degradation impacts, particularly from oil pollution, the direct deforestation impacts of the oil industry and from roads were negligible. The indirect impacts from oil roads to open up new areas for first timber extraction and then colonisation were more important, causing a spontaneous influx of agricultural squatters, who established their farms and gradually 'ate' their way into the forest. In absolute terms, the northern Amazon is the region that has lost the most forest over the past three decades. But deforestation in the southern Amazon and on the Andean flanks was also high, in spite of the fact that there was no oil in these regions. In fact, the expansion of land use was much slower than in the 1980s. Instead, more market-near frontiers were being opened up, especially on the western Andean flanks connecting the Sierra to the Costa, as these zones were better suited to accommodate growing demand from the booming urban economies.

Rather than blaming forest loss on the oil industry, the main story behind the accelerated forest loss in Ecuador involves a single species: cattle. The increase in pastures basically accounted for all net land-area expansion of the agricultural sector during the oil-boom decade, with pastureland expanding at 6.2 per cent/yr in the 1970s (Whitaker and Alzamora 1990: 136). And all this new land basically came from forests. The Dutch Disease certainly hit the agricultural sector, in spite of the various steps that were taken to protect it (import protectionism, subsidised credit, improved infrastructure and technical assistance). These protective measures mostly favoured a capital-intensive agribusiness subsector. In overall terms, agricultural production grew slowly, its share of GDP was halved, rural employment declined and cropped area stagnated. All this alone would have created a scenario favourable to forest conservation. But cattle-ranching did not follow this pattern of stagnation: land-extensive production grew impressively. A number of factors coincided to make this possible:

- growing consumer income induced a shift to meat and dairy products
- meat-import protection made cattle *quasi non-traded*, so high demand raised cattle prices
- growing producer incomes (reduced poverty) permitted investments in cattle
- massive road-building created large forested open-access areas
- land-tenure rules favoured 'homesteading through forest-clearing', followed by extensive use, including ranching as a simple means to discourage squatters and preserve land tenure
- improved funding for land-development agencies supported colonisation
- much subsidised credit with negative real interest rates was earmarked for cattle
- high real rural wages favoured capital-intensive ranching over labour-intensive crops.

It is seldom that a sector has all factors working in its favour, but in the case of cattle this actually seems to have happened. This is no sheer coincidence, given that highland crops were increasingly being replaced by lowland rice and imported wheat, so that rural interests in the Sierra pushed for the development of ranching as an additional source of income. The economic climate was favourable on the demand side (first two factors), but policies also shaped a supply-side package (last six factors) that gave strong incentives to land-extensive and capital-intensive rural investment, a formula that translated directly into the conversion of forests into land for expansionary ranching.

Table 7.4 gives a summary of ten major boom impacts on forests, ranked according to their alleged deforestation size (last column), which is a product of economic significance (column 3) and strength of the link from that production in that sector to deforestation

Table 7.4 Ecuador: oil wealth and deforestation – an overview of impacts

Economic and productive impacts			*Links to deforestation*		*Deforestation impact*	
No.	*Type*	*Intensity*	*Type*	*Strength*	*Type*	*Intensity*
1	New road construction (and transport subsidies)	Strong ●●	Promoting new settlement and agricultural trade	Very close ➔	Opening up frontier areas	Powerful, lasting ▲▲▲
2	Higher urban labour absorption (industry, services, etc.)	Strong ●●	Less rural labour and less demand for cropped land	Close →	Less forest conversion	Strong ▼▼
3	Higher urban incomes and shifts in food demand	Strong ●●	Expands particularly cattle-ranching	Medium ⇨	Forest conversion to pasture	Strong ▲▲
4	Trade policy	Strong ●●	Partially protects agriculture from the Dutch Disease	Medium ⇨	More forest conversion to crops and pasture	Strong ▲▲
5	Loss of agricultural competitiveness	Medium ●	Reducing crop and pasture area expansion	Close ➔	Less forest conversion	Medium ▼
6	Soaring budgets of development agencies	Strong ●●	Supporting colonisation efforts	Medium ⇨	Augmenting and sustaining encroachment	Medium ▲
7	Oil expansion in the Northern Amazon	Strong ●●	Direct (roads, timber); but mostly indirect (colonisation)	Close ➔	Opening up frontier areas (some overlap with impact 1)	Medium ▲
8	Soaring budgets of forestry agency	Medium ●	Augmenting forest control and protected area system	Weak →	Less encroachment and degradation	Weak ∇
9	Directed settlement	Weak ○	Frontier occupation	Medium ⇨	Opening up frontier areas	Weak ○
10	Poverty alleviation	Medium ●	Eliminates low-yield forest loss but increases investments	Variable	Ambiguous	Variable ?

Note
2, 5 and 8 areas – reduces deforestation; 3, 4, 6, 7 and 9 areas – increases deforestation; 10 area – ambiguous.

(column 5). (See Chapter 1 for a discussion of the method.) For instance, the marked expansion in the road network and reduced transport costs (effect 1) had a strong impact on trade, growth and sector distribution (column 3). Road construction also exhibits a close causal link with deforestation, because it is very closely related to frontier expansion (column 5). As a product of the close link from infrastructure to production and from production to deforestation, on the whole road expansion was found to be an extremely powerful source of deforestation, especially in the form of the conversion of forest into pastureland (column 7).

A quick overview provides a hint of why the oil boom in Ecuador caused the rate of forest loss to increase rather than decrease. Most factors (six) accelerated forest loss (3, 4, 6, 7 and 9 areas), three strongly. Only three factors reduced forest loss (2, 5 and 8 areas), one strongly (see last column of Table 7.4), while one (poverty reduction) was ambiguous. The typical market adjustment effects of the Dutch Disease curbed deforestation, *ceteris paribus*. Rapid urbanisation drained agriculture for labour and thus reduced pressures for forest conversion (2). Relative price changes (5) delayed the growth of export agriculture, shrimp-farming and some import-competing crops, but not the growth in cattle-ranching. The latter in turn had to do with the group of factors mentioned above, especially demand shifts accompanying rising incomes (3) and trade protection (7). Oil production itself also caused some deforestation in the northern Amazon, though mostly indirectly (4).

Government agencies also earmarked oil-derived revenues for land-expansion activities that increased deforestation. Here, road construction and transport subsidies (1) were vital, subsidies to colonisation (6) had a medium effect, while the role of directed public settlement programmes (9) was negligible. Combined, these development-budget effects were much stronger than the impact of increased funding for protected areas and forestry regulation (8). Notably, forest-loss impacts from roads (1, 4) and changes in demand (3) proved to be gradual and/or lasting, so that they continued to have an effect throughout the 1980s, pointing to asymmetries in the relationship between oil wealth and deforestation relationship. Basically all other effects were symmetrically reversed during the economic crisis. This illustrates the large impact of the oil cycle on both markets and policy. In particular, agriculture resumed a competitive, leading position in the economy, which continued to exhibit high rates of forest loss in the 1980s.

One more general conclusion is that the external framework – the export and trade incentives – need not be decisive for deforestation. The Ecuadorian experience shows that development strategies and policy packages can set the stage to such an extent that they alone determine whether a boom causes a halt or a spurt in deforestation. The analysis points to road-building, transport subsidies, agricultural trade protection and colonisation agencies as the leading factors in creating forest loss and forest degradation. In Ecuador, the dynamics were fully led by domestic production. Potentially, international trade could have played a larger role. But trade restrictions helped to build and maintain sheltered home-market sectors (timber, ranching) that were highly rudimentary in their technologies, and at the same time extremely wasteful in their use of natural resources (wood, forested land). Both sectors grew pro-cyclically with the domestic market, only being marginally hampered by price competitiveness. While this favoured particular business interests, it was probably a 'lose–lose' policy for society, sacrificing natural forests for the sake of sectors with low economic efficiency. This cast a dark shadow over Ecuador's fairly sound macroeconomic management of the 1970s' oil boom.

Following a prolonged economic and political crisis from 1996, in 2000 Ecuador decided to respond radically to eradicated investor confidence by abandoning its currency, the *sucre*, for the US dollar. As in countries that have pegged their currency to the dollar, inflation did not adjust downwards quickly enough to avoid real appreciation. This was reinforced by very favourable oil prices recently. It is still too early to analyse the consequences, but it is certain that once again an appreciating RER will trigger significant disincentives for 'traded' agriculture, and possibly for some timber production. Will this also reduce the pressures on the forests? This time, a more liberalised economic environment will make it more difficult to use protectionism as a regulatory tool. But in the end, the forest outcome of higher oil inflows and dollarisation will depend largely on the factors outlined in this chapter, namely the policy package that accompanies the reduced competitiveness of the non-oil economy.

Notes

1 See Denevan (1976) for an overview of the debate on pre-Columbian population size in South America. It is likely that post-conquest reduction was particularly dramatic in the Ecuadorian Andes.

2 During 1950–82, Costa population grew from 1,289,000 to 3,947,000 (Benalcázar 1989: 90).

3 The 15.4 million ha figure is an estimate from Wunder (2000: table 4.2), based on material made available by Flemming Skov from the Danida-financed DIVA programme (Centre for Research on the Cultural and Biological Diversity of Andean Rainforests).

4 See Chapter 1 for a critique of the methodology. The net result for Ecuador is that deforestation does not exist for the bureaucrats who send reports to the FAO: forest and woodland area should have *increased* from 15.5 million ha in 1979 to 15.6 million ha in 1994 (ibid.: 8).

5 WRI uses the FAO-FRA figure for 1981–90. However, the 1981–90 and 1981–5 WRI estimates can only be consistent if annual deforestation dropped from 340,000 ha in 1981–5 to a mere 136,000 ha in 1986–90. A reduction may have occurred in the late 1980s, but is unlikely to have been of this magnitude.

6 The gap between the MAG and INEC data reflects different methodologies, rather than any dramatic change from 1984–5 to 1988–9 (Southgate and Whitaker 1992).

7 Information about the low reliability of the post-1990 data was confirmed by various experts, such as Remy Ojara of the Ministry of Agriculture, Vilma Salgado of UNDP, and the statisticians Raul Gaethe and Octavio Recalde of the World Bank – MAG SICA project (Quito, March 2001).

8 I am grateful to the Quito-based research NGO Ecociencia for making these figures available; they differ somewhat from INEC's figures, apparently because of variable definitions.

9 This refers to the forthcoming agricultural census and to the ongoing work of Rodrigo Sierra (Ecociencia/Arizona State University) on national forest-cover changes for different sub-periods.

10 Schmidt, R. (1990): 'Sustainable development of tropical moist forests', FAO Forestry Department, Rome, cited in: Southgate and Whitaker (1994: 36).

11 Rudel (1993: 56–7 and chs 5, 6) describes how penetration into large forest tracts in Santiago–Morona Province (southern Amazon) is linked to road-building and to the financial support from government colonisation agencies that intensified in the 1970s.

12 Torres (2000: table 2). He measured the northernmost area of Napo–Sucumbío Province, but his definition of 'affected forests' – 300 m radius around each well – overestimates forest loss.

13 Case study of the ARCO concession, reported by Kimerling (1991: 55). This includes 1,200 km of seismic lines at 3 m average width, a forest opening which without further disturbance is likely to regenerate back to forest quite rapidly.

14 The detailed calculation is: 392 km $\times$ 0.02 km $=$ 7.84 km^2 (primary); 214 km $\times$ 0.015 km $=$ 3.21 km^2 (secondary); 0.008 km $\times$ 593 km $=$ 4.74 km^2 (tertiary). These add up to 15.79 km^2. We excluded from the calculations Ruiz' category of 'minor feeder roads' (*red vial vecinal*), which are unlikely to have been oil-related. Note also that the assumed clearing width refers to the effects of road-building only, not *de facto* for land colonisation purposes, which classifies as an *indirect* impact.

15 The 1994–5 estimate for forests in the Oriente is 7.56 million ha (Wunder 2000: 102).

16 'Preface by Robert F. Kennedy Jr.', in Kimerling (1991: ix).

17 The detailed calculation is: 392 km × 24 km = 9,408 km (primary); 214 km × 10 km = 2,140 km^2 (secondary); 593 km × 0.4 km^2 = 237.2 km (tertiary). This adds up to 11785.20 km^2.
18 Sierra (2000: 6) finds that 1,228,894 ha had been cleared in 1996, but his study area includes not only the whole of northeastern Ecuador, but also sections of Colombian and Peruvian territory.
19 Because of the high humidity of the Ecuadorian Amazon, this is the locally applied alternative to the slash-and-burn method used in the highlands and in most of the Brazilian Amazon (Thapa *et al.* 1996: 1330).
20 See Southgate and Whitaker (1992: 41); Larrea (1992: 205).
21 This paragraph draws on Parks and Bonifaz (1995), and Southgate and Whitaker (1994: ch.8).
22 CLIRSEN data, reported in Southgate and Whitaker (1994: 91–2). The change from 1969 to 1984 is likely to have occurred mainly in the 1980s, when shrimp production started to grow.
23 See ITTO and INEFAN (1994a), and ITTO and INEFAN (1994b).
24 Tariff rates during the boom were 45 per cent for non-durable and 82 per cent for durable consumer goods, 17 per cent for intermediate and 16 per cent for capital goods (Larrea 1992: 230).
25 Notably, R^2 = 40 per cent, and oil exports are significant at the 1 per cent level in this new regression.
26 The 1993 error apparently had to do with the institutional shift from SUFOREN to INEFAN (J. Meza and H. Thiel, MoE, personal communication, Quito, March 2001).
27 See Wunder (1996a: 370–1) for the cases of three protective forests in the Sierra.
28 Gelb and Marshall-Silva (1988: 182); Mosley (1991: 414).
29 See, e.g. Rosero (1992: A4/A10).
30 The decade-long plans to construct a road through the Sangay National Park from Guamote (Sierra) to Macas (Oriente) have spurred various waves of land claims, abandonments and reclaims along the proposed route, all determined by changes over time in the political and financial prospects of the road actually being built (see Wunder 1995).
31 See Rudel (1993); Pichón (1997); Thapa *et al.* (1996).
32 Apparently, a 'Law of Fallow Lands' entered into force in 1875 which formally linked homesteading to land-clearing requirements (Uquillas 1984: 271); the same principle applied to the subsequent 'Law of Idle Lands and Colonisation' of 1936.
33 See Wood (1972: 603) on the 1960s' IDB-IERAC project of 'support to spontaneous colonisation' in the Santo Domingo area.
34 Cf. Brownrigg (1981: 310) on El Oro Province (Costa), Rudel (1993) on Morona–Santiago (southern Oriente) and Ekstrom (1981) for the south-eastern Andean flanks.
35 Source: World Bank (1992, 1999a). Larrea (1992: 340–1) explains that forty small towns counted in the 1982 census as 'rural' were reclassified as 'urban' in the 1990 census, despite not having grown much on average. This tends to inflate urbanisation figures after 1982.
36 Larrea's figures even exclude two northern Amazon cantons because their extreme growth makes them outliers in the statistical analysis. They include as a minor element the Galápagos Islands, but he is not explicit as to whether other mainland colonisation areas are included too.
37 In 1984, only 210,500 ha of an original forest cover of 400,000 in 1965 were left. Eighty-seven per cent of the deforested areas are under pasture (Rosero 1992).
38 See Wunder (1996b) on forest conversion to pasture in four different study zones of the central and southern Sierra.

8 Papua New Guinea

This last primary case-country is also the most complex one: mineral (oil- and metals-derived) wealth did contribute to low deforestation in PNG, but this was interwoven with many other trends. First, because of mining investments, capital flight and other factors, a smaller share of this wealth entered the country – and out of the part that did, the import content was high, further reducing demand effects. As a unique feature, PNG's mineral boom caused very little urbanisation, due to both income redistribution and a series of structural rigidities and non-market factors (land-tenure, crime rates, legal changes, etc.). The core effect did raise food imports, but more so in the small urban sector than in isolated rural environments. Forest degradation through timber extraction was partially held back by an overvalued currency until the 1990s. As in Gabon, the total neglect of road-building was highly instrumental in keeping forest conversion down.

Deforestation in Papua New Guinea

Vegetation history

Forests cover about three-quarters of the land area of PNG. This makes PNG the country with the second highest share of forest-cover in this book. High rainfall makes moist forests the natural vegetation type on most lands. As in West Papua (Irian Jaya), the other half of the island of New Guinea, closed broadleaved forests predominate (see Map 8.1). More than half of these are lowland rainforests, supplemented by montane and swamp forests, and mangroves on the coasts (Collins *et al.* 1991: 178). Forest composition differs notably from the dominance of dipterocarps in SE Asia: forests in PNG are more diverse. *Pometia, Ficus, Terminalia* and, at higher altitudes, *Nothofagus* species are the most frequent (FAO 2002a).

Papua New Guinea's forest diversity is exceptional; for instance, 200 species of frog, 2,000 ferns, 740 species of bird and 200 mammals have been recorded (Collins *et al.* 1991: 179). The diversity of landscapes, forests and species is matched by a high ethnic diversity: PNG has over 700 linguistic groups. Until recently, there was limited contact both across tribes and with the outside world. The abrupt changes triggered by the sudden encounter with Western civilisation have been well described in a *National Geographic* quote from 1962: 'Here live people who never saw a wheel until it dropped to them from the skies on an airplane' (cited in Connell 1997: 3).

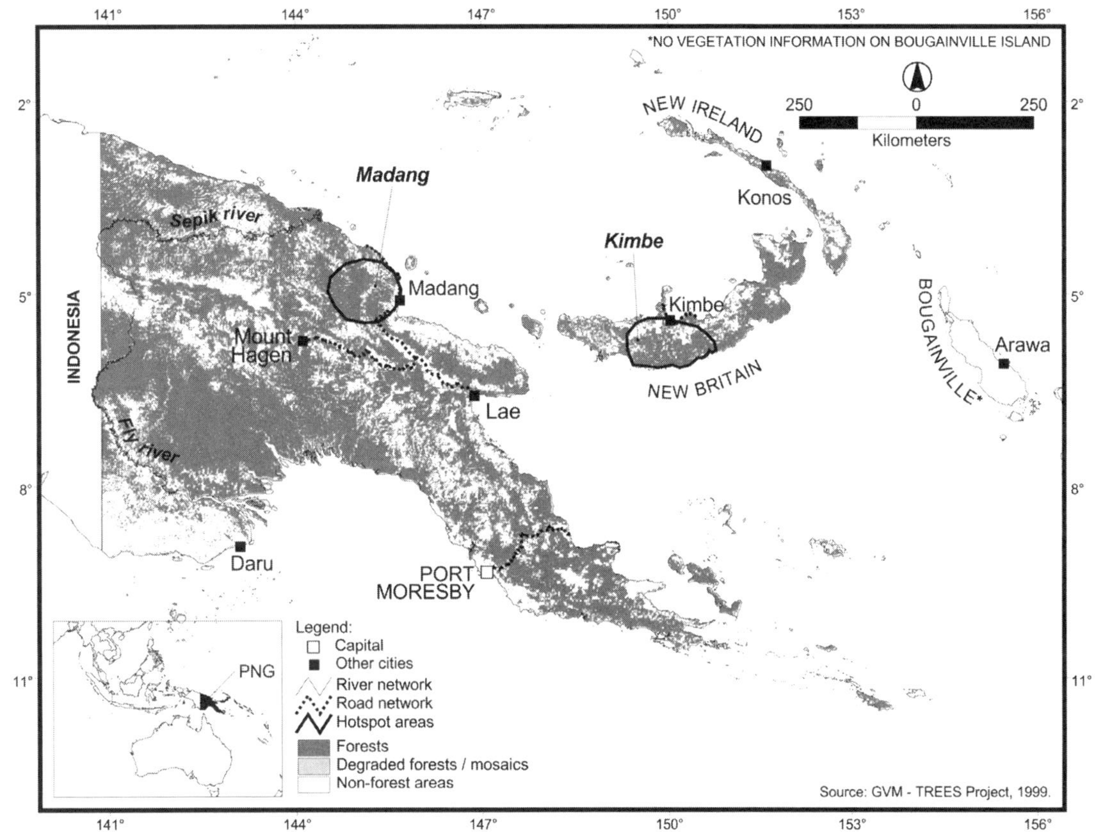

Map 8.1 Papua New Guinea.

Human settlement on New Guinea dates back at least 40,000 years (Filer with Sekhran 1998: 25), though agricultural practices began only 6,000–9,000 years ago. Forest disturbances in pollen records are documented from about 5,000 years ago. It is likely that hunters and gatherers practised extensive shifting cultivation as a supplement to widespread forest extraction, but archaeological research reveals that intensive, irrigated systems were also used occasionally in the highlands (Allen *et al.* 1995: 298–302). A major change in pre-colonial agriculture occurred about three centuries ago with the adoption of sweet potatoes (*Ipomoea batatas*), a species originally introduced to the Philippines from South America. The new staple permitted the colonisation of high-altitude valleys and plateaux, gradually introducing more sedentary systems. The lower parts of the highlands were not suitable for the introduced varieties of sweet potato. More extensive shifting agriculture was practised here, involving the clearing of relatively large areas of forest, though more intensification has occurred over the last hundred years (ibid.: 299–300). More recently, the spread of other introduced varieties like cassava, potato and African yam (*D. rotunda*) has further increased per-hectare food production (Bourke 2001).

What impact did 'pre-contact' agricultural expansion have on the forests? Most of the lowland forests have been converted for shifting cultivation at a variable rate of recurrence (Collins *et al.* 1991: 179). C. E. Lane-Poole, the Commonwealth Forestry Adviser, reported of his pioneer forest expedition in the 1920s that in certain areas 'the natives ... have converted what must have once been high rain forest into savannah forest and pure savannah' (Lane-Poole 1925). Permanent conversion to agriculture and grasslands apparently occurred in those areas where population density rose and/or where ecological conditions were less favourable for forest regrowth. By 1900, the total population was about 1 million. Dutch, German, English and French traders arrived, establishing plantations of copra, the dried meat of the coconut palm (*Cocos nucifera*). Coconut products made up around 80 per cent of the value of PNG's exports throughout the 1920s and around 60 per cent in the 1960s. In land-use terms, however, forest conversion for plantations remained limited to point impacts in coastal areas (Connell 1997: 16–19).

Contact with Europeans, first in the coastal areas and from the 1930s also in the highlands, initially reduced the indigenous population through the spread of new diseases. But after the Second World War, improved health services brought down infant mortality and population growth took off (Allen *et al.* 1995: 300). More people led to more intensive soil uses, and an accelerated shift to the sweet potato, which today, it is the prime staple for 60 per cent of the population, providing 30 per cent of all calories consumed in rural PNG (Bourke 2000: 2). Forest conversion also increased. For example, aerial photographs from 1942 and 1968 of the hilly, dry region around the capital, Port Moresby (south Papua), show that on average 22 per cent of cleared forests converted to grassland or savannah on abandonment. This was due to shortened periods of fallow and frequent burning, combined with the area's dry climate, sloped hills and thin topsoil, which favoured the invasion of aggressive *kunai* (*Imperata cylindrica*) grasses. However, only about 5 per cent of the entire study area was deforested during 1942–68 (Eden 1985).

The rise in population density, which was made possible by improved disease control and increased yields of staple crops, has been fairly unequal across regions. The spread of varieties such as Chinese taro (*Xanthosoma*) and potatoes, and significantly improved techniques in monocropped sweet potato systems, allowed population densities to rise above

100 persons/km^2 in some highland areas, such as Enga Province. In contrast, at lower altitudes, areas dominated by yam and bananas have a population density of around 5 persons/km^2, while those relying only on the extraction of the nutrient-poor sago palms tend to have less than 1 person/km^2 (Ohtsuka *et al.* 1995). Furthermore, many coastal lowlands have only recently become free of malaria, which was previously the leading cause of death in PNG. Settlement has thus been much higher in the traditional malaria-free areas of the highlands.

Papua New Guinea's population was estimated at 4.7 million in 2000, and has grown at 2.2 per cent in the 1980s and 2.3 per cent in the 1990s. The rural population makes up 85 per cent of the total (World Bank 2001a: 279), and even though it has grown at 2 per cent/annum in recent decades, the low initial levels mean that at 10 persons/km^2 national population density is still very low by international standards. Furthermore, rural population growth has not lead to a proportional decline in forest cover. In the early 1980s, around 20–25 per cent of land area was affected by agricultural use, half of which (about 4.6 million ha) was intensive (Freyne and McAlpine 1985; McAlpine and Quigley 1998). Between 1975 and 1996, intensive land use expanded at 0.7 per cent/yr, but total land use only increased by 0.2 per cent annually, a rate far below rural population growth rates of around 2 per cent (Allen *et al.* 2001). This is because 'in PNG, increasing land-use intensity of land already in use, rather than expansion, has been the main response to the need to increase food production' (McAlpine *et al.* 2001). Intensification has been promoted by both physical and social constraints on area expansion and therefore limits to deforestation.

Even in the populated highlands, population density across tribal groups has varied within a large range of 20–414 persons/km^2 (Brown and Podolefski, cited in Harris 1982: 19). This indicates that local agricultural systems have been highly adaptable to varying degrees of land scarcity, using both new varieties and yield-enhancing techniques to respond to growing land shortages.[1] On the other hand, extensive grasslands have been expanding (about 10 per cent of land area), particularly in the lowlands and foothills. Most of these are stabilised disclimax grasslands that are the result of previous anthropogenic impacts, notably a repeated use of fire for cultivation, grazing, hunting or cultural purposes (Wigston 1984: 308). In the mid-1980s, net conversion to grasslands is said to have advanced at a rate of 10,000–20,000 ha/yr (K. J. White, cited in Collins *et al.* 1991: 179).

Why has land intensification predominated over land extensification in a land-abundant country with a growing, mostly rural population? Many authors stress the physical limits to land expansion: about half the land area is on steep slopes, one-fifth of it is seasonally flooded, and seismic and volcanic risks are high. 'For three quarters of the land not to be used, suggests that the constraints to agriculture are severe and that to overcome them will have a high economic cost' (Allen *et al.* 2001). Yet this conclusion begs the fundamental question of why neighbouring West Papua, endowed with similar physical land constraints, has a pace of forest conversion for agriculture that is several times higher than PNG's (Cook 1996).

The answer is that social and institutional obstacles to deforestation in PNG seem at least as important as the physical ones. The country's system of land ownership is unique: local so-called 'resource owners' control 97 per cent of all land and 99 per cent of forested land.

Their land is under customary ownership, which is formally recognised by the state (Taylor 1997: 1–2; Filer with Sekhran 1998: 30–2). Only a residual 3 per cent have been alienated by the state, mainly prior to independence in 1975, for coastal plantations, urban areas and infrastructure. 'Resource ownership' covers a range of intricate arrangements that vary in time, place and across land-derived benefit types, with different layers of individual, family or clan rights (Holzknecht 1996). Customary land is never sold in a formal Western-style manner; hence all externally led land-use initiatives, from logging to mining and plantations, have to be implemented through complex leasing and/or compensation arrangements, sometimes with the state as an intermediary (Louman and Siaguru 1995). But frequently informal claims overlap between different resource owners.[2] This is exacerbated when land becomes scarce, which is one of the reasons why armed tribal conflict over land still repeatedly occurs.

How does the prevailing system of land ownership constrain land extensification and deforestation? Various disincentives are at work, relating both to the type of collective, non-transferable ownership, and to the presence of prolonged conflicts and insecurities concerning tenure. First, customary land ownership is exhaustive, in the sense that no pool of state-owned, quasi open-access forestland exists for private individuals to appropriate, as was the case in Venezuela or Ecuador for example. Frequent, long-lasting inter-clan conflicts over land constitute a major disincentive in establishing perennial plantations.[3] Potential external developers cannot gain secure land rights. While that is less of a problem for loggers who extract value over a limited time-horizon, potential investors in plantations are much more reluctant to make long-term commitments in projects involving forest conversion to other commercial uses (Jones and McGavin 2001). Individual entrepreneurs within communities cannot use land as collateral for bank loans to raise capital for land-converting development. Although collective 'landholder companies' are sometimes formed, this has proved an ineffective tool of land development (Holzknecht 1996: 5). Often the fact that forest areas have remained untouched for decades can be directly related to unsettled disputes over land tenure (Umeazaki *et al.* 2000: 369–70).

An often-neglected additional constraint is the process of intra-clan decision-making over land use. Each village tends to have a range of plots and a mosaic of fallows of different ages and fertility levels. When land becomes scarce, individual households in need of additional areas of cultivation cannot normally decide to clear a plot on their own in a collectively owned forest, but need to obtain collective consent. Indeed, the forest-clearing activity itself, which is mainly done by men (day-to-day gardening is mainly done by women), is often a collective work task (Densley 1977). Obtaining a consensus on clearing extensive areas of forest or 'bush' may not be easy if community members who are less short of land are currently deriving benefits (firewood, fodder, etc.) from the standing forest. Often the individual households will shorten fallows instead and, as they see yields decline, will try fertility-maintenance techniques or new varieties to raise land productivity on existing plots, although intensification requires greater labour inputs than converting forest fringes for cultivation (Bourke 2001).

Current forest loss

Compared to its vast remaining forests, PNG has a slow rate of conversion. But exactly *how low* the rates are, and how they have varied over time, is much less certain. Based on both forest assessment and land-use data, it will be argued in this section that current yearly

Table 8.1 PNG: forest-cover and deforestation estimates

Author	*Forest cover* (in ha)	*Year*	*Annual deforest.* (in ha)	*Relative decline*[a] (%)	*Period*	*Source type*	*Coverage notes*
FAO (1997a)	37,605,000 36,939,000	1990 1995	133,000	0.4	1990–5	Model estimate	All forests (<10%)
FAO (1993)	37,173,000 36,043,000	1980 1990	113,000	0.3	1980–90	Model estimate	All forests (<10%)
WRI (1994)	37,130,000 36,000,000	1980 1990	23,000 203,000	0.0 0.5[b]	1980–5 1985–90	FAO FAO	Natural forests
WRI (1998)	37,145,000 37,605,000 36,939,000	1980 1990 1995	 −46,000 133,000	 −0.1[c] 0.4	 1980–90 1990–5	 FAO FAO	 All forests (<10%)
Myers (1994)	42,500,000 36,000,000	'Original' 1989	350,000	1.0	1989	Various	Deforestation includes logging
McAlpine and Quigley (1998) – FIM	33,065,500 31,750,000[b]	1975 1996	62,619[b]	0.2[b]	1975–96	Air photos 1 : 100,000 Landsat 1 : 250,000	Excludes fragments and mangroves
PNGRIS data base	32,469,400	1975	50,200[a] 22,100[d]	0.1 0.0	1975–96 1975–96	Air photos 1 : 100,000 Landsat 1 : 250,000	Land-use expansion data
Hurst (1990), in Cook (1996)	—	—	21,000	—	1970–90	—	—
Amelung and Diehl (1992)	—	—	22,000	—	1980–8	Model estimate	All forests
Mayaux *et al.* (1998) – TREES	36,157,000	1992–3	—	—	—	AVHHR, Landsat correction	Evergreen + semi-deciduous forests (<70%)
UNEP (2001)	32,422,300	1992–3	—	—	—	AVHHR, USGS database	Closed forests (<40%)
Collins *et al.* (1996)	36,353,000	Around 1975	—	—	—	Maps from 1970s	Rain forest

Notes
a Expansion of 'significant land use' category.
b Own calculations using figures indicated in the specific source.
c Negative figure indicates net reforestation, rounded up to one decimal place.
d Expansion of 'total land use' category.

forest loss over the past two decades has probably been fairly stable in the range of 50,000–70,000 ha. Table 8.1 provides an overview of forest-stock and -loss estimates.

The first two rows show the model-based projections from FAO's FRA 1990 (FAO 1993) and its mid-decade update for 1990–5 (FAO 1997a). They claim that forest cover declined by 113,000 ha in the 1980s (0.3 per cent), accelerating to 133,000 (0.4 per cent) in the first half of the 1990s, that is, about double the range I consider to be the most likely.

Note, however, that in FAO (1997a) the baseline for 1990 forest cover was drastically estimated upwards, by more than 1.5 million hectares, making forest stock actually larger than the original 1980 estimate (FAO 1993). This is a first indication that the FAO may have severely overestimated past deforestation in PNG. As elsewhere, others have reproduced FAO's estimates, such as WRI (1994, 1998). In WRI (1994), the FRA 1990 estimate is juxtaposed to another, extremely low FAO forest-loss estimate (23,000 ha; 0.0 per cent). For this to be compatible with the FRA 1990, forest loss should have increased enormously in the last half of the 1980s (203,000 ha; 0.5 per cent), which is extremely unlikely. Myers's (1994) high estimate for 1989 (350,000 ha) is inflated by the fact that, unlike the FAO, he defines logged-over areas as 'deforested'.

Can other land-use data help us resolve the ambiguities? The Papua New Guinea Resource Information System (PNGRIS), a computer-based database developed gradually with assistance from Australian institutions since the 1960s (Bellamy and McAlpine 1995), recorded basic forest data, but at a very coarse 1 : 1,000,000 map scale (Saunders 1993b). As a condition for a World Bank loan, the National Forest Authority (NFA)[4] developed an interim system of Forest Inventory Mapping (FIM) in 1972–5. They used PNGRIS data, supplemented by aerial photos (1 : 50,000–100,000) and field surveys to provide a 1975 national resource inventory (Saunders 1993b; McAlpine and Quigley 1998: 3; J. McAlpine, personal e-communication, March 2002). The information was updated with Landsat satellite images from 1996, thus allowing for a rough assessment of land-use changes from 1975 to 1996. The summary data in McAlpine and Quigley (1998) (row 6) indicate a yearly forest loss of 62,619 ha (0.2 per cent) between 1975 and 1996.[5]

Note that the PNGRIS-derived forest stocks for 1975 (33 million ha) and 1996 (31.2 million ha) are much lower than most comparable sources, such as FAO (1997a) (36.9 million ha; 1995 figure), the TREES project (Mayaux *et al.* 1998: 36.2 million ha; 1992–3) or IUCN's *Tropical Forest Conservation Atlas* (Collins *et al.* 1991: 36.4 million ha; mid-1970s). The PNGRIS figure is more in line with the recent assessment of closed forests by UNEP (2001) of 32.4 million ha. These stock differences are explained *in part* by the fact that the FIM figure excludes woodlands, mangroves, secondary forest fallows and small forest fragments in landscapes dominated by forest-agriculture mosaics.[6] As shifting cultivation is very important in PNG, this has major impacts on the total stock estimate, because 'this practice of bush fallow cultivation leads to very complex patterns of gardens and vegetation regrowth at varying stages of development' (Saunders 1993a: 1). By counting all fallows as deforested 'light land use', excluding even old secondary forests, the FIM and PNGRIS forest-stock estimates become very conservative, compared to the FAO criterion of 10 per cent tree-crown cover. As McAlpine and Quigley express it: 'The area of *primary* forest is given as a rough indicator of unused land (to which should be added most but not all of the area under light land use)' (n.d., my emphasis).

From the PNGRIS data presented in the section on 'The competitiveness of agriculture and forestry', there was an estimated 50,200 ha of yearly expansion in 'significant land use' (covering intensive uses) and 22,100 ha growth in 'total land use' (including also fallows, secondary forests, etc.) between 1975 and 1996. The balance between the two is made up by a yearly *decline* in 'low land use' of 28,100 ha, indicating intensification: most new, intensive land use is drawn from areas that were already in extensive use. Given the fact that most land classified as 'unused' in PNG is forested, these land-use expansion indicators should also be fairly good proxy indicators of forest-clearing.[7] Note that both of these change processes – that is the expansion of intensive use into previous extensive use areas and the expansion of total area under use – involve deforestation, according to the FAO definition (see the section on 'The competitiveness of agriculture and forestry'). For net deforestation the FIM-derived estimate of 62,619 ha/yr is the most reliable (J. McAlpine, personal e-communication, March 2002). While this is the best available source, two problems still make it a very rough estimate. First, in classifying mixed vegetation types, an intricate system of 'dominant' land-use classes was employed to compile the aggregates.[8] While this may be an excellent procedure *per se*, small-scale forest conversion in a mixed landscape may not be registered as deforestation until it hits a 'dominance' threshold. Furthermore, secondary forests and old fallows, which the FAO would count as 'forest' in its definition, are not included as such. Second, neither the PNGRIS nor FIM map-derived data are always fully explicit in their time references (e.g. Saunders 1993a, McAlpine and Quigley n.d.). This is not just an omission, but expresses the fact that land-use and forest data sometimes come from forestry concessionaires and other sources with highly variable base years.[9] While this may not be a severe drawback in general for a country where land-use change over time is slow, and it is the description of the overall structure that has priority, it does matter for studies like the present book, which have forest change assessments over time as their primary objective.

FAO's FRA 2000 (FAO 2002a) also used the FIM data.[10] However, by adopting the forest definition in FIM that excludes forest patches ('only unused forests are forests'), the FAO made a dramatic reduction of almost 6 million ha in its 1990 baseline stock, from 37,605,000 ha in FAO (1997a) to 31,731,000 ha in FAO (2000b) (see Chapter 1). This is obviously inconsistent with FAO's 'normal' forest definition, which includes secondary and regenerating forests (see Chapter 3). In contrast, the FIM forest definition focuses on primary forests. Indeed, the FAO's own online presentation (as of February 2002) raises doubts about how to interpret the results.[11] Moreover, the FAO also maintained its very high yearly deforestation estimate of 113,000 ha from the previous FRA (FAO 1993) at the same level (113,000 ha) for the 1990–2000 period. This has only been possible by revising deforestation from 1975–85 retrospectively down to zero and assigning all the FIM-observed changes during 1975–96 exclusively to the 1985–96 period. The rationale given for this drastic assumption of an asymmetrical time distribution of forest loss is that logging operations in PNG only increased in the mid-1980s, and logging is supposed to be highly instrumental in deforestation. However, this argument ignores the well-documented fact that deforestation in PNG, as defined by FAO, is led by subsistence agriculture, not logging. The total forest area converted between 1975 and 1996 was 1,315,000 ha, but of this only 355,000 ha, that is, about a quarter, had previously been logged. Conversely, out of a totally logged area of 2,340,000 ha in that period, only 355,000 ha (15 per cent) were converted to other land uses, while the rest (85 per cent) is regenerating (McAlpine and Quigley 1998). Thus the increase in logging may well have accelerated deforestation somewhat after 1985, but only at the margin. Census data and other evidence indicate steadily increasing levels of rural economic activity, as well as between 1975 and 1985, which obviously must also have expanded land use. For instance, in the mid-1980s Freyne and McAlpine (1985) estimated that 200,000 ha of forest, regrowth or grasslands were cleared each year, although the bulk of this in a transitory way.

As a consequence, a drastic change trend in the mid-1980s, as stipulated in FAO (2002a), is unrealistic. Deforestation over the past two decades has probably been both low and reasonably stable, most likely fluctuating in the annual range of 50,000–70,000 ha. As will be further argued below, subsistence food production has been the main driver of deforestation, mining and commercial crops playing secondary roles. This also means that forest-conversion pressures are incremental and fairly well distributed in the populated rural space, though probably with greater extensification in the lowlands than in the already densely populated highlands (see section on 'The competitiveness of agriculture and forestry'). Some fluctuations over time have occurred, for example, with the recent expansion of certain cash crops. There has been a logging-cum-plantation deforestation tandem in the Madang area and on New Britain – both are marked as deforestation 'hot spots' in Map 8.1 (see also section on 'Windfall impacts on government spending'). In the following, the description of macroeconomic and sectoral trends will shed light on the underlying conditions for land-use changes.

The effect of mineral production on forests

In the early 1970s, the economy of PNG experienced a thorough structural change. The country became a significant mineral exporter: 'no country in the Asia-Pacific region has been so transformed by recent mineral exploitation' (Connell 1997: 121). For more than a decade, these mineral exports were dominated by the extraction of copper, gold and silver from the Panguna mine on Bougainville Island. Other mines, producing mostly copper and gold, gradually followed (Burritt 1997: 21–2; Connell 1997: 121–37). In the 1990s, PNG also started to develop its petroleum resources. The rents from both hard-rock minerals and oil have had weighty impacts on the local and national economies. Throughout this chapter, therefore, we shall treat both oil and metals integrally as 'booming sectors', although attention will also be paid to the differences between the two sectors. To start with, this section will look at the environmental impact of extraction activities in both.

Direct mining impacts

Due to the more radical physical impact of hard-rock extraction, the environmental effects of mining have been much more significant than those of oil extraction. However, direct on-site forest-clearing for extraction areas and infrastructure tends to be quite limited. For Porgera, a large gold mine in Enga Province (Highlands), an impact assessment determined that the total area disturbed by the mine over its lifetime, whether originally forested or not, was only 931 ha (Placer Dome 2001). In many cases, active efforts have been made to aid forest regeneration. The Misima gold mine in Milne Bay Province, which will soon be closed, had by the end of 1999 caused 611 ha of site-clearing, plus 52 ha of swamp forest cleared for road-building, but on 188 ha the forest was already in a good state of regeneration (MML 2001).

A much more severe problem is the off-site erosion of mine discharges, the so-called *tailings* (waste rock, sediments, etc.). The most heated controversy in that respect has evolved around the Ok Tedi copper and gold mine, in Western Province, close to the border with Indonesia. During one-and-a-half decades, the company reports having discharged 85 million t of tailings, much more than initially envisaged (Ok Tedi 2002). In this highly sloped, rainy, ecologically unstable area, this massive discharge (between 30,000 and 80,000 t daily) has caused extensive damage not only to the Upper Ok Tedi River, but also to the whole watershed, for hundreds of kilometres downstream. This includes the larger Fly River (see Map 8.1), right up to its fall into the Gulf of Papua (Harper and Israel 1999). The riverbed has been raised by between 5 and 10 m, which has caused flooding and the loss of fertile cultivated land and freshwater and fish resources for downstream dwellers. Pollution has been aggravated by the discharge of cyanide, copper and cadmium into the river. A main concern has been the negative mining impact on agriculture (Government of PNG 1989: 37).

While river and agricultural systems have suffered most from these impacts, forests have also been severely affected. Permanent or increasingly frequent flooding causes physical damage to plants and root-zone waterlogging, which eventually will lead either to deforestation – trees dying and conversion to flood-tolerant vegetation (e.g. wetland grasses and shrubs) – or to forest degradation, in the form of a great alteration in species composition. This phenomenon, known as vegetation *dieback*, has been studied in detail by Marshall and Rau (1999) for the Lower Ok Tedi and Middle Fly River watersheds. Using Landsat images, they found that 47,800 ha had been affected in 1997, of which between 15,860 ha and 22,650 ha of forest had been properly converted,[12] while 23,600 ha were in early stress stages, and 1,550 ha were regenerating. However, the potential impact area is much larger. In the Upper Tedi (close to the mine), dieback is at an advanced stage, but further downstream the process is now advancing too. For instance, 20,000 ha of the Fly River wetlands are thought to have been degraded (W. Kanawi, TNC, personal communication, Port Moresby, 21 November 2001). Based on their own dieback model, Marshall and Rau (1999) expect future impacts within the entire floodplain of 590,000 ha to affect 256,900 ha (with a confidence interval from 188,300 to 378,900 ha).

Hence, the effects from dieback are both gradual and severe. In other mines, they have been observed on a minor scale. Originally, tailings deposits and land flooding were a serious source of dispute with local people at the Panguna mine on Bougainville, which affected the Jaba river watershed. This contributed to the outburst of violent conflict in 1988 on Bougainville, which eventually led to the shutting down of mining operations in 1989 (Connell 1997: 138–9). In Porgera, previously alarming emissions of mercury, zinc and cyanide into the Strickland river system have been somewhat reduced, but apparently the sedimentation of rock waste has kept accelerating.[13] Critics have warned that impacts could become similar to those in Ok Tedi (Shearman 2001), but it seems that Porgera has relatively stable rock-waste dumps to avoid this (C. Filer, personal e-communication, 26 March 2002). Still, the expected long-term dieback at Ok Tedi alone (256,900 ha – see above) corresponds to more than four years of the most likely size of annual deforestation in PNG (50,000–70,0000 ha).

Direct oil impacts

The presence of oil in PNG was first recorded in 1911, but commercial production began only in the 1990s, as PNG went from a phase of oil exploration to being a major oil exporter between 1990 to 1993 (Connell 1997: 132–5). Oil production began in 1992 (53,000 barrels/day), peaked in the following year (126,000 barrels/day), but then fluctuated around 100,000 barrels/day for the rest of the decade (EIA 2001). Although there are off-shore deposits in the Gulf of Papua (e.g. Pandora), the on-shore fields in the forested Southern highlands (notably near Lake Kutubu and Gobe) are the most important ones. A 270 km pipeline transports the oil through a highly forested area to Kikori, on the Gulf of Papua. The same on-shore area has large natural gas deposits, especially the Hides field, with a significant economic potential. A controversial 3,000 km gas pipeline to Queensland, Australia is planned, though not yet built.

As PNG's oilfields are basically all in forested zones, one might expect the impact of oil exploration and production to have been large. However, aggregate on-site forest impacts have been more restricted than those for mining, as production is geographically more concentrated. Kutubu, the main oilfield, was originally almost fully forest-covered. Forest has been cleared for wells, production and support facilities. Off-site clearing has been carried out for roads, and a 40-m strip was deforested for the 280 km pipeline to the south. However, the pipeline was buried in the ground, and vegetation seems to be regenerating well. A company report estimated the total on- and off-site clearing for Kutubu at about 1,300 ha, half of which is now being reforested (Moari and Era 1996). Another assessment reaches similar conclusions, putting aggregate clearing at 1,189 ha (Hartley 1996).

As shown for several of the cases discussed in this book, as well as clearing forest cover, oil production can also have far-reaching pollution impacts that put forest flora and fauna at risk. The current technology at Kutubu is environmentally friendly in the sense that it re-injects both gas and used water into the ground. Both deforestation and pollution impacts have thus been much more restricted than in Ecuador (Chapter 7) or Nigeria (Chapter 9). Yet, environmental groups have focused their resistance to oil development in PNG on the risks of pipeline ruptures and major oil spills in a rugged and ecologically rich yet unstable region. This is the main argument put forward by the Rainforest Action Network against the further development of the Gobe oil and gas field, and against the pipeline across the Torres Strait to Australia (RAN 2002).

Indirect impacts

So far, oil and mining production and infrastructural impacts have been discussed, but forests might also be compromised by the fact that such development enables other activities leading to deforestation or degradation, whether by outsiders or by the local population. For instance, one concern is that oil-related roads may 'open frontier forests to assaults by timber companies, destroying even more of PNG habitat' (RAN 2002). However, most of the oil and mining sites are in remote regions with a difficult topography. For instance, the access road used for the Kutubu pipeline was allowed to revert to 'bush' because the Gulf Provincial Government and local people feared that southern highlanders would penetrate their coastal settlements. Hence, potential timber transport could only occur through Mendi and the Highlands Highway down to Lae, a long access route which even the most desperate logger would not use (C. Filer, personal e-communication, 26 March 2002). The 'assault' is likely to remain limited in most highland areas, where the bulk of oil and mining resources are concentrated.

However, it is not only external actors that can convert or degrade the forest. The local economies could also be expected to receive major stimuli from income opportunities and transfers introduced by oil and mining companies, providing the capital, incentives or access to sell agricultural products, enlarge cultivated area, increase local populations and settlement, etc. A first step here is to assess how great these local benefits have actually been. A second task is then to understand how these benefits have changed the dynamics of local land use. These two questions will be dealt with briefly in what follows. The spending pattern of local oil and mineral incomes will also foreshadow aspects of the macro-economic spending effect, to be discussed later in the chapter.

The particular land-tenure structure in PNG, and the extensive general tradition of paying cash or in-kind compensation between households or interest groups (James 1997),[14] have made it necessary for oil and mining companies to make payments through various channels: royalties, compensation and special development funds or grants. For royalties, there are three different levels of recipient: the state, provincial governments and landowners. The impact of resource rents accruing to the state will be discussed in the following sections. Provincial governments historically laid claim to up to 95 per cent of royalties (Burton, in Toft 1997: 117), and in some cases still receive substantial shares: for example, 70 per cent of royalties from Ok Tedi is received by the Fly River provincial government (Mawuli and Sanida 2000: 18). But few of these revenues have been invested in infrastructure or social development; most of the funds seem to have dissipated into individual consumption. Our interest here is mainly in the benefits accruing to the third level, namely the local landowners. This comprises cash transfers (royalties, compensation payments, etc.), salaries (employment in the mines), new business opportunities (e.g. food sales and services) and infrastructure provided by the companies (roads, schools, health services, etc.).

It seems that cash transfers have been substantial compared to local pre-mine and pre-oil incomes, that they have been rising over time, and that they have often accrued in an unequal way. A debate has developed over whether local cash transfers are fair and equitable, and what broader socio-political impacts they have had (e.g. Polier 1996; Hyndman 1997; Toft 1997; Filer 2001). Our concentration here is only on the aggregate size of these payments and their likely productive impact. Whereas royalties have been the most stable source of income for landowners, on average it is the quest for local compensation payments that has represented the greatest economic benefit (Mawuli and Sanida 2000). In principle, royalties are resource rents, while compensation payments should reimburse the actual opportunity costs incurred by mining operations. In practice, demands for compensation, whether truly 'justified' or not, enable landowners to appropriate an increasing share of rents, pitching demands according to the company's perceived capacity to pay (Filer 1997). Large payments distributed among a relatively small local population can yield high per-capita transfers; in the case of Ok Tedi these reached 25,000 kina in 1984–90, or about the same amount in US$ (ibid.: 17).[15] This includes compensation for companies' forest and 'bush' clearing, which has ranged between K505 and K3,752/ha (Filer *et al.* 2000b: 171–2). Obviously, demands have been strengthened by previous mine closures triggered by violent conflict with local landowners (e.g. Panguna in 1989 or Mount Kare in 1992), so demands now tend to be generally agreed by companies (Connell 1997; Mawuli and Sanida 2000). This has promoted local *rent-seeking* and produced a hand-out mentality among resource owners (James 1997: 101; Filer 2001:14–15).

How much do companies' local cash transfers ('resource rents') and salaries matter to the local economies directly affected by mining? Table 8.2 shows the source distribution of landowners' cash incomes in absolute and relative terms for three major mine sites in

Table 8.2 PNG: average household incomes for landowners affected by mining 1996/7 (12-month average)

Income sources	*Ok Tedi*		*Misima*		*Lihir*	
	*Amount (kina)**	*Percentage*	*Amount (kina)**	*Percentage*	*Amount (kina)**	*Percentage*
Resources rents	2,708	82	1,045	56	19,038	95
Agriculture	177	5	152	8	148	1
Businesses	173	5	221	12	440	2
Wages and salaries	132	4	240	13	102	1
Others	145	4	196	11	182	1
Total	3,407	100	1,854	100	19,910	100

Source: Mawuli and Sanida (2000: 49–56).

Note
* In 1996–7, K1 = US$0.69–0.76.

PNG. Local economies are dominated by resource rents to a striking extent, with an income share of between 56 and 95 per cent. As the absolute numbers show, high shares coincide with very high per-capita rents, as in Lihir. Wages, salaries and business incomes, most of which depend directly or indirectly on the mines, are much more limited. However, they still tend to exceed traditional cash-crop incomes.

Resource rents in the petroleum sector have been of a similar nature, though with a much smaller area of damage and fewer compensation payments for off-site impacts. Total benefits from the Kutubu field up to March 1998 have been estimated at K1.666 billion, three-quarters of which were appropriated directly by the central government through a petroleum income tax. At the sub-national level, provincial governments received 54 per cent of the 1991–8 benefits (Simpson *et al.* 1998). As with mining revenues, very little has actually been invested by the provincial governments (here Southern and Gulf Provinces): instead, much has been wasted or fallen victim to 'financial mismanagement and corruption' (ibid.: 5).[16] About 20 per cent of the sub-national oil funds have gone to landowners. At the Gobe field, Chevron Niugini has used so-called 'mitigation funds' to pay for community development projects, village committees and scholarships. Payments during 1997–9 totalled US$578,065, to which must be added land compensation payments of approximately US$350,000. The total of less than US$1 million represents less than 0.5 per cent of the total project costs, but is still considerable given the size of the local economy.

How have these locally accruing oil and mining revenues been spent? Have they fostered investments that have expanded land use at the expense of forests? Although detailed land-use data are not available, the question can safely be answered in the negative. If the basic theoretical economic rationale is profit maximisation, one might expect local communities to have invested in the cultivation of food and cash crops, taking advantage of their windfall gain, of local food demand by company workers and, in some cases, improved road access. Potentially this response would also have recognised the temporary character of most oil and mining revenues by investing in agricultural development alternatives with a long-term perspective. However, few customary landowners in PNG are long-term profit-maximisers. They therefore reacted to the windfall in exactly the opposite way, which is sometimes called 'full-belly' or 'limited wants' behaviour: 'the landowners are earning sufficient money from compensation payments, thus putting minimal pressure or focus on gardening on a commercial basis' (Mawuli and Sanida 2000: 56). This has hurt cash-cropping in particular. Out of landowners' yearly household expenditure, only 1 per cent was dedicated to agricultural inputs and investments (ibid.: 59).

So, where did all the additional money that was paid to local people go? For twenty large local compensation payments from the Porgera mine, Banks (1993, cited in Filer *et al.* 2000b: 107) found that about half of all landowner receipts were redistributed through the traditional egalitarian *wantok* system to family and clan members outside the study area (see section on poverty). About one-quarter of the funds was invested in local businesses, most of which have been badly managed and produced poor returns. Of the remainder, much went into luxury consumption, such as alcohol, gambling, hotels, prostitution, air travel to distant relatives, or inflated brideprices and the 'purchase' of additional wives.[17] This picture is generally reconfirmed by a larger comparative study of landowner spending in Ok Tedi, Misima, Porgera and Lihir (Mawuli and Sanida 2000: 58–73). Yet, as well as *wantok* redistribution and luxury consumption, much was also spent on consumer durables (refrigerators, televisions, etc.), means of transport (vehicles, motorboats), and children's school fees.

Not more than 10 per cent of resource rents are saved and/or invested, and only a minuscule proportion of that 10 per cent in agriculture. Not only did cash-cropping decline. The Ok Tedi study finds that 'subsistence gardening activities have declined since the inception of the mine ... [with] increasing dependence on store goods' (ibid.: 48). Because of the land-extensive nature of local subsistence farming, this actually significantly reduced the pre-existing pressures for forest-clearing.[18] The Misima mine is surrounded by fertile soils, but 'production of tree crops has been declining since the mine commenced operations' (ibid.: 51). Similarly, around the Lihir mine, 'the interest in tree crops is very minimal now' (ibid.: 56). Finally, the Porgera mine in the Highlands 'causes landowners and people who are recipients of payments to neglect agricultural and farming activities as ways of earning a living' (ibid.: 53).

The picture is thus identical everywhere: mining and oil activities have raised local incomes significantly, which has led people to *reduce* cultivation and pressures to convert forest areas, in spite (or, perhaps, *because*) of increased cash, higher demand and better market access. Local people started to buy more things from outside, which gave a significant economic stimulus to those towns that were closest to the mining sites, such as Mount Hagen for Porgera and Kutubu, Mendi for Kutubu, Kavieng from the Lihir mine, or Kiunga from Ok Tedi (M. Bourke, personal e-communication, 7 March 2002). Although mining may have increased local population growth in some cases,[19] the immigration of non-local mineworkers has been deliberately limited.[20] The indirect local impact of the oil and mineral projects has thus universally been to reduce local pressures on forests.

In summary, the forest effects of the oil and mineral sectors at the local level are composed of different sub-effects of different directions and magnitudes. On-site forest-clearing for the opening of mines or oilfields and related infrastructure is often relatively limited, in the range of 600–1,400 ha for each site. In terms of local land-use dynamics, the mines actually tend to reduce local pressures for forest conversion, because landowners respond to increasing resource-rent transfers by working less in agriculture. On the other hand, the off-site erosion impact of tailings has in several cases caused large downstream forest dieback, resulting over time in both outright deforestation and forest degradation on a scale that constitutes a noteworthy element in national forest loss.

The macroeconomic impact of the mineral booms

Throughout its history, PNG has faced several mineral bonanzas. From 1878 to 1920, and again in the 1930s and 1940s, gold-mining was a prime source of foreign-exchange, reaching a peak of 82 per cent share of exports in 1940 (Connell 1997: 121–3). By 1960, the industry seemed doomed to terminal decline. Yet, colonial exploration in the mid-1960s led to the discovery of low-grade copper deposits at the Panguna mine on Bougainville (O'Faircheallaigh 1984). When new technologies permitted these low-grade deposits to be processed, the share of mining in exports jumped from 1 per cent in 1971 to around 50 per cent in the mid-1970s. The arrival of the revenues from Panguna has been described as 'the greatest single event in the economic history of Papua and New Guinea' (Downs 1980, cited in Connell 1997: 123). Panguna was the country's predominant source of foreign-exchange from 1972 until the mine's forced closure in 1989. With the opening of new gold and copper mines, such as Ok Tedi (1984), Misima (1989), Porgera (1990) and Lihir (1997), the mineral export base was broadened (Connell 1997: 123–32). Significant on- and off-shore oil and gas deposits were found in the 1980s and early 1990s. Oil revenues started at US$312.3 million in 1992, accelerating to a peak of US$696 million in 2000 (IMF 1999; BPNG 2001). Figure 8.1 shows the combined inflow of foreign-exchange revenues from minerals (metals and oil) and foreign borrowing over the last three decades, and compares them with RP changes (RER index: see right-hand scale).

Figure 8.1 reveals several marked phases in PNG's recent economic development. The transformation to a mineral-exporting economy after 1972 was accompanied by a significant real currency appreciation, first in 1971–3, and then, following a price slump in the mid-1970s, again during 1979–81. Note that independence from Australia also implied a change of currency from the Australian dollar to the kina by 1 January 1976, which may have deferred the adjustment of the RER somewhat. PNG's independent government adopted what was called the 'hard kina' strategy, that is, maintaining a strong and stable currency backed by significant international reserves, low foreign debt, and conservative fiscal, monetary and balance of payment policies (Garnaut and Baxter with Krueger 1984). Over the next two decades, this exchange-rate policy triggered a perpetual debate among economists (e.g. Dahanayake 1982; Lam 1984: 199–207; Goodman *et al.* 1987; Jarrett and Anderson 1989; Gumoi 1994; Fallon *et al.* 1995; Gupta 1995; Duncan *et al.* 1998; Mawuli 1998; Duncan and Xu 2000). In the 1970s, the success of the hard kina policy was promoted not only by cautious and consistent macroeconomic policies, but also by favourable copper and coffee prices. Between July 1976 and December 1979, the kina was actually *revalued* four times in nominal terms. This was a conscious strategy to redistribute the benefits of exports to the rest of society and to alleviate inflationary pressures from importables (Dahanayake 1982: 3–5).

This orthodox policy model was faced with serious challenges in the early 1980s. It proved more difficult to hold back spending pressures: at constant prices, government expenditure rose by 7 per cent in 1980 (Garnaut and Baxter with Krueger 1984: 78). At the same time, PNG's terms of trade deteriorated markedly, creating a large trade deficit.

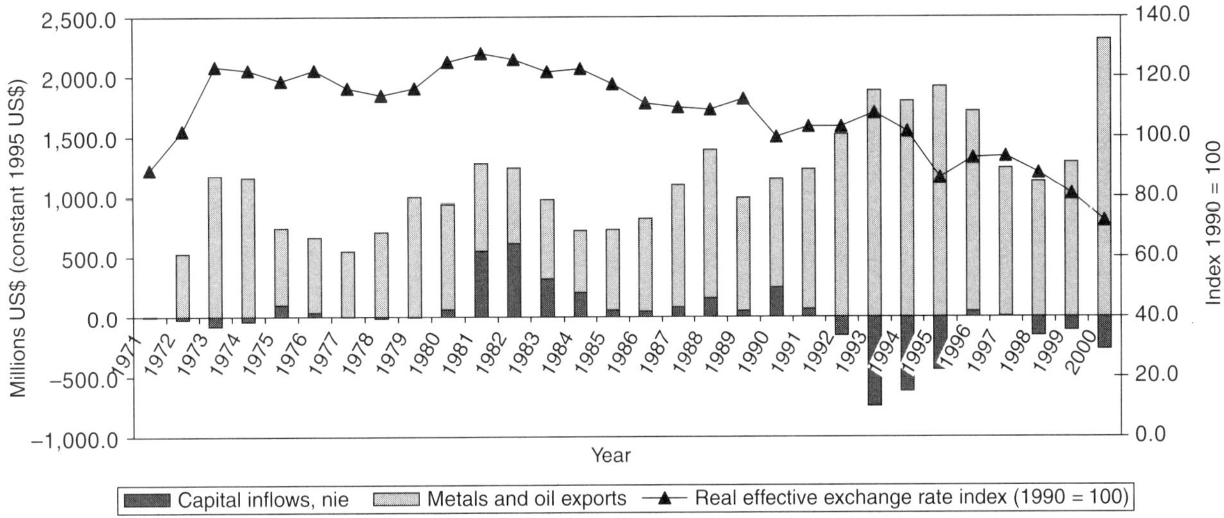

Figure 8.1 PNG: capital inflows, mineral exports and real effective exchange rate 1971–2000, constant 1995 US$.

Sources: BPNG (2001), Connell (1997), Gupta (1995), IMF (1992, 2000); O'Fairchallaigh (1984), UNCTAD (1997), World Bank (1999a, 2001a).

Notes

1 Real effective exchange rate, 1971–8: 1975 weights; 1979: 1985 weights; 1980–2000: 1990 weights.

2 Capital inflows, not including exceptional financing, 1971–89: Other capital nie, 1990–7: Financial account nie, 1998–2000: Capital account balance.

3 Mineral exports, 1971: Metal exports, 1972–5: Metal exports (Bougainville only), 1976–91: Metal exports, 1992–4: Metal exports + Oil exports, 1995–2000: Metal exports + Crude oil exports.

The government was already committed to significant investments in the new Ok Tedi mine. One instrument that helped to balance the equation was the Mineral Resources Stabilisation Fund (MRSF). Established in 1974, its aim was to smooth public spending out of volatile mining revenues, with net savings during high-revenue years and net disbursements to the budget when copper prices or quantities went down. The MRSF was generally managed well for this purpose, although in 1981 a record 81.4 million kina were spent in a 'good' export year, to alleviate the fiscal deficit (Polume 1987: 89). A second escape option was foreign borrowing. From almost no foreign debt in 1980, both debt size and the debt service ratio increased significantly, especially up to 1984 (Goodman *et al.* 1987: 45–8). This helped to reduce the impact of sluggish mining revenues, at least until 1984. However, as Figure 8.1 shows, the RER was generally slow to depreciate in that period. This was obviously related to the fixed kina regime, a policy that relied on sticky domestic prices to fall during busts, which actually happened only gradually.

In 1986–8, copper and especially gold prices experienced a temporary boom (Connell 1997: 147), as is reflected in Figure 8.1 in the record-high mineral export revenues of US$1.2 billion (in fixed 1995 prices) for 1988. However, by 1989 the price boom was over, and worse, violent conflicts led to the closure of the Panguna mine, which caused the loss of 40 per cent of export earnings and 17 per cent of budget revenues (Wesley-Smith 1990: 192). Other mines were beginning to produce revenues (Ok Tedi, Misima, Porgera, etc.), but the shortfall made it necessary for the government to negotiate a structural adjustment package with the World Bank and the IMF in early 1990. Government expenditure was reduced by K25 million, and the kina was devalued by 10 per cent. Macroeconomic adjustment policies succeeded in keeping inflation under control while improving the balance of payments as new mining revenues replaced the Panguna losses, and economic growth picked up again in 1991 (World Bank 1999a).

A number of events in the early 1990s significantly changed the course of the PNG economy for the rest of the decade. Oil exports started to flow from the Kutubu field in 1992 and from Gobe in 1998, which lifted foreign-exchange inflows to permanently higher levels for the rest of the decade. Yet the tools used to stabilise mineral export revenues, like the MRSF, lost efficiency after 1989, as revenues from oil and various different mines accrued in a much more decentralised and uncontrolled manner. At the same time, the government gave in to pressures to relax foreign-exchange controls somewhat (Gupta 1995). It also reversed historic mining policies by pushing for a much higher national equity share in joint ventures, to a degree that was perceived unreasonable by foreign companies and created investor anxieties (Connell 1997: 145–7). Changing fiscal rules continued to be a source of controversy throughout the 1990s (Harden and Sugden 1999: 5).

However, the root problem was that growing mineral revenues made the government increasingly lose fiscal and monetary control. After running a fiscal surplus in 1988 and 1989, the deficit in the following five years accelerated to reach an unprecedented 5.6 per cent of GDP in both 1992 and 1993, mostly financed through an inflationary expansion of the money supply (Duncan *et al.* 1998: 7). The Wingti administration's expenditure in 1992, followed by the highly expansionary 1993 budget of Sir Julius Chan, were radical attempts to 'jump-start' the PNG economy. They initiated a period of extreme fiscal instability (Dorney 2000: 75–9). This was worsened by the costly internal conflict on Bougainville Island, a volcanic eruption that buried the town of Rabaul in 1994, and El Niño impacts in 1997–8. But the prolonged receipt of metals revenues, and the prospect of still greater rents from oil, had come to institutionalise overspending, handouts and an escalating degree of corruption.

Within this high level of public spending, which areas received priority? State-financed prestige-type mega-projects with low economic returns, including exorbitant and non-transparent military expenditure, were undertaken in PNG,[21] but they absorbed a lower share of the windfall than elsewhere. A large share was transferred to private stakeholders, either formally or informally. Public wages and salaries were the main channel of allocation: while public employment stagnated, salary levels grew rapidly, especially from 1980 to 1986 (Blyth 1988), reaching extraordinary levels.[22] In the 1990s, wages and salaries made up about 40 per cent of government expenditure (EIU 1999d: 19). The bulk of the windfall was consumed: at 14 per cent, the investment share in public expenditure was one of the lowest in the Asia-Pacific region (Marsden 1993: 237). A good example of 'informal' transfers to the private sector is the increasing role of extra-budgetary 'slush funds' and other corruption-prone arrangements, such as the Electoral Development Funds.[23] The rise of corruption in the 1990s has been thoroughly documented by Dorney (2000). In general, pressures to appropriate resource rents mounted:

> The floodgates of the State had been forced open by a horde of special interests and local fantasies which sought their Holy Grail by redistributing the "royalties", the "equities", the "compensation" and the many other forms of wealth which they believed the technocrats had hidden in their dungeons of arcane injustice or had simply failed to squeeze from all the wicked foreign devils barricaded in the Chambers of Mines and Petroleum.
>
> (*The Independent*, 15 March 1996, cited in Filer 1997: 223)

The net result was a situation of considerable political and economic insecurity, in which many investors judged that PNG's current set of policies were not sustainable. As a safety measure (and in some cases as a speculative move), capital was taken abroad and foreign investments in PNG were delayed, leading to a continuous outflow of capital. The government used interventions by the Bank of PNG to sustain the overvalued kina, but when international reserves declined uninterruptedly from 1990 until almost reaching zero in 1994, it finally had to give in. After two decades, the hard kina policy was abandoned and the currency was left to float, which caused an immediate devaluation of about 25 per cent with respect to the US dollar.

Although increased domestic inflation and a short-lived economic recovery in 1996 reduced the real impact of the depreciation, renewed crisis in 1997 (cyclone damage and droughts, which caused mine production to plummet), combined with more fiscal over-expansion, gradually led to further devaluation (AusAID 2000: 33–40). At par with the US$ in 1994, the nominal value of the kina plunged dramatically to US$0.7 in 1997, US$0.39 in 1999 and US$0.27 by the end of 2001. In spite of high metals and oil revenues, especially due to rallying oil prices in 2000, the RER index depreciated from index 108 in 1993 to 72 in 2000 (index 1990 = 100). This strong gain in competitiveness eventually also affected sectoral production, land use and forests, as we shall see in the following section.

But how strong a link was there between foreign-exchange inflows and RP or the RER? Comparing visually the correlation between line and bars in Figure 8.1, it seems probable that this link was stronger for some periods than for others (see also regressions below). The upsurge in mineral revenues caused a significant currency appreciation in the early

1970s and again in the early 1980s, but did not follow downward adjustments in the mid-1970s and mid-1980s. But divergences are most evident for the 1990s. The significant real depreciation from 1994 onwards is correlated with marginal reductions in mineral exports, and certainly with the strong outflows of financial capital, but after 1992, when the oil revenues kicked in, this had surprisingly little effect on the RER. Although the entire post-1971 period experienced a boom in mineral revenues, therefore, only the 'hard kina' period up to 1994 was characterised by highly reduced competitiveness.

There are several explanations why the basic boom price mechanism did not work so strongly in the PNG case. First and foremost, part of the metals and oil revenues never had a spending effect inside the country. Mining export revenues had to finance the amortisation of large investment loans, state equity acquisitions, the remuneration of foreign capital, services, etc. Mikesell, cited in Lam (1984), estimates that in the early periods of the boom only one-fourth of export revenues were actually retained in PNG. For the national share of the cake, it seems that wealthy PNG nationals, among which were many politicians, increasingly invested in real estate in Queensland, Australia in the 1990s. In some cases, the value of these investments has exceeded what is registered as official capital outflows.[24] A non-trivial part of the oil and mineral rents in the 1990s never stimulated domestic demand, but went directly abroad. In addition, the government had managed to stabilise fluctuations in mineral export revenues over time, at least prior to 1990, through the MRSF whose independence was severely weakened in the 1990s. Hence, an important general caveat is that domestic absorption effects are far from congruent with the metals and oil export figures shown in Figure 8.1.

Second, the hard-kina exchange-rate policy, backed by capital controls, obviously played an active part in maintaining an excessively appreciated exchange rate prior to 1994. On the other hand, uncertainties as to the new exchange-rate and policy regime, and the accumulated gap *vis-à-vis* the 'real-economy' equilibrium rate, then accelerated the speed of real depreciation in the 1990s. Third, the full indexation of wages up to the 1992 labour market reform was an additional rigidity that favoured a sustained overvaluation of the currency prior to 1992 (Duncan *et al.* 1998; Levantis 2000). Fourth, another potential Dutch Disease factor not considered above was foreign aid and grants. In particular, Australian support for the annual budget was gradually but steadily being reduced from its pre-independence peak of AU$121 million in 1973–4 (Gupta 1995: 29), thus running counter to the trend towards rising mineral rents.[25]

On the whole, therefore, until the 1990s economic policy in PNG was successful in stabilising highly fluctuating mineral-export revenues, but little was achieved in terms of economic development. Non-mining GDP did not grow in per-capita terms for almost three decades. High-cost, low-skill labour, bad roads, excessive regulation and high rates of crime and violence are the factors that are usually cited for this (EIU 1999d: 24; Manning 2000). However, while price competitiveness improved markedly in the 1990s, it had certainly been a major obstacle to traded sector growth earlier. Although year-to-year correlations may be slightly blurred, mineral exports clearly reduced competitiveness (Dutch Disease), especially in the 1970s and 1980s. As other studies confirm (Fallon *et al.* 1995; Duncan *et al.* 1998: 66–71), in the absence of mineral revenues the kina would have depreciated much earlier, allowing agriculture to develop more rapidly. Let us now turn to analysing how this affected production trends in different sectors and thus also changed pressure from these sectors on forested lands.

The competitiveness of agriculture and forestry

Agriculture

In assessing the link between changing competitiveness (last section) and land-use impacts, two preliminary questions arise. First, which of the land- and forest-using sectors in PNG are actually 'T sectors', in the sense of being truly exposed to real changes in the exchange rate? This question will be dealt with more in the section below on trade policy, but we can reveal here that only the export sectors (cash crops, logging) continuously satisfy the requirements. Manufacturing is small and mostly protected (Duncan *et al.* 1998). Food-crop agriculture, the most important economic activity for the bulk of PNG's population, was on the whole a *semi-traded* sector. The factors limiting food-crop tradability relate both to variable access ('natural' protection from outside foodstuffs because of expensive goods transport to remote areas), period (increased versus relaxed import regimes) and products (different substitutability *vis-à-vis* e.g. imported rice and wheat). Forestry is a T sector: the dominant timber exports are 'pure tradables', while the home market is very constrained by low population size, low urbanisation and low consumer incomes.

A second initial question is whether agricultural producers in PNG are actually profit maximisers. To the extent that they are not – as the discussion above on mineral *resource owners'* economic behaviour might indicate – they may not increase production in response to higher prices. For instance, De Silva *et al.* 1987 (cited in Duncan *et al.* 1998: 36) note that '[p]rofit maximisation does not appear to be the objective of the majority of coconut growers'. Increased incomes may not be reinvested but spent on the spot: Connell (1997: 84) shows how, in the Eastern Highlands (1979–84), the seasonal profile of coffee incomes correlated almost perfectly with that of beer sales. In economic jargon, if producers' supply curves were not upward-sloping, this would certainly jeopardise any Dutch Disease mechanism. While cases of *downward*-sloping supply curves are generally hard to document, we should simply keep in mind that supply responses differ across products (see below) and agents (e.g. smallholders versus plantations). They may also have become more elastic over time due to reforms, for example in the labour market: until 1992, agriculture was particularly hurt by fully indexed minimum wages.

A CGE model calculus by Fallon *et al.* (1995) showed that a 7 per cent devaluation would trigger a 6–8 per cent rise in agricultural exports and a 1–3 per cent rise in the volume of traded food crops. The low elasticity of food crops confirms their status as *semi-tradables* – they were to some extent sheltered from import competition. Recent evidence also substantiates the assessment that the 1999 devaluation of the kina made export-crop producers better off, in spite of low world-market commodity prices in US dollars, and that they generally reacted by raising production (Harden and Sugden 1999: 9). It thus seems safe to assume *some* positive price elasticity in both food and especially cash-crop supply, though probably quite variable across sub-products and in lesser quantities than in most developing countries.

One detailed analysis of export perennials compares RER movements with the export quantities of copra, coffee, cocoa and oil palm respectively for 1976–97 (Duncan *et al.* 1998: 35–43). The study finds that copra and coffee are clearly the crops with the highest short-term price elasticity, among other things because marketing arrangements favour a high pass-through rate of price changes to farmers. This is not the case for cocoa, where only 43 per cent of price changes pass through to farmers. For oil palm, the product nature of long-term investment meant that the authors could not detect any short-term price elasticity in econometric terms. However, the sustained expansion of oil palm since 1992 coincides widely with the period of real depreciation. The experience of at least one company, Higaturu Oil Palm, which has planted 14,000 ha in PNG over the last twenty-five years, clearly confirms this: without the sizeable devaluation of the kina, it would not have been possible to compensate for the dramatic drop in US$-denominated palm-oil export prices.[26]

In general, as concluded from another CGE exercise modelling agricultural supply response, 'to be effective, devaluation should be accompanied by complementary policies' (Asafu-Adjaye 1996: 307). But if we accept the hypothesis that real currency depreciation contributed greatly to the expansion of crops like oil palm, how much did that impact on forests?[27] Apparently, at present oil palm covers about 85,000 ha in PNG. This includes both large European and Asian companies (with lease–lease back arrangements where the state mediates between landowners and companies) and about 10,000 outgrowers producing on 2–4 ha holdings (World Bank 1997: 8). For preference, oil palm has been planted in old copra or cocoa plantations with easy accessibility. To some extent, the estate crop is now moving more into logged-over coastal forests (e.g. in West New Britain), causing some deforestation. However, European companies worry about their environmental image, therefore they expand wherever possible into grasslands instead of forests.

Copra has been the traditional anchor of PNG's export economy. The Germans developed 55,000 ha of plantations before the First World War (Lam 1984: 97). These were taken over by the Australians, and the sector was dominated by large-scale plantations until the 1960s, when smallholder production started to expand. Plantation is land-extensive, with poor per-hectare returns. Total coconut area reached 265,000 ha in 1979 (ibid.: 104), but in the mid-1980s and 1990s the sector went into a severe crisis, which was probably in part due to kina overvaluation, and reduced plantation area in absolute terms (Connell 1997: 62). In recent years, including the period of restored price competitiveness following the devaluation, there has been a minor revival in copra production.

Coffee is the smallholder cash crop *par excellence* in PNG, being pursued on a genuine micro scale: in 1975–6, there were 380,385 smallholders cultivating 32,175 ha of *arabica* and 75,037 cultivating 8,431 ha of *robusta* (Department of Primary Industry 1979: 14). The share of large-scale, foreign-dominated coffee plantations was further reduced in the decade following independence, as a deliberate policy of giving the land back to its customary owners. However, it quickly became clear that communities often lacked the technical and managerial skills to make plantation businesses a success. The limited official statistics seem to indicate that *arabica* and *robusta* smallholder plantations expanded between 1975–6 and 1979–80, and possibly further up to 1984.[28] However, in the late 1980s the combined effect of falling world-market prices, a still highly appreciated exchange rate and the financial problems of the price-stabilisation body (the Coffee Industry Fund) meant that producers were being squeezed heavily, production was cut back and no new coffee was planted (Temu 1995; World Bank 1997).

Cocoa, which is cultivated mainly in the island provinces, followed a similar story, but with an earlier peak. First, there was a rapid expansion of plantations from 3,700 ha in

1951 to 50,000 ha in 1965, then a shift to smallholder production until the 1980s. In the expansion phase, cocoa plantations also caused significant deforestation locally. For instance, in Siwai (North Solomons), much forest cover was eliminated for cocoa-growing up to the 1970s. Yet in the years that followed, cocoa faced increasing problems of declining competitiveness, inadequate smallholder management and under-investment. Then the price collapse of the late 1980s coincided with an exhaustion of the Cocoa Industry Fund's capital, further eroding the basis for the industry (Connell 1997: 62–78).

Cocoa, coffee and copra exports have all bounced back since the 1994 devaluation (BPNG 2001: S31–2). There are many product-specific trends, including the emergence of new crops like spices, so for our purposes it may be more illustrative to look only at the aggregate numbers. Figure 8.2 shows the real value of all agricultural exports (in constant 1995 kina) between 1971 and 2000.[29] There was a dramatic export decline from 1971 to 1991, right down to just under one-third of the initial value, but a strong recovery to the initial level following the labour-market reform and devaluation in the 1990s. For most year-to-year changes, we find the expected negative correlation between the RER index and agricultural exports. However, at the end of the 1980s, gradual real depreciation was accompanied by a sustained decline in agricultural exports. As explained above, this was due to a combination of falling world-market prices and the severe financial difficulties of PNG's main commodity stabilisation funds.

Did the severe competitiveness impacts on agricultural exports also help to protect the forests? No time-series for areas planted with cash crops are available, but it seems likely that the answer is 'yes'. In 1971, prior to the mineral boom, the area dedicated to the three major export crops (copra, cocoa and coffee) was 346,180 ha (Lam 1984: 64). In the mid-1990s, a rough estimate of all export crops arrived at almost the same figure, 350,000 ha. Due to the predominance of tree crops, much of this was inter-planted with food crops (World Bank 1997: 8–9). While it is likely that many plantations were left (semi-) abandoned without directly growing back into forests, together with other factors the mineral boom severely disrupted a thriving post-war plantation economy. Most probably, a lot of forest would have continued to be cleared had it not been for the highly appreciated kina.

Certainly, the cash-crop story is important, but how was the dominant food-crop sector affected by RER fluctuations? Probably it did lose some momentum, but much more at the margin of other trends. 'Papua New Guinea's richest assets, after its people, is the food grown and eaten in the villages' (French 1978: 8), but the statistics on food production and land use are dismal. National accounts data seem to show that agricultural value-added in per-capita terms fell between 1971 and 1992, then recovered in the 1990s to its 1971 level (World Bank 2001a). For the period as a whole, this would imply a growth record that is inferior to that of Sub-Saharan Africa (World Bank 1997: 1). However, many of these aggregate production estimates may be imputations and extrapolations: hard knowledge is simply not available. New household survey data from 1996 show that food-crop production may have grown much faster (Gibson 2001). Commercialisation is advancing rapidly: 'For a very large number of people ... food crops are now the most important cash crops' (Connell 1997: 72). The vast majority of farmers combine subsistence farming of food crops with some production for the market. Another factor blurring the distinction is that often more than half of the food-crop tubers are used as fodder for pigs, which are commercial commodities (Stacey and Lucas 1989).

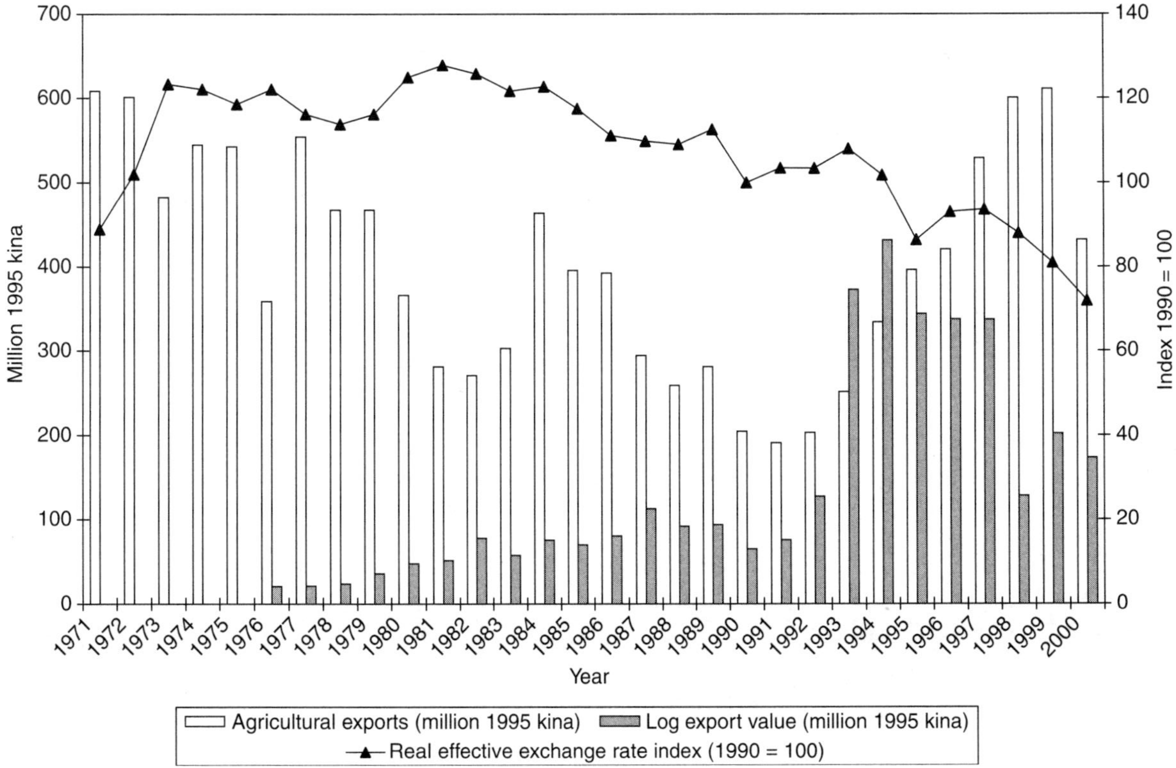

Figure 8.2 PNG: log exports, agricultural exports and real effective exchange rate, fixed 1995 kina, 1971–2000.

Sources: BPNG (2001), Connell (1997), Jarrett and Anderson (1989: 4), NFA data from Peter McCrea. RER from Figure 8.1.

The main competitiveness mechanism slowing down this food-crop production over the last decades does *not* seem to have been a massive penetration of imported foodstuffs (see later). Five-sixths of the calories consumed in rural areas and half in urban areas come from national production (Gibson 2001: 6; see later). While the urban share is high, the aggregate effect has been constrained by the low degree of urbanisation, giving low weight to that sector (section on 'Structural changes in income and demand'). In rural areas, importables like rice and wheat have arrived in those villages where transport access is reasonable, but they have not crowded out local food production on a large scale. In any case, the production-curbing effect on food crops was much less pronounced than for export crops. High imports of foodstuffs already occurred prior to the mineral boom, but one could say that low competitiveness caused by mineral wealth restricted efforts to substitute imports by nationally produced foodstuffs.

How did these production trends affect the size of land areas that were dedicated to food crops? The data is even scarcer than for cash crops. Assessment is also complicated by the fact that most food is produced in shifting systems and in inter-cropped gardens. One national estimate for the mid-1990s is that, in 'net' terms (excluding fallows), 250,000–300,000 ha is cropped yearly (World Bank 1997: 8–9). As already mentioned, intensification predominated over extensification. Although extremely few fertilisers and other external inputs are used (only for vegetables and potatoes), and food-crop extension services are very limited, a variety of local methods for fertility-enhancing intensification have developed spontaneously, including terracing and other forms of soil retention, drainage, composting, crop-rotation, growing nitrogen-fixing fallow species, etc.[30] But did food crops exercise no pressure on forests at all? Using the PNGRIS classifications, we should distinguish between two effects. The *intensification* of areas already in use eliminates some secondary forest that satisfied the 10 per cent crown-cover criterion, so part of 'intensification' is deforestation in FAO terms. This is likely to be significant in the transition from high- to medium-extensive systems: if fallows are already very short, tree cover would not have had time to develop in the first place. The second and more straightforward type is the *extensification* of production areas into previously unconverted forests. The latter effect should be captured unambiguously by the net expansion of total area used.[31]

Table 8.3 shows changes between 1975 and 1996, as recorded in PNGRIS, for different regions and provinces of the country, comparing intensive, low and total land use in 1975 and 1996. Attention should focus on the right-hand side of the table, with six columns that denote the change in land use. The total expansion in intensive (including non-agricultural) uses (column 8) was just over 1 million ha (1,055,000 ha). Column 9 shows how much intensification is due to expanding 'village land use', that is, food-crop production supplemented by smallholder cash crops. On the national scale, this share is about two-thirds, leaving one-third for the expansion of other intensive uses (plantations, mining, infrastructure, etc.). Column 10 documents the major reduction in 'low land use', which on the national scale at 464,700 ha corresponded to more than half the expansion in intensively used areas. Thus the net change in total land use ('intensive' plus 'low' use) over the last two decades, as shown in column 11, was 464,700 ha. Apparently, about two-thirds of this expansion is driven by village-production, especially of food crops. As will be further demonstrated in the econometric exercise below, this part of cultivation was affected less by boom-induced changes in the RER. It is thus probably the remaining one-third of (non-village) land-use

Table 8.3 PNG: land-use intensity in 1975 and 1996 ('000 hectares)

Region/Province	*Land area*	*1975 land use*			*1996 land use*			*Land-use change 1975–96*					
		Intensive	*Low*	*Total*	*Intensive*	*Low*	*Total*	*Intensive, total*	*Intensive, village use (%)* [a,b]	*Low*	*Total*	*Total, per year*	*Intensive, per year*
	(1)	(2)	(3)	(4)	(5)	(6)	(7)	(8)	(9)	(10)	(11)	(12)	(13)
Southern Lowlands													
Western	970,650	275.7	517.4	793.1	390.4	478.1	868.5	114.7	61.8	−39.3	75.4	3.6	5.5
Gulf	338,470	88.2	291.9	380.1	104.8	286.9	391.7	16.6	84.9	−5.0	11.6	0.6	0.8
Central	299,540	487.3	153.4	640.7	577.0	128.8	705.8	89.7	49.1	−24.6	65.1	3.1	4.3
Milne Bay	141,250	398.0	171.1	569.1	424.7	158.4	583.1	26.7	50.2	−12.7	14.0	0.7	1.3
Northern	225,100	203.9	221.9	425.8	293.0	186.3	479.3	89.1	46.8	−35.6	53.5	2.5	4.2
Total S. Lowlands	1,975,010	1,453.1	1,355.7	2,808.8	1,789.9	1,238.5	3,028.4	336.8	54.7	−117.2	219.6	10.5	16.0
Highlands													
Southern Highlands	256,980	612.9	98.3	711.2	661.7	81.2	742.9	48.8	67.4	−17.1	31.7	1.5	2.3
Enga	118,390	328.0	46.9	374.9	357.7	38.6	396.3	29.7	56.6	−8.3	21.4	1.0	1.4
Western Highlands	88,970	413.5	34.7	448.2	447.3	22.0	469.3	33.8	92.3	−12.7	21.1	1.0	1.6
Simbu	60,220	222.9	28.6	251.5	233.3	28.5	261.8	10.4	82.7	−0.1	10.3	0.5	0.5
Eastern Highlands	110,060	431.0	122.9	553.9	450.8	106.9	557.7	19.8	24.7	−16.0	3.8	0.2	0.9
Total Highlands	634,620	2,008.3	331.4	2,339.7	2,150.8	277.2	2,428.0	142.5	66.2	−54.2	88.3	4.2	6.8

Northern Lowlands													
Morobe	335,250	630.7	593.8	1,224.5	721.0	546.8	1,267.8	90.3	73.9	−47.0	43.3	2.1	4.3
Madang	287,320	617.6	987.0	1,604.6	714.6	893.8	1,608.4	97.0	52.4	−93.2	3.8	0.2	4.6
East Sepik	437,200	287.0	612.2	899.2	325.0	580.1	905.1	38.0	80.0	−32.1	5.9	0.3	1.8
West Sepikd[b]	360,100	303.0	518.1	821.1	331.5	495.6	827.1	28.5	186.3[b]	−22.5	6.0	0.3	1.4
Total N. Lowlands	1,419,870	1,838.3	2,711.1	4,549.4	2,092.1	2,516.3	4,608.4	253.8	79.2	−194.8	59.0	2.8	12.1
Islands													
Manus[c]	20,980	20.3	154.4	174.7	68.6	106.2	174.8	48.3	95.9	−48.2	0.1	0.0	2.3
New Ireland	96,150	143.1	310.0	453.1	170.5	283.3	453.8	27.4	44.2	−26.7	0.7	0.0	1.3
East New Britain	151,090	214.7	162.5	377.2	264.6	115.3	379.9	49.9	90.4	−47.2	2.7	0.1	2.4
West New Britain	207,530	92.9	474.7	567.6	176.8	415.8	592.6	83.9	70.8	−58.9	25.0	1.2	4.0
North Solomons[d]	93,290	252.8	262.5	515.3	365.2	219.4	584.6	112.4	n.a.	−43.1	69.3[d]	3.3[d]	n.a.
Total islands	569,040	723.8	1,364.1	2,087.9	1,045.7	1,140.0	2,185.7	321.9	77.8	−224.1	97.8	4.7	15.3
PNG Total	4,598,540	6,023.5	5,762.3	11,785.8	7,078.5	5,172.0	12,250.5	1,055.0	68.2	−590.3	464.7	22.1	50.2

Sources: PNGRIS data, unpublished, Port Moresby; McAlpine *et al.* (2001: 13).

Notes

a Village land use can be distinguished from non-village activities in the 1996 land-use figures. This column represents change in village use as a percentage of change in total intensive land use.

b From McAlpine *et al.* (2001). Total intensive area change 1975–96 differs from corresponding PNGRIS data.

c Manus land-use figures doubtful according to McAlpine *et al.* (2001).

d North Solomons 1996 land-use figures were extrapolated from the rate of land-use change of total islands from 1975 to 1996.

expansion, plus the cash-crop share within village use, which have been most depressed by low agricultural competitiveness.

Through division by the number of years, columns 12 and 13 denote the yearly rates of change for both total and intensive uses. On the national scale, the average annual expansion of total land use was 22,100 ha, while intensive uses expanded at 50,200 ha/yr. Both these figures are potential deforestation indicators, and were therefore used as such in Table 8.1 above. As a percentage of the area already under use in 1975 – 0.83 per cent for intensive, 0.18 per cent for total use – these annual expansion rates are very low, underlining further that PNG has been a case of low land-extensification rates.

Forestry

The high diversity of tree species, frequent difficulties with topographical access, remoteness from the largest markets, and policies discriminating against log exports were all structural factors working against a strong logging sector (Lane-Poole 1925; Collins *et al.* 1991; Thompson 1995). Thus, until 1975 logging was minimal. But cheaper transport, acceptance of a larger number of species and the exhaustion of timber resources elsewhere gradually made timber in PNG more interesting. A negligible export level of 35,000 m^3 in 1965 was expanded to 430,000 m^3 in 1971 and 1,020,000 m^3 in 1982. At the peak of the timber boom in 1997, log exports passed 3 million m^3. Obviously, in the early 1990s this was further stimulated by the increased price competitiveness after the real devaluation. Figure 8.2 above shows that, during the peak of 1993–4, the value of log exports even came to exceed the aggregate value of agricultural exports. For the period as a whole, the figure also lends support to the idea, to be more formally tested below, that a depreciating RER stimulated log exports. About 97 per cent of exports are raw, unprocessed logs, the balance being wood chips from the Gogol Valley (Madang Province). Malaysian firms control 85 per cent of exports, and, Rimbunan Hijau, a single company with its subsidiaries, controls about half of the export volume (Filer with Sekhran 1998: 49). Traditionally, more than half of the logs have gone to Japan and a quarter to South Korea (Filer 1997: 210–1), but China has recently increased its share.[32] The timber sector has recently attained great importance in the national economy. During the 1990s, the industry employed 10,000–12,000 people and annually paid landowner compensation of K40–50 million (US$17.4–21.8 million) and log-export taxes of K140–150 million (US$61–65.4 million) (EIU 1999d: 28).

A traditional society with customary, often forest-dependent landowners, rapidly growing export demand, a state with little regulatory capacity, and the extreme market power of single companies with little environmental pressure from their Asian home country – this set the scene for a highly conflictive form of forestry development in the 1980s, in social, economic and ecological terms. Malpractices in the forestry sector led in 1989 to the Barnett Inquiry, with a report that exposed both fraud by timber firms and corruption in the form of government officials being paid off by companies to avoid observing the rules (Barnett 1990). One concern was logging firms' mis-classification of species and widespread use of 'transfer pricing', that is, pretending to sell logs substantially below true market prices to reduce tax payments (Grynberg *et al.* 1988). In the aftermath of Barnett, international institutions like the World Bank, the WRI and the International

Institute for Environment and Development (IIED) all became involved in a debate that demanded higher transparency in PNG's timber sector (Filer 1997; Filer with Sekhran 1998; Filer with Dubash and Kalit 2000; Seymour and Dubash with Associates 2000). Exports were stimulated by the post-1994 real devaluation, but the Asian crisis in 1997 severely reduced volumes. Also, sustainable harvesting rates became part of PNG's World Bank-led structural adjustment programme (EIU 1999d: 28–9). The public image and political leverage of the Malaysian firms active in PNG were also severely damaged. Log taxes were recently reinstated by the Mourata government, a concessions moratorium was launched in 1999, and it is now official policy to phase out log exports. However, timber taxes still remain an important source of government revenue.

What impact did this logging boom have on forests? Timber-harvesting in PNG causes forest degradation, not deforestation. The lowland rainforests that have been exploited for exports are not sufficiently rich in commercial species even for careless extraction to cause openings that would decrease crown cover to less than 10 per cent. But the increasing number of species harvested over time has also increased the degradation impact of selective harvesting (Kwapena 1985: 96–9). Incipient efforts have been made to map the areas that are affected by forest-cover disturbances (Estreguil and Lambin 1996). In terms of logging intensity, merchantable quantities were typically set at 50–60 m^3/ha officially, but actual extraction has only been half that, at 20–30 m^3 (Collins *et al.* 1991: 178; Sekhran and Miller 1994: 51). A heated debate has developed as to whether selective harvesting at current production levels, without active regeneration efforts and with few environmental safeguards, is 'sustainable'. The NFA claims it is, because the area cut remains small in relation to the remaining forest stock; other assessments say it is not sustainable, based on alternative ecological criteria and measures both of the potentially workable and the currently worked area.[33] This is not the place to resolve this debate. Let us simply note that logging has been concentrated in accessible coastal and island areas where, due to the type of intervention, lowland forests have often been degraded biologically, and it is dubious whether current harvesting rates allow sufficient time for the regeneration of commercial timber stocks.

Pulpwood has been harvested since 1973 in the Gogol Valley for the Japanese-financed JANT pulp-and-paper mill in Madang. Unlike selective logging, this type of clear-felling operation does cause deforestation, and it may sometimes even endanger forest regeneration.[34] Yet it is highly misleading to represent Gogol as an archetypal example of deforestation in PNG in general, as De'Ath and Michalenko do (1980). About 30,000 ha, almost half the project area, have been clear-felled (Davidson n.d.: 77). Lamb (1990: ch. 5) found rapid regeneration of secondary vegetation in Gogol, except for some severely compacted and waterlogged areas, and those where agriculture or plantations had permanently replaced natural forests. Conversion occurred to a much lesser extent than originally envisaged: only 10,000 ha were put into agriculture, two-thirds into smallholder gardens dominated by food crops. Species composition in the regenerating forest may differ from the natural forest, but in terms of wood supplies JANT operates 'sustainably' on about 13,000 ha of acacia plantations established both directly by the company and with trees purchased from outgrowers.[35]

To sum up, the PNGRIS data show that 2,340,000 ha were logged between 1975 and 1996 (111,429 ha/yr), but only 355,000 ha of this were also converted to other land uses.

The remainder, more than four-fifths of logged-over forests, is under regeneration. At the same time, only a quarter of the converted area was logged first (McAlpine and Quigley 1998). Logging roads can sometimes stimulate new clearing, but often they are too rudimentary and not built to last (see section on roads later). This shows that, while logging provided some access for conversion, for example, for plantation development in West New Britain (one of the deforestation hot spots shown in Map 8.1), even these indirect deforestation effects were quite localised. Logging greatly profited from the depreciation of the kina, which was one of the factors promoting its expansion.

Trade policy impacts

A final question in this section is whether PNG's trade policies protected the land- and forest-using T sectors from the long-term impacts of a highly appreciated kina or whether, on the contrary, they punished these sectors by exposing them to further competition and decline. In general, from independence to the early 1980s, PNG's trade regime was extraordinarily liberalised, but protection increased during the 1980s. A general tariff of 10 per cent was introduced, later raised to 17.5 per cent (Whalley 1982; Fallon *et al.* 1995: 77). During the 1980s tariffs and quotas became more variable, creating a differentiated structure of effective protection rates (Robson 1985: 20–1; Bosworth and Anderson 2000). Also certain food imports, like canned beef, fish, poultry or sugar, and forest products, such as imported plywood, were completely banned during particular sub-periods. Manufacturing, a small sector often processing primary commodities, has generally been highly protected. Over time, the government also came to rely more on trade (import and export) taxes, which made up 28 per cent of its tax revenue base in 1998 (Bosworth and Anderson 2000: 19–20). It has been claimed that PNG's protective trade policy on balance had a negative bias against agriculture (Jarrett and Anderson 1989; World Bank 1997). If true, this would imply that trade policies aggravated declining competitiveness, thus restricting land-use expansion and forest loss.

But in order to understand the impact of specific trade policies on forests, it is necessary to examine the situation more closely. For export crops, for instance, price stabilisation schemes were applied to the main export crops. It has been claimed that they functioned as an implicit tax on producers: the benefits of risk stabilisation were allegedly inferior to the costs of transfer payments to the funds (Temu 1995: 6–11), while forced savings reduced producer incomes and reinvestments in export crops (Jarrett and Anderson 1989: 63–73). On the other hand, as official documents stress, the aim of this policy was both macroeconomic (to stabilise inflation) and to 'help to cushion and protect farmers from the effects of wide variations in world market prices' (Government of PNG 1989: 25). Given the proven low capacity of small rural producers in PNG to save and invest, the latter argument cannot be readily dismissed. In any case, the functioning of commodity stabilisation funds has generally been much less corrupting and confiscatory than for either of the two African cases considered in this book.

What about the trade protection of specific land- or forest-using goods – did it promote deforestation? Effects of this type are probably very limited indeed. Most of the protected goods, like cement, fish or poultry, did not have any direct land-use implications. What about forest products? Transitory plywood and furniture protection through import quotas (e.g. Whalley 1982: 40) may have stimulated some additional harvesting of timber. In the 1970s, a legal obligation on log-producers to process a certain share of logs into sawnwood and plywood prior to export would have reduced their profits and extraction levels (Jarrett and Anderson 1989: 55–6), though doubts have been raised if producers actually complied with the regulations (L. Tacconi, personal e-communication, 20 July 2002).

In agriculture, fruit and vegetables have sometimes enjoyed tariffs of up to 75 per cent, but production has been both small and highly land-intensive. In land-use terms, probably the most important case has been sugar. The national Ramu factory at Gusap has been massively protected from import competition; there is now a complete import embargo on sugar. This has sustained employment for about 10,000 people, but also held sugar at internationally uncompetitive price levels, to the detriment of PNG consumers (Bosworth and Anderson 2000: viii). Already by the early 1980s, sugar plantations occupied about 6,000 ha in Morobe Province, mostly in the Ramu Valley. However, originally this land was almost exclusively grassland rather than forest. Similar considerations apply to cattle-raising (see consumption pattern section below). Imports of certain cuts and types of beef have been prohibited during subperiods (Jarrett and Anderson 1989: 52–4), but the additional production stimulated by this policy has been concentrated in the Markham and Ramu Valleys, which were dominated by grasslands (T. Gumoi, personal communication, Port Moresby, 23 November 2001). In this sense, import protection for beef and in particular sugar certainly had a significant land-use impact, but probably a negligible effect on forests.

On the whole, over the past three decades there have not been any good examples of protected industries that have caused significant deforestation. The initially very liberal trade regime became slightly more protectionist, which may have hurt agriculture in particular, although there was some liberalisation of log exports. Export-price stabilisation policies may or may not have reduced cash-crop production. In the early 1990s, many quantitative restrictions were converted to tariffs, which in some cases were prohibitive. In 1998 the average import tariff was 20 per cent, but with a range from 0 to 125 per cent.[36] However, in 1999 the average tariff level was reduced to 9 per cent, as part of the WTO process (Bosworth and Anderson 2000: 22–3). On the whole, there may have been some general trade-policy bias against agriculture, and possibly in some sub-periods against timber exports. By reducing output, compared to a counterfactual laissez faire situation, trade policy thus mildly protected forests.

A quantitative view

The regression results in Table 8.4, using time-series for the last three decades, provide an opportunity to summarise the main points in this section. The first two equations concern the determinants of changing competitiveness. As was suspected, a simple relationship for the whole period, linking mineral (oil and metals) revenues and financial capital inflows to the RER, gives poor results. The parameter for mineral exports has the unexpected negative sign, while that for capital inflows is positive but insignificant. Several explanations for this mismatch were mentioned above. Mainly the policy of fixed exchange rates actively kept the kina overvalued in the period up to 1994, when the regime shift to a floating rate, and then initiated a sizeable downward adjustment, in spite of rising oil revenues. The equation in row 2 tests for the hypothesis that RER levels were fundamentally different

Table 8.4 PNG: relating mineral wealth to relative prices and traded sector production. Regression results, 1971–2000

Independent variables/dependent variables	*Mineral exports (1995 US$)*	*Capital inflows (1995 US$)*	*RER (1990 = 100)*	*Time dummy* [a,b]	R^2	*F-value*	*Years*
1 RER (1990 = 100)							
Coefficient	−0.01052*	0.01370			0.3195	6.3378	1971–2000
T-value	−1.72484	1.22713					
2 RER (1990 = 100)							
Coefficient	0.00278	0.01876**		−26.04710[a]***	0.6747	17.9742	1971–2000
T-value	0.55923	2.36774		−5.32811			
3 Agricultural exports (fixed 1995 kina)							
Coefficient			−2.94580**	−245.03536[b],***	0.3803	8.2857	1971–2000
T-value			−2.15146	−3.70438			
4 Agricultural per capita value added (fixed 1995 kina)							
Coefficient			−0.12409	−44.40779[b]	0.4598	10.6407	1971–98
T-value			−0.50397	−4.58228***			
5 Industrial timber production (million m^3)							
Coefficient			−0.03842***		0.3068	11.9508	1971–99
T-value			−3.45699				
6 Log exports (million m^3)							
Coefficient			−0.03174***		0.3159	12.9299	1971–2000
T-value			−3.59582				

Notes

a Dummy 1: Hard-kina policy: 1971–94 = 0, 1995–2000 = 1.

b Dummy 2: Time: 1971–89 and 1993–2000 = 0, 1990–1992 = 1.

* Parameter *T*-value significant at the 10 per cent level.

** Parameter *T*-value significant at the 5 per cent level.

*** Parameter *T*-value significant at the 1 per cent level.

before and after 1994 by introducing a time dummy variable. It takes the value zero for 1971–94 and one for 1995–2000, that is, it allows for a difference in model-estimated RER levels either side of 1994. The new model explains about two-thirds of RER variation (R^2 = 67.5 per cent), as compared to less than a third before (R^2 = 32 per cent). The variable for capital inflows is now significant at the 5 per cent level, and mineral exports have the theoretically expected positive parameter sign, though not at a significant level. The time dummy is highly significant, indicating that exchange-rate policies had an important independent role to play.

Were production quantities in agriculture strongly affected by changes in competitiveness? For the 'pure' tradables, agricultural exports (in fixed 1995 kina), the initial regression (not shown in Table 8.4) did not exhibit any significant linkage. But an analysis of the model residuals proved that this was due to three influential 'outliers' – the overall results were biased by a systematic under-performance of agricultural exports during 1990–2. As explained above, this was due to a combination of low world-market prices, the financial crisis in commodity funds and accumulated rigidities prior to the 1992 labour-market reform. Again, introducing a dummy variable in the equation in row 3 – allowing for a lower export level for these three years – much improves model properties. It also confirms that the RER was a highly significant (at 5 per cent level) determinant of agricultural exports.

Is the same true for the entire agricultural sector? As food-crop production grew greatly with population growth, we use per-capita agricultural value-added (in fixed prices) as the dependent variable in the equation in row 4. The dummy variable proves highly significant, indicating the dismal agricultural output in the early 1990s, but unlike exports the value-added in agriculture as a whole is not significant.[37] This reconfirms the observation above, and from other studies, that agricultural exports were 'pure tradables', while food crops were 'semi-tradables', in the sense that they were only partially exposed to foreign competition. Due to the relatively liberal import regime for food crops, they were probably fairly exposed in urban areas, but that sector only has about 15 per cent of population. In the dominant rural zone, exposure to lower competitiveness was reduced by 'natural protection' – the costs of transporting goods to remote areas with undeveloped roads (see the section on 'Windfall impacts on government spending'). Unfortunately, we do not have annual land-use figures to document how production affected the size of cultivated area and, ultimately, of forest-covered area.

How was log production over time affected by price competitiveness? Two time-series are used: FAO's data from the FAOSTAT database (equation in row 5), and log export data from the NFA database (equation in row 6). The first of these should include an imputed estimate of production for the domestic market, which is minor in the case of PNG.[38] The second refers only to exports, and is probably the most reliable source of the two.[39] Both regressions confirm that price competitiveness is a key determinant of log production, explaining just under one-third of its variation (R^2 = 30.7 per cent; R^2 = 31.6 per cent). The numerical value of the parameter in the equation in row 6 is about four-fifths the size of that in the equation in row 5 (both significant at the 1 per cent level), a difference which should be due to the inclusion of domestic production in the latter.

We can conclude that price competitiveness, as measured by the RER, was affected by financial capital in- and out-flows and by the size of mineral exports, but the relationship was less clear cut in PNG than in the other cases discussed in this book. Notably, shifting exchange rate regimes over time played a role of their own. Among the potential tradables, the 'pure' T (exports or export-dominated) sectors, timber and cash crops, were clearly hurt by the real appreciation and benefited from its reversal in the 1990s. This is less the case for 'semi-traded' food crops, which were partly sheltered from import competition. Trade protectionism played a minor role in that; much was simply due to the economic dualism keeping the subsistence sector isolated from exposure to market forces. Though statistics on cultivated areas are very sparse, they certainly confirm that the area devoted to cash crops was stagnant, and for some products even declining, over three decades. Deforestation for cash crops and forest degradation from timber extraction are thus the two activities that were mostly held back by the boom-led real currency appreciation. Let us now see how public budget priorities affected the production environment in different sectors.

Windfall impacts on government spending

Agriculture, forestry and conservation

Public funding for renewable natural resources has faced its ups and downs, but has not been a strong priority in PNG for the past three decades as a whole. Government expenditure on agriculture, forestry and fisheries rose substantially from 1975 to 1979, and continued a moderate rise throughout the 1980s (Goodman *et al.* 1987: 234; Blyth 1988: 14). On the other hand, the fall in the 1990s was much more dramatic than overall budget cuts, especially in US dollars: from K68.1 million (US$71.3 million) in 1991, the budget fell to K38.3 million (US$29 million) in 1996 and K38 million (US$14.1 million) in 2000.[40] In particular, in the 1990s the Department of Agriculture and Livestock (DAL) become 'chronically underfunded' (EIU 1999d: 21).

For forestry, there have also been cuts, though less severe. The NFA budget was reduced from K21–22 million at the end of the 1980s to around K17 million now, in spite of the large expansion in logging levels in the early and mid-1990s and the fiscal resources that became available from oil revenues. This also triggered staff reductions to around 460 people, of which about 100 are based in the capital (G. Amos, personal communication, Port Moresby, 23 November 2001). The budget cuts have severely reduced the NFA's capacity to monitor logging practices in the field, which may have worsened forest degradation.

In terms of forest conservation, the Department of Environment and Conservation (DEC) experienced a rapid growth in funding in the first half of the 1990s from foreign donors, NGOs and contributions from mining companies. This was used to expand both urban-based activities (administration, research, teaching at university, etc.) and a range of integrated conservation and development projects in the field. For instance, public spending on biodiversity-related activities alone (including donor financing) grew from K1.2 million in 1992 to K4.5 million in 1994 (Sekhran and Miller 1994: 365). However, the trend was not sustained. In recent years, the DEC has been downgraded from a 'department' to an 'office', and its total staff has been cut from around 300 to 80 people. One view is that the objectives of forest conservation are being left more than ever to foreign donors and NGOs (W. Kanawi, TNC, personal communication, Port Moresby, 21 November 2001). Another view is that, even when fully staffed, the DEC never achieved much in terms of field results, so the recent cuts have made little difference (L. Tacconi, personal e-communication, 20 July 2002).

The above has focused on budget allocations, but an institution's implementation capacity depends on both the size of its budget and the efficiency of its spending. Especially in agriculture, efficiency has been critical. For instance, under the pro-growth strategy of the Wingti government in the late 1980s, increased budget allocations to DAL and other agencies seem to have had little effect, as the latter 'were not ready or able to spend the extra money' (Dorney 2000: 83). Conversely, the impact on agricultural spending efficiency was equally negative under the fiscal austerity imposed in the 1990s. It was noted that 'the agriculture ministry has become largely moribund' (EIU 1999d: 9), as a result of declining funds and mismanagement. Most of its functions were transferred to provincial governments and commodity industry corporations. But, as one of the representatives of the latter noted, there was a trend for the DAL 'to divest itself of functions with no corresponding reduction in expenditure and staff' (cited in Dorney 2000: 88). In many cases, government agencies faced with allocation cuts chose to safeguard public employment, while the expenditure necessary to make public workers productive were cut drastically. While that may sound like a socially laudable approach, it gives rise to a range of paradoxical situations: for example, agricultural extensionists cannot leave their offices because there is no money either to have their pickup repaired or to buy the petrol needed for field trips.[41]

Did the public sector directly interfere in land-using activities? PNG differed from many other oil and mineral boom countries by not using a large share of the windfall for investments and recurrent deficits in agricultural parastatals. In 1991, four of the most important public enterprises (in transport, electricity and communication) were all making minor surpluses. However, none of these was involved in business that posed threats to forest cover or quality.[42] On the other hand, strongly subsidised agricultural credit schemes received a lot of attention, but with poor results across the board. In the late 1970s, about K7 million were spent on smallholder cattle projects, but with few results to show for the money (Dorney 2000: 87). The Agricultural Bank of Papua New Guinea (ABPNG) was created in 1984, but in practice lent mostly to better-off farmers (Connell 1997: 80–5). By 1984, accumulated capital grants disbursed by the government through ABPNG already amounted to K29.3 million (Jarrett and Anderson 1989: 40). Yet the lack of collateral in land was a main problem for all credit schemes, inviting default on the concessional financing from government and donors alike. In other words, a fair share of the money was wasted rather than used productively. Nevertheless, subsidised credit must have had some benefit on cash-crop production: coffee and cocoa were the largest recipients in the 1980s, followed by oil palm and cattle (Jarrett and Anderson 1989: 41). At the margin, this must have alleviated somewhat the strong crisis in the cash-crop sector and the contraction in area planted. Government spending thus supported processes that would, in and of themselves, slightly accelerate forest-clearing.

Roads and transport infrastructure

However, there is evidence that these mildly pro-deforestation budgetary trends were more than reversed by the neglect of roads in PNG. PNG's road network is extremely badly developed. Its road density of less than 0.05 km/km^2 of land area is the lowest registered by the

International Road Federation for any country in Asia or Oceania (IRF 2000: 29). This makes it necessary to rely more on river and coastal shipping in PNG, as well as on a sophisticated network of air transport. But air transport is far too expensive to be used in expanding either timber or most agricultural production. The underdevelopment of roads (and the lack of a railway system as a potential terrestrial supplement) has thus been a prime obstacle to cash-crop expansion, and a main factor in protecting forests from degradation and conversion.

Moreover, growing metals and oil revenues have not helped to improve the road situation – quite the contrary. The Australian colonial rulers expanded the road network gradually, from 4,333 km in 1950 to 9,879 km in 1960 and 12,878 km in 1965. Over the decade from 1956 to 1965, the number of registered vehicles also tripled from 624 to 1,801 (Department of Territories 1967: 18–20). There was a further expansion up to independence, and by 1986 total road length was 19,736 km, 6 per cent of which was surfaced (IRF 1988: 20). But in the following decade registered road length actually *declined* to 19,445 km in 1995, rising slightly back to 19,600 in 1996 (UNDP 1999e; IRF 2000: 29). Eighty-five per cent of these are unsurfaced, and less than half are 'national roads' that are passable for more than 90 per cent of the year. In other words, road use is unreliable, especially in the wet season (EIU 1999d).

In part, this unfavourable trend was linked to shrinking budget allocations in the 1990s. The funds for transport and communication fluctuated around K100 million during 1980–6 (Blyth 1988: 14), then rose to K129.8 million (US$135.9 million) in 1991, but plummeted back to K61.8 million (US$46.8 million) in 1996 and only K45.2 million (US$16.7 million) in 2000.[43] Yet again the problem is not only one of the absolute amount of funding, but also of how the money is spent. First, road expenditure has been concentrated around urban centres, neglecting rural areas (EIU 1999d). One example of such spending is the costly Poreporena highway, which has been built in Port Moresby. Second, road maintenance is generally very poor. It is estimated that thorough road repairs would cost about K150 million (US$42 million), but the government currently spends less than 15 per cent of that (K20 million). Deteriorating roads are estimated to add K200 million to the yearly operating costs of road users (EIU 1999d: 15). One particular example is the Highlands Highway down to Lae, the country's main seaport. Its progressive deterioration has raised transport costs for cash-crop producers, in particular coffee-growers, often to an extent that inhibits commercialisation.[44] Third, as official budget figures show, expenditure in the Department for Works and Implementation are dominated by salaries, while the share allocated to 'Roads and bridges construction' (K0.5 million in 1999, K0.2 million in 2000) is negligible. It can thus be no surprise that a recent private-sector survey ranked deficient infrastructure as the third most important business obstacle overall in PNG, and 90 per cent of all respondents classified road services as 'below average', 'poor' or 'very poor' (Manning 1999: 15–19).

Obviously, a lack of good-quality roads has greatly restricted the scale of commercial agricultural production in particular. In many cases, the only alternative available to smallholders in the absence of road access is to carry heavy loads, such as coffee bags, on foot to regional markets (Ellis 1999: 16). There is a large amount of wastage of perishable crops in PNG, due to difficulties in marketing production (EIU 1999d: 27). However, a side-effect is also that rural food-crop consumption was less affected by import competition, because bad or lacking roads make it too expensive to transport goods into remote rural areas (see above). Some of the few roads built recently have been for logging, but most have been transitory. Case studies confirm that these logging roads have little land-use impact (Klappa 1999), though the construction of more permanent roads does. As in the case of the Bewani road in West Sepik, new roads are likely to induce smallholders to move closer to them, establish new cash-crop plots (in this case cocoa gardens) and market their food-crop surpluses in town (Kocher-Schmid 1999).

Where permanent roads have been built, in most cases this has also accelerated forest-clearing. For instance, ribbon-type settlement and forest-clearing patterns have emerged around the new highway from the upper Markham valley to Madang (McAlpine *et al.* 2001: 8). The flip side of this is that the reluctance to build more roads (and to maintain the existing ones properly) has held back forest-clearing. The high spatial correlation between road density and population density (UNDP 1999e: 126) indicates that roads are key determinants for settlement and the expansion of rural economic activity.[45] The policy since independence to neglect roads has indirectly protected forests from greater pressures for agricultural conversion. Had policies in PNG continuously favoured investing mineral revenues in the expansion of the road network, and had ambitious plans such as the Trans-PNG Highway been implemented, forest cover would probably have been reduced much more.

Directed settlement

Historically, the state has had a quite limited role in distributing forested land and in determining settlement patterns. Whereas in the other countries considered in this book governments have approved of resettlement from the remote interior to areas near roads, this has not happened in PNG. A main restriction has been that, due to the land-tenure situation in PNG, no 'idle' public land has been available to relocate people to, and the state has had little jurisdiction in terms of getting people to move. In colonial times, experiments were made with directed settlements around small-scale plantation schemes producing, for example, cocoa, but there was greater success with the larger-scale plantations established after 1967 for oil palm, tea and rubber (Hulme 1982). The land reform programme after 1974, which aimed to return alienated land to customary landowners, could in some cases also involve resettlements. The reform was most successful with coffee, a typical labour-intensive smallholder crop, while experiences were less satisfactory with copra, rubber and cocoa plantations (Eaton 1983).

Have there been important examples of government-promoted resettlement schemes after the mineral boom set in? In the 1970s some unproductive, unclaimed swamp lands around the Highlands Highway were drained and improved by the state, which thus became their rightful owner and reallocated them to settlers from densely populated regions on an individual smallholder basis (P. Senat, personal communication, Mount Hagen, 31 November 2001). But although this may have involved some conversion of the original vegetation, the scale of these operations was very limited. The most important contemporary examples of resettlement schemes concern oil palm, where since independence the government has assumed an intermediary role between landowners and private companies such as the Commonwealth Development Corporation (CDC) in Higaturu (Williams 1982) or the Hoskins Oil Palm Scheme (Hulme 1983). In the latter case, for instance, this involved not only the conversion of, at the time, 4,775 ha for company plantations, but also the state-directed resettlement of smallholders and conversion of 5,280 ha to oil-palm outgrower blocks. In the former case, resettlement was also involved. However, in both cases most converted areas were former cocoa and copra plantations rather than forests. In other words, some resettlement may have contributed to forest conversion, but to a negligible extent.

To sum up the arguments in this section, PNG's shifting budgetary priorities and spending policies had contradictory impacts on forests. They did provide subsidised cash-crop credits, and they reduced the staff and budgets available for timber regulation and forest conservation, especially in the 1990s. This worked partly towards more forest-clearing and degradation, even though much money was wasted. On the other hand, the government certainly did not initiate large-scale projects requiring land conversion, either through parastatals or large resettlement projects, and it neglected agricultural extension and research. Yet all of these effects were minor in scope. The most important and consistent impact over the past three decades has consisted in the fact that it built extremely few new roads, especially in rural areas, and maintained the existing ones very poorly. This greatly restricted farmers' options in expanding the cultivation of commercial crops by converting forest.[46]

Structural changes in income and demand

Poverty alleviation

Unlike the other countries discussed in this book, welfare-related changes from the mineral boom were neither massive nor abrupt in PNG, so the derived structural changes that might have caused land-use changes and affected forests also remained more gradual and restricted. There are a couple of reasons for this. First, poverty levels apparently have not changed much for most of the population in recent decades. World Bank data for 1979 recorded 75 per cent of the rural population and 10 per cent of the urban population as being under the respective poverty lines, indicating a large rural–urban income gap.[47] Non-mineral GDP was stagnant during most of the period, and private per-capita consumption during 1980–98 even slightly decreased at 0.3 per cent per annum (World Bank 2001a: 277). But these figures may underestimate advances in food-crop production: many very cash-poor areas have increased subsistence production. In addition, some rural areas with transport access have made progress by marketing food crops. A lack of market access is the main regional poverty trap in isolated fringes between the highlands and lowlands, on remote islands and in the more inaccessible highland regions.[48] The mineral boom does not seem to have reduced the welfare gaps between provinces.[49] On the other hand, poverty in Port Moresby has grown during the 1990s (World Bank 1997, cited in O'Collins 1999: 37). Rural–urban poverty gaps may have narrowed somewhat over the last two decades. At the national level, most social indicators continue to lag considerably behind those in neighbouring countries in the Pacific region. Life expectancy at birth is fighty-eight years (1998) and female adult illiteracy is 45 per cent (1998), though infant mortality fell from 7.8 per cent in 1980 to 5.8 per cent in 1998 (World Bank 2001a: 277, 287).

Second, in spite of increased inequality, extreme poverty may have been held down by redistribution. The inequality of consumption, with a Gini coefficient of 50.6 in 1996, is higher than in neighbouring countries (O'Collins 1999: 37; World Bank 2001a: 283). Also, most mineral revenues probably accrued to relatively small groups of politicians, civil servants, 'Big Man' local leaders and on-site compensated resource owners. But the traditional *wantok* system, with its pros and cons, has provided a fairly efficient security net that safeguards most people from extreme deprivation:

> Absolute poverty, in terms of hunger over an extended period, and homelessness, are not observed in Papua New Guinea because of the existence of a dependable network of social relationship – especially a kinship system that lends support to the needy, and shares wealth. ... On the positive side, it provides hardship alleviation. On the negative side, it undermines the economic advancement of entire kinship groups, and retards progress towards escaping temporary poverty because it adversely impacts on household savings. ... Because of a strong network of *wantok* relationships, no one really feels poor.
>
> (Mawuli 2001: 11–12)

While some observers speak about the emergence of a new 'hidden poverty', Mawuli's description still seems widely applicable. For instance, Allen (1999: 43) reports that urban-to-rural remittances are probably the largest cash source in the country's most remote regions. Likewise, the above discussion on the distribution of mining- and oil-development compensations showed that rural-to-rural remittances can be important too. It also indicated that the people who were made better off by mineral wealth generally spent the money quickly and worked less in agriculture. Less poverty thus also reduced their pressure on the forests. *Wantok* redistribution is widespread: the 1996 household survey estimated that 90 per cent of households either received or made annual transfers (Igua 2001: 85). So, although mineral revenues and derived incomes accrued in concentrated form, their welfare effects were spread over a larger number of recipients, and 'losers' from the boom's negative side-effects were probably widely compensated. As we shall see in the following sections, redistribution mechanisms also gave people less incentive to change economic behaviour (e.g. move location, job or profession), thus leaving the structure of the economy more stable and less responsive than it would otherwise have been.

Rural–urban migration

One of the areas of relative stability has been the spatial distribution of population. PNG's official 17 per cent urbanisation rate for 1999 (World Bank 2000b) may represent a underestimate, especially the supposed 252,000 residents of the capital, Port Moresby (M. Manning, personal communication, Port Moresby, 22 November 2001). Yet, even with a notable upward correction, PNG would still clearly remain the least urbanised of all the countries analysed in this book. In all other cases, urban spending sprees, construction booms and a growing service economy have induced a much greater proportion of the population to settle permanently in cities and towns. Why has that not happened to the same extent in PNG?

One potential explanation is that, as a very recent phenomenon, urbanisation started at too low a level to advance more. Indeed, urbanisation was only 2.7 per cent in 1960, growing to 10.2 per cent in 1971 and 17 per cent in 1997 (World Bank 1999a, 2000b). The speed of urbanisation was particularly rapid from 1966 to 1971. Still, in 1971 only three urban centres (Port Moresby, Lae and Rabaul) had more than 20,000 inhabitants (Skeldon 1978). Nevertheless, looking at the rate of change over time, the difference in percentage points between 1960 and 1971 was the same as in the three following decades. In other words, the speed of urbanisation actually *slowed down* with the mineral boom. This is a remarkable deviation from the standard mineral-boom pattern.

A second explanation would be that the spending of metals and oil revenues, and the derived demand effects on urban income generation, were less strong in PNG: that is, pull motives for rural–urban migration remained moderate. The sections on macroeconomic adjustment and poverty alleviation lend some support to this interpretation. Data on the structure of employment further reinforce this picture. Government employment (which has a clear urban bias) initially went up, but then stagnated in the mid-1980s when more money was spent on salary increases. The share of the labour force in industrial and service activities (almost exclusively urban) even declined, from 22.7 per cent in 1980 to 18.2 per cent in 1990 (Marsden 1993: 231–2; EIU 1999d: 43). Rigidities in the formal labour market up to 1992 are partly to blame, but the data show that the urban informal sector mostly lagged behind, in part because specific policies discouraged its growth (the prohibition of petty services, street vendors, etc.). After independence, the net exodus of 10,000–12,000 predominantly urban-based expatriates had a weighty impact, directly in terms of reduced urban population, but more importantly because the mass departure of skilled labour and capital heavily reduced urban business opportunities (Skeldon 1978: 6–8). In other words, urban labour absorption of a rapidly growing population slowed down. As expected, this was less pronounced in the capital, where the publicly accruing mineral rents were administered, than for provincial towns (Lae, Madang, Goroka, etc.) that depended on the declining cash-crop trade (ibid.: 10–12). In the 1990s, accelerating crime rates and deteriorating government services were further obstacles stressed by private business(wo)men in explaining slow urban growth (Manning 1999). Consequently, urban unemployment has risen sharply, discouraging accelerated migration to the cities.

Third, push motivations for rural out-migration may have been modest, meaning that people have preferred to stay in rural areas, with some rural–rural migration. Obviously, this is not confirmed across the board. According to census data, some predominantly rural provinces had a high 'push'-type of net out-migration between 1980 and 1990, such as the densely populated and poor Simbu Province (−40,973). Other highland provinces and East Sepik Province have also faced net out-migration. Net migrant recipients were mainly the Capital District (+84,487), but also the thinly populated Provinces of Western Highlands (+33,990) and West New Britain (+22,817) where oil-palm development is quite strong (UNDP 1999e: 73). But in spite of the dismal performance of cash crops, especially in the 1980s, agriculture as a labour-intensive sector was the default activity to fall back on. The labour force that was engaged in agriculture, forestry and fishing, which declined from 90 per cent in 1960 to 70 per cent in 1980, actually *grew* back to 75 per cent in 1990.[50] In other words, there was some rural–urban as well as rural–rural migration, but migration flows were generally small compared to total population size, and rural employment and self-employment remained vital.

Finally, figures for permanent migration may also underestimate the real degree of geographical mobility between urban and rural areas. There seems to be a reasonably high degree of circular migration, where people use their kinship networks to live and work in urban areas for some time, before eventually returning to the countryside. They may have come in the first place to supplement rural family income, as a strategy of diversification, or to save enough money in the city for a high brideprice. However, strong cultural ties make them go back – it is 'the village' that is the centre of the universe, not the city. Furthermore, those who move and do not come back run the risk of losing customary land rights and other village benefits – and to be able to go back and cultivate the land is an important safeguard against poverty (Dorney 2000: 76). In political terms, many observers were concerned about the rapid pace of urbanisation in the early 1970s (Skeldon 1978). Consequently, economic and socio-cultural factors combined are needed to explain why a strong drive towards urban areas is lacking in PNG.

The structure of consumption

The first question in this section is whether boom-derived income changes altered the consumption of agricultural and forest products during the last decades. The second question is to what extent any such changes in demand had significant land-use and forest impacts. For instance, it has been claimed that higher cash incomes have permitted the spread of new tastes favouring massive rice and wheat imports, which allegedly reduced food security by cutting back local production of staples (Spencer and Heywood 1982; Igua 2001). If true, this shift in tastes towards imported cereals would have interacted with price competitiveness in retarding some crop intensification and some extensification into forested areas, which would otherwise have been necessary to produce these additional food crops locally.

However, this so-called 'paradox at the heart of food policy in Papua New Guinea' – booming cereal imports crowding out local food crops – is an example where speculations and hasty conclusions have led to the structural stability of the PNG economy being underestimated. Gibson (2001: 65–6) shows that, although rice imports grew by 86 per cent and wheat imports by 94 per cent from 1962 to 2000, at 2.26 and 2.47 per cent, the annual growth of both was below the rate of population growth. It is true that rice and wheat have become important food crops, especially in urban areas, and rice has entered into the consumption basket of rural households, but clearly as a minor supplement to domestic food crops. The currency devaluation accelerated the levelling-off of wheat imports in the mid-1990s and the outright decline of rice imports after 1997, pointing back to competitiveness factors (section on 'The competitiveness of agriculture and forestry'). But in his analysis of urban household survey data from the late 1980s, Gibson (1995: 62–4) finds that among staple foods the income elasticity for sweet potatoes was actually marginally higher than that for imported cereals. Preferences for rice or wheat over sweet potatoes or bananas were due more to cost factors, making local staples more expensive than their imported substitutes (Igua 2001). The view that consumers in PNG see traditional starchy staples as inferior goods to be abandoned when incomes rise is thus not confirmed; there even seems to be a slight preference for traditional staples over imported ones. This is in stark contrast to, for instance, Cameroon, where preferences were clearly biased towards importables.

Luxury foodstuffs, with high income-elasticities from the household survey are mainly alcohol, chicken and meat (Gibson 1995: 62). As we have seen in the two Latin American cases, a growing appetite for beef stimulated by higher incomes can have disastrous consequences for forest cover. Did the same happen in PNG? There are several reasons why this question must be answered in the negative. First, cattle-ranching in PNG does not dominate the livestock sector, where raising poultry and pigs are traditionally very important. Both are less land-extensive types of production, although a lot of sweet potato production is fed to pigs. Second, customary rules among smallholders do not go along well with a cowboy culture: ownership disputes, a lack of finance, technical requirements and law and order problems are all disadvantageous for an investment-oriented cattle sector. This is exactly why cattle numbers have declined sharply since independence, from a peak of 153,000 head in 1976 to 75,000–90,000 head now. Large ranches now take up 113,000 ha of grazing land, while smallholder farms occupy 78,000 ha (Igua 2001: 51). In addition, since the devaluation of the kina, three out of four urban areas have seen a major decline in beef consumption in favour of fish and lamb (ibid.: 73–4). Most important of all, grassland availability is seldom the factor that limits cattle production. As mentioned above, large cattle projects have been concentrated in the Ramu and Markham Valleys, where grasslands already predominated before the ranches came. Igua (2001: 51) estimates that an additional 300,000 ha of customary land would potentially be suitable for cattle-ranching, with few competing land uses. Thus, even if the cattle sector in PNG suddenly overcame all its management and production problems, its deforestation effect would still be negligible for many years to come.

To sum up, boom-induced changes in consumption patterns played a negligible role in impacting on the forests. First, this was because the economic structures that could have triggered these changes, such as growing incomes and urbanisation, remained fairly stable. For instance, had urbanisation been more rapid, it is likely that PNG would have experienced the same construction boom as other mineral boom countries, increasing the demand for timber on the domestic market, but the size of food imports replacing local staples might also have been larger. Second, preferences in the form of a continued inclination to buy local food crops also remained fairly robust in the face of competing importables. Third, for some luxury goods like beef, where demand definitely did increase, the domestic supply response was actually negative. Finally, even if that supply response had been large and positive, the increased land demand might have been satisfied without much forest conversion.

Synthesis and conclusion

Papua New Guinea is a country where more than four-fifths of the population live and farm in rural areas, where population has grown by close to 3 per cent over the last three decades, and where annual forest loss over the same period has probably been 50,000–70,000 ha, corresponding to 0.1–0.2 per cent of the remaining forest stock. This is not a usual pattern for a tropical developing country. The main aim in this chapter has been to determine why forest loss was so low and how much the mineral wealth from the metals and oil sectors has contributed to this trend, compared to other factors that were more specific to the country and its institutional setting.

Papua New Guinea became a high mineral exporter in the early 1970s, with metals exports first from the Panguna mine on Bougainville, and later from other mines that went into production in the 1980s and 1990s. Since 1992, oil exports have become an important additional source of revenue. While mineral export revenues have often fluctuated from year to year, for our purposes it is best to conceptualise the whole period after 1971 as a 'boom period'. Mineral revenues jumped up from zero to become the dominant export revenue for the entire period. Furthermore, a commodity fund, the MRSF, successfully smoothed budget spending, in spite of year-to-year fluctuations in revenue. Other channels of mineral-wealth absorption were the provincial governments and local landowner clans. Notably, PNG's mineral wealth had less impact per US$ of registered export revenue than in other cases in this book because a non-trivial share left the country before having a demand effect, either as amortisation of mining investments, fly-in-fly-out experts' salaries or as capital flight by wealthy national recipients of mineral riches.

From the part of wealth that did enter PNG's economy, the government wage bill was a main route for distribution. This was supplemented by the social sectors (health, education, etc.) and subsidies, for example agricultural credit, though corruption and wastage began to claim an increasing share over time, especially in the 1990s. Primary transfers may have accrued rather concentrated, but it is likely that a lot of redistribution occurred through the *wantok* system. For both public and private agents, savings and investments from mineral revenues were very low; much of the windfall was consumed. For urban recipients, the import content of this consumption was high (vehicles, fuel, luxury items, etc.), which further reduced the derived impact on the PNG economy.

Because of this and other circumstances, non-mineral production basically did not grow in per-capita terms. For many people in rural areas, life changed little through the oil and metals booms, except perhaps that their sales of food crops to urban areas grew. Economic dualism remained accentuated and the urban economy – that is, construction, private services and other 'modern-sector' candidates – did not 'take off'. As the investment climate deteriorated further – for example, because of growing law-and-order problems, but also unwise fiscal, monetary and foreign-investment policies – capital flight increased in the 1990s and the economy deteriorated. In productive terms, one can thus ask to what extent PNG's oil boom in the 1990s 'went up in smoke'. This is an important question, as there is probably more Dutch Disease to come in PNG. The Liquified Natural Gas project, with a pipeline to Australia and huge investments in Hides, Kutubu and Wewak, would create additional export revenues of up to K1.7 billion (Bosworth and Anderson 2000: 78–9). Thus, the sound absorption of mineral rents will probably continue to be a cornerstone of economic policies and management in PNG.

How did the mineral boom actually alter the macroeconomy, and how did this affect land use and forests? Table 8.5 gives a summary of ten major boom impacts on forests, ranked according to their alleged amount of deforestation (last column), which is a product of economic significance (column 3) and the strength of the link between that sector's production and deforestation (column 5) (see Chapter 2). Factors that reduced deforestation are marked as areas 1, 2, 4, 6–8 while factors that increased forest loss are areas 3, 5, 9 and 10. As we are looking at the long-term effects of the mineral boom over almost three decades, the reasonable counterfactual 'no-boom' baseline scenario is not a static economy, but one where pre-existing trends would likely have continued – basically a growing agricultural economy with rapid population growth, producing food and cash crops.

Table 8.5 PNG: mineral wealth and deforestation – an overview of long-term impacts

Economic and productive impacts			*Links to deforestation*		*Deforestation impact*	
No.	*Type*	*Intensity*	*Type*	*Strength*	*Type*	*Intensity*
1	'Hard-kina': loss of export-crop competitiveness	Very strong ●●●	Stagnating/ declining cash-crop area	Medium ⇨	Fewer forest conversion pressures	Strong ▼▼
2	Neglect of rural roads	Strong ●●	Limits agricultural extensification	Close ➔	Protects forest frontiers from conversion	Strong ▼▼
3	Mining and oil production impacts (regional)	Strong ●	On-site clearing, off-site tailings, local economic stimulus	Medium (average) ⇨	On-site: small; off-site dieback: large; stimulus forest-protecting	Medium (regional) ▲
4	Higher urban labour absorption (government, services, etc.)	Weak ●●	Less rural labour and less demand for cropped land	Weak →	Less forest conversion	Weak ▽
5	Growing budgets for agricultural development	Weak, fluctuat. ●	Supporting agricultural production	Weak →	Potentially augmenting area expansion	Weak, fluctuating △
6	Growing budgets for forestry and conservation	Weak, fluctuat. ●	Improving forest control and protected areas	Weak →	Less forest degradation	Weak, fluctuating ▽
7	Trade policy	Weak, variable ●	Average exposed agriculture more to Dutch Disease	Weak →	Fewer pressures for forest conversion	Weak, fluctuating ▽
8	Poverty alleviation	Weak ●	Reduces 'full-belly' forest loss but can increase investments	Medium variable ⇨	Average: lower 'full belly' cash- and food-crop conversion	Weak, variable ▽
9	Higher urban income shifts food demand	Negligible ○	Preference for local staples over imported cereals	Weak →	Higher food-crop demand and forest loss	Negligible ○
10	Directed settlement	Negligible ○	Moving people to plantations	Medium ⇨	Occupying forested areas	Negligible ○

Notes
1, 2, 4, 6–8 area – reduces deforestation; 3, 5, 9 and 10 areas – increases deforestation.

According to Table 8.5, six factors reduced deforestation while four expanded it. However, the most important observation is perhaps that so few of the ten mineral-boom factors that were common in other countries had a significant impact in PNG. Basically, two forest-protecting impacts dominated the picture: our core effect of competitiveness loss through the 'hard-kina' (1), and road neglect (2). First, the near wipe-out of agricultural exports in PNG, falling in 1971–91 to one-third of pre-boom value, was strongly influenced by real appreciation. Generally speaking, coffee, cocoa and copra plantations probably did not grow back into forests, but were under-utilised, abandoned or converted to food-crop uses. It seems certain that, in the absence of a mineral boom, the export-crop sector would have been much more expansionary, and additional land demand would eventually have caused more forest conversion. Second, the striking neglect of rural road-building compared to the pre-boom period was a great obstacle not only to agricultural exports, but also to food-crop commercialisation in urban areas. Note that both factors not only restricted deforestation, they also hampered logging. Although logging had only a limited impact on deforestation proper, the delay or reduction in the growth of log exports also decreased forest degradation significantly.

Unlike the other countries analysed in this book, most of PNG's oil deposits are located on-shore. But the environmental impacts of the hard-rock mining sector were more serious than those from oil production (3). On-site forest-clearing for mining and oil sites has been very limited, and negligible on the scale of national deforestation. Oil pipelines, access roads and other infrastructure had negligible direct and indirect deforestation impacts. The income-derived land-use impact from locally paid royalties and compensations actually protected forests, as communities unambiguously reduced cash- and food-crop cultivation alike in response to their sudden mineral wealth. But the single downstream impact of mining tailings being distributed through river systems proved to be momentous, triggering flooding and gradual vegetation dieback that may eventually cause damage and loss of sizeable forest areas in the watersheds affected.

Basically the other seven effects were all either weak or negligible. For instance, boom-led urbanisation (4) was limited to some expansion in the capital. This low level was due to cultural ties to the village and the land and circular migration, as well as the very limited employment generation in urban areas. Budgetary effects on agriculture (5) and forestry (6) shifted over time, were not clearly connected with mineral revenues, and sometimes had no clearly defined impact on the forests. For instance, agricultural agencies in PNG were much less geared towards the task of frontier expansion than in the Latin American cases, Ecuador and Venezuela. Trade policy (7) also had variable impacts, both over time and across products. On balance, it may have discriminated against agriculture, but the effect was not very pronounced. One traded land-using sector that was heavily protected, sugar, exploited grassland reserves rather than forests. Poverty alleviation (8), in principle an ambiguous factor, seems predominantly to go hand in hand with less pressure on forests in PNG. Yet, the overall extent of pre-boom *acute* poverty was apparently limited, just as was the extent of poverty alleviation from the mineral boom. The consumption structure (9) also changed very little, partly because changes in household incomes and urbanisation were not large, and partly because of the strength of preferences for local foods, such as sweet potatoes. Where preferences did change, for example towards higher beef consumption, the domestic supply response was actually negative, for a variety of socio-cultural reasons. And public resettlement schemes (10) were *a priori* limited by the restrictive nature of PNG's customary land-tenure system.

At this stage, the sceptical reader might think that institutional and non-economic factors seem to have had a much greater impact on land use in PNG than the mineral boom. For instance, isn't the land-tenure system in PNG a much more powerful determinant of low deforestation? Isn't the law and order situation a greater obstacle to land-use development than the appreciated RER? Doesn't the *wantok* redistribution system reduce people's 'push' and 'pull' incentives, so that they become less sensitive to changes in RPs? Wasn't the mass departure of expatriates after independence a more important blow to urban businesses than any urban public-spending spree could compensate for? Certainly, there is some support for all of these arguments. For instance, the survey of major business obstacles mentioned above ranked as the top four factors crime and theft, corruption, poor infrastructure and policy instability (in that order). Issues relating to price competitiveness, such as inflation (rank 5), taxes (rank 6), trade regulations (rank 11) and price controls (rank 15), all came further down the list of perceived major obstacles (Manning 1999: 15).

The present author is not claiming that mineral wealth had causal primacy in holding back forest loss and degradation in PNG. On the contrary, it should be recognised that in no other case considered in this book were the cultural, institutional and political factors such strong independent determinants of the land-use picture. However, there are a couple of caveats. For instance, the survey of business obstacles was carried out in 1996–7, and 71 per cent of the respondents were from Port Moresby. The results may therefore not be fully valid for the pre-1994 hard-kina period, nor for the agribusiness sector, which should have the largest land-use impacts. A second observation, sustained by the other cases in this book, is that certain political aspects, although apparently independent, are actually strongly linked to the prolonged presence of mineral rents. This is certainly true for corruption, but also for the state's political autonomy. For instance, one may conjecture that the pronounced 'indigenisation' of the economy would not have been implemented so rapidly or so radically had there not been large mineral revenues to bolster the budget. Instead, the political goals of economic self-determination would have had to make room sooner for some pragmatism in order to offer a good investment climate. In this sense, the mineral boom was not only a factor that influenced land-use decisions and reduced deforestation pressures at the margin – it also helped to shape the enabling framework within which other, non-economic factors operated.

Notes

1 On land intensification, see Freyne and McAlpine (1985), Allen *et al.* (1995), Ohtsuka *et al.* (1995), Bourke (2000, 2001), Allen *et al.* (2001) and the section on agriculture.

2 Recently, a World Bank project attempted to register all customary land claims centrally. However, this was interpreted by many as a hidden strategy to take over the land, and the project had to be abandoned after experiencing violent resistance (T. Gumoi, personal communication, Port Moresby, 1 December 2001).

3 In several contemporary cases in the Highlands, raiding clans have cut down their neighbours' coffee trees on a massive scale as a form of tribal retaliation (P. Peng, personal communication, Mt Hagen, 30 November 2001).

4 I would like to thank Mr Ripa Karo of NFA for providing me with FIM data and information.

5 In fact, McAlpine and Quigley (1998) do not present a forest stock estimate for 1996. The estimate of 31,750,000 ha in Table 8.1 was calculated by subtracting the yearly estimates of permanent forest conversion from the 1975 estimate.

6 'In part' is emphasised here because some stock differences are hard to explain, such as the large and troubling discrepancy between the TREES and the UNEP figure. Both are supposedly derived from the same NOAA-AVHRR satellite imagery, for the same year (1992–3), although TREES uses a Landsat correction procedure. In fact, the UNEP criteria for closed forest (<40 per cent crown cover) should be more inclusive than the TREES one (<70 per cent crown cover). This should make the TREES forest stock *smaller* than the UNEP one, not almost four million ha larger.

7 This claim is supported by the fact that 'total land use' and 'forest area' combined made up 97 per cent of total land area in 1996. See also the similar assumption in the forest land-use map by Saunders (1993b: 3): 'Areas with no land-use class indicated undisturbed forest and were denoted by the appropriate forest type code.'

8 For instance, in the case of FIM, a calculation algorithm was used with an automatic distribution of 65–35 per cent ('dominant' to 'sub-dominant' vegetation cover), abstracting from the actual distribution when it could not be determined at a detailed scale. The same applies to the so-called 'resource mapping units' in the PNGRIS: majority use is counted in the numerical calculation, implying that the clearing of small fragments may not be registered in a change assessment (R. Kiele-Sapak, PNGRIS, personal communication, Port Moresby, 21 November 2001). On the national scale, this would either under- or over-estimate actual deforestation, depending on the distribution of 'majority' and 'minority' forest fragments in the total of mixed vegetation classes.

9 See the footnote above on the insecure date of the forest-stock estimate in McAlpine and Quigley (n.d.), as well as in the corresponding printout I received in Port Moresby. Also, most of the forest figures reported generally as 1975 data in McAlpine and Quigley (1998) seem to date back to air photos that had already been taken in 1972 (ibid.). In addition, selective updates to the 1975 map seem to have been done for some areas with significant levels of land development, such as the Upper Ramu and Markham valleys (Bellamy and McAlpine 1995: 115).

10 This section was written as a postscript, which is why the FRA 2000 results are not included explicitly in Table 8.1. However, the general results of FRA 2000 are discussed in Chapter 1.

11 While the FAO statistics report a forest cover of 30.6 million ha or 66.1 per cent of land area, the text section under 'Resources' states that 'almost 80 percent of the country is under forest'.

12 The difference between the upper and lower estimate is vegetation dieback that may be due to increased flooding caused by El Niño, rather than by mine discharges.

13 According to the mine's own report, the category 'waste rock mined – to erodible dump' increased from 18.4 million t in 1999 to 37.2 million t in 2000 (Placer Dome 2000l).

14 For instance, compensation consisting of a large monetary payment plus a thousand pigs was demanded from those who were thought to be responsible for the recent death of a political leader from a Highlands clan.

15 K1 = US$0.27 in February 2002, but in the 1980s the kina was approximately a par *vis-à-vis* the dollar. In 1984–90, the kina exchange rate fluctuated between US$1.00 and US$1.15 (IMF 1992).

16 Corruption is just as widespread at the provincial as at the national level. In 1984, the administration in Enga Province was the first to be suspended for financial mismanagement, to be followed by several others (EIU 1999d: 8).

17 For instance, Bonnell, in Toft (1997: 137–42), found that the incidence of polygamous marriages in Porgera had more than doubled because men had increased purchasing power from compensation payments, leading them to marry additional wives, who increasingly come from outside Porgera.

18 In the early 1980s, Ok Tedi subsistence farming of taro, sweet potato and banana had expanded temporarily into the lower altitude foothills, but 'gardens ... were exhausted after a single harvest, which placed pressure on Foothill Rain Forest near the roadside villages' (Hyndman 1997: 5).

19 For instance, in the case of Porgera, it was estimated from census data that population growth between 1980 and 1990 was some 80 per cent more than one would expect from natural population growth (Burton 1992, cited in Mawuli and Sanida 2000: 42). But even in Porgera, until now this has not been enough to expand local agriculture: 'there was little development in the agricultural sector at Porgera' (Porgera Social Monitoring Programme, 1996 Annual Report: 43).

20 Notably, the now widespread practice of the *fly-in fly-out* of technicians and mineworkers is used deliberately to limit the immigration of mine employees (James 1997). This limits not only potential conflicts and negative socio-cultural impacts, but also the local economic opportunities that would result from higher service demand, petty trade, etc. related to larger settlement.

21 Most importantly, the handling of the Bougainville Island separatist conflict would probably not have been as costly without booming oil revenues. By the mid-1990s, it had absorbed almost US$500 million. Much of this was allocated outside the budget, was spent inefficiently and in some cases very probably diverted into private pockets (Dorney 2000: 180–3).

22 In 1990, only 0.2 per cent of PNG's population were government employees, compared, for example, to 0.3 per cent in Thailand or Sri Lanka. Yet public expenditure on consumption reached 24 per cent of GDP (Thailand: 10 per cent; Sri Lanka: 9 per cent). This is because the average public wage in PNG is six times the GDP per-capita level (Philippines: 2 times; Sri Lanka: 1.8 times) (Marsden 1993: 237).

23 Originally created in the 1970s as the Prime Minister's Discretionary Fund, in 1984 all members of parliament were allocated K20,000 per annum, increasing to K100,000 in 1990 and K300,000 in 1994, amounting to a total expense of K32.7 million for that year. Following pressure from the World Bank and local students, the fund was abolished at end of 1995 (Connell 1997: 277–8). However, budgetary lines with little or no transparency attached to them continue to thrive. In President Mourata's recent 2002 budget, as presented on 28 November 2001 in the *Papua New Guinea Post Courier* (p. 2), no less than K311 million (US$86.5 million, or 27 per cent of the national government's departmental expenditure) is allocated under 'Miscellaneous: Dept. of Finance and Treasury', a budget line that in the past has been used for discretionary allocations of a non-transparent nature (T. Gumoi, personal communication, Port Moresby, 1 December 2001).

24 In particular, many PNG nationals hold real estate in Cairns, Australia. In a high-level scandal, it became known that PNG's Public Officers Superannuation Fund (POSF) had bought a large property from a real-estate agent at about double the value it had been purchased for just weeks earlier. The operation thus served to appropriate public funds and to take them abroad (Dorney 2000: 280–1).

25 As a more technical factor, it has been claimed that PNG's inflation indexes, contained in the RER index, over-represent imported goods, thus understating the degree of adjustment from nominal re/devaluations (Duncan *et al.* 1998: 68; AusAID 2000: 39–40).

26 *The National*, 8 November 2001, Special Feature section on 'Higatura Oil Palm Silver Jubilee: Development over 25 Years'.

27 I owe the information in this paragraph to personal communications from Ian Rowland (Port Moresby, 1 December 2001) and Leo Mandeakali.

28 According to three different versions of the *Handbook of Agricultural Statistics* (Department of Primary Industry 1979, 1983, 1986). It is not clear whether the rise in smallholder areas simply corresponds to the reduction in large holdings or to a rise in total cultivated area. The two latest *Handbooks* of 1981 and 1984 report only selected provinces, and some of the figures for these two years are identical, indicating that no new assessments exist for these provinces for 1984.

29 The numbers were deflated by implicit GDP prices, not by individual commodity prices.

30 The extensive literature on land-use responses and intensification in PNG includes Freyne and McAlpine (1985), Allen *et al.* (1995, 2001), Ohtsuka *et al.* (1995), Bourke (1997), Bourke *et al.* (2000), Umeazaki *et al.* (2000) and McAlpine *et al.* (2001). Additional information was gathered from a range of interviews with farmers and extensionists during the field trip to Mount Hagen.

31 In a way, this conceptual division is mirrored by highlanders' own distinction between 'village fields' (located close to settlements), which are most prone to intensification, and 'bush fields', which are more remote and often represent an extensified land use (Stacey and Lucas 1989).

32 Among non-timber forest products, rattan is the prime export good (annual value is about K200,000). A large variety of species, such as wild tubers and vegetables, nuts, game, medicinal plants, etc. are harvested on a subsistence basis. Flour made from the sago palm (*Metroxylon sagu*) is an important staple food, especially in Gulf and Sepik Provinces (Saulei and Aruga 1994). There is no evidence to suggest that NTFP harvesting was systematically influenced by the mineral boom.

33 See, for example, Nadarajah (1994) and Sekhran and Miller (1994: 53) for discussions.

34 For instance, lasting damage may occur due to the large-scale removal of nutrients, topsoil or hydrological disturbance. However, total clear-felling in Gogol only seems to have occurred on flat areas, some trees being left standing in the hilly zones (Lamb 1990: ch. 5).

35 Information from field trip to Gogol, 26 November 2001, Mr Dewe Enn (NFA Madang).

36 This refers to the unweighted average of Most Favoured Nation (MFN) countries.

37 Using absolute instead of per-capita value-added, and estimating the relationships for different sub-periods, both proved not to change the overall results of the equation in row 4.

38 As in the other chapters, a functional form was tried with non-agricultural GDP included as an independent variable, as a proxy indicator of urban incomes, but the results were insignificant.

39 According to Peter McCrea, NFA, Moresby, 22 November 2001. I am also grateful to Peter for providing me with this unpublished long-term time-series on log exports from NFA.

40 Data for the 1990s were provided by T. Gumoi, based on government figures from *Estimates of Revenue and Expenditure* (various issues), Waigani, Port Moresby.

41 This situation was confirmed in various interviews in both Madang and Mount Hagen.

42 This refers to the so-called 'commercial statutory authorities', namely the Electricity Commission (ELCOM), the Post and Telecommunication Corporation (PTC), Air Niugini and the Harbours Board. From 1981 onwards, they operated with small but sustained surpluses (Whitworth 1993).

43 Data for the 1990s were provided by T. Gumoi, based on government figures from *Estimates of Revenue and Expenditure* (various issues), Waigani, Port Moresby.

44 P. Peng and E. Kavanamur, personal communications, 30 and 31 November 2001, respectively, Mount Hagen. See also UNDP 1999e: 125.

45 In principle, causality could also be the reverse: roads are 'endogenously' being built into those areas where people live and demand them. However, that caveat seems weak due to the fact that road-building has stagnated recently where population and settlement have continued to grow.

46 One might object here that more agricultural credits, extension and other support might have helped farmers to *intensify* more instead, easing the pressure on forests. While this may be true in part, it should be remembered that farmers actually managed to intensify food-crop production fairly well, *without* noteworthy public assistance. In turn, a larger road network would unambiguously have enabled them to expand cultivated areas.

47 Per-capita poverty lines were US$275 in rural areas and US$400 in urban areas (Tabatabai 1996: 41).

48 Allen (1999: 41–3), and Heywood and Hide (1992), cited in Allen (1999: 43).

49 The provincial welfare ranking, according to UNDP's Human Development Index, remained basically unchanged between 1972 and 1980 (Fernando 1992: 8).

50 See EIU (1999d: 43) and Marsden (1993: 231–2). The two sources disagree on the 1990 figure, which is 79 per cent in the former case and 75 per cent in the latter, probably due to different sector definitions.

9 More tales of oil wealth and forests

Mexico, Nigeria and Indonesia

The preceding five chapters have analysed the relationship between oil wealth and forests in the five primary cases. All were small or medium-sized countries, which allowed us to deal in a comprehensive way with highly problematic forest statistics and a multitude of opposite impacts from macroeconomics to land use. However, this choice, governed by analytical convenience, has also meant that certain large, more complex, but important tropical mineral exporters have been ignored up to this point. The present chapter will try to remedy this by giving a brief, *ad hoc* account of the oil wealth link to forests in three of the largest countries, one in each of the main tropical continents: Mexico, Nigeria and Indonesia. For each country, we shall start with the oil sector proper, then turn to the macroeconomic/policy sphere, and finally discuss how land use and forests were affected on the national and regional levels.

Mexico

Shifting oil wealth

Mexico has been a petroleum exporter since 1901, and during sub-periods of the country's modern history the sector was of prime importance. In 1921, Mexico even became the world's second largest oil producer, satisfying 25 per cent of world demand. However, by the early 1970s, the sector accounted for less than 2.5 per cent of GDP; it was an important, but in no way a dominant sector (Gavin 2000: 167). Although some exploration and infrastructural development occurred even in the 1960s, up until the mid-1970s production remained moderate at around 500,000 barrels/day, almost all of which was destined for the domestic market. This means that in the mid-1970s, Mexico was *not* a specialised oil or mineral exporter, for example, in the classification of Sunderlin and Wunder (2000: 330), which is based on average exports in 1974–6.

Together with the discovery of oilfields in Campeche, the 1973 price boom was a key factor in the decision to accelerate oil exports. In 1976, under the López-Portillo administration, Mexico opted for a more aggressive, export-oriented petroleum strategy. Production grew continuously from 653,000 barrels/day in 1974 to 2,748,000 in 1982, and remained high throughout the 1980s. Net exports rose even more dramatically, from 72,000 barrels/day in 1976 to 1,608,000 in 1984. The world price (in fixed 1988 US$) jumped from US$9 in 1972 to US$22 in 1973 and US$38 in 1980 (Gavin 2000: 171).

During the second oil boom, Mexico thus moved into the category of high mineral exporters, based on average export composition in 1979–81 (Sunderlin and Wunder 2000: 330). The oil development that made possible this rise in production took place both off-shore (Bay of Campeche) and on-shore (Chiapas and, mainly, Tabasco State). In the on-shore case, there is a large degree of spatial correlation between tropical forests and oil deposits (O'Brien 1998: 83 and see below).

In the 1980s, the world-market price for Mexico's oil ultimately fell back to US$27 in 1983 and US$12 in 1988, though thanks to the high production levels, oil-export value and foreign exchange inflows remained higher throughout the 1980s than in the mid-1970s. Gavin (1999: 172) quantifies the size of the windfall at 3.8 per cent of GNP in 1987, compared to only 1.5 per cent in 1978. Thus, although the oil-price boom was temporary, the rise in production gave it a semi-permanent feature. However, 'the macroeconomic boom that began with the oil boom ended much earlier than did the oil boom itself' (Gavin 2000: 179). As explained in what follows, this mainly had to do with foreign borrowing and with the fiscal policies adopted.

The macroeconomic response

Mexico's is the third largest economy in Latin America. Traditionally, it has had sizeable, widely protected home-market sectors (industry, food crops), but there have also been strong, strategic trade ties to its northern neighbour, the US. Prior to the oil boom of the mid-1970s, Mexico experienced a period of stable growth based on import substitution, combined with a moderate rise in both agricultural and manufactured exports. Mexican agriculture also did fairly well in the 1950s and 1960s, with growth being accompanied by an incremental modernisation. Medium-sized and larger farmers successfully adopted high-yield varieties, leading some to speak about 'the Mexican miracle' in agriculture (O'Brien 1998). Continuing redistribution of land through the land reform that began in the 1920s gave ownership of about two-thirds of rural forest and cropland to communities (*ejidos*), though the distribution of capital and technology remained unequal. As *ejido* members usually could not sell the land, use it as collateral or move away from it without losing usufruct rights, this landholding structure is unfavourable to rural change and innovation (L. Snook, personal communication, Bogor, 17 April 2002). A sizeable peasant sector thus continued to rely fundamentally on traditional techniques of shifting cultivation such as the *milpa*, a low-input cropping system used for the main staple, maize, in association with beans, squash and other vegetables (Collier 1998; Ochoa-Gaona and González-Espinosa 2000: 19).

In the first oil-boom years (1977–9), Mexico faced an accelerated, demand-led economic growth that benefited all economic sectors to some extent. Foreign exchange inflows from oil were accompanied by significant external borrowing, in anticipation of further rises in oil incomes.[1] Mexico's external debt rose steeply from US$3 billion in 1973 to US$35 billion in 1979 and US$ 60 billion in 1982 (Karl 1997: table A7). Foreign borrowing in the late 1970s financed a public spending spree and a large fiscal deficit. As we would expect from other Dutch Disease countries, economic growth was highest for typical urban sectors like construction, services and the largely import-protected manufacturing sector. The economy's NT sectors became increasingly over-heated and there was a strong real currency appreciation of the *new peso*, from index 67 in 1976 to index 152 in 1981 (1979 = 100) (Feltenstein 1992: 275). However, when in 1982 real interest rates in the US rose sharply and foreign financing decreased drastically, Mexico, now a large-scale debtor, became a showcase for the emerging foreign debt crisis in the developing world. The Mexican economy went into a severe recession, and the RER depreciated back to pre-boom levels.[2] In this sense, the relative price story in Mexico was affected at least as much by financial capital flows as by oil exports. The sharp contraction and reversal of relative price changes caused the economic boom to end in the mid-1980s.

What were the large oil revenues and the accompanying financial capital inflows spent on? Initially, in the 1970s large amounts were invested in the energy sector itself, that is, in petroleum infrastructure, but also in a series of hydroelectric projects. Second, the Mexican governments of the period chose to subsidise the domestic consumption of energy heavily, thus insulating the economy from the cost effects of rising oil prices. In 1980, the average price of energy products in Mexico was only one-fourth of the corresponding world-market price. The value of the matching transfer from the public to the private sector rose gradually to a staggering 5.4 per cent of GDP in 1982, before it was dismantled in the fiscal adjustment to the economic crisis (Gavin 2000: 191–2). Third, a variety of other budgetary priorities was pursued, jointly raising the share of government spending in GNP from 12.1 per cent in 1972 to 20.8 per cent in 1981 (Scherr 1989: 552). Among these were also categories with clear (positive or negative) implications for land-use expansion:

- Urban spending – higher public employment, wages and other investments in urban infrastructure – similar to what occurred in other oil-boom countries.
- Rural investments: in particular the road network was extended, and electricity and other infrastructure were improved (Collier 1998: 118).
- A land-colonisation and frontier-opening strategy in the southern, tropical zone, started in the 1960s, was greatly strengthened by oil revenues. This included a geopolitical policy of strengthening settlement along the Guatemalan border (Harcourt and Sayer 1996: 200).
- Other subsidies (foodstuff, electricity, guaranteed farm-producer floor prices, agricultural credits, etc.). For example, large food subsidies aimed at compensating 'losing' agricultural producers were administered through CONASUPO, the giant national food-marketing agency (Scherr 1989: 551).

The Mexican policy response to the boom was thus quite balanced across sectors and, compared to most cases in this book, quite favourable to agriculture. As in Ecuador, it also actively promoted the *extensification* of cultivated area. For instance, in Quintana Roo state the government financed large-scale forest conversion for crops and pastures, using heavy machinery for clearing, ploughing, etc. One author has observed that since the crisis of the mid-1980s, 'the subsidies for these programs have waned' (Edwards 1986: 127). The same happened in Chiapas state (Collier *et al.* 1994: 399), in Oaxaca and in the Uxpanapa project in Veracruz (L. Snook, personal communication, Bogor, 17 April 2002). There was thus a direct link between, on the one hand, booming oil revenues and, on the other, funding for colonisation programmes and agricultural subsidies.

What happened to forests?

The national picture

According to Harcourt and Sayer (1996: 195), forests covered 51.5 million ha, or 27 per cent of Mexico's land area in the early 1990s. That is about half of what is thought to be the country's 'original' forest cover (97.2 million ha). Of present forests, about half is conifer and oak forest, while the other half is made up of tropical rain, seasonal and montane forest. Deforestation figures are just as uncertain for Mexico as for the other countries described in this book. Estimates range from FAO's low-end 615,000 ha/yr for 1980–90 (FAO 1993) and 631,000 ha/yr (1.1 per cent) for 1990–2000 (FAO 2001c: 394) to the high-end 1,500,000 ha/yr (Toledo 1988, cited in Harcourt and Sayer 1996: 199). A main source of divergence is in the definition of 'forests', for example, with respect to old secondary fallows and agroforestry systems with significant cover of shade trees. One of the most reliable sources, Masera *et al.* (1992, 1996, cited in Harcourt and Sayer 1996: 199; O'Brien 1998: 42) found from satellite imagery a yearly loss of about 800,000 ha/yr in the late 1980s and early 1990s. This corresponds to a deforestation rate of 1.6 per cent/yr, though the yearly loss in the tropical biome (237,000 ha) is even higher (2.4 per cent). Unfortunately, the survey periods coincide poorly with the oil boom, which makes it impossible to test our time-series hypothesis fully. But there is little indication that national deforestation fell significantly during the boom years, that is, from the mid-1970s to mid-1980s.

Pasture and cropland expansion have clearly been the main source of deforestation in Mexico, which is why agricultural land-use figures can be an illustrative supplement to forest statistics (Barbier and Burgess 1996: 209). Combining agricultural census data with satellite imagery, Liverman *et al.* (1997: 6; figure 3) found that, between 1970 and 1990, the area of closed forests was halved. At the same time, cultivated pastures increased by 75 per cent and cropland by 34 per cent. Probably, there was little change between the 1970s and the 1980s. Using national agricultural census data, Deininger and Minten (1999: 315) identify a similar trend between 1980 and 1990. In Rudel and Roper's (1997a: 58–9) binary classification based on regional land-use reports, Mexico is labelled a high-deforestation country in both the 1970s and 1980s. Furthermore, deforestation was stronger in the southern, tropical zone of Mexico than in the temperate part of the country. In some regions, deforestation is contained by large labour outmigration to the US (T. Rudel, personal e-communication, June 2002). Up to three-quarters of Mexican forest conversion for agriculture may occur in the tropical part (Masera 1992, cited in Barbier and Burgess 1996: 203–4).

Why does oil seem to have had have such a small 'forest-protecting' land-use effect at the national level in Mexico? The direct and indirect regional effects of oil production are one candidate for explanation. But, in spite of the significant spatial overlap between petroleum deposits' and tropical forests, this effect was not strong. It is true that oil roads, together with timber roads, helped to provide access to immigrating settlers in Tabasco and Chiapas States (including Guatemalan refugees), and thus had some indirect deforestation effect. But the direct clearing effect of oil production was as limited as in the other cases in this book. For an exploration area in the Marqués de Comillas forested region

(Selva Lacandona, Chiapas), O'Brien (1998: 84) estimates that only 0.07 per cent of the study area was cleared for oil roads and drilling wells.

Rather, a main explanation is that agriculture declined in relative terms but was not hit by the oil boom in a disastrous way; it also received capital injections and policy incentives to expand cultivated areas. Agriculture's share of GDP declined during the boom from 10 per cent in 1977 to 7 per cent in 1982, before rising back again to 9 per cent in 1987. This is a slight boom-led acceleration of the previous declining trend from 14 per cent in 1965 (Karl 1997: table A12). The agricultural export sector truly suffered, with sustained reduction until 1983. Only then did the overvalued RER start to depreciate and exports recuperate (Feltenstein 1992: 275). But the decline in agricultural exports seemed to be of little concern to the government. For part of this period, exports of beef and sugar were actually deliberately restricted through trade policy to increase domestic food supply and alleviate inflationary pressures (Scherr 1989: 551), as well as to maintain food security (Collier 1998: 118). But for the semi-traded food-crop sector, many public interventions (roads, floor prices, credits, etc.) helped to maintain or increase production.

Maize is the most important food crop in this case; this staple accounted for more than half the cropped area during the period, mainly being cultivated under land-extensive production systems in tropical areas. After a decline up to 1980, when maize imports from the US reached 25 per cent of total Mexican consumption (Collier 1998: 118), trade and price policies shifted radically, and domestic maize production increased significantly (Scherr 1989: 550). Panel data indicate that maize production is quite sensitive to changes in maize prices and fertiliser prices, both of which were manipulated by policy interventions in favour of producers (Barbier and Burgess 1996: 239). In another panel study at the municipal level, government agricultural subsidies were found to be a significant factor in accelerating deforestation (Deininger and Minten 1999). The land-tenure structure of the *ejidos*, mentioned above, made the supply of maize less elastic and, in the specific situation, less competitive *vis-à-vis* imports. However, for motives of political stability and food security, the food-subsistence sector was widely protected after 1979.

Another sector that deserves special attention regarding land use is cattle. Pastures for cattle-ranching and dairy production account for around 60 per cent of deforestation in Mexico (Harcourt and Sayer 1996: 200). The cattle sector was favoured by the rising incomes during the boom, promoting shifts in consumption patterns towards meat and dairy products. Panel data indicate that the number of cattle is significantly correlated with rising per-capita income, and also with the size of publicly subsidized credits (Barbier and Burgess 1996: 239). A middle class that was already large before the boom created a favourable scenario for this effect to operate forcefully (Scherr 1989). This structural change in consumption and production was probably as important for land use and forests as it was in Venezuela and Ecuador.

The regional picture

For a large country like Mexico, it may be preferable to look not only at national, but also at regional land-use adjustments to shifting degree of oil wealth. In this section, we shall examine in more detail the state of Chiapas, which in terms of land area is almost half the size of two of our study countries, Gabon and Ecuador. As we shall see, production and employment structures in Chiapas were significantly affected by oil-related development, and evaluation at a sub-national level allows us to identify a variety of land-use responses. The more specifically tailored time spans for these land-use studies will also allow us to juxtapose them to the effects of oil wealth. We shall look at three levels of aggregation: the state of Chiapas, a large sub-region with extensive forests (Selva Lacandona), and a couple of micro-level study areas in the central highlands of Chiapas.

Chiapas is a thinly populated state, with only about 3 per cent of the Mexican population, but it is important in terms of agricultural production. In the early 1990s it was Mexico's largest coffee-producing state, and third in terms of both maize and cattle (O'Brien 1998: 99). Oil production, mostly in the neighbouring state of Tabasco, significantly changed the structure of employment in Chiapas in the 1970s, both directly and indirectly. Higher salaries attracted unskilled rural workers from Chiapas to work in the oilfields (in Chapter 2 this was called the booming sector's 'resource movement effect'). Other off-farm employment was related to the spending of oil and foreign-borrowing inflows, notably the building of several hydroelectric dams and a general urban construction boom. But job opportunities also opened up in transport and other services. Collier (1998) thoroughly documents the impacts of these changes on land use in Chiapas. In many rural areas, the formerly all-dominant farming activities played a much lesser role, as many men migrated seasonally from their villages to construction projects and to the cities. In some villages, maize cropping was abandoned entirely (ibid.: 124). Variable remittances from off-farm labour also led to an increasing social differentiation among households in rural areas.

What were the consequences for forests? In northern Chiapas, a comparison of Landsat images between 1972 (before the boom) and 1980 (the peak of the oil boom) shows a decline in closed forests, but an even larger rise in tall secondary fallows, most of which are probably to be classified as 'forests' according to FAO criteria. The area under pasture and crops both declined (Collier *et al.* 1994: 401). This indicates that logging and other extraction continued to open up natural forests, but this was more than counteracted by increases in secondary forests on abandoned agricultural land. In the Selva Lacandona, one of the largest remaining tropical forest zones in Mexico, the situation was slightly different. Aided by road construction and government colonisation programmes, immigration pressures (including from Guatemala) continued in this frontier area, and 32,000 ha/yr were cleared between 1969 and 1988 (O'Brien 1998). In a 34,000 ha study area in the central highlands of Chiapas, land-use trends differed across both sites and sub-periods. For 1974–84, that is, more or less the period of the oil boom, the rate of deforestation (1.58 per cent) was found to be lower than in the post-boom period 1984–90 (2.13 per cent). For the smaller (3,200 ha) study area of Huistán, however, a densely populated, established agricultural zone that is well connected to urban markets, forest loss was actually greater during the boom period (1.84 per cent) than after the boom (1.10 per cent). For the remote and unpopulated frontier area of Chanal (3,800 ha), forest-clearing almost stopped during the boom (0.46 per cent), but exploded in the post-boom period (3.42 per cent) (Ochoa-Gaona and González-Espinosa 2000).

These divergent results underline the spatial variability of land-use responses, even within Chiapas state. Those rural areas that could produce easily for growing urban markets, and those that were being opened up by new roads and government colonisation projects, faced increasing or at least an unchanged pace of deforestation during the oil boom. Those that met neither of these two conditions typically lost much of their labour force to booming off-farm activities that could pay higher wages, thus allowing local secondary forests to grow back on abandoned or only sparsely used crop and fallow land. Similarly, when agriculture experienced a general revival towards the end of the 1980s, land-use adjustment was asymmetrical. Additional land demand was satisfied from both forests and extensive fallows. But many peasants had acquired skills and capital that allowed them to preserve links to the off-farm economy (Collier 1998). Also, richer farmers increasingly replaced shifting *milpa* production with more intensive cropping systems, using more fertiliser and herbicides (Barbier and Burgess 1996: 209). The capital for these developments was partly provided by boom savings from off-farm employment. But poorer farmers without capital became increasingly marginalised, which also provided one root cause of the *zapatista* rebellion in Chiapas (Collier 1998).

Conclusion

Mexico's national deforestation statistics do not allow us to test explicitly whether the country conforms to the core hypothesis of this book or not. The sparse and debated forest statistics and land-use figures that exist give decade-wise estimates for the 1970s and 1980s, respectively. For our purposes, we would have needed data from the mid-1970s to the mid-1980s, compared to the periods before and after. What we can say on the basis of the sparse existing evidence (FAO, regional reports and some satellite imagery) is that, on average, national deforestation was probably high in each of the past three decades, that is, the 1970s, 1980s and 1990s. In this sense, Mexico seems to be a case that *deviates* from the core hypothesis of this book. On the national scale, the sequence of an economic boom followed by a severe bust was most likely not particularly beneficial to forest cover and quality. Yet the regional results from Chiapas state indicate variability: greater boom-led deforestation pressures dominated in some forested areas (newly opened frontiers, periurban zones), while diminished pressures occurred elsewhere (areas of rural exodus to the booming off-farm economy).

The partial-effect framework described in Chapter 3 can help us understand why this outcome occurred. On the forest-protecting side, the core factor, namely competitiveness loss in agriculture and forestry, was quite prominent in Mexico, and it cut back export cash-crops strongly. The urban drive, fuelled by the usual public spending spree, construction boom, etc., also successfully drew agricultural labour into the cities. But the urban bias was generally mild. The government gave special attention to rural development: 'an historically high proportion of all government expenditure was directed to rural areas during the oil boom' and, more specifically, 'a major program of infrastructure construction outside urban areas was undertaken' (Scherr 1989: 553). This emphasis was not favourable to forest conservation.

Several other policies of macroeconomic management also worked clearly against forests. Extensive rural road construction, massive energy and transport subsidies and government-subsidised settlement programmes all helped to lower transport costs and to integrate remote forested areas more into national markets, thus augmenting conversion pressures. Timber and oil roads had similar indirect effects, though their direct impact on forest loss was negligible. Trade and pricing policies protected food crops, notably the heavily land-using staple maize, which would have declined dramatically had it been a genuine 'tradable' commodity that was fully exposed to import competition. And a substantial middle class income-led shift in consumption towards meat and dairy products heavily promoted cattle-ranching – the end-use of more than half the deforested areas. The Mexican oil-wealth story thus considerably resembles that of Ecuador and, in part, Venezuela: strategic policy responses favoured land-expansionary development patterns, and a domestic 'hamburger link' was the key single pathway to sacrificing forests.

Nigeria

Shifting oil wealth

Booms and busts

With 124 million people (1999), Nigeria is the most populous country in Africa (World Bank 2001a: 275). Petroleum was discovered by Shell-BP in 1956, following half a century of exploration. Oil production became important in the 1960s, but the Biafra conflict and civil war between 1967 and 1970 delayed further expansion. The main reservoirs are located in and around the Niger Delta, in both on-shore mangroves and shallow off-shore basins, and since 1990 exploration has increasingly moved to deep, off-shore areas. Unclear boundary demarcations in the Niger Delta have caused disputes relating to several strategic areas. Cameroon and Nigeria both claim zones on and off the Bakassi peninsula (see also Chapter 6 on Cameroon), and ownership of the Zafiro oilfield is disputed with Equatorial Guinea. Shell continues to be the most important company, but it has been joined by a series of other multinationals over the years. In 1977, the giant parastatal, the Nigerian National Oil Corporation (NNOC), was created (Mbendi 2002).

By 1970, petroleum exports already made up 58.1 per cent of the country's export value. Unlike Mexico, Nigeria thus took full economic advantage of the first oil-price hike in 1973–4 (UNCTAD 1999: 128). Oil revenues jumped from 1.3 billion naira in 1973 to 3.9 billion naira in 1974 (in fixed 1973 prices). After a slight drop in the late 1970s, they rose back to a peak of 4.9 billion naira in 1980. Until 1983, real revenues then dropped to less than half their value (2.0 billion naira). The oil price had slackened, but even more importantly, Nigerian production quantities almost halved due to insufficient previous exploration (Oyejide 2000: 420–4). However, after the bust in the 1980s, production resumed at previous levels of around or above 2 million barrels/day throughout the 1990s. Petroleum's share of exports was 96.9 per cent in 1980, 93.6 per cent in 1990 and 95 per cent in 2001 (UNCTAD 1999: 128; Mbendi 2002). Exxon Mobil's and Esso's major deepwater discoveries in 1999 will help to ensure high or rising levels of oil production in the future. In addition, Nigeria also has major natural gas reserves, which are currently still under-exploited. In other words, oil exports boomed in 1973–81, they fell off in the 1980s but resumed high levels in the 1990s. Over the last decade, mini-booms and -busts have occurred, as a result of fluctuating world-market prices. But Nigeria's structural transformation into a highly specialised petroleum exporter has proved long-lasting.

Environmental effects

The environmental and human-rights aspects of Shell's and other companies' oil operations in Nigeria have been one of the main targets for international and local protests against the oil industry. Attention was lifted to new heights when Ken Saro-Wiwa and several colleague activists from the indigenous Ogoni people who were opposing Shell's practices in the Niger Delta were executed by the Nigerian regime in 1995 (Bravo 1999). Many of the oil production and exploration areas are in the 700,000 ha of delta mangroves. These represent the main remaining forested biome in Nigeria and about two-thirds of all the mangroves left in Africa, and are an essential ecosystem both for local people and for fisheries off West Africa's coast. No deforestation estimates are available, but operations and infrastructural development in the mangroves have been quite intense: 349 drilling sites, 700 km flowlines, 400 km pipelines, 22 pumping stations and 1 terminal (Bassey 1999).

However, forest degradation impacts in the ecologically fragile mangroves have been much more serious, and have been aggravated by poor practices. According to Rowell (1995, cited in Bassey 1999), Shell alone has spilt 1.6 million gallons of oil in twenty-seven incidents between 1982 and 1992, which account for no less than 40 per cent of Shell's global oil spills. Probably equally important are extensive ecosystem changes from water flow disruptions, including involuntary damming, tidal changes and the exchange of fresh and salt water, which are caused by road construction, dredging and the use of heavy machinery. Other direct effects include pollution from drilling mud and the effects of explosives. Indirect effects are also present, such as accommodating 6,000 temporary workers for a gas plant, who practise hunting and other forest extraction (Bassey 1999). Gas flaring is higher than anywhere else in the world; current ongoing efforts are aimed at eliminating the current levels of 75 per cent flaring, while 12 per cent is being reinjected (Mbendi 2002). So, on the aggregate, the oil companies' environmental practices in a particularly sensitive forest ecosystem have been deficient; direct and indirect deforestation effects have been limited, but forest degradation has been serious.

The macroeconomic response

Few developing countries have been studied so much from the Dutch Disease point of view as Nigeria. One reason for this is that the policy response in the 1970s was so disruptive that some observers talked about a particular 'Nigeria Disease' (Collier 1988). Some of the elements in this were an extraordinary government spending spree, the financing of fiscal deficits through the money supply (inflation tax), an extreme real currency appreciation and the complete sacrifice of export agriculture. Particular public-spending priorities were the following:

- Investments in infrastructure (physical, social)
- Prestige projects
- Investments in 'modernisation' and 'indigenisation'
- Public salaries
- Military expenditure.

Probably the single most important element was investments in infrastructure. After the civil war, Nigeria launched a drive towards command-and-control economic management and centralisation. Yet at the same time there was a fragile power balance between different federal states and ethnic groups, as well as high regional inequalities in the amount and quality of past public investments. Growing oil wealth thus presented an opportunity to appease conflicting stakeholders, but it also triggered strong political pressures from vested interests. The urge to deliver new schools, roads, hospitals, etc. rapidly to a variety of regions triggered a public construction boom. Investments also included more extravagant prestige projects. Notably, this included the multi-billion dollar decision to build an entirely new capital, Abuja, in the middle of the savannah, after the model of Brasília and Canberra. In general, many costly projects were directed at urban areas, such as the new highway system put in place in Lagos, the largest city and the capital until 1986. These projects were usually accompanied by a large degree of inefficiency and corruption that multiplied their costs. Probably a sizeable share of the oil windfall did not even go through the public coffers, but was siphoned off directly from the oil industry into the pockets of the ruling classes and their entourage (J. Fiebach, personal e-communication, 24 April 2002).

The government's modernisation strategy also led it to make the costly long-term commitment to build a huge steel complex and to promote the auto-assembly industry (Pinto 1987: 432). The increased tendency by the state to pursue nationalistic goals was also reflected in strategies to *indigenise* the economy, for example, by introducing limiting quotas for expatriates, minimum purchase requirements of Nigerian goods, increase Nigerian ownership, etc. (Gelb with Bienen 1988: 236–7). While a disproportionately high share of the windfall was invested, public wages also rose sharply. In 1975 alone, average public wages were doubled in an attempt by the government to buy political support from key stakeholders (Ezeala-Harrison 1993: 199). Military expenses were not as high as during the civil war, but remained a major burden on public budgets.

Consequently, as the boom progressed, civilian and military governments alike were no longer able to hold back spending and prioritise effectively among different project proposals. Fiscal discipline was lost after 1975, and in particular public capital spending went from 3.6 per cent of non-mining GDP in 1970 to 29.5 per cent in 1976. That absorption alone was *more* than the revenues from oil; the balance was made up by drawing on reserves and expanding the money supply, obviously driving up domestic inflation (Gelb with Bienen 1988: 241). After a short-lived fiscal cut-back in 1978, oil prices soared again the next year and accelerated spending. Nigeria was cautious in its foreign indebtedness until 1981, when external debt constituted only 8.2 per cent of GDP, but this changed considerably in the 1980s. The downturn in both oil prices and quantities after 1981 was widely compensated by higher borrowing to sustain elevated levels of spending (Pinto 1987: 428). Thus, the debt share of GDP reached 112.8 per cent in 1987 (Karl 1997: table A-14; Ezeala-Harrison 1993: 200). In only six years, Nigeria had entered the circle of highly indebted countries.

In land-use terms, it is notable that none of the specific spending categories gave incentives to expansionary land uses and forest conversion, except perhaps for transport subsidies and some expansion of the road network. Notably, public investments in agriculture only assumed 3 per cent of the disbursements in the 1970s (Pinto 1987: 432). Indeed, the emphasis on services, construction and industry reinforced a pre-existing urban bias that would increasingly draw labour out of agriculture. This also diminished pressures to convert forest, relative to what would have happened without a boom. The only specific toll on forests was that the construction boom accelerated timber demand. To satisfy domestic interests, in 1976 the federal government imposed a ban on the export of logs and semi-processed timbers (Sayer *et al.* 1992: 231). But even before this, the exhaustion of high-value timber species and oil-wealth effects together had eliminated timber exports through economic rather than legal forces. The share of timber in total exports declined from 4.1 per cent in 1960 to 0.7 per cent in 1970 and 0.1 per cent in 1975 (Ezeala-Harrison 1993: 198).

Notwithstanding specific budgetary effects, the 'core effect' of declining non-oil competitiveness was perhaps stronger in Nigeria than in any other high-absorbing oil country. The real exchange rate already started to appreciate with rising oil production in the late 1960s, but the price booms accelerated that trend. From index 1968 = 100, the rate appreciated to index 131 in 1973, 257 in 1980 and to a remarkable 357 in 1985 (Collier 1988: 769).[3] A parallel market for foreign exchange developed during the period, with an increasing black-market premium, so a range of transactions was implemented at a lower naira rate.[4] Still, it is fairly obvious that a competitiveness loss of such vast dimensions would have a disastrous impact on domestic tradables. Not surprisingly, the agricultural export sector, which was dominated by smallholders, was basically wiped out, and it failed to revive later. Prior to the oil era, Nigeria was the leading world exporter of both cocoa and palm kernels, while rubber and groundnuts were also important. The total, real value of these exports was halved, from 549 million naira in 1973 (in fixed 1975 prices) to 270 million naira in 1980, and hit rock bottom at 63 million naira in 1984 (Gelb with Bienen 1988: 250). The share of agricultural exports (including timber) was 80.9 per cent in 1960, progressively declining to 30 per cent in 1970, 7.2 per cent in 1975 and 2.4 per cent in 1980 (Ezeala-Harrison 1993: 198).

It is remarkable that no devaluation of the naira was carried out before 1986 to realign RPs in the face of the economic crisis. In fact, however, cheap imports were thought to be more important to political stability than agricultural exports (Scherr 1989: 546). Although the massive devaluation in 1986 brought back some growth momentum into agriculture in the late 1980s, the renewed oil mini-booms in the 1990s triggered sharp RER fluctuations that impeded sustained cash-crop export growth.[5] Over the past half-decade, the share of non-oil exports has fluctuated around only 5 per cent. Fiscal imbalances continue to emerge, due partly to a decentralised federal system of external debt contraction and legislation that mandates the direct distribution of oil revenues to the states, without assigning responsibility for macroeconomic stability to the lower levels of government (IMF 2001).

On the whole, agriculture's share of GDP was almost halved, from 48.8 to 26.5 per cent, in only four oil-boom years (1970–4). That share rose back to 36.3 per cent in 1986, after the crisis (Ezeala-Harrison 1993: 195). But in the 1990s, agricultural growth was mediocre, not exceeding population growth of about 2.8 per cent and the near-zero per-capita growth of the entire Nigerian economy, which raises serious doubts as to whether oil has really made Nigerians better off.[6] Yet agriculture has at least grown in absolute terms over the whole period, which would not seem warranted at all from the drastic RP change working against it. The reason is that the food-crop sector, by far the largest component of Nigerian agriculture, has enjoyed significant trade protection from importables, at least during extensive sub-periods. Trade policy fluctuated greatly, being used as a stop–go policy to respond to shifting external conditions. But on average, 'the domestic price of food was far higher than the world price converted into domestic currency' (Collier 1988: 767). Following Taylor *et al.* (1986), therefore, Nigerian food crops should rather be treated as a *semi-tradable* sector, producing goods that were *imperfect* substitutes for imported foodstuffs. Although food imports grew rapidly during the oil boom, the initial level was negligible. Even at the height of the oil boom (1979–81), Nigeria met 91 per cent of its calorie consumption from domestic sources (Collier 1988: 762). There was thus some crowding-out of locally produced crops, but not on a massive scale.

What crops are we talking about, where are they grown, and what were the underlying land-use trends? Basically, sorghum and millet are the traditional food crops in the non-forested north, while yams, cocoyams and cassava dominate in the south, where many of the remaining forest reserves and fragments are located. Maize and rice are grown throughout the country, and together with wheat (which is entirely imported) they are also subject to direct import competition. Between 1978 and 1982, import duties on maize, rice, wheat and sorghum were raised drastically. Yet, quantitative restrictions via import licensing were the main trade-policy tool (Scherr 1989: 551–2). Port congestion was an additional technical barrier to food imports, though smuggling was another way of partly circumventing restrictions (Oyejide 2000: 440–1). On the other hand, in remote areas with increased transport costs, locally grown food crops also enjoyed a 'natural' protection that made them quasi non-tradables (Scherr 1989: 553).

Poor statistics on food crops do not allow us to make a definite statement, but on the whole food-crop production probably rose at a rate slightly inferior to population growth in the 1970s and early 1980s. Collier (1989: 774) estimates that the total labour force allocated to food crops rose from 17.1 million (72 per cent of the total labour force) in 1970 to 21.1 million (63.3 per cent) in 1983. He also finds widespread evidence of return migration from the cities to the countryside during the pronounced crisis in the last half of the 1980s (ibid.: 778). Thus the oil boom very likely slowed down food-crop expansion, while the crisis probably reinforced it. As land productivity probably remained constant, these production conclusions are likely to translate fairly well into agricultural land demand. Scherr (1989: 555) emphasises that the boom-crisis effect on food crops was probably strongest in the forested southern part. This is because this zone was closer to both large urban areas and the oilfields. Being also more labour-intensive, southern agriculture suffered from the tripling of real rural wages in the 1970–83 period. When many men migrated to the cities, more emphasis was placed on cassava, which is traditionally grown by women in Nigeria. As the absolute decline of export crops (notably of cocoa) also was strongest in the south, oil wealth should have had a dampening effect on forest-clearing in Nigeria. More specifically, we would expect deforestation to have slowed during the boom in the 1970s and to have accelerated somewhat in the 1980s, while probably having been slower again in the 1990s.

What happened to forests?

About four decades ago, Nigeria was a forest-rich country and an important exporter of tropical timber. In 1951, total forest size was about 14 million ha, 6 million of which were closed forests (Sayer *et al.* 1992: 232). Up to the present, forests have been reduced and fragmented drastically. The most heavily forested zone is still in the south, in a coastal belt up to 250 km in width that contains mangroves (about one-third of forest cover), swamp and lowland forest. Going north, a 'derived' savannah zone (man-made by repeated burning for shifting cultivation; Aweto 1990: 129) borders the forest zone, followed by drier Sahelian savannah (FAO 2002b). Some semi-deciduous high forests are still found, but most trees outside the forest zone are in woodlands and different open or fragmented formations.

This also means that contemporary forest-cover estimates vary dramatically with the definition of 'forests'. While some observers state that the country today is 'moderately well forested' (FAO 2002b), others refer to studies saying that already by the end of the 1960s 'no significant areas of rain forest remain outside the legally constituted forest reserves, constituting about 2 per cent of the total land area' (Soladoye and Ola-Adams 1990: 190). Stock estimates show an equally stark contrast. At the high end, the FAO's figure for 2000, using a 10 per cent crown-cover criterion, is 13.5 million ha (FAO 2001c: 391), corresponding to 14.8 per cent of land area. UNEP's 1992–3 estimate of closed forests (40 per cent crown cover) is 8.5 million ha, or 9.3 per cent of land area (UNEP 2001: 41). The TREES estimate of closed evergreen and semi-deciduous forests (same year, also 40 per cent crown cover) is 5.6 million ha, or 6.2 per cent of land area (Mayaux *et al.* 1998: 44). Finally, at the low end, the IUCN Conservation Atlas estimates closed forests in 1989–90 at 3.9 million ha, or only 4.2 per cent of land area (Sayer *et al.* 1992: 232).

The causes of land expansion and the clearing of natural vegetation in Nigeria are, in ranked order, food crops, export crops, livestock grazing and firewood collection, with road-building and logging as main enabling factors (Osemeobo 1988, 1992; Aweto 1990; Soladoye and Ola-Adams 1990; Sayer *et al.* 1992). Grazing and firewood are relevant sources only in the northern region, where the climate is drier and less forest is left. In the forested southern zone, food crops and cash crops are the main sources of deforestation. Yet the area used for food crops (and its expansion over time) is much larger than that for cash crops, because of the higher weight in agricultural production and because food crops are mostly produced in shifting, rather than sedentary, cultivation systems. This implies that the production of food crops has been the major deforestation factor in Nigeria, mainly because the production system has proved inadequate under circumstances of a high population density and rapidly increasing growth rates (Osemeobo 1988: 18–23; Aweto 1990).

How much deforestation has there been, and how has it evolved over time? Of course, the ambiguity as to what counts as forest makes it very difficult to compare deforestation rates from different sources. The FAO (2001c: 391) reports a yearly deforestation of 398,000 ha for the 1990s. But closer inspection of its sources (FAO 2002b) shows that this figure is extrapolated from a study comparing land use and vegetation in 1978 and in 1993–4,[7] so it is actually more representative of what was happening in and around the 1980s. The FAO's and other's estimates for the 1970s were 300,000 ha/yr (FAO 1981; Soladoye and Ola-Adams 1990) and 260,000 ha/yr (Osemeobo 1988). Although different methodologies may apply, this picture of 25–30 per cent less forest-clearing in the oil-rich period from 1970 until the early 1980s would be consistent with the situation described in the previous section: extensively cultivated food crops – the main source of deforestation – grew more moderately in this period of high oil wealth and low competitiveness, while export crops declined rapidly. This is supported by contemporary observations of visitors to the coastal forest zone, who reported that relatively large tree-crop plantations for export (cocoa, oil palm, rubber, etc.) had been either abandoned or were being converted to food crops (Sayer *et al.* 1992: 235; J. Fiebach, personal e-communication, Bogor, 14 April 2002).[8] These are indications that reduced land demand also reduced the pressures on forests.

Conclusion

Nigeria as a country is in a relatively advanced stage of forest transition. At 136 people/km^2 (1999 figure) its population density is similar to the average for Western Europe, the population continues to grow at a rate close to 3 per cent (World Bank 2001a: 275, 279) and large amounts of forests have disappeared. Compared to the little forest that is now left, relative loss rates are above 2 per cent, placing the country in the high-deforestation category for the 1970s, 1980s (Rudel and Roper 1997) and 1990s (FAO 2001c). However, deforestation is much more moderate in comparison with the sizes of the population and the economy, and in relation to the continued growth in both (see the discussion in Chapter 10). Yet, with an agricultural production dominated by smallholders, and the main food crops being grown in shifting cultivation systems with only a limited degree of land-saving technological change, the trend is to expand cultivated area in line with population growth and food demand. These are the basic long-term features that were at work over the entire period, and any assessment of the impact of oil wealth must be balanced against this.

However, at the margin of these trends, shifting macroeconomic conditions induced significant land-use changes. From the mid-1960s onwards, Nigeria went through long-term structural change from a specialised cash-crop exporter to an oil-dependent economy. This shift was greatly accelerated by the oil-price booms in the 1970s. The country indulged in a public-spending spree that was prolonged by a foreign-borrowing spree, government spending mainly being in construction and other non-traded goods. As fiscal and monetary control was lost and wages rose, the economy's traded sectors completely lost price competitiveness. The naira real exchange rate appreciated to sky-high levels because inflation was rising and no devaluation was made before 1986. Non-oil exports were the main victims of this policy, in particular cash crops such as cocoa, rubber and oil palm. This significantly reduced land demand and forest-conversion pressures in the southern coastal zone, where more cash crops are grown, and where most of the remaining forests are located. For the mangroves in the Niger Delta, where most of the petroleum is produced, there was some deforestation and serious forest degradation. But on the whole deforestation declined due to the urban bias in development, which induced many male labourers to go to the cities to take part in the boom. With the crisis in the 1980s, there was return migration to rural areas and a renewed impetus for agriculture, but in the 1990s oil revenues have again risen, though with unstable oil-price levels.

As in Mexico, the land-use impact was moderated by the fact that the main food crops were able to continue their growth, though at a lower rate, thanks to a high level of import protection that turned them into semi-tradables. In both countries, the social impact on smallholders and national food-security concerns provided the political justification for constraining the rise of food imports. The fate of export crops shows what could have happened to food crops had they not been trade-protected: they would basically have been wiped out by the vastly elevated levels of real wages and RERs.[9] In that case, we might have seen the core hypothesis confirmed not only on a relative scale in the sense of a reduced deforestation rate, but also in absolute terms in the sense of widespread forest regrowth on abandoned agricultural land.

Indonesia

Shifting oil wealth

Booms and busts

Turning our attention now to SE Asia, Indonesia has the third largest tropical forest area on the planet (FAO 2001c). However, the world's largest archipelago contains a variety of forest types, from the very densely populated, forest-scarce inner islands (such as Java, Bali, Lombok and Madura) to rapidly advancing frontiers of forest-clearing (Sumatra and Kalimantan) and remote, still widely forested regions like West Papua (Irian Jaya). Indonesia is also a significant petroleum exporter (oil and gas), and has been a member of OPEC since 1962. Royal Dutch started to develop petroleum activities as early as 1885. In the 1950s, Caltex began developing giant fields in central Sumatra, but oil production only took off after a production-sharing agreement was signed with foreign companies in 1967. This followed the transition from President Sukarno's 'Guided Democracy' to President Suharto's pro-business 'New Order' regime. Production expanded rapidly from 1 million barrel/day in 1972 to 1.7 million barrel/day in 1977. In the 1980s, petroleum production was slightly reduced, to 1.4 million barrel/day in 1988. But following the discovery of a number of minor fields, production has resumed levels fluctuating between 1.5 and 1.7 million barrel/day in the 1990s. Currently known oil reserves are about 5 billion barrels, around 80 per cent of which is on-shore; 70 per cent of this is in the central and northern parts of the island of Sumatra (Riau and Aceh). In addition there are large natural gas reservoirs, and Indonesia is a large exporter of LNG products, which in 1998 even surpassed the value of crude oil exports (Warr 1986: 314; Gelb with Glassburner 1988: 197; Amigransa *et al.* 1997: 48–50; Bravo 1999; EIU 1999e: 61).

Like Nigeria, Indonesia fully enjoyed the benefits of the first oil-price hike in 1973–4. The export share of petroleum was 32.8 per cent in 1970, rising to 71.9 per cent during the second price hike in 1980. However, unlike in Nigeria the share then declined to 44 per cent in 1990 and 25.8 per cent in 1996 (UNCTAD 1999: 122). This had to do with a number of policies that promoted the growth of other exports (see below). But also Indonesia never acquired an extreme dependency on oil. Oil exports were only US$62 per capita at the peak in 1979, compared to an average of US$778 for five comparable high-absorbing oil economies (Nigeria, Algeria, Ecuador, Venezuela, and Trinidad and Tobago). On the other hand, with a per-capita income of only US$200 in 1974, Indonesia was also clearly the poorest of the oil countries (Gelb with Glassburner 1988: 197). On these grounds, therefore, we should expect booming oil incomes to have an important impact on budgets and absorption, but triggering fewer structural changes in employment than in countries where the per-capita size of the boom was ten times greater.

Environmental effects

Most of Indonesia's oil is found in forested areas. Has expanding oil production itself had a large impact on forests? As for Mexico and Nigeria, it seems that direct deforestation impacts have been very limited. With respect to forest degradation, the environmental record appears mixed, although the topic is under-researched.[10] For a positive example, the French firm TOTAL has responded to earlier environmental impact assessments in a way that adopts best-practice technologies to minimise both clearing and degradation effects in its operations in the mangroves of the Mahakam delta in East Kalimantan. In stark contrast to the Shell example in Nigeria, previously cleared sites seem to have been reforested, and current pollution impacts are negligible; TOTAL has received an environmental award in recognition of its achievements (IPIECA 2002b). Caltex, the largest producer, has also taken steps to reinject a greater share of produced water and to apply new drilling methods that reduce direct deforestation impacts (IPIECA 2002a). However, the same source also acknowledged that the oil industry has played a prime role in providing roads, bridges and other infrastructure to central Sumatra since the 1950s, thus providing key support to economic development that enabled deforestation. For some companies, such as Saga's involvement in the 94,000 ha Merang block in Jambi (Sumatra), direct degradation impacts were apparently understated, especially pollution with mercury, zinc and cyanide (Loraas and Eraker 1999).

The macroeconomic response

In many multi-country comparisons of oil countries, Indonesia is referred to as the rare success story, demonstrating timely and adequate policy responses to both boom and bust (Neary and van Wijnbergen 1986; Pinto 1987; Gelb 1988; Scherr 1989; Bevan *et al.* 1999b). In general, there are three reasons for this. First, boom-led fiscal expansion expended most of the windfall, but Indonesia did not borrow further against its oil revenues to exacerbate or prolong its public spending spree – it actually reduced its foreign debt slightly. The GDP share of public external debt declined from 33.5 per cent in 1973 to 21.1 per cent in 1982, and in 1982 the debt-service share in exports was very low (8.3 per cent) compared to other oil countries (Warr 1986: 289). Second, while Indonesia did face strong monetary expansion and inflation, the government significantly devalued the rupiah three times in order to realign relative prices and protect T sectors from decline. Third, agriculture had a very high priority in government policy, which was less urban biased. These three factors in combination meant that the costs of adjustment to the oil bust of the 1980s were reduced, and economic growth and other welfare measures improved markedly for the period as a whole, more than for other oil-exporting countries. Growth rates had already been high in the pre-boom period of 1967–73, which was characterised by economic recovery. They accelerated during the boom (1970–82) to 5.4 per cent per capita (Warr 1986: 288). But even after the oil boom proper, between 1980 and 1998, private per-capita consumption grew at an average rate of 4.6 per cent. Other welfare measures also improved markedly: the mortality rate of children under 5 years fell from 12.5 to 5.2 per cent, life expectancy at birth was raised to 65.5 years (World Bank 2001a: 276), and primary education levels improved nationwide, in particular for females (Hill 1992: 357).

However, behind this rosy picture of sustained growth based on sound economic management and fiscal discipline, fierce political struggles arose over the control of oil revenues, with a shifting balance of power and alternate policy directions over time. On the one hand, the technocrats of the so-called 'Berkeley Mafia', a number of US-trained PhD economists centred on the Ministry of Finance and the National Development Planning Agency (BAPPENAS) who came to dominate economic policy-making for decades, pressed for cautious spending and continuous economic diversification. On the other hand, some interest groups pushed for much higher spending on government projects, favours and benefits. These ranged from the military to government bureaucrats, parastatal enterprises, native *pribumi*[11] capitalists and President Suharto's family and friends. Winters (1996) shows how Indonesia's economic policy under Suharto generally followed a stable pro-growth, pro-competitiveness path that accommodated the interests of 'mobile capital', that is, of foreign companies and ethnic Chinese entrepreneurs. But during the specific periods of peaking oil revenues, free-market, open-door interests and the technocratic 'economists' were temporarily sidelined, and vested interests managed to influence state policies more towards the goals of nationalism, indigenisation, politicisation and tight regulation, which also involved deficit spending (Bevan *et al.* 1999b: 241–3). An illustrative example was Pertamina, the state oil company. Its top man, General Ibnu Sutowo, a close friend of Suharto, ran his own development budget by controlling some of the oil revenues and having autonomous access to foreign borrowing. Mismanagement and a company debt growing to US$3 billion eventually led to the insolvency of Pertamina in 1974 (Winters 1996: 81–90; Ascher 1998), and the rescue package cost the Indonesian state US$1 billion – a non-trivial share of the entire oil windfall for 1973–4 (Winters 1996: 106). Yet politically, this event also helped the 'economists' to restore control (Bevan *et al.* 1999b: 389–92). During the second price boom of 1979–80, a new competing planning unit under the State Secreteriat known as 'Team 10' distributed most of the oil revenues in a way that was characterised by non-competitive tenders, patronage, inflated project costs and corruption (Winters 1996: 125–42). Yet the economists from BAPPENAS managed to use currency devaluation as a main tool to correct macroeconomic imbalances.

Some categories stand out as the main public-spending priorities oil-boom period (1973–81):

- Physical and social infrastructure in both urban and rural areas
- Agricultural investments and subsidies
- 'Strategic' investments and prestige projects
- Public employment, administration and the military.

Like most oil countries, Indonesia invested heavily in infrastructure. In 1970, more than four-fifths of the population lived in rural areas, so while a lot of the spending was urban, it was considered politically vital to improve living conditions in the countryside. A major focus was on primary schools, but physical infrastructure also received additional funding (Gelb with Glassburner 1988; Scherr 1989: 556). Protests over high consumer prices for rice had been one of the main factors undermining Sukarno in 1966–7. Furthermore, the unresolved issues of food production and poverty alleviation represented a latent threat to Suharto's regime and to foreign donors' interests. To make the country self-sufficient in the production of rice therefore became a political goal for the Suharto government (Sunderlin 1993: 60–96). Large investments were accordingly made in rice intensification programmes, especially on densely populated Java and Bali, which the government also supported by subsidising fertiliser and pesticide prices. At its peak in 1974–5, these subsidies corresponded to 11.5 per cent of total public expenditure and 2 per cent of GDP, though they had to be cut back in the 1980s when oil revenues declined (Osgood 1994: 218–19; Bevan *et al.* 1999b: 360–1) Total public spending on agriculture rose to 22 per cent in 1979–80, a share unrivalled by any of the countries in this book (Scherr 1989: 556).

In other aspects, Indonesia's spending was closer to the norm of high-absorbing oil exporters. The country tried to establish or revive steel and aluminium industries, which were seen as strategic for economic development, although they turned out to be less successful than expected. It also spent money on investments of a more extravagant nature, such as a satellite and other telecommunication equipment that was purchased for US$ 1 billion in 1975 (Winters 1996: 106). Finally, public-sector employment, administration and the military[12] also increased their expenditure in Indonesia, as they did in other oil countries. Their combined share in a rising budget grew from 6.3 per cent of GDP in 1971 to 9.2 per cent of GDP in 1980–1 (Bevan *et al.* 1999b: 358–61).

What did these spending priorities mean for land use? A main observation is that there was less of an urban or regional bias (Scherr 1989: 552). Rural areas and agriculture were clearly among the budgetary beneficiaries of the oil boom. The mix of policies favouring both agricultural expansion and land-intensification makes it hard to predict what land-use impact agricultural policies had. But probably more important than the specific interventions was the handling of the 'core effect' of real currency appreciation. Indeed, competitiveness declined, especially following the first oil-price hike in 1973. In that year, the oil boom coincided with a severe drought that increased domestic food prices, and temporarily the rupiah appreciated in real terms by more than 20 per cent. Another period of appreciation came in response to the second price boom, from 1979 to 1982. However, in both cases significant rupiah devaluations interrupted this trend, in 1978 and 1983 respectively. When oil prices dropped in 1986, a third large devaluation brought the rupiah down to almost half its real value in 1973 (Warr 1999: 334–5). The RP of non-tradables thus rose temporarily, but the devaluations forced it rapidly back to the downward trend followed since 1967, accompanied by a large reduction in import protection levels, which opened up the economy.

One special characteristic is thus that the nominal exchange rate in Indonesia was actively used to protect the price competitiveness of the traded, exposed sectors of the economy, even in advance of balance-of-payment difficulties (Bevan *et al.* 1999b: 256–7). For more than a decade after 1986, with reduced dependence on oil and diversified foreign-exchange inflows, the real exchange index remained stable, within a range of 10 per cent variation up and down. But in 1997 the Asian financial crisis hit Indonesia, being followed by a 32 per cent drop in the value of petroleum exports in 1998 (EIU 1999e: 61), and much political instability in the aftermath of President Suharto's fall. The RER dived abruptly from 103.80 in July 1997 to 33.68 in January 1998 (December 1987 = 100). It then slowly stabilised at a level between sixty and seventy over the following two years.[13] On the whole, therefore, we are looking at a rupiah real depreciation during 1967–71, fluctuations in the 1970s, strong resumed depreciation in 1983–6, stability in 1987–97 and a marked depreciation followed by high volatility during 1998–2001.

What was the production response to this pattern of RP changes? The most obvious trends prior to the oil boom (1967–72) were a relative decline in agriculture and growth in manufacturing. The boom delayed somewhat the growth of manufactured exports that occurred in other ASEAN countries, which picked up in Indonesia in the 1980s and 1990s. As expected, the GDP share of NT sectors (services and construction) rose rapidly during the 1970s, while their growth was dampened in the 1980s and 1990s. Timber production experienced a rapid rise in the 1970s, with harvesting rates doubling between 1970 and 1975.

To return to agriculture, did RER changes feed directly through to producer prices, or did trade policy moderate the effect? In Indonesia trade policy had a less disruptive influence on price competitiveness, and over time net effective protection rates became gradually less biased against agriculture (Timmer 1994: 2–4). In the semi-traded food-crop sector, there were intermittent quantitative restrictions on maize imports, whereas rice prices and imports remained heavily controlled (Scherr 1989: 551). During the 1970s, according to one source, 'agricultural performance was on the whole impressive, especially for food crops' (Gelb with Glassburner 1988: 215), although another source says that 'the rate of decline in agriculture's contribution to GDP was quite high by international standards' (Warr 1986: 307). The quotes would seem contradictory, but the first refers to absolute and the second to relative performance. Warr (1999: 341–2) finds a statistically significant negative correlation between the GDP shares of oil and agriculture (the latter expressed as a deviation from long-run trends of decline). In other words, the temporary loss of competitiveness accelerated the relative decline of agriculture within a fast-growing economy under rapid structural change. Correspondingly, agriculture responded favourably to the real depreciation after 1986, especially cash-crop exports. But in absolute terms, even from 1971–2 to 1980–1, production volumes still grew annually at 1.8 per cent for rubber, 6.6 per cent for palm oil and 13.9 per cent for coffee. Output of the main food crop, rice, almost doubled from 13.1 million in 1970 to 22.8 million in 1982 (Gelb with Glassburner 1988: 210–16). Maize production increased by 50 per cent and cassava by 25 per cent. Total agricultural production grew at 3.8 per cent annually (1.5 per cent capita) during 1970–82. As a result, the rural population also declined by only 4 per cent during that period (Scherr 1989: 547), an outcome that is singular among the high-absorbing oil countries. In particular, rural population density in the Inner Islands was already so high that rural out-migration to booming urban businesses did not greatly raise real wages, and there was no disastrous labour-cost impact on agriculture (Scherr 1989: 554).

The forestry sector had a larger impact in Indonesia than in any other country in this book, both for the economy and for forest conditions. On the one hand, huge commercially rich dipterocarp forests on the Outer Islands were, prior to the 1970s, basically untouched. They thus provided a large, untapped economic potential, which was aggressively exploited by the Suharto regime. After 1967, the New Order government distributed over 60 million ha of timber concessions to private companies (Barr 2001: 19–36). Unlike for agriculture, the oil-wealth impacts were insufficient to hold back timber expansion, while the sequence of protective rupiah devaluations may have played a role and an urban construction boom raised domestic demand (Osgood 1994: 217), probably the main factor was that concessions and other incentives offered for investors made it too favourable for the competitiveness effect to stop the process. New roads and other infrastructure facilitated the expansion, especially in Kalimantan. Timber rents were simply too large, and the quest for harvesting them caused forestry to 'run ahead' of land conversion for crops and thus become a leading sector in deforestation. By 1979, Indonesia was the world's largest producer of tropical logs, it exported roughly three-fourths of its production, and the annual export value reached US$2.1 billion, corresponding to a world-market share of 41 per cent (FWI/GFW 2002: 24–6). In the 1980s, a ban on the export of raw logs promoted a successful diversification into the plywood sector; in the 1990s the nation's pulp and paper industry expanded sevenfold (Barr 2001: 19–24).

For Indonesia's mild and tightly managed version of the Dutch Disease, we can therefore mainly say that oil wealth temporarily held back agricultural expansion, compared to what would have happened otherwise. But nor did it impede absolute growth in agriculture or the rise of logging. Depending on changes in land productivity, we would also expect cultivated area to have continued to expand during the whole period, though at a lesser rate in the 1970s and at a higher rate in the 1980s and 1990s, especially after 1997. As most of Indonesia's original vegetation was forest, we would also expect these fluctuations in cultivated area expansion to be reflected in corresponding deforestation patterns.

What happened to forests?

Assessing the Indonesian forest situation is made particularly difficult by the heterogeneous situation within the archipelago, as described above. But also, only a few countrywide forest assessments have been made over time, and transparency concerning existing government data has been low. In a historical context, in 1950 162.3 million ha were forested, corresponding to 84 per cent of land area. Around 1985, that had been reduced to 119.7 million ha (63 per cent of land area). For the early 1990s, an inventory estimated forest cover (including bush and scrub) at 121.2 million ha (64 per cent). Finally, in 1997 a forest area of 100 million ha (50 per cent of land area) was recorded.[14] However, not even these four basic assessments, on which most other recalculations and reinterpretations are based,[15] are compatible, due to varying scales and definitions and missing data. For instance, forest cover did not really expand between the mid-1980s and early 1990s, because the inclusion of scrub in the latter (National Forest Inventory, NFI) drove up the estimate; rather yearly deforestation during these years was probably 1,200,000 ha (0.9 per cent) (Scotland *et al.* 1999). Direct overlay of the 1985 and 1997 maps shows that almost 17 million ha is not comparable, meaning that almost 18 per cent of the 1997 forest-stock estimate is in fact not matched by comparative data (FWI/GFW 2002: 10). Also, the 1997 map in Holmes (2000) was subjected to little ground truthing, making it more preliminary than the 1985 assessment. At the same time, two other estimates with different forest definitions, coverage and methodologies, were published for 1997, one significantly lower, one much higher than Holmes's figure.[16] What makes it further difficult to estimate yearly deforestation rates is that many assessments are based on imagery from several years rather than a single year.[17]

Taking into account these common deficiencies, one can attempt to correct for several inconsistencies. Like Sunderlin and Resosudarmo (1996) and Muhamad (2002), one can also try to integrate evidence from studies of sub-national coverage. By taking these two steps, at least a tentative picture of changes over time emerges. First, taking the historical 1950 and 1985 figures at face value, average annual forest loss between these two years was 1.2 million ha (0.8 per cent). That is about average at the global deforestation level, but moderately low in per-capita terms, that is, compared to Indonesia's large and growing population. We do not have forest cover figures to isolate the pre-boom period of 1950–70 within this long time-span. Sumahadi *et al.* (1997) compared Landsat images from 1972 and 1990 (excluding Java) and found a deforestation rate of 1.0 per cent, which would seem to imply that deforestation was higher in 1972–90 than in 1950–72.[18] However, Sunderlin and Resosudarmo (1996) reviewed a variety of estimates for roughly the same period, reporting estimates that lie in the wider range of 0.6–1.2 per cent.

While the evidence on pre-boom deforestation is inconclusive, there is some support for my suggestion that forest loss was somewhat lower in the oil-rich 1970s than in the 1980s and 1990s. For example, the Ministry of Forestry and FAO (1990) estimated that smallholder forest conversion rose from 300,000 ha in the 1970s to 600,000 ha/yr in the 1980s. The same source sets total forest loss for 1982–90 at 1.3 million ha (1.2 per cent). Hasanuddin (1996) reports a very high annual deforestation estimate of 2.4 million ha for 1982–93 (based on Landsat images), but this also defines logged-over forests as 'deforestation'. Richards *et al.* (1999) find from the TREES maps that forest loss during 1990–5 was 1.9 million ha, though this estimate is for 'Insular SE Asia' and therefore includes Malaysian Borneo (Sabah and Sarawak). Admittedly, the problems related to the forest statistics mean that one cannot provide more than tentative support to the idea of a slow-down in deforestation rates in the 1970s, and acceleration in the 1980s and 1990s.

As with the other countries considered in this book, therefore, it may be beneficial to look at agricultural (census and survey) area data to consolidate the picture of forest conversion to alternative land uses. From the statistics on five major crops (rice, maize, cassava, potatoes and soybean) published by Arif *et al.* (1996), we can calculate that average annual area expansion for all the crops was only 87,255 ha for 1971–80, but tripling to 269,460 ha for 1981–92. This is a partial but still strong indication that the expansion of cultivated area was dampened significantly during the oil boom. In the 1970s, the total expansion of food crops was dominated by rice, the main staple, expanding at 68,074 ha (0.8 per cent)/yr. This is very low compared to the doubling of rice production over the same period, achieved overwhelmingly by adopting 'Green Revolution' land-intensive technologies (see above). It was thus a period of strong food-crop intensification. In the 1980s, rice yields continued to rise, but area expansion (143,457 ha/yr) now had a much stronger role to play. Soybeans (71,311 ha/yr) and maize (56,192 ha/yr) also expanded much more than in the 1970s, where they had been static. Most of this expansion occurred in the Outer Islands, and mostly under land-extensive systems of production. For instance, the area under soybean in previously dominant Java stagnated, while in the Outer Islands it grew by 6.8 per cent annually during 1980–98 (Erwidodo and Hadi 1999: 29). For maize and cassava, the trend was similar though area expansion was less pronounced (ibid.: 39, 51). The land-use data, which are probably on average of a higher quality and consistency than forest statistics, thus provide additional support for the hypothesis that forest conversion picked up significantly in the period following the oil boom.

What were the factors that determined the shift in emphasis over time from food-crop intensification to extensification? On the one hand, subsidies for inputs needed for intensification were cut, and a sharply depreciated rupiah made imported inputs more costly. On the other hand, various incentives and opportunities for extensification were increased in the 1980s. Expanded industrial timber extraction was one factor. In spite of the 1985 ban on the export of unprocessed logs, there was an accelerated rise in log production in the 1980s. Whereas the 1970s were characterised by booming log exports, in the 1980s Indonesia became the world's main plywood producer (Barr 2001: 24–8). The expansion of forest-based industries was aided by policies designed to stimulate foreign investment and to capture rents destined for patronage allocations that would help the Suharto regime to maintain political support (Ascher 1998; Ross 2001: 166–89). While selective logging of Indonesia's dipterocarp forests normally causes degradation rather than deforestation, it indirectly provided access to settlers and investors who converted the forest to agricultural uses (e.g. Barbier *et al.* 1994: 245). This has also been a main cause of the recurrent fires, as in 1982–3 (burning a 2 million ha land area), 1991 (500,000 ha), 1994 (5 million ha) and 1997–8 (10 million ha) (FWI/GFW 2002: 53–8). Note, though, that these figures refer to the total amount of vegetation burnt, not only to forests. In contrast to selective logging, the further diversification of forestry into pulp and paper from the late 1980s onwards caused direct deforestation, basically because this process harvests all the wood from the forest. Huge investments in the industry, the rapid expansion of production capacity, and a forest plantation scheme that lagged behind these other trends meant that, in 1988–2000, around 90 per cent of the raw material for pulp and paper came from natural forests, and only 10 per cent from plantations. Pulpwood caused an accumulated loss of approximately 900,000 ha of natural forest (Barr 2001: 60), though part of this was reforested by monoculture tree plantations. Finally, in the 1990s the plywood industry shifted increasingly from selective harvesting to staged clear-cutting, which satisfies FAO's deforestation criterion (C. Barr, personal e-communication, 4 September 2002).

Another land-extensification factor was the tremendous expansion of the government's transmigration programme in the 1980s. 6,570 families had been moved yearly in 1950–79, but in 1980–4 the average figure rose more than tenfold to 73,200 families (Sunderlin and Resosudarmo 1996: 7). This periodisation undermines the hypothesis that oil revenues helped push forward the transmigration programme; larger World Bank credits in the 1980s may have been important sources of funding. To the extent that Indonesian transmigration mostly moved people from densely populated, non-forested areas to thinly populated areas with large forest cover, this contributed to accelerated forest loss. One government estimate is that 1.7 million ha of agricultural land were opened up between 1969 and 1993 (cited in FWI/GFW 2002: 48–9). However, there are caveats implying that the deforestation impact of transmigration may frequently have been overestimated.[19] As with logging, the expansion of transmigration projects in the 1980s was also associated with much more road-building, which decisively enabled additional land-clearing to take place. This is observed in case studies, but it is also indicated first and foremost by the fact that the length of the Indonesian road network more than doubled in the 1980s and 1990s.[20]

In addition to forest conversion for food crops, several tree and estate crops also became much more important in the 1980s and 1990s, both in the form of industrial plantations and for smallholders. Indonesia is the world's largest producer of palm oil and natural rubber, the third largest of cocoa and the fourth largest of coffee. Obviously, these largely export-oriented sectors were greatly assisted by the depreciation of the exchange rate in the 1980s and 1990s. Oil palm, the largest 'consumer' of converted forestland among these crops, basically existed in the 1970s only as state-owned plantations covering a relatively small area. But both industrial and smallholder production (often in outgrower schemes) progressively picked up after 1983–4. The highest yearly expansion rates occurred in the last half of the 1990s: in 1995–8, planted area grew at 220,000 to 270,000 ha each year. The total area under oil palm is now passing the 3 million ha threshold (Casson 2000: 9, 40). Rubber covers about 4 million ha, 80 per cent of which are occupied by smallholders. Copra plantations make up at least 2.7 million ha, coffee 1.1 million ha (FWI/GFW 2002: 49–50). The area planted to cocoa, a mere 17,498 ha in 1975 and 37,082 ha in 1980, expanded tenfold in the 1980s to 357,480 ha in 1990, and further to around 600,000 ha by 1996 (Quane 2002). Smallholders in Sulawesi planted most of this cocoa, in most cases clearing forests to do so (Murray Li 2000). Simply adding up the area on an (incomplete) list of estate crops and perennials yields a conservative accumulated total of 11–12 million ha, the bulk of which was established in the 1980s and 1990s.

In the previous subsection, it was expected that the 1997 financial crisis, the post-1998 political uncertainty and marked capital outflows, combined with relatively low oil-export revenues, would lead to higher deforestation and forest degradation. This was basically because the strongly devalued rupiah was expected to favour increased timber exports and conversion to agricultural 'tradable' crops. But potentially return migration could also contribute: large numbers of people being laid off in crisis-hit urban sectors would return to the countryside and put more pressure on natural resources. From 1997 to 1998, 'employment in the agricultural sector rose by 15.2 per cent as redundant urban employees returned home to work in family fields' (EIU 1999e: 14).

Again, Indonesian forest statistics do not provide us with definite answers on land-use impacts, but some indicators may confirm the trend of a recent rapid increase in forest-clearing. Different interpretations of the study by Holmes (2000), cited above, put annual forest loss between 1985 and 1997 at somewhere between 1.2 and 1.8 million ha (Muhamad 2002: table 1). According to an interview with a senior official from the Ministry of Forestry and Estate Crops (MoFEC), preliminary analysis of a new forest inventory shows that current deforestation 'may well reach 2 million ha'.[21] Holmes himself suggested, based on a comparison of his data with the NFI data, that forest loss since 1996 must have been 'well over 2.0 million ha per year' (Holmes 2000: 4). His observation is supported by a recent synthesis assessment (FWI/GFW 2002). Using their own primary data, Sunderlin *et al.* (2000) found, in a 1999 sample survey of 1,050 farmers in six of Indonesia's Outer Islands, that land-clearing increased slightly from 1996–7 to 1997–8 and greatly in 1998–9, that is, in the second year of crisis.

These observations need further consolidation, and it would be premature to attribute additional forest-clearing after 1997 exclusively to the effect of the devaluation of the rupiah on price competitiveness. The forest fires of 1997–8 obviously played a role, too. So did the fact that many forest concessions came to an end in the late 1990s, and both companies and the Forestry Department were looking for ways to extract more revenues by pressing for a reallocation of these areas to 'conversion forest' (R. Dennis, personal e-communication, 14 June 2002). Even the economic crisis itself had contradictory impacts.

For instance, Casson (2000) finds that the severe shortage of financial capital reduced investments in new oil-palm plantations, thus reducing area expansion. Similar financial constraints applied to the pulp and paper industry (Barr 2001). Sunderlin *et al.* (2000) found from their smallholder survey that both farmers who had benefited from devaluation *and* those who had become worse off expanded their cultivated area. They also planted more cocoa and rubber, their traditional cash crops, in spite of the fact that rubber prices had actually gone down, while pepper had become much more profitable. These results, which seem counter-intuitive at first sight, underline the fact that capital-shortage effects and risk-averse product diversification under conditions of high uncertainty and volatility may be just as useful as the real exchange rate in understanding changing land-use strategies. Return migration to the Outer Islands also seemed more limited than expected.[22] Another very important factor after 1997 was the collapse of government authority and the inability of the security forces to prevent unauthorised forest exploitation and colonisation. After 1997 government support for road construction collapsed, with beneficial consequences for forests (W. Sunderlin, personal communication, Bogor 23 May 2002). In other words, a closer examination reveals that the causes of the accelerated forest loss were much more complex.

Conclusion

Indonesia is an extraordinary oil-country case in many respects. It was the poorest, but also the most populous exporter country at the beginning of the oil era in the 1970s. The per-capita size of its oil boom was much smaller than for either Mexico or Nigeria. It managed the boom more prudently than any comparable developing oil country, basically by applying three policy principles:

- maintaining fiscal balance (including no net increase in foreign debt);
- spending greatly on agriculture; and
- protective currency devaluations.

This combination of limiting preconditions and restrictive policy responses meant that Indonesia had a very mild attack of the Dutch Disease, or what Indonesians sometimes joke was a 'Double-Dutch Disease', the first one being the legacy of Dutch colonial rule (Warr 1986: 288). Compared to other oil countries, basic economic structures like rural–urban migration, sectoral employment etc. were affected more marginally. In the 1970s, the economy basically continued its long-term path of export-led growth and structural change, though with some modifications. Among the changes that had most relevance for forests, one was that some export crops were hampered by the temporarily appreciated rupiah, while new ones (e.g. oil palm, cocoa, soybeans) only took off in the 1980s, once price competitiveness had been restored. A second change was in the semi-traded food-crop sector: rice production doubled, while the area of rice cultivation rose less than 10 per cent, reflecting the strong government-supported programme of intensification. Both these changes curtailed expansion into forests, compared to what was to happen in the 1980s. Although rice intensification occurred mostly in the Inner Islands, where little natural forest is left, the scale of the rise in food production and the lower national price of rice meant that pressures to convert land for food cultivation also diminished greatly in the Outer Islands. While there are tentative indications of this slow-down in the forest statistics, this is better reflected in agricultural land-use figures. On the downside for forests, the oil boom and RER appreciation did not impede the beginning of the timber bonanza in the 1970s, although the subsequent expansion of forest-based industries in the 1980s and 1990s was to be even stronger. Logging, transmigration, new roads and cash crops turned into the new motors of land extensification and deforestation in the 1980s and 1990s. The government actively supported this process through a deliberate strategy of cashing in natural resource rents and diverting them to other sectors.

Can we thus conclude that Indonesia's persistent efforts over the last three decades to maintain a competitive exchange rate have been good for the economy but bad for forests? Looking at the aggregate picture, this is probably close to the truth. The strategy of openness and export-based growth was indeed both very successful in development terms and harmful to forests, at least in combination with the other policy incentives for land extensification that were applied. However, one important caveat is that Indonesia also had a sizeable urban labour-intensive sector that depended on exports: manufacturing. The competitive rupiah was thus equally crucial to absorb labour in the cities – labour that otherwise might have depended on the exploitation of natural resources, including forest conversion and degradation (Kaimowitz *et al.* 1998: 61). This distinguishes Indonesia from all other country cases in this book, where the urban economy was largely dominated by NT sectors with a clear stake in maintaining an overvalued currency.[23]

As in Mexico and Nigeria, food security had a high priority in Indonesian policy. Food crops are also produced largely by smallholders, so their decline would have strong social impacts. For these reasons, they were at least partly import-protected (semi-tradables) in all three countries. In the larger economies analysed in this chapter, home-market sectors obviously make up a large share of total land use, and trade policies thus become an important influence over land use and forest outcomes. While in Nigeria, and especially in Mexico, food crops continued to cause some deforestation because food production grew with food demand and was sheltered by import controls, this effect was minimised in Indonesia. This was not because food crops were so much less protected, but because the massive intensification of rice cultivation 'divorced' rising production from deforestation fairly effectively.[24] Obviously, a precondition for this process was the rich volcanic soils of Indonesia's inner islands, which were able to support these intensive systems. During the boom there was no policy-led drive for massive road-building, establishment of pastures or land extensification (as in Mexico), nor an oil sector causing as significant a degree of forest degradation as in Nigeria's mangroves.

On the other hand, rural areas were not sidelined in terms of infrastructure and investments, as happened in the two other countries, where there was a clear urban bias. Export crops were in no way sacrificed, as they were in Mexico and Nigeria, where most of the core land-use effects manifested themselves in terms of a drastic decline in cash crops. In Indonesia, this did not happen to any comparable extent. This had fundamentally to do with the 'core effect' of a more controlled and limited appreciation of the real exchange rate. Especially in the 1980s and 1990s, Indonesia became a more open economy, thus also resembling Mexico and Nigeria less. This implied that the cash-crop export sector kept expanding its demand for cultivated land, thus also putting pressure on forests.

Finally, as a note of caution on Indonesia, one should not forget the high level of uncertainty that surrounds deforestation, where 'opaque statistics make it virtually impossible to determine exactly how much forest remains' (EIU 1999e: 31). In particular, Indonesian deforestation is characterised by the riddle that 'the numbers don't add up': the recorded total of converted land uses is significantly lower than the forest area that seems to have disappeared when interpreting the satellite imagery. For instance, Dick (1991) systematically estimated separate deforestation contributions from seven different forest-loss sources and agents, but his total of 623,000 ha/yr during 1979–89 is dramatically lower than any other deforestation estimate for the 1980s. The recent Forest Watch state-of-the-art assessment (FWI/GFW 2002: xi–xii) is concerned with the same question: more than 30 million ha forests are supposed to have disappeared in Indonesia between 1980 and 2000, but where has all this forest gone?

This is not the place to get to the bottom of the puzzle, but at least we can make some guesstimates. The description above showed that estate crops and perennials may account for a fair share (at least 12 million ha),[25] forestry (creating plantations and pulpwood harvesting) may explain about 3 million ha,[26] and food crops might perhaps add another 7–8 million ha (extrapolating the expansion rates calculated above). Contained in the latter would be the expansion of shifting cultivation, set at 4 million ha for 1985–97 by FWI/GFW. On the whole, it becomes clear from this back-of-the-envelope calculation that at least one-quarter of total forest loss is not explained by any new land use. This is even more the case when considering that some double-counting between forestry and agriculture sources has occurred, as when an area is first cleared for pulpwood, and then planted for oil palms, but with both deforestation sources being counted simultaneously.

To some extent, the numerical discrepancy may reflect the fact that the pace of conversion is actually falling short of clearing and creating idle lands. It is obvious that forest-based industries and estate-crop companies are more important (direct and indirect) deforestation actors in Indonesia than in any other country analysed in this book. They gradually acquired this role over the last decades of a Suharto regime that as a basic rule attempted to please 'mobile capital' through cautious macroeconomic strategies, price competitiveness, generous company operating-conditions and overall political stability.[27] The management of the oil boom is indicative of that strategy. One land-use implication is that deforestation is currently 'running ahead' of the demand for converted land. A concessionaire's speculative motives in gaining control over rich timber rents may lead to forest-clearing well in advance of implementing any intended alternative land use. The showcase is when companies receive permits to convert forest areas to oil palms, but never actually plant after harvesting all valuable wood from their concession area. However, it would still seem to require quite heroic assumptions about large pre-clearing on behalf of estate-crop and timber-plantation concessionaires to explain the numerical gap between the alleged 1980–2000 deforestation and alternative land uses from this source alone (FWI/GFW 2002: xii).[28] Other candidates for explaining the area difference are abandoned agricultural areas (e.g. those covered by *alang-alang* grasslands), and notably forests that have been burnt reiteratively so that they are no longer 'counted' as forests.

Still, to some degree a *fata morgana* caused by poor data consistency and inadequate interpretation might easily apply as well. As Sunderlin and Resosudarmo express it (1996: 3), 'assessments of the extent and causes of deforestation in Indonesia are at best "semi-educated guesses" '. Apparently, it does not take too many changed assumptions or reclassifications of the RePPProT or Holmes maps to alter the size of national deforestation substantially. Potentially, the large deforestation numbers may be due in part to the inconsistency of a small number of deforestation figures. Thus there seems to be great scope for improved analyses of satellite imagery over time – country-wide or at a detailed case-study level – which could fill in missing cells in the land-use transition matrix, and help resolve the puzzle.

Summary notes

To supplement the primary case studies in Chapters 4–8, in this chapter, summarised descriptions of the cases of Mexico, Nigeria and Indonesia have provided three additional 'big-country' and, for two of them, 'forest-poor' perspectives on the land-use and forest impacts of oil wealth. In distilling the analyses, there seem to be in particular three observations to emphasise before turning to the concluding chapter:

1 the vital importance of protective trade policies, oriented towards food security and smallholder protection, in sheltering domestic food crops from declining competitiveness (in all three countries);
2 land extensification leading to higher forest pressures from these food crops in two countries (Mexico and Nigeria), but with intensification slowing it down in the third (Indonesia);
3 a complete sacrifice of cash crops that partially alleviated deforestation in Mexico and Nigeria, versus currency devaluation and rural investment in Indonesia that favoured a continuously aggressive export-led expansion of forest-based industries, estate and cash crops, at the expense of forest cover.

Notes

1 Apparently, at the time oil reserves were overestimated, which helped to justify the highly expansionary fiscal and foreign indebtedness strategy (M. Musacchio, personal communication, Bogor, 25 April 2002).
2 Feltenstein's RER index is based on wholesale prices and a (non-specified) basket of trading partners. Alternatively, the bilateral peso–US$ index in Gavin (2000: 186), which is based on the consumer price index, shows a less drastic real appreciation and a more drastic real depreciation after 1981.
3 Collier calculates a bilateral US–Nigeria index, not a trade-weighted one, and uses Nigerian consumer prices and US wholesale prices.
4 The parallel market premium was 21 per cent in 1973, rising to 42 per cent in 1978 and 82 per cent in 1983 (Oyejide 2000: 443). A large spread between multiple rates was preserved throughout the 1990s, but by the end of 2000 it had been reduced to about 10 per cent (IMF 2001: 54).
5 For instance, government revenues from petroleum rose sharply between 1994 and 1997. Thus the naira RER appreciated by about 50 per cent between 1995 and 1998, but then depreciated by 10 per cent in 1999 (IMF 2001: 4–7).
6 In 1980–98 as a whole, private per capita consumption in Nigeria has *declined* by an average of 4.2 per cent/yr, a negative record that is only rivalled by war-torn countries such as Angola and the Democratic Republic of Congo (World Bank 2001a: 276–7).

7 The original source cited by the FAO is Geomatics International Inc. 1996. 'The assessment of landuse and vegetation changes in Nigeria between 1978–1993/5. Environmental Management Project'.

8 In more detail, Fiebach reports from the Ife-Calabar road (southeastern Nigera) that 'widespread forest regrowth ... seems to tally with what I could observe riding to Calabar in 1983 – many miles of "recuperating" rather thick, lush "bush", on soils that had been used for palm kernel and cocoa export/large scale crop agriculture' (J. Fiebach, personal e-communication, 24 April 2002).

9 Of course, even under the most liberal import regime, food crops in remote areas would still have been 'naturally' protected by high transportation and distribution costs, as mentioned above.

10 For further study references, I contacted a Jakarta-based environmental NGO specialising in oil and mining impacts, Jaringan Advokasi Tambang (JATAM). They responded that very little quantitative information existed on the forest impact of oil companies in Indonesia (T. Glynn, personal e-communication, 3 May 2002).

11 This term is normally used for the class of indigenous Indonesian investors, as distinct from the country's ethnic Chinese investors (Winters 1996: 42).

12 Beyond spending on the armed forces proper, the military played an important role in the civil economy, running its own projects and operations, as with its link to the state oil company, Pertamina. It was widely accepted that the military had the right to appropriate a share of oil revenues without interference from the state. This extra-budgetary allocation was accomplished mainly by sub-declaring the amount of oil revenues received by Pertamina (Ascher 1998).

13 I am grateful to Vikram Nehru, Lead Economist at the World Bank office in Jakarta, for providing this unpublished time series from their database.

14 These estimates come from, in chronological order (see FWI/GFW 2002 and Muhamad 2002):

   1 Hannibal (1950 reproduced in FWI/GFW 2002: 8);
   2 RePPProT (1990 – cited ibid.: 10–2 and in Muhamad 2002: table 1), reflecting different years around 1985); figures revised in 1996 by WCMC;
   3 National Forest Inventory data from the GoI and FAO, reflecting different years in the early 1990s and revised by Scotland *et al.* (1999);
   4 Holmes (2000), adopted by the World Bank and the Government of Indonesia (GoI).

15 For instance, the FAO's figure for FRA 2000 reproduces the figures from Holmes (2000); see the country data at the FAO's site *http://www.fao.org/forestry/fo/country/index.jsp?lang_id = 1&geo_id = 82* (accessed May 2002); however, it does not reproduce the cautionary notes by Holmes himself on data comparability over time.

16 MoFEC (1997, cited in Muhamad 2002: table 1) estimates forest stock at 93,424,614 ha, but this excludes Java, Bali and Nusa Tenggara Provinces. The TREES project, using low-resolution NOAA-AVHRR data, calculates national forest cover at 110,829,000 ha (Stibig and Achard 1999). Both estimates are for 1997.

17 One explanation for this is that it is virtually impossible to obtain cloud-free imagery for the same year for all of Indonesia (R. Dennis, personal e-communication, 14 June 2002).

18 In fact, the absolute estimate in Sumahadi *et al.* (1997) is only 840,000 ha, but this compares with a 1990 forest stock of only 83.2 million ha, that is, a more exclusive forest definition has been used. Two other problems are the low resolution of the Landsat images and the long time span over which they were collected (Landsat MSS originate from 1972–82; Landsat TM from 1986–90).

19 Note in particular (Sunderlin and Resosudarmo 1996: 7–9; Levang 2001):

- that the total number of transmigrant families is disputed;
- that distinguishing them analytically from 'spontaneous' migrants is difficult;
- that they were not *always* moved out of densely populated areas into forests (sometimes the reverse applied); and
- that the average amount of forest cleared by transmigrants at their new destination is also disputed.

20 For the former, see, for example, the case study by Angelsen (1997: ch. 4) from Seberida (Sumatra). Nation-wide road statistics show a continuous expansion in the total road network in the 1980s and 1990s, from 156,907 km in 1983 to 219,009 km in 1986 (IRF 1988: 19) 244,164 km in 1991 (IRF 1994: 19) and 342,700 km in 1997 (IRF 2000: 20).

21 Oesman Yoesoef, personal communication to N. Z. Muhamad, 12 November 2001, cited in Muhamad (2002: 1).

22 According to Sunderlin *et al.* (2000: 27–8), only 3 per cent of the population had migrated to the study villages in the two years following the crisis. The discrepancy in respect to the account of high return migration caused by the crisis in EIU (1999e), mentioned above, could be due to limitations in the representativeness of the village sample (see discussion in Sunderlin *et al.* 2000: 28). But it is perhaps more likely that a large part of the return migration went to rural areas on Java and Bali. To the extent that the latter explanation applies, this would probably imply little for forest cover, as the forest frontier on these Inner Islands is closed, and by far most of the deforestation pressures occur in the Outer Islands.

23 Obviously, countries like Mexico or Venezuela also had a significant urban-based manufacturing sector, but with much higher trade-protection and orientation towards import substitution, rather than exports.

24 This statement ignores the possibility that the intensive and more profitable cropping systems may eventually have spread to new areas in the 1980s and 1990s, and thus still have contributed to more forest-clearing. For a thorough discussion of the relationship between technological advances in agriculture and deforestation, see Angelsen and Kaimowitz (2000).

25 This figure referred to the accumulated area, which of course also includes those established before 1980, implying an overestimate. On the other hand, the list did not include all perennials and estate crops, which implies an underestimate.

26 This comes from the 900,000 ha assumed to have been cleared for pulpwood and 2,000,000 ha estimated to be in timber plantations (FWI/GFW 2002: xii). Of course, the latter includes areas that are covered by trees of at least 5 m height and 90 per cent crown cover, which are to be counted as 'forests' according to the FAO definition used in this book. This means that plantation-led deforestation has been over-estimated.

27 This statement is probably less true for the forestry sector, because the log export ban in the 1980s was implemented in part to push foreign ('mobile') capital out of the sector (C. Barr, personal e-communication, 4 September 2002).

28 For instance, in FWI/GFW (2002: xii) it is stated that while 9 million ha have been allocated to industrial timber plantations, only 2 million ha have actually been planted with fast-growing species, so '7 million ha of former forest land are lying idle' (ibid.). But this assumes that all 9 million ha had been fully cleared in the first place – hardly a realistic assumption. A similar caution applies to the 7 million ha allocated for estate crops, where 3 million ha are supposedly lying idle (ibid.).

10 Comparison, conclusions and recommendations

It is now time to pull together the different pieces in this final chapter of the book. The plan of the five sections is to move gradually towards the formulation of recommendations: which policies, strategies and actions might improve land-use patterns in- and outside our group of study countries, and which are likely to prove blind alleys? First, different background factors and preconditions will be evaluated across the sample to frame the validity of the conclusions. Second, we shall compare the analytical results, including a synthesis of the ten partial links between oil wealth and forests from Chapter 3. We will examine the five primary cases and, where relevant, the secondary cases from Chapter 9. Third, we shall turn to an explicit policy analysis, including some cross-cutting political-economy factors in oil countries. Fourth, conclusions and policy recommendations, from both a forest-conservation viewpoint and an economic development perspective, will be outlined. Finally, a section looking beyond mineral wealth, with parallels to related issues (debt relief, structural adjustment, etc.), will close the book.

Comparing background variables

Economic structures

To grasp fully the implications of the results from the previous chapters, it is first necessary to understand the range of different country characteristics. What types of country are we covering in the extended sample? How do they cluster in terms of different economic and forest preconditions? Table 10.1 compares various basic economic and other indicators for our study countries, including the three secondary cases from the last chapter.

In terms of both land area and population, the three secondary cases (Indonesia, Mexico and Nigeria) are the largest, but even Venezuela, with its 88 million ha land area and 24 million people, ranks as a relatively sizeable developing country. PNG and Cameroon are medium-sized in terms of land, but the latter has a much higher population. Ecuador and Gabon are smaller countries in terms of both land and population. Economic development levels in 1970, that is, prior to the first oil boom, were subject to a huge spread. Nigeria and Indonesia had extremely poor economies, at an average GDP per capita of less than US$300 (in constant 1995 prices). Cameroon, Ecuador and PNG were lower middle-income countries in the US$500–1,000 range. Mexico, Gabon and Venezuela were all

Table 10.1 Comparison of economic structures in the study countries

	Year	*Gabon*	*Venezuela*	*Cameroon*	*Ecuador*	*PNG*	*Indonesia*	*Mexico*	*Nigeria*
General									
1 Land area ('000 ha)	1999	25,767	88,205	46,540	27,684	45,286	181,157	190,869	91,077
2 Population (million people)	1999	1.2	23.7	14.7	12.4	4.7	207.0	96.6	123.9
3 GDP (million constant 1995$)	1999	5,279	76,180	9,640	17,610	4,741	199,121	349,000	30,958
4 GDP/capita (constant 1995$)	1970	3,390	4,306	508	879	870	298	2,295	264
5 GDP/capita (constant 1995$)	1999	4,369	3,213	656	1,419	1,008	962	3,613	250
Trade									
6 $\{(\text{Exports} \times \text{imports})^{0.5}\} \times 100/\text{GDP}$ (%)	1999	41	17	16	19	42	24	29	31
7 Fuel exports (current million $)	1980	1,912	18,068	429	1,563	480[a]	15,774	12,081	25,189
8 Fuel exports/GDP (%)	1980	45	26	6	13	19[a]	20	5	39
9 Fuel exports/merchandise exports (%)	1980	88	94	31	63	47[a]	72	67	97
10 Fuel exports/capita (current million $)	1980	2,767	1,197	50	196	156[a]	106	179	354
11 Forest products export value (% of merchandise exports)	1999	13.0	0.3	28.1	1.3	9.0	9.9	0.2	0.3
Land use and sector distribution									
12 Population density (people/km^2)	1999	5	27	32	45	10	114	51	136
13 Population, rural (%)	1999	46	13	52	38	83	61	26	57
14 Agriculture, value added (% of GDP)	1999	7.8	5.1	43.5	12.2	29.6	19.5	5.0	32.1
15 Population density, rural (people/km^2)	1999	2	3	16	17	9	69	13	77
16 Population growth rate (%)	1980–99	2.0	2.4	2.8	2.3	2.3	1.8	2.0	2.9

Sources: FAO (2001), FAOSTAT (2002), World Bank (2001a).

Note

a Mineral exports instead of fuel exports (PNG only).

already well beyond the US$2,000 threshold: Venezuela came first, with US$4,113, that is, more than ten times the per-capita income of Nigeria and Indonesia. Three decades later, the ranking had not changed dramatically, but there was a widely different economic performance. In 1999, Nigeria had become an even poorer country (US$250), PNG and Cameroon had progressed only marginally, while Indonesia had entered the middle-income group, thanks to rapid economic growth. The more diversified Mexican economy and the two oil countries *par excellence*, Gabon and Venezuela, are still the most prosperous, but Venezuela has faced a sharp drop in real per-capita GDP since 1970. The average growth record of our eight countries over three decades is disappointing.

In the section on trade indicators, the first line shows the geometrical average of imports and exports as a share of GDP, which can be seen as a measure of the openness of the respective economy. With a trade share of more than 40 per cent of GDP, PNG and Gabon have extremely open economies dominated by import and export flows. Nigeria, Mexico and Indonesia are moderately open (share 24–31 per cent). Cameroon, Venezuela and Ecuador (share 16–19 per cent) are more inward-looking economies in which a more pronounced import substitution has taken place over time. A second trade-related issue relates to the size of the oil boom. Lines 7–10 describe the absolute and relative size of fuel exports in 1980, a year that marked the height of oil wealth in most of these countries. Note that the PNG figure shows mineral exports, since the country only became an oil exporter in 1992. In absolute terms, Nigeria, Venezuela, Mexico and Indonesia had the highest windfalls, in that order. But for economic structure, the degree of oil-export dependence is more important (line 9). In Nigeria, Venezuela and Gabon, oil was overwhelmingly dominant; in Ecuador and Mexico the export share was about two-thirds, while it was less than a third in Cameroon. For the oil share of GDP, there is a variation from 5 per cent (Mexico) and 6 per cent (Cameroon) to 39 per cent (Nigeria) and 45 per cent (Gabon). In per-capita terms, Gabon and Venezuela had outstanding windfalls (US$2,767 and US$1197, respectively), Nigeria's was also high (US$354), while Indonesia (US$106) and Cameroon (US$50) came bottom, though with some caveats for the latter.[1] The large variation in the size of these indicators makes it desirable to distinguish between high-rent economies (Gabon, Venezuela), medium-rent cases (Nigeria, Ecuador) and countries where mineral rents were less prominent (Mexico, PNG, Indonesia and Cameroon).

Turning towards sectoral and land-use issues, national population densities can give us a first indication of pressure on land resources. There is a huge spread from Nigeria and Indonesia (136 and 114 persons/km^2, respectively), to PNG (10 person/km^2) and Gabon (5 person/km^2). However, urbanisation rates also differ widely, from the extremely rural PNG (17 per cent urban) and the intermediate cases of Nigeria, Indonesia, Cameroon and Gabon to the highly urbanised economies in Latin America, Mexico (74 per cent) and Venezuela (87 per cent). As argued above, perhaps *rural* population densities give a more accurate reflection of general pressures on forests than do total densities. These are even more polarised, again with Nigeria (77 person/km^2) and Indonesia (69 person/km^2) at the top, then with a big jump to Ecuador, Cameroon and Mexico (13–17 person/km^2), PNG (9 person/km^2) and the rurally depopulated Venezuela and Gabon (3 and 2 person/km^2, respectively). However, the two high-density countries differ radically in terms of population growth: Nigeria still has a very high rate of 2.9 per cent, while in Indonesia the demographic transition has brought it down to 1.75 per cent. The other countries lie in between these extremes, with, for example, the Latin American cases at the lower and the African ones at the higher end of the spectrum. In other words, land scarcity is highly variable within the sample, which is also bound to influence land-use patterns.

The forest situation

As discussed in Chapter 3, different forest statistics give highly variable estimates, and none of them is fully free of disadvantages. In Table 10.2, I use the 1991–2 TREES estimates from Mayaux *et al.* (1998) as a source of comparison. As can be seen, there is a wide disparity in forest abundance. Gabon and PNG are extremely forest-rich: four-fifths of their land area is forested. Indonesia, Ecuador and Venezuela follow in the next group, with three-fifths of forest cover, although the TREES estimates for the first two countries are probably a bit high (see Chapters 7 and 9). There is then a large gap to Cameroon, with a third of its land area being forested, although this is concentrated in the humid forest zone in the south. Mexico (16 per cent) and Nigeria (6.2 per cent) are forest-scarce countries according to the TREES classification, though they have large areas of woodland and large transition zones dominated by tree cover. An additional forest-scarcity perspective is obtained by dividing forest area by population. Again, Gabon (17 ha of forest per capita) and PNG (8 ha/capita) clearly head the sample, followed by Venezuela, Ecuador and Cameroon (1–2 ha/capita). The three secondary cases from Chapter 9 are much more forest-poor: Indonesia (0.5 ha/capita), Mexico (0.3 ha/capita) and Nigeria (0.05 ha/capita). Yet, as we have seen, for Indonesia this average conceals the fact that some of the Outer Islands remain very forest-rich, while on Java the figure is similar to Nigeria. It also hides the fact that some countries were originally almost entirely forested (Gabon, Ecuador, PNG, Indonesia), while others (Cameroon, Nigeria, Venezuela, Mexico) have large regions that are not naturally forested. But a tentative classification would label Gabon and PNG as extremely forest-rich, Venezuela, Ecuador and Cameroon as moderately forest-rich, and Mexico and Nigeria as forest-poor; Indonesia contains a spectrum of land areas that fit under each of these three categories.

Turning now to deforestation figures, Table 10.2 first reproduces the range of likely forest loss in the last decade, as estimated in the country chapters (line 5, as summarised in Table 1.2, Chapter 1). In line 6, we simply calculate the midpoint of that range, in order to have a single number to work with in the following.[2] For the three secondary cases, no detailed analysis of land use was made, so for lack of alternatives we use FAO's deforestation figures from the FRA for 2000, although we must remember that FAO's estimates were significantly different from those obtained for several of the primary case studies. Yet, using the TREES forest stocks for comparison, we can make some estimates of deforestation. The ranking between the countries below singles out all the three secondary cases as high-deforestation countries (Nigeria with an extreme 7.1 per cent), Cameroon, Ecuador and Venezuela just at or below the 1 per cent threshold, and forest-rich Gabon and PNG as low-deforestation.

Comparative global analyses of deforestation and its determinants have almost exclusively focused on these *relative loss rates*, that is, current deforestation divided by remaining forest stock (see Chapter 3). This choice has been driven by the conservationist concern for how quickly we are losing important biomes: the question is, how much time do we

Table 10.2 Comparison of the forest situation in the study countries

	Gabon	*Venezuela*	*Cameroon*	*Ecuador*	*PNG*	*Indonesia*	*Mexico*	*Nigeria*
1 Forest area (TREES) (million ha)[a]	20.7	49.3	17.4	17.0	36.1	110.8	30.5	5.6
2 Forest area (% of land area)[a]	80.3	55.8	37.3	61.3	80.0	61.2	16.0	6.2
3 Forest area (ha/capita)[a]	21.2	2.5	1.5	1.6	9.3	0.6	0.4	0.1
4 Forest products export value (million $)[b]	333	46	351	50	240	5,578	267	33
5 Deforestation 1990s – likely range ('000 ha/yr)	0	250–400	150–200	<180	50–70	n.a	n.a	n.a
6 Deforestation 1990s – point estimate ('000 ha/yr)	0	325	175	150	60	1,312	631	398
7 Deforestation rate 1990s – (point estimate as % of forest area)	0.0	0.7	1.0	0.9	0.2	1.2	2.1	7.1
Ranking	8	6	4	5	7	3	2	1
8 Deforestation 1990s – point estimate/capita (ha/1,000 people)	0.0	13.7	11.9	12.1	12.8	6.3	6.5	3.2
Ranking	8	1	4	3	2	6	5	7
9 Deforestation 1990s – point estimate/GDP (ha/million constant 1995$)	0.0	4.3	18.2	8.5	12.7	6.6	1.8	12.9
Ranking	8	6	1	4	3	5	7	2
10 Combined points from 7–9	24	13	9	12	12	14	14	9

Sources: FAO (2001), FAOSTAT (2002b), Mayaux *et al.* (1998), World Bank (2001a), own calculations.

Notes
a Year 1991–2 map data.
b Year 2000 data.

have left before we exhaust forest stocks? While this accounting procedure is understandable from that perspective, it is much less clear why relative loss rates should be the most appropriate unit in analysing the causes of forest loss.

For the sake of illustration, imagine two countries called 'Sylvania' and 'Steppelandia'. They are completely identical in terms of their small population size, low level of economic development, similar political system, size of economy, growth rate and land area, say 1 million ha each. Sylvania is situated in a naturally forested zone and has 90 per cent forest cover (900,000 ha), whereas Steppelandia extends more towards drier areas dominated by savannahs and steppe, and has only 10 per cent (100,000 ha) forest cover. Last year, Steppelandia extended its cultivation areas into the forest zone by 2,000 ha, amounting to a national deforestation rate of 2 per cent. The striking socioeconomic resemblance between the two countries might lead us to expect that forest loss has been similar in Sylvania. Yet for their deforestation rates to come to 2 per cent, Sylvania would need to clear 18,000 ha. That is nine times as much with respect to the size of Sylvania's economy, population and every other comparative indicator. While in absolute terms Sylvania may be more prodigal in using its abundant forest resources, it is unclear what (economic or other) forces should make the two deforestation rates equal: indeed, it does not make much sense to expect equal loss rates.

Consider now that a civil war breaks out in the forests of Sylvania, which eventually leads to the country being divided. The rebels, an ethnic group of highly entrepreneurial farmers, come to control 90,000 ha of cleared land and 10,000 ha of forest. They form a new country, 'Agrolandia', which has 10 per cent forest cover. The rest of Sylvania (condescendingly called 'Trans-Sylvania' by the Agrolandians), which is dominated by forest-dwellers, is left with 890,000 ha of forest and only 10,000 ha of farmland (98.9 per cent forest cover). In the first year after Independence, Agrolandia clears 2,000 ha of forests (20 per cent). Rest-Sylvania does not have the funds to make a forest assessment but, compared to Agrolandia, how much would we expect the country to clear? Obviously, we would not expect it to amount to 20 per cent (178,000 ha)! The actual answer seems to have little to do with the respective remaining forest stocks. We should rather look at how population and economic sectors are distributed in the two now separate countries in order to predict forest-loss patterns.

This example demonstrates that deforestation rates alone can only serve as a partial comparison of forest loss intensity. In particular, this caution applies when juxtaposing countries at different stages of their forest transition, that is, a forest-poor, population-rich country (like Nigeria) compared to its direct opposite (e.g. Gabon). We should thus also look at forest loss compared to the size of population and of the economy. Line 8 in Table 10.2 shows deforestation expressed in per-capita terms. Somewhat surprisingly, this criterion shows Venezuela to be the highest forest converter, compared to its small population density (13.7 ha/yr/thousand people),[3] closely followed by PNG, Ecuador and Cameroon. For population-rich countries like Mexico, Indonesia and especially Nigeria, per-capita loss rates are significantly inferior. This indicates that in the forest-rich oil countries, people also depend more on expanding land use into forests than in those where a lot of forest has already disappeared.

Finally, another approach is to compare deforestation with the size of the economy. How land-extensive and 'forest-using' is the economic system in a given country?

How dependent is the economy on advancing its land-use into forested areas? Table 10.2 shows forest loss as a ratio to GDP in 1999. The results open up a novel perspective. The three most deforesting economies according to that criterion are Cameroon, Nigeria and PNG, the latter in spite of having a low relative loss rate: they 'consume' between 12 and 18 ha of forest for each million dollars of their gross domestic product. The three countries share certain features: they have a low degree of urban economic development and of 'value-added' diversification, they have experienced low expansion of cash crops over the last decade and, perhaps most importantly, they depend greatly on slash-and-burn agriculture to expand their food production to feed a rapidly growing population. The next, intermediate level (5–10 ha/million US$ of GDP) is occupied by Ecuador and Indonesia, which on the whole are less dependent on shifting cultivation but have developed specific, heavily land-using commercial sectors like cattle (Ecuador), estates and cash crops (Indonesia). At the low end of the scale (0–5 ha/million US$ of GDP) we have Venezuela, Mexico and Gabon, countries with a higher income level, greater economic diversity or maturity and a certain bias towards urban development.

On the whole, one should at least keep in mind the fact that most forest-rich countries also lose more forests with respect to their population and economy size. Only Gabon (with its zero deforestation) was continuously classified as the lowest forest-loss country according to all three indicators, as shown by the sum of combined ranking points (line 10). Nigeria and Cameroon come out as highly deforesting according to this simple index, while the other countries are clustered within a closer range (12–14 points), reflecting the fact that the deforestation ranking of each depends greatly on the selected indicator.

Comparing the outcomes

Macroeconomic versus deforestation trends

How do we compare?

Having sketched both basic economic factors and forest-cover indicators in the last two subsections, we can now start to compare the changes over time in the two equations. Hypothesis 1 claimed that oil- and mineral-exporting countries had more forest left, and were losing it less rapidly than comparable non-mineral countries; support for that hypothesis was presented in Chapter 3. Second, a particular inter-temporal hypothesis in Chapter 3, also referred to as the *core hypothesis* throughout the text, linked changing oil wealth to changing pressure on forests:

> **Hypothesis 2**: *Over time, tropical forested countries that specialised in high-rent mineral exports will have lower deforestation and forest degradation during boom periods than during bust periods of low mineral export revenues.*

This section is thus central to the book, as we shall be using it to examine how land-use/deforestation trends compare with the big macroeconomic picture in primary and secondary study countries, thus allowing us to evaluate the core hypothesis. Which cases confirm the hypothesis, and which ones do not? If the country was affirmative, was that the case in absolute terms (implying that forest cover grew in absolute numbers) or in relative terms (i.e. the relative rate of forest loss slowed down)? And what was the main reason for this? Which factors were working for and against forests? Table 10.3 summarises the main trends on forests and macroeconomics, first for the primary countries and then, more tentatively, for the secondary cases.

As a preliminary observation, care is needed in juxtaposing the macroeconomic and land-use cycles, because the timing of both proved to vary considerably across countries. On the forest side, the main constraint is the scarce availability of data, which means that the two types of cycles do not fully coincide. Even on the economic side, booms are not unanimously dictated by the fluctuations in international oil prices and differed greatly between countries. First, this has to do with variable national production levels of petroleum, which are influenced by the exhaustion of existing reserves and by new discoveries, but also by shifting government policies with respect to multinational oil companies. Second, the timing of the boom was frequently affected by financial capital flows. Many countries exacerbated or prolonged their oil booms by borrowing against revenues (e.g. Nigeria and Venezuela), while in others sudden capital outflows could end or dampen a boom in spite of high mineral revenues (e.g. in PNG or Mexico). Third, prices proved to be sticky in their adjustment to a bust: most countries maintained overvalued exchange rates that also delayed the incentives for land-use changes. In PNG and Cameroon, overvaluation endured and worked strongly against cash-crop agriculture. Indonesia is the counter-example where the currency was devalued swiftly, permitting a fast revival of the T sectors. Let us now examine the cases one by one.

Primary cases

Gabon provides a textbook confirmation of the core hypothesis. With its transformation to an oil exporter in the early 1970s, this low-population country became the peak rent economy in our sample. In spite of the mini-bust in 1986–9 and its highly fluctuating oil revenues in the 1990s, it is appropriate to view the entire post-1973 period as an era of high oil wealth. Although these oil revenues were very unequally distributed within Gabon, they still caused massive structural changes. With the total neglect of rural areas and road-building, and the massive stimulation of urban employment and rent-seeking, there was an exodus of young people from the countryside, and abandoned agricultural fields grew back into forest. Although we do not have national forest inventories for sub-periods, the two point estimates plus a number of case studies give us a quite solid basis for saying that high oil wealth led to an *absolute net expansion* of forest area. Within this picture, there may well have been local deforestation processes in periurban areas, especially during mini-crisis periods, when some people lost their urban jobs. But the long-term trends will obviously still dominate, at least until oil revenues enter into a phase of rapid decline.

Venezuela's historical transformation from a specialised agrarian exporter prior to the Second World War showed many similarities with the Gabon story and was another case that confirmed our core hypothesis in absolute terms: there was a marked net forest regrowth in abandoned agricultural areas from 1920 to 1950. In the 1950s, Venezuela invested greatly in road-building, promoting the expansion of the cattle sector in particular,

Table 10.3 Comparing macroeconomic and deforestation trends

Country	*Macroeconomic cycles*	*Deforestation cycles*	*Evaluation of the core hypothesis*
Gabon	1960–73: pre-boom 1974–85: boom 1986–9: mini-bust 1990s: fluctuations	1960–90s: forest regrowth Recent: probably periurban clearing under mini-crisis years	*Absolute confirmation* (short and long run)
Venezuela	1920/30s: oil transformation 1956–8: mini-boom 1974–83: boom 1984–: crisis and mini-booms	1920–50: forest regrowth 1950–80s: slow loss 1980s/90s: more rapid loss	Pre-Second World War: *absolute confirmation* Post-Second World War: *relative confirmation*
Cameroon	1960–78: pre-boom 1979–85: boom 1986–94: bust, fixed CFA 1995–: devaluation, recovery	1973–85: slow loss 1986–94: high loss After 1994: probably high loss	*Relative confirmation*
Ecuador	1960–73: pre-boom 1974–81: boom (rising) 1982–5: boom (declining) 1986–95: bust 1996–: mini-booms	1975–90: high loss 1990s: probably lower loss	*Rejection* – absolute and relative
PNG	1972–94: mineral boom (rising), fixed overvalued kina 1995–: oil boom, financial capital outflows, devalued kina	Whole period: probably stable, low loss After 1994: perhaps loss acceleration	*Relative confirmation* (hesitant)
Mexico	1975–85: boom 1986–90s: crisis 1990s: recovery, lower oil exports	Probably stable, high-forest loss (1970s, 1980 and 1990s)	Probably *rejection*
Nigeria	1972–82: boom 1982–9: bust (gradual decline) 1990s: mini-booms and busts	1970s: probably lower loss; 1980/90s: loss accelerating	Probably *relative confirmation*
Indonesia	1973–82: boom 1983–7: bust 1988–96: recovery 1997–2000: financial crisis	1970s: probably lower loss; 1980/90s: loss accelerating, especially after 1997	Probably *relative confirmation*

which accounted for almost all of the country's deforestation. Our regression results showed that cattle-ranching as a semi-tradable sector expanded pro-cyclically with urban incomes, with a higher demand for protein-rich food among a large urban middle-class leading the sustained expansion. On the other hand, cropped area remained constant. From the regressions, it became clear that cropped area was negatively correlated with both rising urban incomes and declining price competitiveness: that is, crop cultivation was a true Dutch Disease victim. While Venezuela still had quite low forest-loss rates during 1950–80, the trend was an accelerating one. With the deep economic and political crisis of the 1980s and 1990s, the speed of forest extraction and clearing increased, even in the areas south of the Orinoco river that had previously been too remote from urban centres. This process was closely related to a depreciating RER. Since the Second World War, Venezuela has been a case that confirms the core hypothesis in relative terms, although roads and cattle increasingly emerged as confounding factors.

Cameroon clearly had the most pronounced economic cycles in our sample of countries. The prosperous period of 1978–85, with simultaneously high international prices for the main exportables (oil, coffee and cocoa), combined with access to foreign capital, turned into misery in 1986 when the external environment reversed sharply – and in an equally simultaneous manner. Furthermore, during 1986–94 the fixed value of the CFA franc prolonged the crisis. Since the 50 per cent devaluation in 1994, there has been a slow recovery. In land-use terms, specific case studies that had been tailored to compare boom and bust, with good coverage of the entire humid forest zone, allowed us to evaluate the core hypothesis exhaustively. The boom (1979–85) had a clear urban bias and accelerated rural–urban migration, which led to a slow-down in deforestation. In the crisis period, with its overvalued exchange rate (1986–94), cash crops in the humid forest zone (mainly cocoa) were stagnant. But a significant return migration to rural areas and an upsurge in slash-and-burn food production (notably plantains) was sufficient to cause a strong rise in deforestation in the forest zone. Since the devaluation of 1995, cash crops have experienced a slow revival, and deforestation continues to be high. Unlike in neighbouring Gabon, oil wealth was thus not high enough to *reverse* forest loss and cause forest area to expand during the beam. Rents were simply insufficient to fully compensate for other factors, notably rural population growth. But there was an abrupt reduction in the *pace* of deforestation. Cameroon thus provides a strong *relative* confirmation of the core hypothesis: forest did not outright increase in size during the oil-wealth period, but the rate of loss clearly slowed down.

The case of Ecuador was rather different. The size of the boom was greater than in Cameroon, but much smaller than in Venezuela and Gabon. Due to rising production, oil revenues remained quite high until the mid-1980s. A bust followed, combined with a strong political crisis in the 1990s, so although there have been mini oil bonanzas, the entire post-1985 period is best characterised as an economic downturn cycle. Nevertheless, deforestation actually accelerated during the oil boom, stayed high for a decade after, and then probably decelerated. This is because the government deliberately spent a large share of oil revenues on things that promoted more extensive land uses, notably new roads (especially connecting the highlands with the lowlands), but also transport subsidies and land colonisation programmes and support. As in Venezuela, a second factor was cattle. A relatively large urban middle-class used part of their new purchasing power to buy cattle-derived proteins, thus creating the level of demand to match the supply-side factors that enabled the large-scale extension of pastures. In Ecuador, these contrary factors were stronger than the core-effect ones – and policy was decisive in creating that balance. Although there was an upsurge in agriculture after the devaluation, as the core hypothesis would suggest, on the whole this country case was contrary to the hypothesis, in both absolute and relative terms: forest cover decreased in absolute terms and also in comparison to pre-boom deforestation rates.

Papua New Guinea became a high-mineral exporter from 1972 onwards. Copper and gold predominated until 1992, when significant petroleum production began. The size of these mineral rents with respect to the economy as a whole was somewhat greater than in Ecuador. Firm macroeconomic management succeeded in stabilising their impact over time until the 1990s, but then fiscal control was increasingly lost. The fixed kina exchange rate introduced in 1975 greatly hampered non-mineral exports until 1994, but then a floating kina implied large real devaluation. In other words, the whole period is characterised by fairly high mineral revenues, but with different economic management cycles. In terms of land use, cash crops never became a large source of deforestation: even after the kina devaluation rural violence, land-tenure arrangements and infrastructure problems were too great as impediments. PNG is thus an example where improved price competitiveness was not enough to revive the cash crop sector. An exception is the oil-palm sector, which caused some deforestation. As in Gabon, government policies ignored rural infrastructure. An overwhelming share of mineral revenues went into consumption, some being wasted, and little being invested to enhance production. On the national level, deforestation was clearly driven by the shifting cultivation of food crops. The rigid land-tenure structure favoured land intensification, so deforestation for agricultural extensification remained restricted. Forest-loss rates were stable and low, although as we saw in Table 10.2 deforestation was actually high with respect to the small size of the population and the economy. This is because of the high dualism in the economy: few value-added and service sectors have developed that might have drawn people out of food-crop agriculture and into urban activities to delink deforestation from population growth. On the whole, mineral wealth in PNG did not stop deforestation in absolute terms. It does seem to have slowed forest loss down, in particular by curtailing the cash-crop sector. Yet none of the other cases in this book was as complex as PNG: factors such as land tenure, crime and political instability were at least as important as mineral wealth in shaping land use.

The secondary cases

For the three additional countries considered in Chapter 9, only a summary assessment of land-use trends was made, so the conclusions presented here are somewhat more tentative. For instance, for Mexico no national forest assessments were available for sub-periods that would have allowed us to juxtapose them directly to the boom-and-bust cycles. Mexico had booming oil revenues until 1985, although the international debt crisis had already affected the country before, and in fact came to be known as 'the Mexican crisis'. In the 1990s, there was a recovery. It seems that, irrespective of booms or busts, forest-loss rates were relatively high throughout the period, although per-capita and economy-wise forest loss was comparatively much lower (Table 10.2). The reason for this is what one could call 'the Latin American syndrome': just as in Venezuela and Ecuador, the roads–cattle nexus enabled rapid expansion of land use, supplemented by fuel subsidies and government-sponsored land-colonisation projects. This is not to say that Dutch Disease land-use effects were absent in Mexico. On the regional scale within Chiapas state, one finds precisely the deforestation-curbing effect envisaged by the core hypothesis, but in other zones there was an acceleration. Most crops declined during the boom due to a lack of peso competitiveness, but the main land-extensive food crop, maize, was import-protected during most of the period and thus continued to expand. On the whole, the core hypothesis probably does not apply to Mexico, mainly due to Mexican policy responses to oil wealth.

Like Mexico, Nigeria is also a more forest-scarce country, but the absolute size of the oil boom was about double that of Mexico's (Table 10.1). Since Nigeria's per-capita GDP in 1970 was less than one-fifth that of Mexico, oil revenues had a very strong economic impact during 1972–82. Furthermore, in the early 1980s foreign borrowing prolonged the heyday of the Nigerian economy. After the crisis, Nigeria's oil revenues resumed reasonably high but fluctuating levels in the 1990s. Nigeria's forest statistics do not provide us with a clear picture, but land-use figures and other observations seem to indicate that there was some slow-down in deforestation in the 1970s. The oil boom certainly wiped out the cash-crop export sector and probably slowed down food-crop expansion, but that was not enough to eliminate pressures on forests entirely. As in Mexico, key food crops remained partially import-protected (semi-tradables) and thus expanded during the boom, driven by the country's high population growth and the rise in incomes. As most of these are produced under land-extensive shifting cultivation, forests were still converted to make room for new areas of cultivation, but at a slower pace. Consequently, it seems likely that Nigeria provides relative confirmation of the core hypothesis.

In Indonesia, the timing of boom-and-bust periods was rather similar to Nigeria, but the macroeconomic management of these booms was much more cautious, and recovery from crisis was much more rapid and sustained. None of the countries in our sample has had such high long-term economic growth in recent decades. The tentative assessment of the forest situation was also similar to Nigeria's, with high deforestation, but probably lower clearing rates during the booming 1970s than during the 1980s and 1990s. Although the end result was similar, the factors behind this result were very different. In Indonesia, export crops kept growing, due to 'protective' devaluations of the rupiah that were explicitly aimed at sheltering the cash-crop sector from the Dutch Disease. On the other hand, food crops only marginally expanded their area during the boom, mainly because rice production was intensified substantially. With the devaluation in 1986, forest-based industries and new cash crops like oil palm, cocoa and soybeans led the country back into a high-deforestation scenario, in particular in response to the 1997 crisis and the mega-depreciation. In spite of the weak forest data, Indonesia is very probably another case that confirms the core hypothesis in relative terms.

A brief look at the last column in Table 10.3 shows that about half the cases confirmed the core hypothesis in relative terms. The two cases that did so absolutely for all or part of the period were also those with by far the highest mineral rent. In other words, it probably takes a very pronounced degree of oil wealth as a necessary condition to put the deforestation process completely into reverse. In a country like Nigeria, even a fairly large windfall was not enough to halt deforestation. In most countries, one should expect oil wealth to slow down forest loss; the example of Cameroon showed that this slow-down can be quite significant. Finally, the last two cases, both in Latin America, showed a pattern that was not supportive of the core hypothesis. In both cases, both preconditions (a sizeable meat-eating middle-class) and policy responses (road-building and colonisation programmes) contributed to this counter-hypothetical outcome. We shall return to this later on, but first let us consider the different partial factors that were supposed to explain variable forest outcomes. The first of these was the extent to which the oil and mining sectors *per se* had a decisive impact on forests.

The impacts of mineral production on forests

How much did oil-production interventions proper contribute to forest loss and degradation in the study countries? Table 10.4 summarises the findings regarding the effects that had originally been outlined in Table 3.3 (Chapter 3). This comprises the direct and indirect impacts of petroleum exploration and production, and those of hard-rock mining, which were important for those countries where minerals represented a significant source of economic rent (e.g. PNG and Venezuela). The first questions to ask in every country are to what extent oil and mineral production actually coincide with land area (on-shore versus off-shore production) and, if they are predominantly land-based, to what degree that land area was covered by forests. The answers are shown in columns 3 and 4. The next columns (5 and 6) then show the main effects identified, their specific location and likely magnitude.

Petroleum

In Gabon, most oil production is on-shore. In the coastal production zone, it affects both forests (moist forests, mangroves) and savannahs. But the direct deforestation impacts, which have been well studied in Gabon, prove to be very limited. For instance, 1,405 ha (5.7 per cent of the site, and 0.06 per cent of the permit area) have been cleared for the Rabi field, which alone produces more than half of Gabon's oil. In the exploration phase, clearing typically lies in the range of 0.2–0.5 per cent of the areas affected. Most clearings are small and generally seem to regenerate just as well as historical disturbances caused by shifting cultivation. The accumulated national direct deforestation effects totals approximately 10,000 ha, or 0.05 per cent of forest area. Indirect clearing for oil workers' food production, new settlement, etc. is perhaps marginally greater (10,000–15,000 ha). This also includes forest degradation, like increased hunting for bushmeat commercialisation. But both effects remain quite limited, due to Gabon's low rural population pressure.

In Venezuela, most oil production is also on-shore, but it is concentrated in savannah areas. Only a minor share comes from deciduous forests. More recently, exploration has been carried out in the mangroves of the Orinoco delta. Direct deforestation effects are negligible. Historically, some oil roads have helped to open up forests in the Llanos region. Oil incomes were decisive in the economic development of Maracaibo, which has led to considerable deforestation in Zulia state. The cases of Mexico and Indonesia, not studied in depth here, seem similar, namely restricted on-site effects and some indirect impacts. In PNG, oil deforestation effects are also negligible, for example, 1,200–1,300 ha in the main production area of Kutubu, and good signs of regenerating vegetation. But the indirect effects

Table 10.4 Comparison of forest impacts from mineral production

Production type	*Country*	*On-/Off-shore?*	*In forests?*	*Main effects on forests*	*Extent and location*
Petroleum	Gabon	Mostly on-shore	Rainforest	Deforestation, direct	National: up to 10,000 ha over time (0.05% of forest area)
			Mangroves		
			Savannah	Deforestation, indirect	Perhaps 10,000–15,000 ha
	Venezuela	Mostly on-shore	Savannah (mostly)	Deforestation, direct	Negligible
			Some deciduous forest	Deforestation, indirect	Historical – Llanos and Maracaibo regions
			Some mangroves		
	Cameroon	Off-shore	No	None	None
	Ecuador	On-shore	99% in rainforest	Deforestation, direct	3000–6500 ha (0.04–0.09% of Amazon forest)
				Deforestation, indirect	Strong road-related timber extraction and colonisation
				Degradation	Strong pollution
	PNG	Mostly on-shore	Mostly forest	Deforestation, direct	1200–1300 ha (Kutubu)
				Deforestation, indirect	Negative (clearing reduced)
Hard-rock minerals	Gabon	On-shore	Mostly savannah	None	Zero
	Venezuela	On-shore	Mostly forest	Deforestation and degradation	Up to 40,000 ha/yr cleared
	Cameroon	Undeveloped	No production	None	None
	Ecuador	On-shore, small	Mostly forest	Not studied	Not studied
	PNG	On-shore	Mostly forest	Deforestation, direct, on-site	600–1400 ha/site
				Deforestation, direct, off-site	Huge for Ok Tedi: long term 188,300–378,300 ha
				Deforestation, indirect	Negative (clearing reduced)
				Degradation, off-site	Variable

were actually negative. Higher local oil (and mining) revenues, paid as compensation to local landowners, generally made local people buy more food from outside and reduce clearing for agriculture. This extraordinary reaction had to do with the limited economic alternatives in isolated mountain regions, but also with a 'full belly' reaction (valuing leisure more than higher incomes) on behalf of local inhabitants.

Whereas in Cameroon oil was produced off-shore without impacts on forests, Ecuador was where the forest impacts of oil were the largest. No less than 99 per cent of Ecuador's oil production comes from the Amazon forest, and although the direct effects have been limited to 3,000–6,500 ha (0.04–0.09 per cent of the forest in Ecuador's Amazon region), indirect deforestation effects have been much greater. This was due to oil roads providing access to forest areas, combined with Ecuador's high rural population pressure in the highlands, favouring out-migration to forested frontiers. Yet, note that relative forest loss was actually higher in the generally more accessible and market-integrated southern Amazon region, where until now no oil development has occurred. This makes it likely that some colonisation pressure would indeed have occurred even if no oil had been found in the northern Amazon region. Forest degradation from spills and other forms of pollution, including those from the 500 km pipeline crossing the Andes, has been severe. Rudimentary drilling-mud deposits and other careless practices have had severe off-site impacts on rivers and lakes, affecting both wildlife and the indigenous population. Similarly severe impacts in our country sample seem to have occurred only in Nigeria, in the sensitive mangroves of the Niger delta.

Mining

Hard-rock mining causes greater impacts on average, although this was not a major topic of interest in the case studies in this book. First, this is because the direct forest effects in most countries were negligible. In Gabon, manganese- and uranium-mining occurs in a limited space in savannah areas, and does not affect forests. Cameroon is rich in minerals, but mining activities have not yet been developed. Ecuador has some mining in forested areas, for example of gold, but forest impacts are known to be limited, and were not studied in Chapter 7. For Venezuela, the case was different. Gold- and diamond-mining are emerging sectors in the forested south that do have important forest-loss and degradation impacts. In PNG, mining was the exclusive source of mineral wealth up to 1992, when oil production began. The on-site clearing effects of these mining projects seem to have been quite limited (600–1,400 ha/site), but there have been more significant off-site pollution and erosion impacts on downstream watersheds. This is extreme in the case of the Ok Tedi mine, where models predict that tailing erosion is causing vegetation dieback that in the long run may affect an area between 188,300 and 378,900 ha (Chapter 8).

In none of the countries analysed was petroleum a main direct deforestation factor, although in a couple of cases (Ecuador, Nigeria) pollution and other degradation impacts were severe. Mining causes more significant yet still restricted site impacts, but off-site impacts may be extreme in particular cases. In most countries, better practices have reduced forest-clearing and pollution effects over time. First, this underlines the fact that, from a conservationist point of view, one should not extrapolate effects from single disastrous projects to all oil and mining industries: good practices do make a significant difference. Second, oil and mining as global sources of deforestation are clearly subordinate, say, the shifting cultivation of plantains, new oil-palm plantations, or the establishment of new pastures for cattle.

These conclusions must be qualified in part. First, while the impact of these activities is limited on an *aggregate* (national or global) scale, they may be vitally important to local people who have an important reason to try and avoid or limit them or, as in PNG, obtain sizeable compensations. Second, the evaluation of forest impacts from oil and mining operations always depends on how much they have actually been studied: there may well be forest loss and degradation impacts in our study countries that have simply not been documented. On the other hand, the state of knowledge and of environmentalist attention also depend greatly on the geopolitical context. It seems to be no coincidence that the environmental struggle against oil companies in Ecuador – close to the US and dominated by US companies – has been more prominent than in Gabon, a more isolated and closed country. Within countries, it also seems obvious that the attention given to multinational companies like Shell (in Nigeria) and Texaco (in Ecuador) far exceeds that directed towards national (mostly state) oil companies, like Petroecuador or Pertamina in Indonesia. Even when the latter employ inferior technologies that are far less benign in their forest impacts, internationally funded environmental NGOs have much less political leverage to influence them than Western companies, which in most cases need to worry much more about their corporate image.

Notwithstanding these caveats, a key finding of this book is that the impact of oil and mineral *rents* is far more important for forests than the effects of oil and mineral *production*. In Chapter 3, this was termed the *derived* or economy-wide impact. The remainder of this chapter will give attention to these derived impacts. We shall start with a key relationship for our core hypothesis: the linkage between petroleum revenues and the RER.

The link between oil wealth and competitiveness

One of the lessons of the macroeconomic assessment above was that booms differ greatly in their extent and duration, so that a simple comparison of boom and bust periods with deforestation phases (as in Table 10.3) may only provide a partial account. Unfortunately, we do not have year-to-year time series for either deforestation or alternative land-use expansion (except for Venezuela). But at least we can trace the effects more precisely through the stages of relative-price and production-quantity adjustments. The purpose of the next three subsections is to give a quantitative-comparative and qualitative summary of the key linkages that are inherent in the core hypothesis:

1 From mineral exports to competitiveness (RER).
2 From competitiveness to agricultural production, land use and deforestation.
3 From competitiveness to timber production, forest land use and forest degradation.

Table 10.5 summarises the quantitative evidence regarding the first of these points. The results show the parameters from selected regression analyses in the country chapters (4–8).[4] We can see that for four of the five countries, petroleum exports had the expected and statistically significant effect of producing an appreciation of the RER. For the fifth

Table 10.5 Comparing petroleum exports and capital inflows' effect on the RER

	Petroleum exports (million constant 1995 US$) [a]	*Capital inflows (million constant 1995 US$)*	*Time dummy*	R^2 (%)	*F-value*	*Years*
Real effective exchange rate (1990 = 100)						
1 Gabon						
Coefficient	0.01	0.03		31.0	6.3	1968–98
T-value	2.32**	3.43***				
2 Venezuela						
Coefficient	0.01	0.00		63.0	15.2	1977–97
T-value	5.44***	0.65				
3 Cameroon						
Coefficient	0.05	0.03		92.1	58.1	1986–98
T-value	8.90***	8.47***				
4 Ecuador						
Coefficient	0.03	0.01		14.5	2.1	1970–97
T-value	1.86*	1.05				
5 PNG						
Coefficient	0.00	0.02	−26.05[b]	67.5	18.0	1971–00
T-value	0.56	2.37**	−5.32***			

Sources: Chapters 4–8.

Notes

a Mineral exports for PNG.

b Dummy: Hard-kina policy in PNG: 1971–94 = 0, 1995–2000 = 1.

* Parameter *T*-value significant at the 10 per cent level.

** Parameter *T*-value significant at the 5 per cent level.

*** Parameter *T*-value significant at the 1 per cent level.

country, PNG, mineral exports were used as the independent variable, but although the estimate (including a time dummy describing the shift in exchange-rate policy) yields the expected positive parameter, it is not statistically significant. Even for Ecuador, the mineral export variable is only significant at the 10 per cent level, and the equation has a very low R^2. In many oil countries, foreign borrowing and other capital in- and out-flows were an additional instrument related to the boom-and-bust cycles. In three of the five countries (all but the two Latin American cases), this variable had a significant effect on the RER. As expected, higher capital inflows had the partial effect of causing an additional real currency appreciation.

Although data problems were identified for each of the cases, much could be done to optimise the RER model for each country by including other variables, change specifications, etc. However, the objective of this exercise was rather to produce a simple, comparative test of one of the main core-hypothesis relations. This allows us to compare the parameters from the regression equations. For instance, an additional US$100 million in oil-export revenues (in constant 1995 prices) would cause the RER to appreciate by 1 per cent in Gabon, but by 5 per cent in Cameroon. For capital inflows, the parameter was similar for the two countries. Note, however, that the estimate for Cameroon was based on a shorter period (1986–98) than for Gabon (1968–98). This coincides with the crisis period, when Cameroon was apparently particularly sensitive to changes in the availability of foreign exchange.

One additional factor that did prove significant was the management of the *nominal* exchange rate. In the Dutch Disease model, currency devaluation has no impact on RPs or competitiveness, as domestic inflation would pick up immediately so that the RER would stay exactly at its equilibrium level.[5] In the real world, domestic prices proved sticky, and overvalued exchange rates could persist for years without falling to their new 'equilibrium'. This means that, during bust periods, repeated devaluation played a vital role in protecting T sectors in some countries (Indonesia), while fixed-currency regimes had the opposite effect in others (PNG, Nigeria, and Cameroon).

The link between competitiveness and agricultural expansion

The spatial expansion of agriculture, that is, the growth in areas dedicated to crops and pastures (including temporary fallows), is not only the single most important source of deforestation – in many cases it is the *only* factor that has any significance in terms of alternative uses of deforested land. Agriculture is the big land-use competitor of forests. This produces a somewhat uneasy implication: most of what is good news for agriculture tends to be bad news for forests, and vice versa. And policies that are biased against agriculture may, unintentionally, become forest-conserving. The extreme example is Gabon, where until now forest conservation has had a very low priority among policy-makers; yet no government has been as 'efficient' as the Gabonese in achieving it. The key to efficiency has been the total neglect of agriculture in national economic development, allowing much vegetation cover to return to its default state: forest.

There were some non-agricultural sectors and projects with a deforestation impact, which were specific responses to the need to generate (or save) scarce foreign exchange during periods of bust and economic crisis. This included the expansion of shrimp farming into mangroves (Ecuador, since the 1980s), the construction of a transiting oil pipeline servicing a neighbouring country (Cameroon, recently), the expansion of artisan mining with small economic rents (Venezuela, in the 1990s) and some hydroelectric projects (several countries). These are examples of T-sector activities that typically developed in a post-boom period. Yet none of them had land-use impacts that compared to those of agriculture.

In quantitative terms, how dependent was agricultural production on changes in competitiveness? How did policy affect competitiveness? How much land expansion resulted from rising production? And to what extent was this land drawn from forests? Table 10.6 starts with the first of these questions, showing regression results for the five primary study countries. Slightly different indicators for the dependent variable were used, according to the absolute versus relative confirmation of the core hypothesis.[6] In all five equations, the appreciation of the RER thus had the expected negative effect on agricultural production; in four of these the effect was highly significant. As in Table 10.6, the equation did not yield a significant coefficient for PNG, even when using the time dummy for the shift in exchange rate policy.[7] The size of the estimated parameters is not fully comparable, due to the different underlying value-added measures.[8] Nevertheless, in all the cases the RER on its own can explain 40–60 per cent of the variation in agricultural value added. This underscores the idea that the link from competitiveness to agricultural production was a strong

Table 10.6 Comparing RER effects on agricultural production

	RER (1990 = 100)	*Time dummy*	R^2 (%)	*F-value*	*Years*
Agricultural production					
1 Gabon[a]					
Coefficient	−2.29		44.3	23.8	1966–97
T-Value	−4.88***				
2 Venezuela[a]					
Coefficient	−6.66		56.8	25.0	1977–97
T-Value	−5.00***				
3 Cameroon[b]					
Coefficient	−0.33		50.6	26.6	1971–98
T-Value	−5.16***				
4 Ecuador[a]					
Coefficient	−4.38		56.3	33.6	1970–97
T-Value	−5.79***				
5 Papua New Guinea[c]					
Coefficient	−0.12	−44.41[d]	46.0	10.6	1971–98
T-Value	−0.50	−4.58***			

Sources: Chapters 4–8.

Notes

a Agricultural value-added (million constant 1995 US$).
b Agricultural value-added (% of GDP).
c Agricultural per-capita value-added (fixed 1995 kina).
d Time dummy: 1971–89 and 1993–2000 = 0, 1990–2 = 1.
* Parameter *T*-value significant at the 10 per cent level.
** Parameter *T*-value significant at the 5 per cent level.
*** Parameter *T*-value significant at the 1 per cent level.

one, probably stronger than that from oil revenues to competitiveness. Note, however, that regressions for agricultural *exports* (not shown in Table 10.6), the genuine tradables, had generally higher elasticities and were more significant than total agricultural value-added.

A common confounding factor in the linkage between competitiveness and total agricultural production was the presence of shifting trade policy regimes. The aggregate impact of different trade interventions in terms of *effective* protection rates for different products proved to be quite variable across countries and products. Furthermore, when protection or de-protection occurred, they did not always have a forest impact. Some general observations regarding the impact of trade policies on agriculture can nonetheless be noted:

1 With a couple of exceptions (Indonesia, Venezuela), there was an overall bias against agriculture.
2 The bias was strongest against export crops (through taxes, pseudo price stabilisation, etc.).
3 Subsidies for imported inputs (fertilisers, insecticides, etc.) sometimes pulled in the opposite direction.
4 Some protected food crops became semi-tradables (quantitative restrictions, prohibitive import duties).

5 Import restrictions on meat and powdered milk made Latin American cattle quasi non-tradable.
6 Over time, the countries reluctantly followed the global trend towards trade liberalisation.

To the extent that trade policies overall were biased against agriculture, one would expect that they also restricted deforestation. This seemed true in particular for export crops, which were often sacrificed by policy-makers, for example, through the impact of price stabilisation schemes that ranged from heavy taxation to outright confiscation (Gabon, Cameroon, PNG). But the cases showed that generalisations cannot be made from the overall sectoral effect to forests – trade-policy impacts at the product level must also be examined. In spite of a general trend towards trade liberalisation (6), 'stop–go' trade policies were often used to respond to particular episodes (inflationary surge, balance-of-payments crisis, etc.), meaning that the effect over time was inconsistent.

In land-use and forest terms, factors 4 and 5 in particular had large impacts, due to their land-extensive character. Their specific deforestation impact could thus be much stronger than any overall bias against agriculture. First, in Latin America the restriction on the imports of meat and dairy products made it possible for an often inefficient cattle sector to profit from income gains and expand behind the walls of protection by 'consuming' huge land areas for nutrient mining: Ecuador was the most glaring example in this respect. The cattle sector actually became more *land-extensive* over time, with a declining average number of head per ha of pasture (indicating technological *regress*), especially in agricultural frontier areas. Cattle thus became the dominant source of deforestation in all three cases from that continent overall. Had the import regime been more liberal, deforestation arising from this sector would have been more limited, probably allowing for much larger beef and dairy imports from areas with a natural comparative advantage in their production, such as Argentina or southern Brazil.

Second, food crops were sometimes naturally protected by inaccessible transport in remote areas (e.g. in PNG), but otherwise key products were protected because of food-security concerns (e.g. maize in Mexico, rice in Indonesia). With growing domestic markets, the analytic dichotomy between food and cash crops is becoming anachronistic: in many periurban areas of, for example, Cameroon or PNG, food crops were found to be farmers' main source of cash. The deforestation problem with food crops grown in their traditional swidden systems – such as the bulk of maize produced in *milpa* (Mexico), or plantains and tubers in *esep* fields (Cameroon) – is that they require simultaneous fallows. Despite fallow shortening over time, this multiplies the forest impact of any expansion in cultivated area. In addition to direct product-specific protection, these food crops were normally semi-tradable, profiting from the import restrictions for imperfect substitutes. To take the most prominent example, had the import of rice been continuously liberalised in all our study countries, there is no doubt that the penetration of this staple would have been much faster, and aggregate deforestation would have been more limited, in particular in periurban areas.[9] On the whole, food-crop import-protection was an important factor in accelerating deforestation.[10]

If the land-expansion impact of cattle was sometimes more than proportional to the rise in production, for food crops the situation was more often the reverse: some land intensification occurred. However, this was most obviously the case where land scarcity had already reached a chronic stage (as on the inner islands of Indonesia), where sudden rapid growth accelerated competing demands for land (as in periurban areas of all the study countries), or where inflexible land-tenure arrangements forced a 'closure of the frontier' (as in the highlands of PNG). In areas where forest was still (perceived as being) abundant, there was no Boserupian land intensification. Yet some intensification was also directly induced by oil wealth, including the fact that an appreciated exchange rate made it cheaper to buy imported inputs and that in addition the price of these inputs was often subsidised. Although these subsidies were more usually directed at cash crops, there is evidence that they were also actually used for food crops (e.g. in Cameroon). Finally, in some countries (Cameroon, Nigeria, Mexico, Venezuela) a non-trivial share of agricultural expansion (cattle, sugarcane, wheat, etc.) was achieved at the cost of savannahs and grasslands rather than of forests. For the other countries where forest cover predominated, almost any agricultural land-use expansion had to be drawn from the stock of forested land. In all countries, a significant share of the cleared land was temporarily abandoned after cultivation. In other words, as an overall measure of deforestation, the expansion of agricultural production area is a good yet in most cases conservative indicator.

The link between competitiveness and forest degradation

The international timber trade has been one of the main targets of forest-conservation campaigns in the past. Yet, as explained above, few tropical forestry operations imply clear-felling, so, given the deforestation definition adopted in Chapter 3, timber extraction would typically cause forest degradation rather than deforestation. However, as one of the main interventions into tropical forests, it could easily be the most important degradation factor, both directly, and through its indirect, access-providing impact in opening up forests to other destructive uses. Thus the questions to be asked here are similar to those in the previous section. How much was timber production actually obstructed by an appreciating RER during mineral booms? Was trade policy as important a factor as for agriculture? Was the rise in affected forest area as large as the rise in timber production? And to what extent has timber production in our study countries actually caused forest degradation?

Table 10.7 can clearly answer the first of these questions in the affirmative. In all five countries, the RER was a significant determinant of industrial timber production. For three countries, the parameter was significant at the 1 per cent level, for two of them at the 5 per cent level. As for agriculture, timber exports (not shown in Table 10.7) are even more sensitive to RER movements.[11] The situation was quite variable between countries. Some of the Latin American economies were never large timber exporters; one of the African countries (Nigeria) used to be a large exporter but exhausted its stocks; another kept its traditional timber-exporting role (Gabon), while yet another expanded it massively, in part thanks to a depreciated currency (Indonesia). Several of our countries more recently became large timber exporters (PNG, Cameroon), but did so on a large scale only after their currency had been devalued. This indicates the key role of price competitiveness for timber markets. It might also document the ability of new players like the Malaysian conglomerates effectively to 'shop around' among different continents and countries where they have concessions, accelerating harvesting where costs are currently low, while scaling back where the exchange rate is overvalued.

Table 10.7 Comparing RER effects on industrial timber production

	RER (1990 = 100)	*Non-agricultural GDP (million constant 1995$)*	R^2 (%)	*F-value*	*Years*
Timber production					
1 Gabon[a]					
Coefficient	−20.51		71.6	78.2	1966–98
T-value	−8.84***				
2 Venezuela[b]					
Coefficient	−2.46	0.03	83.0	43.8	1977–97
T-value	−2.80**	4.91***			
3 Cameroon[b]					
Coefficient	−36.66	0.56	70.7	30.1	1971–98
T-value	−3.75***	7.43***			
4 Ecuador[b]					
Coefficient	−10.31	0.42	77.6	41.6	1970–92
T-value	−2.30**	6.43***			1994–7
5 PNG[b]					
Coefficient	−38.42		30.7	12.0	1971–2000
T-value	−3.46***				

Sources: Chapters 4–8.

Notes

a Total timber production ('000 m^3).

b Industrial wood production ('000 m^3).

* Parameter *T*-value significant at the 10 per cent level.

** Parameter *T*-value significant at the 5 per cent level.

*** Parameter *T*-value significant at the 1 per cent level.

Even for the total production equations in Table 10.7, in all countries (except PNG) between 70 and 83 per cent of the variation in timber exports is explained by the equation containing the RER. However, note that for three of the countries this model includes non-agricultural GDP, an indicator of urban incomes, as an explanatory variable. For Gabon (with its very small total population) and PNG (with its very limited degree of urbanisation), this variable was not significant. But for Venezuela, Cameroon and Ecuador – all middle-income countries with a large or growing urban middle class – oil-wealth periods triggered an urban construction boom that raised demand for furniture, construction timbers, etc. It is thus not surprising that non-agricultural GDP was highly significant (at the 1 per cent level) in all three countries.

This result merits some further discussion. It has recently been shown that in a large, semi-closed economy like Brazil, 86 per cent of all timber produced in the Amazon is absorbed by the domestic market (Smeraldi and Veríssimo 1999). It has probably received less attention internationally that even in middle-sized developing economies with a greater dependence on trade, the dynamics of domestic timber demand can also play a decisive role. Cameroon, for instance, is seen by most observers as a timber producer for the world market, a country where domestic timber demand is nothing more than an addendum. Nevertheless, the significance of domestic production here and in other countries was documented in the country chapters. This too is explained partly by tradability: when timber is a genuine tradable commodity, then the RER will be the only significant determinant (as in Gabon and PNG). Conversely, non-agricultural GDP should not have any influence: excess demand from higher incomes is allegedly satisfied completely by imports. In turn, if a commodity is a non-traded or quasi non-traded commodity, then the RER coefficient should be estimated positively. The timber sectors in our three countries obviously occupied the middle ground of what were called 'semi-tradables' in this book: they were affected negatively by RER appreciation, but still profited from the growth in domestic urban demand.

The explanation for the status of semi-tradables is again to be found in the sphere of trade policies. First, as a bulky commodity one might expect timber to enjoy natural import protection in any tropical developing country. Yet in fact several of our study countries implemented import restrictions during certain sub-periods, especially with respect to forest products with a higher value-added. Second, it was also common in our case sample to implement time-limited bans on the export of logs and other export restrictions in order to stimulate the domestic processing industry, but also to saturate the domestic market for wood products without causing additional inflationary pressures. This is not the place to discuss the overall adequacy of log-export restrictions as a development strategy – a huge debate in itself. Suffice it to say that these export restrictions made the timber industry more home-market oriented, depressed price levels for raw logs, thus stimulating higher urban consumption, and encouraged industries with a high wastage of wood in all value-added stages, from the forest to the final domestic consumer. Ecuador was a notorious example, and Cameroon does not seem to have done much better. Like protected cattle-ranching and semi-sheltered food crops, domestic timber production was a third example of a sector that over-expanded its wasteful use of natural resources, thanks to trade-policy interventions. This was clearly to the detriment of the forests.

Did this accelerated timber harvest cause a proportional rise in harvested forest area? The country evidence indicated not, since normally the increase in area affected was less than proportional. In some countries, the internal discussions confused harvested area with concessions areas: granting the latter was more a process of gaining control over land for future harvesting, not necessarily for current production. The reason that the rise in harvested area was less pronounced was that a greater quantity of species was being harvested per hectare. Again, this trend in production resulted mostly from a growing scarcity of the traditionally most valuable timber species, leading gradually to a greater acceptance of secondary species. In some cases, new technologies also favoured the extraction and processing of these secondary species. While in a way this trend has 'saved' some forest area, it also frequently meant that the environmental impact per hectare harvested was greater.

This leads us to the assessment of forest degradation trends and their link to the oil-wealth situation. If logging was significantly curtailed by the RER appreciation from mineral booms, does that mean that oil wealth also protected forests from degradation? For many countries, the answer is 'yes'. In particular, currency devaluation in response to an economic downturn seems to be a highly efficient way to stimulate logging for export. The assessment is complicated by what exactly is to be seen as 'degradation'. The sustainability of selective logging depends heavily on what it is that we are seeking to sustain. Is it, at one end of the scale, the full range of biodiversity and ecosystems or, at the other end, a sustainable harvest of wood? If the latter is the criterion, some observers make the case that forest-rich countries like Gabon and PNG are currently practising sustainable forestry, while those from the other end of the spectrum are vehemently arguing the opposite. On the other hand, in a country like Indonesia there can be little doubt that current forestry operations are not sustainable and that they are having a forest-degrading effect; regardless of what criterion is applied. The depreciated rupiah seems to have been one important factor in this process.[12]

However, in no way is logging the only source of forest degradation. For a checklist of other effects, we can return to Box 3.4 in Chapter 3. Factors like oil production, mining and shifting cultivation have already been discussed. Others like overgrazing were not relevant, as there were no countries with very dry forests in our core sample. Fire was an important cause in Indonesia, but being a secondary case study, our analysis of it was not detailed enough. Firewood harvesting was an activity that seemed to be reduced by various oil-wealth-related effects: rising incomes, accelerated urbanisation and subsidies for fossil fuels. On the other hand, firewood seldom constituted a serious degradation concern in humid forests, except for periurban forests where harvesting rapidly gathered speed with rising urban markets.

What about the harvesting of non-wood products? One type of degradation that did stand out as a serious problem was over-hunting stimulated by the bushmeat trade, especially in Central Africa. 'Empty forests' created by this phenomenon of defaunation frequently jeopardise long-term regeneration processes. While there was no clear link to oil wealth from the bushmeat demand side (see later), there seemed to be a clear supply-side connection to logging as the main access provider. If logging was obstructed by oil rents, this also partially reduced defaunation. Finally, many other non-timber forest products (forest foods, medicinal plants, etc.) are being used 'counter-cyclically': during periods of macroeconomic crisis, their consumption increases, which in some cases may lead to over-harvesting (see Chapter 6 on Cameroon). Correspondingly, during periods of oil wealth, harvesting pressures often seem to diminish – a pattern that conforms to the core hypothesis.

The role of budgets for agriculture, forestry and conservation

The RER mechanism was the most important, but not the only way oil wealth affected forests: funding allocations and specific policies with land-use implications also had an impact. The literature on SAPs has claimed that cuts in public budgets undermine forestry regulation and conservation on the ground, which contributes to forest degradation. Conversely, under an oil boom one would expect the same agencies to receive more money, which would thus enhance their capacity to manage and protect resources. As a simple set of hypotheses, one should then expect that:

1 More funding for agriculture helps agriculture expand, at the cost of forests.
2 More funding for forestry helps regulate timber extraction, reducing forest loss and degradation.
3 More funding for conservation helps protects forests, reducing forest loss and degradation.

If we look first at agriculture (1), the overall deforestation impact from oil-induced budgetary increases was quite limited. This was due to the weakness of several links in the causal transmission structure from funding priorities to forest impacts (see Figure 2.1, Chapter 2):

I. Agriculture received a fair, but usually less than proportionate share of the oil money Agriculture had a variable priority across countries, from 'very high' in Indonesia to 'medium' in Venezuela, Cameroon, Ecuador and Mexico and 'low' in Nigeria, PNG and, as an extreme case of negligence, Gabon. In addition, spending priorities were often discontinuous over time within countries. Agricultural (and rural-area) funding was usually much less than the corresponding share of GDP. However, in the quest of the oil-rich state to please all interest groups, agriculture was never totally neglected. The 'leakage' in this first transmission was thus moderate.

II. A large share of agricultural budgets was not designed to cause direct land-use changes Much of the money going to agricultural agencies went to increased ministerial employment in the capital, expensive equipment (new offices, cars, etc.), or even a new tractor factory, which in Venezuela at one time consumed half of the agricultural ministry's budget. While these budgetary allocations *could* eventually cause changes on the ground, the link is indirect and likely to be weak. This was an important filter in diminishing the 'deforestation efficiency' of agricultural spending.

III. The implementation of agricultural projects and policies was highly inefficient Even those projects that were actually designed to change land use seldom managed to achieve these changes to the extent planned. Adverse factors cited in most countries were corruption, bad project planning, insufficient market studies, and most of all biases towards the large scale which generally ignored the smallholder sector (see next point). Yet even subsidised credit schemes designed particularly for smallholders largely failed (e.g. a cattle programme in PNG), or else earmarked funds were commonly diverted by the recipients to more profitable non-agricultural uses (e.g. in Ecuador). Inefficiency was an important filter. But a notable exception was the various land-colonisation support programmes in the three Latin American cases, as well as credits for cattle-ranching, which seem to have enabled significant deforestation to take place.

IV. If implemented, agricultural projects and policies were often capital- and land-intensive Much of the agricultural budgets in the mineral-rich countries were spent on large-scale, capital-intensive parastatals, which it was hoped would be vehicles of modernisation that would do a better job than allegedly primitive smallholder systems. In reality, more than anything they proved to be vehicles of rent-seeking, producing 'white elephants' and, at best, expensive commodities at high cost. While inefficiencies were thus very large (see III), production methods on *newly established* plantations or large-scale farms were also very input-rich, mechanised, using much money but little land per output unit. Even the funds that actually came through to the implementation stage thus led to little expansion in area. In addition, nation-wide policies like fertiliser subsidies (at one point the largest single item in Cameroon's agricultural budget) also in some cases made *pre-existing production* more land-intensive, thus reducing pressures for land expansion.

V. If implemented, agricultural projects frequently extended into non-forest areas Finally, as another yet minor caveat, among those oil-financed agricultural projects that did come through all the filters and actually claimed new lands, a fair share did not cause damage to forests. In countries where non-forested 'land reserves' existed (Cameroon, Gabon, Venezuela), the limited expansion tended to first go into savannah zones or previously cleared areas – normally not out of a strong concern for forest conservation, but because access and soil preparation often proved to be easier than for forest conversion.

With few exceptions in which there was targeted frontier expansion, agricultural budget expansions thus did not manage to accelerate deforestation significantly. But the corresponding funding for forestry and conservation agencies was equally ineffective in stopping it. The picture thus mirrored the agricultural one in a number of respects. Perhaps factor I was more important here: forestry and conservation did not receive so much money in the first place. In particular, forest conservation remained almost completely donor-driven in Central Africa (Gabon, Cameroon) and thus detached from the oil cycles. In PNG there was domestically funded expansion in the early 1990s, but it was sharply reversed towards the end of the decade. The Latin American cases were the exception: in Ecuador and in particular in Venezuela, much of the protected area system was financed and established in the oil-rich 1970s.

In forestry, the funding situation was probably more counter-cyclical. During the period of peak oil revenues, countries like Gabon and Cameroon basically neglected both timber regulations and taxes: there were plenty of rents to capture elsewhere in the economy. Only during periods of crisis or mini-busts did forestry agencies receive a higher priority. In PNG, funding and personnel for the National Forest Authority declined markedly in the 1990s, despite record mineral export revenues. However, even in the presence of solid budgets, for many countries management efficiency on the ground remained a highly controversial issue in forestry. There often seemed to be an excessive degree of centralisation of power, institutional discontinuities over time, a disproportionate influence of large timber firms on policies, instances of corruption at the local level, etc. These factors were not all present in the entire country sample, but nevertheless they were widespread. It may be that the pattern of *fluctuations* in funding – booms and busts in budgetary allocations causing asymmetries – is disruptive of institutional efficiency. However, it also raises the question if it can be taken for granted that cuts in public funding always have upsetting impacts on natural-resource management objectives. Much seems to depend on the extent to which these institutions were effective in reaching their aims in the first place.

The role of expenditure on transport

Roads, other transport investment[13] and subsidies formed one area where the priorities of countries differed most, and where spending choices had a catalyst effect on forests. The construction of infrastructure proper through forested areas causes negligible *direct* impacts. For instance, the much-debated building of roads in the northern Ecuadorian Amazon region caused direct clearing of maximum 4,000 ha of forest (about 0.05 per cent of Ecuador's Amazon forest). For the Transgabonese railway in Gabon, the total direct effect of this mega-project was a permanent forest loss of 2,000–2,500 ha (0.01 per cent of Gabon's forest). But, due to the strong *indirect* link between roads and deforestation, they proved to be a decisive variable for the fate of forests.[14] Slightly simplified, we can distinguish between the following types of situation:

Scenario 1. Near total neglect of rural roads: static structure – no deforestation This is the situation we found in Gabon and PNG. Gabon has a total road density of 0.03 km/km^2: one of the absolute lowest in the world. In both countries, the road network has deteriorated since independence, and in particular after the start of their respective mineral booms. This was most evident in PNG, where the Australian administration had undertaken a significant extension of the road network in the 1960s. Road budget allocations shrank in the 1990s, and the bulk of what is received is spent on personnel costs, rather than on maintenance and new construction. Only urban road-building has received adequate attention. Consequently, in both countries there is also a very low proportion of vehicles. As an alternative, air services are highly developed in both countries as the preferred means of passenger transport. Freight (of logs, minerals, etc.) in Gabon is undertaken using river transport and the Transgabonese railway, which has had only negligible deforestation effects.

Scenario 2. High oil rents – incipient road construction: rural out-migration – to the benefit of forests This seemingly paradoxical situation was found in pre-Second World War Venezuela, in some remote areas in Gabon, and in Cameroon at the peak of the oil, coffee and cocoa boom. High rents and high labour opportunity costs mean that there is much more money to be made in the cities than in the countryside. In spite of the impact roads have in lowering transport costs to the market, this alone is not sufficient to make agricultural expansion worth while because the gap with urban earnings is too large. In such a situation, easier access to the cities accelerates rural–urban migration: roads are sufficient to 'pay the way out' for people – or help them integrate with the urban economy in a variety of ways (Waters 1997) – but not to make profitable the commercialisation of crops and forest products. Roads may also help to 'pay the way in' for cheap imported foodstuffs that out-compete local staples (e.g. in Gabon). Both these effects reinforce agricultural abandonment and forest re-growth. The example underscores the fact that, while roads in general are strongly associated with forest loss, they are not always a *sufficient* instrument for deforestation.

Scenario 3. Moderate oil rents – strong road building: high forest loss – now and in the future This situation was found in all the Latin American countries: in Venezuela, most of the process had already been finished in the 1950s; in Mexico, it accelerated with the 1970s boom. In Ecuador, an economic latecomer in the region, road expansion during the oil boom was strongest, and had the most severe forest impact. Cameroon, with its strong agricultural and especially logging interest groups, was another country where the road network increased strongly during the boom, especially in the 1981–6 period, and in particular in the humid forest zone. In all of these countries (including Indonesia), roads eventually promoted strong agricultural expansion. In all of them, part of the deforestation effect was also lagged, that is, forest loss around the new roads occurred only gradually over the next decade or two. Lagged deforestation was particularly strong when a strong bust followed a boom, thus turning relative prices sharply in favour of agriculture. Roads that were built from oil revenues then became vitally important for the post-boom revival of agriculture, as happened in Ecuador. The forest-loss effect of roads was generally stronger in frontier forests than in forest fragments. Deforestation was also generally stronger in periurban areas, where the increasing demand for food crops from growing cities directly stimulated the clearing of forests.

In the third scenario, it is notable that roads initially enabled forest degradation processes, skimming the most high-value products at an early stage of intervention. Indeed, logging was a main motive for road construction in Cameroon (especially in East

Province) and historically in Venezuela (especially in the Llanos region and the Andean *piemonte*). Road-building also enabled other degradation processes, notably defaunation (providing a stimulus to bushmeat supplies), usually in combination with logging. In Ecuador, road expansion seems to have been driven much more by agricultural expansion (except for the timber-producing Esmeraldas Province), while log extraction provided a convenient side benefit.

Finally, we should not forget the negative effect of fuel subsidies on forests. In Ecuador, at the height of the boom in 1980, these constituted 7.3 per cent of GDP. In Venezuela, 8 per cent of the whole oil bonanza was spent on subsidising energy costs, and even in the 1990s the consumer price of gasoline was cheaper than that of bottled drinking water! Some believe that a price distortion of this magnitude would help forests by promoting the use of fossil fuels and thus alleviating pressures from harvesting firewood. But this dynamics is overshadowed by a much stronger effect of energy subsidies, namely that remote areas suddenly become economically integrated into national markets because transport costs are held artificially low. Unlike roads, subsidies constitute a non-spatial, nation-wide effect that lasts only as long as the subvention is in place. This makes it more difficult to discern the impact on forests. But there is no doubt in the mind of this author that a price distortion of the size that occurred in Latin America in the early 1980s is bound to have had a strongly expansionary land-use impact, which eventually complemented the impact of road-building.

The role of directed settlement

Government-sponsored colonisation schemes that move thousands of people into forested zones where their only livelihood option is to clear forests to establish new farms in ecological conditions they are often unfamiliar with, and thus likely to damage – this is the conservationist's nightmare vision of a publicly financed directed settlement programme. With Indonesian transmigration in mind, this factor obviously had to be included as a separate comparative point when looking at the fate of forests in oil countries. Would oil wealth finance more directed settlement? If yes, would this be another government policy tool reversing the forest-protecting effect of oil wealth, just like road building? The country chapters have clearly shown that this was actually not the case. And although the nightmare vision of directed settlement certainly exists in the real world, this was again only one of three main scenarios found in the study countries:

Scenario 1. 'No empty space – no room for large-scale resettlement' One scenario was that all those land areas where people might possibly want to live were already occupied. Furthermore, native people would actively claim the land and be able to resist efforts by others to grab it, or even to make temporary economic use of it. This applied in particular to PNG, but the situation was similar in much of the humid forest zone of Cameroon, where customary land property rights were well-articulated, too. In addition, there were strong ethnic and cultural barriers to resettlement. In these cases, there might still be small-scale resettlements for which the state acted as a catalyst, such as in the case of plantations of tea, rubber, cocoa and most of all, oil palm in PNG. The state could also act on lands that nobody else wanted, for example, draining swamplands and using them to resettle people from over-populated regions. But that was the exception rather than the rule, and forest impacts were negligible.

Scenario 2. 'Open space available – get people into the forest' This scenario of 'bringing people without land to land without people' was valid for the main objectives of the Transmigrasi programme in Indonesia, and for several more sporadic schemes in Latin America, both now and in the past. These programmes had a *net impact in increasing deforestation*, by helping to open up new forested frontiers.[15] The rationale of these programmes was some combination of:

- a 'social escape-valve' function: relieving socio-political pressures in very densely populated areas (such as Java or the Interandean Valley);
- economic development: pressures to increase cultivated area to raise national food supply (e.g. the historical settlement of European farmers in Venezuela);
- geopolitical motives: safeguarding thinly populated border areas, which in some cases have mineral riches, against squatter invasions from the neighbours (e.g. in Ecuador the Cordillera del Condor, on the border with Peru; Venezuela's 'Conquest of the South', on the Amazon border with Brazil).

However, in the Latin American cases, directed land-colonisation schemes in most cases had a poor success rate. They have now increasingly been replaced by *government-assisted* schemes of spontaneous colonisation (see 'Agricultural budgets' section above).

Scenario 3. 'Open space available – but get people out of the forest' This scenario, in contradistinction to the previous one, implied the directed resettlement of scattered populations living inside forests to concentrated roadside settlements – the creation of a 'linear space'. Using incentives and force, this state instrument was exploited in population-scarce areas to:

- obtain an increased supply of labour for plantations and encourage the 'modernisation' of production;
- produce economies of scale in the provision of social services (schools, medical assistance);
- increase political control over scattered ethnic groups.

The most dramatic case was the policy of *regroupement* in Gabon between 1940 and 1970, but this was far from being an isolated case. In Cameroon, a similar process occurred historically, in a slightly softer version. In Venezuela, some of the early missionary settlements in the south had similar features, and even within the Indonesian transmigration programme, a considerable number of people were also resettled *out of* the forests, to concentrate them near provincial centres (P. Levang, personal communication, Bogor, 10 September 2000). The *net impact on forests was reduced forest-clearing*, as a result of the introduction of more sedentary agriculture with shortened fallow and permanent plots, replacing land-extensive shifting cultivation systems in their place of origin.[16] For instance, this was a main component behind the story of forest re-growth in Gabon. Indirectly, in some cases redirected settlement to roadside villages also accelerated out-migration to urban areas.

How were these resettlement scenarios related to the oil cycles? The connection was weak at best, perhaps even counter-cyclical to periods of oil wealth. It is possible that oil money paid for a minor share of the (forest-protecting) *regroupement* process in Gabon, though this was mostly completed before 1970. In Cameroon, the process was definitely concluded prior to the oil boom, at least that part of it that affected the humid forest zone. In Venezuela, oil historically ended some (deforesting) efforts at colonisation because both settlers and the government lost interest in agriculture. In Indonesia, transmigration only became a strong factor in the 1980s, that is, when oil revenues were reduced. On the whole, directed settlement, whether it actually protected or sacrificed forests, was probably more important in bust than in boom periods, and also more of a historical than a contemporary phenomenon.

The role of poverty alleviation

The core hypothesis in this book is that (macroeconomic) oil wealth protects forests. It would *a priori* seem a strange 'micro-macro paradox' if on the one hand that hypothesis holds, while on the other there is no microeconomic relation between a nation's (poor) individuals becoming better off from oil wealth and the fate of forests. The findings above indicate that, in spite of highly unequal income distribution in some of the study countries, there is a clear positive correlation between oil booms and the alleviation of absolute poverty. This is generally true for urban poverty and, except perhaps for the highly PNG and Gabon, also in most rural areas. In other words, oil wealth does 'trickle down' to the poor, or at least some of them. The symmetric causality is also true: during busts, urban poverty rose significantly, as was clearest in the long-term crisis-hit Cameroon and Venezuela.

Yet the onward link from poverty alleviation to deforestation is a much more complicated matter. Initially, it depends on whose poverty is being reduced. Which constraints will be alleviated for the new wealthier labourers, producers and consumers, respectively? How will their improvement change economic behaviour and its impact on forests? The full answer to these questions includes factors that will also be treated in the next two subsections, that is, how changing income and poverty levels have affected migration and consumption patterns. But for the time being we should again distinguish between two different basic scenarios:

Scenario 1. Rural safety nets: reduced poverty reduces both deforestation and forest degradation
This scenario of an unambiguous, or at least clearly dominant forest-protection effect, was probably a key factor in three of the five primary cases: Gabon, Cameroon and PNG. In Gabon and Cameroon, the positive effect of poverty alleviation on forests mostly worked through rural–urban migration (see next section). High oil rents drove up urban wage levels, drawing people to the cities, as remuneration levels in labour-intensive agriculture were no longer competitive. Conversely, urban bust, unemployment and higher poverty either partly pushed people back to the countryside to resume slash-and-burn agriculture, as their economic default option (as in Chapter 6 on Cameroon), or they would go into the informal urban sector and practise 'weekend farming' in periurban areas (one common response to mini-crises in Gabon). In PNG, there was also growing urban unemployment

during the crisis in the 1990s, but a strong customary system of income redistribution and remittances (*wantok*) reduced 'poverty-push' responses, as absolute deprivation remained limited. The village always remained the vital safety net. Furthermore, people in rural areas of PNG in receipt of direct oil or mining compensations responded unanimously in a 'full-belly' way by scaling down agriculture and forest-clearing to produce more leisure time. Forest degradation from the over-exploitation of 'inferior' forest products (certain open-access NTFPs) also typically declined in all three countries with higher alternative labour-remuneration and lower poverty levels. Conversely, people fell back on these options during bust periods of economic crisis. There is thus a negative correlation between poverty levels and deforestation and/or forest degradation overall. In this scenario, two contrary but subordinate factors are a general rise in income increasing overall demand for land or causing extraction of forest products that depend on higher income levels, such as timber for the domestic construction industry.

Scenario 2. Rural investments and safety nets: ambiguous poverty impacts on deforestation and degradation This is very much the Latin American scenario (Venezuela, Ecuador), though also with some application elsewhere in the country sample. One typically finds all the effects sketched in Scenario 1, for example, historically a reduced poverty-push migration in Venezuela and less forest degradation in both countries, as a result of the oil boom. However, one additional and strong effect of decreased poverty was that higher urban incomes stimulated not only the demand for construction timber, but also for cattle meat and dairy products. At the same time, better-off rural producers also have more money to invest in cattle. As Latin American deforestation is so dominated by forest conversion to pastures, this easily became the dominant effect of reduced poverty, Ecuador being an example. Some of these elements could also be found elsewhere as a 'deforestation-for-investment' scenario of cash crops (estate crops in Indonesia, cocoa in Cameroon's forest zone): impoverished, capital-scarce producers have less ability to respond to market opportunities (e.g. rising cash-crop prices), thus reducing their ability to clear forests.

The role of rural–urban migration

First, a key argument behind the core hypothesis is that most oil wealth tends to be spent in urban areas. As Professor Roland Pourtier remarks of Gabon: 'The rent was converted to towns'.[17] A second factor is that most NT sectors, which benefit from the Dutch Disease, are found in urban areas: government and private services, urban construction, protected industries, etc. all expand. As a result of this urban bias, more people move from rural to urban areas to take advantage of the boom. This implies that they are giving up livelihoods that depend mostly on the conversion and cultivation of the land, in favour of activities that add economic value to products or services, without directly using land resources as an input. This premise implies that, although certain periurban forests may suffer from this demographic reallocation, in net terms urbanisation is unambiguously a good thing for forest conservation: rural–urban migration will help to bring down deforestation, other things remaining equal.[18]

Many observers would doubt this basic premise, arguing that growing cities leave a sizeable 'ecological footprint', and that urban sprawl by itself (residential areas, urban infrastructure, etc.) occupies much space. The country chapters showed that, because of the transport costs involved, growing urban centres are typically supplied with a larger share of foodstuffs from more intensive, periurban systems; also, a certain share of food supply comes from imports. Regarding the urban sprawl proper, this proves to be negligible relative to total forest loss. For Venezuela, one source stated indignantly that Metropolitan Caracas was growing at a speed of one entire hectare per day. But even if this expansion was fully drawn from forests, the loss of 365 ha/yr would correspond to a mere 0.09–0.15 per cent of the estimated likely range of yearly forest loss in Venezuela. Another example is Indonesia. In recent decades, *Jabotabek* (Greater Jakarta) has been one of the fastest-growing urban agglomerations in the world. Between 1969 and 1989, satellite images show that forests in this area receded from 73,266 ha (10.9 per cent) to 43,691 ha (6.5 per cent). Yet, even calculating with the most conservative national deforestation figure of 600,000 ha annually, the roughly 2,000 ha/yr of forests lost to urbanisation correspond to only 0.3 per cent of Indonesian deforestation.[19] In other words, urban deforestation may be important regionally – and these changes may be conspicuous to an urban conservationist observer – but it is important to keep in mind the proportions: on the whole, moving people from rural to urban areas is a land-saving undertaking that is bound to reduce deforestation.

Growing urbanisation over time is a structural change that is basically occurring in all developing countries. Our study countries are at very different stages of this process, ranging from 17 per cent in PNG to 85 per cent in Venezuela. But to what extent did oil and mineral wealth accelerate the ongoing drive towards urbanisation? In the predominantly rural PNG, the speed of urbanisation actually *slowed down* during the mineral boom, due to both anti-urban policies and specific factors that hurt the urban economy (the exodus of foreign businesses after independence; growing crime rates, etc.). Yet in all other countries mineral wealth significantly accelerated urbanisation. Conversely, economic crisis decelerated the process. In most cases, rural–urban migration was driven by pull factors from urban rent allocation. In Gabon, the pull was created directly by the state through projects and employment. In Venezuela, it was more due to a general expansion of both public and private businesses, though the state remained the leading sector, and spent its oil rents in the capital in particular. In some cases, rural push factors also contributed, such as the labour-saving mechanisation of agriculture in Ecuador, but rent-determined urban pull was dominant.

A caution regarding data is necessary. Shifting trends in urbanisation do not always show up in official statistics, nor in the international ones that copy the official information. In some cases, such as Cameroon, no demographic census has been done in recent times. The official 'data' are thus mechanical extrapolations from old census data that do not capture the real trend changes. A minor additional problem is that statistics based on rural versus urban *residence* sometimes underestimate the growing role of the urban economy, and the corresponding decline in rural activities. In PNG, there was plenty of circular migration for temporary urban jobs. In Ecuador, rural residents commuted in large numbers from rural areas to the cities to take advantage of off-farm boom-led opportunities. In Central Africa, many urban migrants were in the most productive age groups; those remaining in the countryside were elderly, received remittances and had little capacity to clear the forest. Nevertheless, the statistical figures still provide a clear overall message: the pro-cyclical pattern of urbanisation strongly alleviated rural population pressures and deforestation. For instance, in Venezuela long-term rural population growth remained at less than 1 per cent, or about a third of the national rate of expansion. In thirty-eight villages of the humid forest zone of Cameroon, population growth was merely 0.75 per cent during the boom period, but 4.6 per cent during the bust period. This reversal of demographic change translated strongly into galloping rural deforestation in the humid forest zone (Sunderlin and Pokam 2002).

The role of the structure of consumption

Obviously, changes in competitiveness shift demand, in particular between domestically produced and imported goods. Yet, beyond these price-induced changes, a sustained higher income from a boom causes *structural* changes in demand. Luxury goods come to occupy a higher share of a richer household's budget, while inferior goods will decline. At the macro-level, the drivers of change are the two factors analysed in the previous two sections, namely higher incomes and urbanisation. As we have seen, the force with which these drivers operated was highly variable among the primary study countries, so we should also expect the changes in demand structure to vary much. Ranked from low to high degree of change, the situation since 1970 has approximately been as follows:

- in PNG, per-capita income and urbanisation were both rather stagnant;
- in Venezuela, incomes and urbanisation were high, but the former diminished in the 1980s and 1990s;
- in Cameroon, per-capita income and urbanisation rate both followed a roller-coaster trajectory;
- in Gabon, both income and urbanisation accelerated greatly from low pre-boom levels;
- in Ecuador, per-capita incomes rose strongly and enduringly, and urbanisation further accelerated.

Higher income and urbanisation both tend to change lifestyles and consumption habits, which can ultimately affect forests, for instance through the demand for forest products. Urban households use less firewood, both because more of them can afford to buy fossil fuels, and because storing firewood is awkward in a densely populated urban environment. On the other hand, urban households may use more charcoal for speciality foods (broilers, barbecues, etc.). As discussed above, urban construction booms were a frequent response to oil wealth, greatly accelerating national timber demand. However, apart from forest products, wealth-triggered shifts in food consumption were even more important for forests. Two factors proved to be vital, though aiming in opposite directions:

1. Animal product demand rises with higher income – 'protein factories' take their toll on forests
Growing urban areas with higher incomes demand more protein foods, that is, meat, dairy products, fish, etc. In Latin America, there is a strong preference for beef consumption, which, together with demand for dairy products, greatly stimulated cattle-ranching, the main cause of deforestation in this part of the world. Ecuador, with its large increase in income and its trade protection of the sector, was the clearest example. Livestock

production rose by 4.6 per cent/yr during 1965–81, which declined to 1.9 per cent annually in the 1982–9 crisis period, underlining its pro-cyclical character. Pasture areas increased 60 per cent during the boom (1972–82). In Venezuela, the rise of the middle class from 1940 to 1980 had a similar effect, but with the crisis, per-capita meat consumption declined after 1983. It was also moderated by a more liberal yet erratic trade policy that allowed for larger imports of meat during sub-periods. This contrasts with the examples outside Latin America. In PNG, the structural-change drivers were rather limited to begin with, but for the increase in protein consumption that did occur, pigs, poultry and fish were at least as important as beef – and they had a much more limited land-use impact. There was also some increase in beef consumption, but most of it came from imports, while some came from ranches established on already cleared land in the Markham and Ramu valleys. In Gabon, meat consumption rose sevenfold during 1973–84, but the share of domestic production fell simultaneously from 12 to 4 per cent. A few cattle ranches expanded, but only into savannah zones. In Cameroon, the situation was exactly the same: the boom raised demand, but increased supplies came from the savannahs and from imports. In Cameroon's humid forest zone, there were both pathological (tsetse fly) and cultural impediments – cattle as a 'capitalist agriculture' did not fit into traditional egalitarian structures. The latter was also an obstacle in PNG and Gabon. The deforestation impact of cattle in PNG, Cameroon and Gabon was therefore close to zero. However, in the two Central African countries, there was an accelerated forest degradation impact from defaunation caused by higher bushmeat consumption. There is tentative evidence that bushmeat is not an 'inferior' NTFP sought after in times of crisis, but rather a 'luxury' meat category that is slightly preferred when incomes rise.

II. Staple foods: substituting local food crops for imported cereals alleviates forest pressures Consumer preferences for staple foods vary with income and urbanisation, and staple foods vary greatly in their land-use impact. Again, in Ecuador substantial structural change triggered the strongest impacts, with a massive boom-accelerated switch from land-extensive maize, barley, potatoes and cassava to imported wheat and, in particular, rice, either imported or produced in a land-intensive manner in the Guayas lowlands province. In Venezuela, the long-term increase in incomes and the low income elasticity for food crops also favoured the massive imports of staple foods; cropped area remained stagnant for decades. In both Cameroon and Gabon, booming incomes and urbanisation induced a strong penetration of imported wheat and rice.[20] The corresponding reduction in the domestic production of plantains, bananas and different tubers was probably a key factor in reducing deforestation pressures. In Cameroon, the strong crisis then partially reversed that process, with lower incomes triggering a return to land-extensive domestic food crops. Only in PNG did the situation differ fundamentally, as consumers actually had a preference for domestic food crops over imported substitutes such as wheat and rice. In household consumption surveys, the income elasticity for sweet potatoes, the main domestic food crop, was slighter higher than for wheat and rice. This means that imports of the latter have expanded over the last four decades, but at a rate that is inferior to population growth.

In most cases, however, the wealth-induced increases in incomes and urbanisation triggered both a shift away from domestic staple crops and an increase in the demand for protein-rich food, especially beef. Where the former mechanism helped to alleviate forest pressures, the latter increased them. The balance between these two land-use effects was very different between countries. In Latin America, where the beef effect dominated, the expansion of pastureland became a main driver of deforestation. In the African cases and in PNG, staple-crop substitution was dominant. Slash-and-burn production of food crops was a main forest-loss factor, which was reduced by oil wealth.

Conclusions and perspectives: what determines the forest outcome?

Summing up

This book set out to investigate the core hypothesis that oil and mineral wealth helps protect forests through its macroeconomic impacts. While it is difficult to summarise all the observations made in the above discussions, ten main lessons stand out:

1 Most of the eight countries studied conformed to the core hypothesis in relative terms ('oil wealth *reduces* pressures to degrade and convert forests'). Some did so even in absolute terms ('oil wealth *reverses* pressures to degrade and convert forests'), while two of them did not conform to the hypothesis ('oil wealth *does not reduce* pressures to degrade and convert forests').

2 Absolute conformers (Gabon, pre-Second World War Venezuela) were characterised by very high oil rents both per capita and with respect to the size of their non-oil economy. The negative cases were both from Latin America, and a roads – colonisation – cattle nexus drove their contrary outcome.

3 Regarding deforestation, conversion to agricultural land (including pastures, fallows and degraded croplands) was the dominant proximate cause in all countries. Depending on this country, most of this land expansion occurred at the expense of forest cover, though savannahs were also targets of conversion in some cases. Agricultural production, especially for exports, was clearly held back by oil wealth. This happened through a fairly solid link between oil revenues and competitiveness, a strong link between competitiveness and cash crops and a more variable link to semi-traded food crops. Expansion of agricultural land could not be measured or analysed on a year-to-year basis. Where land scarcity induced intensification, land expanded at a slower rate than production. Still, it was clear that in oil countries, as in most of the tropics, the bad fate of agriculture was mirrored by the good fate of forests. The by far strongest causal force in tropical deforestation is agriculture. Indeed, those who largely blame deforestation on other, readily identifiable villains (loggers, miners, urban developers, etc.) miss the key reason why forests are disappearing.

4 Regarding forest degradation, the expansion of selective logging (whether 'unsustainable' or not) was crucial, through both direct impacts and access-providing effects (e.g. for bushmeat extraction, conversion). Yet timber production was even more sensitive to fluctuations in competitiveness than agriculture. During oil booms with real appreciation it declined, while during busts with currency devaluation it expanded. This effect was stronger for exports, while domestic timber consumption rose in response to the oil-induced urban construction booms. Thus, through this

mechanism, oil wealth also partially protected forests from degradation by reducing the incentives for logging and for over-harvesting of wood and non-wood forest products.

5 Trade policy played a changeable and ambiguous role in land use, but for a number of key sectors protectionist policies stimulated a significant over-expansion that had large costs in terms of forest-clearing and degradation. First and foremost, cattle-ranching in Latin America expanded into marginal areas with poor returns and declining carrying capacity, often because restrictions on the imports of meat and dairy products sheltered domestic production from competition, at least during sub-periods. Certain land-extensive food crops were also allowed to expand (or saved from decline), at times by erratically applied import prohibitions or prohibitive tariffs, in spite of an appreciated RER. Finally, log-export bans or impediments, and some import restrictions, promoted the over-expansion of highly inefficient home-market timber sectors that consumed a lot of wood for low-value products.

6 Oil production in itself is a negligible direct source of deforestation, compared to national land use. Its direct degradation impacts are variable, and have in many cases declined over time through better practices. The same is true of mining, though its effects can be more significant: there are some examples of severe forest loss caused by mining. Like logging, oil and other minerals can provide access for other forest degradation and conversion. Again, this effect is variable between countries.

7 Government spending of oil revenues generally had an urban bias protecting forests. Some potentially damaging forms of spending, like directed settlement projects inside forest areas, were not related to greater oil wealth. Others like agricultural spending did rise with oil wealth, but did not greatly affect forests due to inefficiencies and the restricted expansion of land use. Yet the same is often true of the reverse situation of inefficient spending on public forest and conservation agencies. On the whole, much of this funding came to be part of the urban nexus of employment, rent-seeking and inefficiency, which drew resources away from productive sectors, including those exerting pressures on forests.

8 A minor exception to this pattern of government spending without direct implications for forests was some of the Latin American forest colonisation projects. A major exception was oil-wealth spending on roads (complemented by transport subsidies), which as a policy choice proved to have a catalyst effect on the forest outcome. Generally, there was a correlative pairing of scenarios, with 'no roads – no deforestation' versus 'substantial road-building – high forest loss', both currently[21] and over the next decade after road completion.

9 Urbanisation was greatly accelerated by high oil rents and was an important part of the 'oil-wealth package' that reduced the pressures on forests in most cases. As such, it was linked to poverty alleviation: rural–urban migration was a major route out of poverty, and the higher levels of remuneration for urban labour reduced the incentives for rural forest- and land-dependent activities. Conversely, in times of crisis the village, together with the forest-consuming production of food crops, became a rural safety net when urban incomes fell. At the same time, higher incomes also shifted demand away from land-extensive staple crops, such as tubers and plantains. This suggests a clear link between reduced poverty, changing consumer preferences and

reduced pressures on forests, which indeed was confirmed in PNG and in the Central African cases.

10 However, there were also some confounding factors, mostly but not exclusively in Latin America. Greater urbanisation and higher incomes also increased the demand for construction timber and, first and foremost, protein-rich foods (in particular beef), which require extensive pasture areas. At the same time, poverty alleviation also alleviated constraints on rural capital, thus promoting higher investments in cattle and cash crops. Wherever this contrary effect became significant – because of any combination of profitable investment opportunities in deforestation, an urban middle-class with greater purchasing power, or a special taste for hamburgers, charcoal broilers or precious-wood furniture – the overall effect of poverty alleviation on deforestation became ambiguous.

Preconditions make a difference ...

The above discussion has shown that, while oil wealth in the majority of cases promoted forest conservation, a number of factors contributed to different scenarios with variable forest outcomes. The 'Latin American syndrome' seemed to cluster most of the contrary effects, although there was no perfect geographical congruence in the division of scenarios. But, how much of the variability in forest impact could be attributed to country- or even continent-specific preconditions, compared to different policies adopted autonomously in oil-rich forested countries?

A first observation is that the only two 'absolute conformers' to the core hypothesis were also the two countries where the size of the oil rent was greatest in relation to the population and to the non-oil economy. Venezuela (historically) and Gabon were the only two countries where the transformation to a rentier economy was so great that some agricultural areas were abandoned and previous deforestation trends completely reversed, allowing national forest cover actually to increase. Thus one obvious deduction in regard to the forest impact of oil wealth is: 'size matters'!

Yet, as most sensible people would suspect, size is not the *only* thing that matters – indeed, 'culture' is also important. For instance, a strong taste for beef and a cowboy tradition triggered pasture expansion in Latin America, while lacking social acceptance of a cattle-led capitalist type of agriculture among the Central African forest people was the main obstacle impeding that scenario there. Another example is the preference for domestic food staples in PNG, compared to their 'inferiority' in all other countries, that is, the fact that when incomes increase they lose relative significance to imported cereals (rice, maize or wheat). Yet, this had land-use implications that differed markedly between PNG and Central Africa, with much more land extensification in the latter case. The case of PNG demonstrated better than any other country discussed in this book how important a range of socio-cultural factors outside the economic sphere potentially can be, in particular land tenure, social organisation, labour mobility and labour-supply responses.

Obviously, 'technique' is another decisive precondition for impact. In particular, agricultural technologies were important. If cattle were the main culprit in Latin American deforestation, then shifting cultivation had this leading role in all the other countries except for Indonesia, where cash and estate crops have come to dominate over the past two decades. The forest impact of slash-and-burn techniques, in turn, depended greatly on increasing population pressures over time, although an oil boom could temporarily de-link the two factors. This happened in Cameroon, where the shifting balance between urban jobs (with a zero impact on direct deforestation), cash crops (with a relatively low intensity of deforestation) and food crops (with on average a high intensity of deforestation) jointly determined the forest outcome. Expanding food crops at the expense of cash crops, as happened during the crisis prior to the FCFA devaluation, therefore increased deforestation. In addition to technology factors, the product mix thus also played a complimentary role.

While the sedentarisation of agriculture (i.e. zero fallow length) will reduce deforestation, a reduction in fallow length with land-extensive shifting cultivation will not have the same effect, at least not with the deforestation definition used by FAO, and applied in this book (see Chapter 3). Consider the following example.[22] A household with a nine-hectare plot cuts and cultivates one hectare at a time, using it for two years before moving on. The abandoned hectare achieves its 10 per cent tree cover of 5 m height (and thus technically becomes 'forest') in four years after cultivation ceases. Thus, of the total of nine hectares being used by this family, 33 per cent (3 ha) will be deforested at any given moment, and the other 67 per cent will be secondary forest at some point in its eighteen-year succession. Assume now that the family's plot is reduced to only 5 ha, which is also farmed in 1 ha plots for two years before moving on to the next plot. Does this 'land intensification' reduce the incidence of deforestation per household? The answer is 'no': the maximum succession/fallow length is reduced to ten years, and 60 per cent (also 3 ha) will be deforested at any given moment. Because the initial stage of land use was highly extensive, marginal intensification through reduced fallow length occurs at the expense of secondary forests, so the additional clearing will be fully reflected in higher deforestation, as measured by the FAO criterion. This type of process was of great relevance for some regions in Central Africa, PNG and probably also in Nigeria, which is why forest loss was extremely high in these countries compared to the size of the economy, as shown in Table 10.2 above.

Finally, historical backgrounds were also decisive, that is, the development stage of each country prior to the arrival of oil wealth, with regard to both its economy and its forests. For instance, in accounting for the cattle-led effects on forest loss in Latin America, it was important that a relatively large urban middle-class was already in place, which was not the case in PNG or Central Africa. Otherwise, the demand expansion would simply not have been sufficient to be able to dominate the land-use picture to such an extent. It is also important to consider where each country was located on its historical forest-transition curve, because this set the stage for the underlying trends. For instance, in 1970 a country like Nigeria had already gone a long way towards the extinction of all its natural forest cover. With a current forest cover of just 6.2 per cent of land area, a population of 124 million growing at 2.9 per cent annually, a density of 136 persons/km^2, and a rural population basically relying on shifting food-crop cultivation, it is unlikely that any macroeconomic policy, nor the largest imaginable oil boom in the world, could *reverse* deforestation outright. The underlying pressures on land are simply too strong for agricultural areas to be abandoned and let forests grow back, as indeed happened in low-population Gabon. The best one could hope for in Nigeria is a slow-down in deforestation, which is what actually seems to have happened during the 1970s.[23]

Policies may count for even more: four observations on political economy

Can the analysis of this book point to policy recommendations that improve land use in our study countries, judged from a set of criteria at the interface of development and conservation objectives? This is only possible if, in addition to the deterministic preconditions listed in last subsection, oil-wealth governments had additional room for manoeuvre to actively shape the country-specific outcome of macroeconomic, land-use and forest variables. In principle, this is a debatable premise.[24] But if the lessons from the present book had to be summarised in two words, they would probably be: 'policies count!' In fact, the rent-rich petro-state had a much greater ability to steer free of both internal and external pressures than the average developing country. They were little concerned about the forest outcome of these policies, but had a strong ability to shape the economic development path, which then indirectly came to have a number of derived effects on forests. Before proceeding to an analysis of policy instruments in boom periods, it is worth seeking to understand the nature of these particular political-economy phenomena in petroleum countries in greater depth, using four interrelated observations.

First, booming oil revenues greatly enhanced autonomy for policy-making *vis-à-vis* external influences. A country like Indonesia depended heavily on donor loans, assistance and foreign investment inflows in the first pre-boom decade under Suharto, which also gave foreign economic interests much greater leverage regarding the design of macroeconomic policy. The oil revenues then basically turned the tables in the late 1970s and early 1980s, markedly improving the Indonesian negotiating position with respect to foreign donors and the Bretton-Woods institutions (Sunderlin 1993; Winters 1996). As the World Bank's top official at the time in Jakarta dryly expressed it: 'They didn't need our money and they didn't listen to what we had to say'.[25] At the height of their mineral booms, Venezuela, Nigeria, Indonesia and PNG all engaged in different campaigns to indigenise the economy by strengthening domestic capital at the expense of foreign interests. During the boom, oil wealth meant that the size of the cake was not a great matter of concern, so political attention was focused more on how to divide it. This would change drastically in those countries where the crisis of the 1980s hit hard, and the enabling conditions for economic growth became top of the agenda. Nevertheless, while Indonesia reacted rapidly to a changed external environment, most oil-rich countries deferred adjustment and resisted outside pressures. The most extreme case was Venezuela. The governments of Herrera Campins and Lusinchi negotiated with the IMF throughout the 1980s and obtained concessions in terms of debt rescheduling. But since no comprehensive conditionalities had been imposed, there was a prolonged period of 'non-adjustment':

> the institutions of the petro-state and pacted democracy conspired to encourage the short-term preference of governments to avoid necessary adjustments which might have eased eventual crisis – even at the expense of future economic productivity and political stability.
>
> (Karl 1995: 34)

A second observation is that the internal power of central governments was also greatly strengthened by oil wealth. In Indonesia, oil created an autonomous source of revenues, which diminished the degree of accountability to the regions, compared to a situation in which the central government had to collect the majority of taxes from local sources (Sunderlin 1993). In Ecuador, power was traditionally divided between the highlands and the coast, but shifted strongly during the oil boom towards the highlands, in particular the capital Quito, where oil rents were distributed. During the bust of the 1980s, it then partially shifted back to coastal Guayaquil, especially when the central political figure León Febres-Cordero assumed power (1984–8). He implemented a package of liberalisation policies that favoured coastal agro-export interests, at the same time pleasing the World Bank and the IMF with his policies of sharp devaluation, removal of export taxes and a shift towards a more open economy. But oil could also dramatically increase state power over, and abuse of, political opposition groups. There is cross-country evidence that oil wealth has an at least moderately negative correlation with the degree of democracy (Ross 2000). Case studies may be more illustrative, as in this description of the government in recently oil-blessed Equatorial Guinea: 'Before oil was discovered, the government often yielded to foreign pressure to allow some degree of political openness. ... Now that they have money, they don't care what the outside thinks. ... Oil has made them more arrogant, more intolerant'.[26]

Third, in Indonesia (Aceh Province), Nigeria (Biafra), PNG (Bougainville) and elsewhere, resource-rich federal states or regions within these countries could also contest the rights of central government to appropriate these underground riches. This could fuel armed conflict, civil war and a strong drive towards regional independence. It reflects a systematic trend among countries that are rich in natural resources (Collier 2000). Collier argues that most civil wars over the last three decades can be explained by the presence of spatially unevenly distributed rich resource rents from primary commodities, which provide the incentives, and often also the financial basis, for rebel separatist movements.[27] Cases include oil and other minerals, but also conflicts over rich timber resources (Doward 2000; Onishi 2000).

As a fourth observation, the qualitative analysis in the case studies in this book showed that the oil countries are extremely prone to corruption. In oil-rich states the electorate does little to support the government through taxes, so the political classes are less answerable to the electorate for their actions, which eliminates a key constraint on the politician's behaviour.[28] There is also some quantitative evidence for the strong oil–corruption link. The Berlin-based NGO Transparency International produces a global-comparative Corruption Perceptions Index each year, based on a survey of business people, academics and risk analysts, who each classify countries on a scale from 10 (highly clean) to 0 (highly corrupt). Table 10.8 shows the comparative performance of our study countries for the 1999, 2000 and 2001 surveys, except for PNG and Gabon, which were not included for any of these years. The results speak for themselves. For comparison, we also show Denmark and Finland, both of which scored top with practically 10 in all three years. Except for mid-placed Mexico, the other five oil countries figure continuously in the bottom third of the sample, ranked according to increasing degrees of corruption. Nigeria, Cameroon and Indonesia are scored among the most corrupt countries in the world. It seems likely that oil can explain much of this through its promotion of a rent-seeking mentality: 'Quickness, adaptability, and improvisation were valued over constancy, continuity, and discipline', as Coronil (1997: 319) notes for Venezuela, which at the individual level

Table 10.8 Comparing corruption incidence in the study countries. Corruption Perceptions Index 1999–2001

Country	*Rank*			*CPI score*[a]			*Standard deviation*			*Surveys*		
	1999	*2000*	*2001*	*1999*	*2000*	*2001*	*1999*	*2000*	*2001*	*1999*	*2000*	*2001*
Denmark	1	2	2	10	9.8	9.5	0.8	0.8	0.7	9	8	7
Finland	2	1	1	9.8	10	9.9	0.5	0.6	0.6	10	9	7
Mexico	58	59	51	3.4	3.3	3.7	0.5	0.5	0.6	9	8	9
Venezuela	75	71	69	2.6	2.7	2.8	0.8	0.7	0.4	9	8	9
Ecuador	82	74	79	2.4	2.6	2.3	1.3	1.0	0.3	4	4	6
Indonesia	96	85	88	1.7	1.7	1.9	0.9	0.8	0.8	12	11	12
Nigeria	98	90	90	1.6	1.2	1.0	0.8	0.6	0.9	5	4	4
Cameroon	99	84	84	1.5	2.0	2.0	0.5	0.6	0.8	4	4	3
No. of countries	99	90	91									

Sources: Transparency International 2000, 2001, 2002.

Note

a Scores range from 10 (highly clean) to 0 (highly corrupt).

meant that '[t]o be someone, one had to be clever, daring, and rich; and to be rich, one had to have the power to stand outside the law and above social constraints' (ibid.: 360).

From these four observations, the following picture emerges. The rich petro-state has considerable freedom to pursue its own, independent policies. It needs to conform much less than other countries to either external economic and political interests or internal pressures. In that sense, it has political room to deviate considerably from pre-established national and international norms, according to the ideological preferences of whoever happens to be in power. On the other hand, high and prolonged rents also change the character of the society and its people. Demands for personal favours and handouts will be more strongly articulated by individuals and vested interest groups. To please the maximum number of voters and supporters, the petro-state may completely lose fiscal control and the ability to react adequately to a bust. It may also be faced with general economic myopia, the increasing moral deterioration of its institutions, and violent conflicts over rich natural resources.

What does all this mean for land use and forests? Governments used their room for independent manoeuvre for things that had contradictory impacts on forests. The main controversies may have occurred in the areas of trade policy, exchange-rate policy and forest administration. For instance, Ecuador long resisted pressures by the IMF and the World Bank to liberalise imports of food crops, which sustained land-extensive production and promoted deforestation. On the other hand, policies also sustained an urban bias, export taxes and an overvalued exchange rate that delayed the expansion of export agriculture, to the benefit of forests. In the 1990s, diverging views about forest-sector practices induced the World Bank to push for domestic reforms through 'green conditionalities' – that is, an environmental agenda – as part of its structural adjustment programmes. In PNG, this process was much more successful than in Cameroon. Indonesia probably occupied an intermediate position (Seymour and Dubash 2000). Obviously, in all three countries the process could only be initiated during a bust period, when loans were urgently needed.

Thus country experience is probably variable; state autonomy was good for forests in some cases and bad in others. My personal estimate is that, on average, the forest-protecting consequences of autonomy may have predominated slightly, principally because a sustained urban bias and currency overvaluation were so strong, damaging the economy while at the same time conserving forests. In that sense, even the society-wide rent-seeking and corruption of the oil countries are in most cases part and parcel of a forest-protecting cluster that kept people busy in the cities, and away from productive, land-using activities. This is because the 'unproductive' petro-state's rent-seeking behaviour draws resources out of 'productive' activities that would have implied forest-clearing. This partial mechanism would lead to the controversial conclusion that corruption can help to protect forests. The counter-argument, put forward for Indonesia, is that rent-seeking and corruption can also 'infect' forestry-sector practices, that is, accelerate institutional disintegration and promote the ruthless over-harvesting of timber (Ascher 1998; Ross 2001). But looking at the causes and extent of forest loss in all the study countries in this book, the latter was probably a subordinate effect.

What policies counted most?

It was claimed above that policies accompanying oil wealth were fairly independent, and that they mattered a lot for forest outcomes, although no clear statement was reached as to whether that autonomy generally worked for or against forest conservation. But certainly not all policy instruments carried the same weight. Road-building was the single policy that clearly had the greatest impact on the use of forest lands. It also varied immensely across the country sample, belying the idea that roads emerge as a response to universal pressures for local development. Ecuador, the main case deviating from the core hypothesis, had a massive road-expansion programme because national infrastructural integration was made a specific government priority, and the oil revenues gave the government the room for manoeuvre to implement this policy. At the other end of the scale, Gabon and PNG let their rural road networks deteriorate. For PNG, it became obvious from interviews that rural producers were effectively constrained by the impassability of some roads in their commercialisation of agricultural produce. This implies that better road access or maintenance would have been a sufficient condition to clear more land for agriculture, independently of any other preconditioning factor.

Another instrument with a high leverage on forests was trade policy. General trade liberalisation or protectionism had ambiguous impacts, but trade restrictions proved detrimental to forests in three areas: cattle, food crops and timber. Consider the case of Ecuador. Imagine that, counterfactually, the country had a more modest road-building programme below the Latin American average, that its credit programmes had not been widely earmarked for cattle and, first and foremost, that beef and dairy imports had been fully liberalised throughout the entire period after 1970. Even considering other preconditions that were favourable to cattle in Ecuador, it seems obvious that most of the huge expansion into marginal areas that occurred in reality would have been curtailed. Some of the cheaper-quality meat-cuts and some fresh dairy products would probably still have expanded, due to the 'natural' protection provided by the transport costs for imported substitutes. But imports of milk powder and of higher qualities of meat would clearly have soared, as they did in other countries of the sample.

For the sake of symmetry, consider now the case of Gabon, where meat consumption rose sevenfold during the oil boom, almost all of which came from imports. Imagine now hypothetically that some preposterous Gabonese dictator had got (away with) the following brilliant ideas:

- to prohibit imports of meat and dairy products entirely;
- to use half of the budget for public investment to build roads connecting Libreville with the provinces;
- to earmark half of all agricultural credits exclusively for cattle-ranching;
- to initiate a large-scale cattle-assistance programme, including the eradication of the tsetse fly.

Combined with the pre-existing increase in demand, this hefty cocktail of supply-side and demand-side policies with a touch of Latin-American inspiration, or even just a subset of it, would have created huge economic rents for cattle-ranching in Gabon. What would have been the domestic production response? Obviously this question is difficult to answer, but surely in the short term some expansion of production would have occurred into areas of savannah. In the medium term, as demand would have risen further and idle savannah areas became scarce, it is almost unthinkable that all the additional demand for protein would have been diverted continuously to poultry and fish, and that nobody would have thought of converting some forest into pastureland. In spite of inherent cultural impediments, one would perhaps first expect immigrant farmers from neighbouring countries with cattle-ranching experience (e.g. from northern Cameroon) to take up this economic opportunity, while others might have followed at a later stage. The counterfactual claim made here is that a set of very large and sustained economic incentives of the type applied in Ecuador would tend to remove most unfavourable preconditions in the medium term.

In addition to road-transport and trade policies, a third vital instrument was government spending on urban areas, which promoted migration and relative price shifts in favour of non-tradables. A fourth, related policy was the management of the nominal exchange rate. At the one extreme, this meant the 'hard-kina' policy in PNG of letting the currency appreciate continuously, and at the other Indonesia's repeated rupiah devaluations, which were aimed at restoring competitiveness in the agricultural, land-using sector as quickly as possible. Yet it is just as important to identify the policies that did *not* have an important impact on forests. Notably, public budget allocations for spending on agriculture, forestry and conservation were subject to a series of filters that in most cases greatly divorced them from having any (intended or unintended) impact on forests. Too often, the link was simply excessively indirect. In cases where the land-use link was more direct, as with public resettlement schemes, both the link to oil wealth and the ultimate effect on forests was ambiguous.

Policy recommendations

The 'improved Ecuadorian recipe' for forest destruction

Finally, in order to distill policy messages from the foregoing, we are now ready to identify the clusters of factors that protect forests, and to compare them to those that are likely to increase the pressures on forests. This exercise is not only relevant for the oil countries we have been discussing: it will also provide more general lessons about the links between macroeconomic factors and forests, thus adding to the theoretical framework described in Chapter 2. The best way to make this exercise the most realistic is to start out in the real world. Choosing as the first case study a country that was highly forest-destructive in managing its oil wealth, namely Ecuador, a first step is to see which ingredients actually 'worked' in effectively destroying forests. A second step is to add policy instruments from other countries that may add spice to the cocktail and maximise the negative impacts on forests. A ten-point formula for governments might read as follows:

1 Make rural road-building and road-maintenance a major government priority.
2 Spend a lot of money in the countryside, especially in agricultural frontier areas.
3 Make gasoline consumption ridiculously cheap through subsidies.
4 Devalue the exchange rate frequently to raise the competitiveness of agriculture and logging.
5 Give generous forest concessions to encourage the entry of forest-based 'hit-and-run' industries.
6 Protect the most land-extensive sectors (cattle-ranching, food crops, etc.) from import competition.
7 Earmark credits with negative real interest rates for the same land-extensive sectors.
8 Resettle people inside forested areas, or assist them to go there 'spontaneously'.
9 Make secure private land tenure contingent upon the continuous clearing of forests.
10 Abandon all family-planning programmes in favour of a pro-natalist strategy.

As can be seen, half of this list contains genuine Ecuadorian elements (1, 3, 6, 7, 9), while the added spices could, for example, come from Indonesia (2, 4, 5, 8). Point 10 on population policies is probably a controversial one, but it is nevertheless included because it is a main underlying 'slow driver', especially in respect of the expansion of land-extensive food crops. Demographic transition has been initiated in some densely populated countries (Indonesia), while in others it has not (Nigeria). In some countries, higher per-capita incomes have accelerated the transition (e.g. Mexico), but in others in the same continent, population growth remains relatively high (Venezuela, Ecuador). This indicates that there is room for a policy that in the long run will also have an impact on forests. Finally, note also that some of these policies are frequently recommended as 'good' development policies – road-building (1), public investments in rural areas (2) and currency devaluation (4) – while the others would be criticised by most people as being problematic in both environmental and economic development terms. Unfortunately, the analysis indicates that these three 'good' policies, if implemented strongly, can cause a lot of forest loss in their own right.

The 'improved Gabonese recipe' for forest conservation

Conversely, Gabon offered the opposite basic scenario of a set of policies that were highly favourable to forests. In Chapter 4, this was called the 'Gabonese recipe' in achieving maximum forest conservation. In the following list, the Gabonese elements have been slightly modified and supplemented:

1 Neglect demands for new road-building and road-maintenance in rural areas.
2 Draw as much labour as possible out of rural areas by spending all your money in the cities.
3 Sell gasoline domestically at prices that produce similar rents to exported fuels.
4 Let the RER appreciate markedly, and then keep it systematically overvalued.

5 Tax logging companies heavily so as to keep their operations at the margins of profitability.
6 Over-tax export agriculture by confiscatory so-called 'price-stabilisation schemes'.
7 Intensify food crops and/or liberalise food imports.
8 Force rural people to settle in concentrated roadside agglomerations.
9 Waste your agricultural budget on agro-industrial 'white elephants' and ignore smallholders.
10 Nourish a rent-seeking environment in which few people find it worthwhile to produce.

The ten-point list 'reverses' the main elements of the Ecuador-inspired scenario above, but it also replaces a number of points that were deemed to be more important for the forest-protecting cluster, such as the rent-seeking mentality (10) (present in all countries) or the possibility of the land-saving intensification of food crops (7) (most prominent in Indonesia). The true Gabonese scenario comprises points 1, 2, 6, 8, 9 and 10. Factor 4 (overvalued currency) was also present, but much more articulated in PNG. Factor 5 (logging taxes) reflects best the situation in PNG in the late 1990s. Probably no country in the sample followed recommendation (3), to sell gasoline domestically at fully competitive prices, but the degree of subsidisation was much less in some countries (e.g. PNG) than in others (e.g. Venezuela). Again, note that only (3), (5) and (7) are elements that would be considered desirable as development policies: most of the rest would be characterised as bad for the welfare of a country's citizens. Viewed as a strategy of conservation, it is striking that, of the ten points, only one (5) is directly related to the forest resource and its protection. All the other elements are indirect, yet strong measures of protection. Moreover, the countries that applied them did not do so because they cared much about preserving their forests – mostly the contrary! These were 'blind' conservation strategies, adopted for completely different purposes, which proved to have powerful side-effects that relieved the pressures on forests. This stresses the dominance of extra-sectoral factors, that is, issues originating outside the forests and the forestry sector, in determining outcomes that were decisive for forest cover and quality.

Policy recommendations: some notes for the Minister of Environment

Bearing in mind this comparative picture of preconditioning factors and policy impacts, what policy changes could possibly be considered to reduce excessive forest-clearing? What recommendations could be made to governments in mineral-rich countries, as well as, to some extent, to those in other tropical countries facing similar structures and constraints? While we approach the problem from the forest-conservation side, thus taking into account 'Northern' environmental concerns, our recommendations would certainly need to be relevant for the synergies and trade-offs *vis-à-vis* national development objectives.

Let us imagine that we are preparing a set of annotated points for consideration by the Minister of Environment in an oil-rich country.[29] We assume, rather optimistically, that 'our' minister has the ear of the senior ministers in the development sphere (Prime Minister, Finance, Agriculture, etc.), and that they are willing to take environmental viewpoints into account if these can be integrated sensibly into the country's development strategy. The Minister of Environment him-/herself has good backing from international donors, who are currently funding a series of projects that together make up 80 per cent of the Ministry's budget. These donors also directly or indirectly finance a number of local environmental NGOs, which have helped create an incipient domestic constituency for forest conservation. Our minister sees his responsibilities mainly within three areas of action:[30]

(a) advocate certain domestic policies that would favour forest conservation to other government bodies, with the aim of finding broader political support;
(b) serve as an environmental advocate and educator in respect of the general public, strengthen environmental values and awareness in civil society, and recast social thinking on the benefits for future generations and the precautionary management of finite resources under growing environmental risks, in order to advance sustainable production programmes, certification, encourage lower rates of population growth, etc.; and
(c) serve as a bridge between his government and other governments and international organizations to facilitate the inflow of environmental aid and technical assistance, obtain compensation for environmental services, promote research into the country's wild resources, but also advocate the reduction of externally created pressures on these resources in international forums (irresponsible conduct by multinational companies, trade-induced unsustainable harvesting, superfluous consumption patterns, etc.).

Our terms of reference, however, would mostly be to contribute to (a), that is, to help identify more rational and sustainable policies at the national level. Referring to (b), the minister would stress that the conservation of forests and their various ecological functions is generally a wise policy that in the long run benefits nature and human beings alike as part of a single integrated system, so that 'development versus the environment' becomes a misleading dichotomy. However, in discussions with economic interest groups and other ministers who do not necessarily share this view, it would be good to know which policy proposals would be rewarding both for forests and for people's short- and medium-term aspirations for development. We would call these 'win–win' policies, and they would constitute our most forceful arguments for acting on the offensive, with policy proposals that actively improve forest conditions.

We know that it would be much more difficult to receive political attention for proposals that benefit the environment at any significant cost in hampering development here and now ('win–lose' policies). But there could also be policies where this trade-off is limited, that is, where conservation benefits come at the cost of restricting economic benefits that are only marginal. To emphasise this in a way that sounds attractive to our dialogue partners, we would call these 'win more (forest) – lose less (development)' policies. In order to be proactive, we would argue that the international community would probably be willing to assist us financially – that is, with respect to point (c) – if these policies were implemented in an effective way. On the other hand, we know that economic interests will have an overwhelming primacy in any negotiations involving the government, so a good protective approach will in many cases be crucial. Thus we would also want our minister to influence development policies so as to minimise unavoidable forest losses in quantity and quality terms – 'lose less (forests) – win more (development)'.

In the following, let us look briefly at the situation country by country, using the five primary cases only, as the three secondary ones were not really treated in sufficient depth to make specific policy recommendations. We will then close with a couple of considerations that provide global perspectives on these main suggestions.

Gabon

Gabon would probably be the country where the negotiation strategy should be the most defensive and protective, in the sense of finding adequate reactions to forest threats that will almost inevitably emerge. This is because of the scenario of declining oil reserves, under the assumptions that large new oil strikes are very unlikely, and that no alternative rent sources are available that could take the place of oil. In other words, although per-capita incomes remain high and pressures on forests rather limited, living standards will decline notably. Dutch Disease effects will decrease, but slowly because the currency is tied to the euro. This will make any national devaluation impossible, and will probably delay the adjustment of relative prices. But there will be a gradual reorientation towards land-using sectors in Gabon, especially agriculture. Given this scenario, the environmentally most important policy responses would be:

1. Limit the decline of the urban economy through good macroeconomic management, safeguarding as much employment as possible to avoid mass return migration to the countryside. Try to develop alternative urban-based activities in the areas of services and value-added, taking advantage of those fields where Gabon has (or could rapidly obtain) a good supply of educated labour (win–win).
2. In spatial terms, accept increasing periurban deforestation as a necessary evil, but try to rationalise the greater pressures predicted towards the occupation of rural space, so that forest loss is minimised and that particularly biodiversity-rich forests are not compromised (lose less–win more).
3. Even under a recession, keep a liberal import regime for meat and food crops, especially for the most land-extensive ones. Dedicate resources to increase yields so as to make food-crop cultivation more land-intensive (argue that this is a win–win strategy).
4. If put under pressure, accept the idea of additional investments in rural roads, but try to steer most of it towards the improvement of existing roads (lose less–win more).
5. Tax logging companies heavily so as to maximise current rent per timber unit and make the speed of extraction sustainable, arguing that a stable long-term source of foreign-exchange inflows is needed as oil revenues decline.

Venezuela

In the other high-rent economy, Venezuela, defensive-protective approaches are also needed, but because of the general political and economic crisis rather than an accelerated decline in oil revenues. Capital outflows have occurred, and following the political turmoil of coup attempts against the populist Chávez government, political instability abounds.

Oil prices fluctuated greatly in the 1990s, but attempts at stabilisation have been ineffective. Prime factors influencing the environmental outcome would be to:

1 Safeguard labour absorption in the urban economy through more rigorous macroeconomic (especially fiscal) management, avoiding excessive currency devaluation. Increase investments in human resources (win–win).
2 Accept as a necessary evil some economic diversification of the petro-dependent economy, including some expansion of mining and logging, but try to restrict the most damaging types (e.g. artisan gold-mining or high-impact timber-extraction) (lose less–win more).
3 Liberalise imports for food crops, timber and especially meat and dairy products, so that these become genuinely 'traded sectors'. If these sectors are greatly damaged to the detriment of economic development, compensate by supporting their technological advance, but in ways that make them more land-intensive and less resource-using. This will be seen by some as 'win more–lose less', but argue that more resource-efficient domestic production and lower consumer prices are ultimately in the best interests of the economy.
4 Strongly resist additional investments in roads, unless this is absolutely vital for high-value sectors such as oil and mining. In these cases, take steps to control the indirect effects of spontaneous colonisation. Argue that existing road networks in the northern part are already highly developed, and that land resources here are sufficiently abundant so that agricultural expansion south of the Orinoco is not desirable (this may also be seen as 'win more–lose less').
5 Eliminate vigorously perverse tenure-related schemes, such as the economic compensation of illegal squatters and 'homesteading' rewards for deforestation. These schemes only favour individual vested interests, so their removal should clearly be a win–win measure for society.

Cameroon

Compared to the two previous cases of high-rent economies, Cameroon is a much poorer country, with harder trade-offs between environmental and development objectives. Following a deep and prolonged recession after 1986, the country has experienced a slow economic recuperation since the FCFA devaluation in 1994. In this situation, the following would seem a desirable route to follow from an environmental viewpoint:

1 Support measures that help revive the urban economy through good macroeconomic management and a path of stable economic expansion (win–win).
2 In spatial terms, accept increasing periurban deforestation as a necessary evil, but try to intensify food crop production (lose less–win more). Perhaps controversial to some: help revive the Human Forest Zone cash-crop sector, notably cocoa, because it provides better income alternatives per hectare for the rural population than the alternative of growing land-extensive food crops for subsistence and for expanding food-crop markets.
3 Maintain a liberal import regime for meat and food crops, especially rice. Provide credits and incentives for domestic rice production in the northern part of the country, reducing pressures to expand production of tubers and plantains from the HFZ (arguably win–win).

4 Limit additional logging roads, especially in the extensively forested East Province. Where this is impossible, control indirect impacts, especially bushmeat extraction. Try to obtain international compensations for the areas that are kept protected (win more–lose less).

5 Tax logging companies heavily, even though this may slow down extraction rates, arguing that fiscal revenues have to be maximised over time, not only in the present. This will also work towards reducing the important access-providing impact of logging (argue that this is win–win).

Ecuador

Apparently, the political and economic crisis in the last half of the 1990s reduced Ecuadorian deforestation, in the sense that few new roads have been built and the cattle sector has slowed down. In addition, Ecuador has recently adopted the US$ as its national currency. We expect that dollarisation will lead to a significant real appreciation in the medium term – Ecuador will become a much more expensive country. Because its economy fluctuates more than that of the currency-issuing US, in boom periods (e.g. with high oil prices) there will be higher demand and inflationary pressures leading to real currency appreciation *vis-à-vis* the US. This will be slow to reverse during busts because of price stickiness. It may take deep depressions to bring prices down, as happened in Argentina, except that that country maintained the ultimate escape clause of devaluation. The good thing about dollarisation is that it keeps foreign investment flows high because currency security eliminates the risks of devaluation. These additional capital inflows will contribute to real pressures on appreciation. Ecuador will probably rely more on oil and other high unit-value exports (mining, shrimps, perhaps bananas, etc.), but the heyday of booming agricultural exports may be over. If Ecuador is effectively tied simultaneously to trade liberalisation through the WTO process, this may also hurt import-competing agriculture. This scenario may be bad for employment and income distribution, but good for forests. Key policy measures would be somewhat similar to those for Venezuela:

1 Keep urban services and value-added sectors growing and stabilise inflation levels, although at an appreciated level of the RER (win–win).

2 Liberalise the trade of food crops, timber and, especially, meat and dairy products. Eliminate ear-marked cattle credits and other selective cattle incentives. Support technological changes to make cattle-ranching more land-intensive. Given widespread inefficiencies in domestic production, one could make the convincing case of win–win, at least in the medium term.

3 Resist a new wave of road investments on the basis of the disastrous land-use consequences this had in the 1970s and 1980s. Given the tight fiscal situation of the government, this may not be a large development sacrifice currently.

4 Eliminate tenure rules that provide 'homesteading' deforestation rewards, and cut the budgets of agencies supporting 'spontaneous' frontier colonisation. The latter may induce losses in agricultural incomes, but these will be small (win more–lose less).

5 Continue ongoing efforts in forestry-sector reform, including the fight against corruption and efforts to diminish logging impacts in the highly threatened remnants of Chocó (Esmeraldas Province).

Papua New Guinea

In spite of high oil and mineral revenues in the 1990s, PNG faced an economic crisis that was accompanied by a significant (nominal and real) depreciation of the kina. This created the potential for economic diversification towards cash-crop sectors, which currently remain under-exploited due to non-supportive framework conditions (a lack of roads, high crime rates, an inflexible land-tenure system, etc.). Thus deforestation remains closely tied to food-crop expansion, which is driven by population growth but is expanding at a less rapid rate, due to land intensification. In various ways, PNG has proven to be a paradox with distinct mechanisms at work, but nevertheless some of the policy recommendations concerning PNG are similar to those for the other countries:

1. Support reforms that help the urban economy to grow (reduction in crime rate, liberalisation of commerce, improved education etc.), accelerating urban labour absorption in services and value-added industries that reduce the chronic dualism of PNG's economy (win–win).
2. Keep the import of wheat, rice, dairy products and meat liberalised. This does not involve a radical policy change with respect to the status quo, but would rather be a measure ensuring that increasing domestic food-crop demand is at least balanced by some growth in imports.
3. Try to channel existing pressures for road investments into road improvements, for example, for the deteriorated Highlands highway down to the port of Lae, rather than new roads through forests.
4. Safeguard the present inflexible and non-transparent land-tenure system, although it causes frequent tenure conflicts and prolonged insecurities, because it provides an effective obstacle to the sustained expansion of cash crops. This is a true win–lose scenario, but due to the strong political resistance to reform, our minister is not likely to be exposed to much pressure for change.
5. Improve monitoring of logging impacts, and try to eliminate the most damaging types of extensive land use, especially in *piemonte* zones (e.g. the burning of forests for hunting).

A few reasons for optimism

The policy recommendations for our five countries have perhaps not provided an overly optimistic picture: many bad development policies are good for forests, and many of the sound policies in macroeconomic terms come at a forest cost. For instance, it is almost impossible to argue completely against new road-building in PNG or Gabon, both of which are forest-rich countries that are under pressure to diversify their economies, although we know that this is the most certain way of causing greater forest loss and degradation. Reducing overvalued exchange rates to diversify the economy also seems to make sense for most specialised primary commodity exporters, but we know that this will increase the pressures on forests in terms of logging and expansion of cash crops.

On the other hand, there seem to be some win–win exceptions to this pessimistic outlook. In some countries, forestry-sector reforms that reduce the incentives for over-harvesting would be beneficial for both the forest as a resource and for the economy. In others, the elimination of perverse land-tenure arrangements has the potential to reduce both unproductive speculation and deforestation – yet in PNG the anti-developmental character of tenure arrangements was a main protector of forests. In other words, the role of these factors varies across countries. But the following two win–win policies seem to have more general validity.

First, for all five countries the low-deforestation route depends on a scenario in which oil rents sustain urbanisation. But this can only be achieved continuously if these rents can be reasonably well managed and stabilised from the macroeconomic point of view. The example of Cameroon showed clearly that pronounced fluctuations – sharp boom-and-bust cycles – are bad not only for the economy but also for forests. This is predominantly a question of effective domestic policies, but the international environment also has a role to play. Notably, easy foreign borrowing in the 1970s, in some cases banks pushing third-world governments to accept new loans, exacerbated macroeconomic cycles, which would turn into bust in the 1980s with lower oil prices and sharply rising real interest rates. This, in its turn, would lead to SAPs with sharp currency devaluations promoting the rapid expansion, and in some cases over-expansion, of cash crops.

As we have seen, a sharp devaluation or drastic currency depreciation following a financial crisis also greatly promoted logging exports. Pronounced macroeconomic fluctuations thus implied an extra cost in terms of deforestation and forest degradation. Furthermore, areas once cleared for cash crops did not symmetrically return to forest once these cash crops become less profitable. As became clear in the case of Cameroon, farmers would keep a larger portfolio of land uses (e.g. under-utilised ageing cocoa plots) precisely to maintain a diversified range of future responses to minimise the livelihood risks from sharply changing external conditions. Similar risk-minimising land-use responses may have occurred in Indonesia following the financial crisis in 1997. For the rural household, it made sense to diversify, but if counterfactually the (expected) external fluctuations had been less pronounced, the household could have obtained more income from specialising on the most profitable crops on a smaller area and/or using less labour. The diversifying over-shooting of deforestation is thus not good for either the household economy or for forests. The average return to both household labour and land would be higher if risks and fluctuations were reduced, thus reducing the need to diversify and allowing the household to specialise more on the most rewarding activities.

To address this problem, it would be ideal if measures existed to reduce excessive fluctuations in capital flows and international commodity prices. In the current absence of such measures, it seems that the Bretton Woods institutions' encouragement of sound macroeconomic policies – striving for fiscal balance, avoiding foreign-borrowing sprees – are also beneficial in environmental terms. Too many times, myopic governments in oil-rich countries have deferred necessary economic adjustments to a bust, passing on the costs of accentuated crises to the population, and therefore favouring short-term strategies in the management of natural resources. If mechanisms existed for oil countries to enter into binding agreements on the use of oil windfalls – for example, reserving a share for external-debt repayments – this would arguably be to the benefit of the country as such.[31] A more stable growth in the urban economy, based on a stabilised flow of rents, would also be favourable to forest conservation.

Part and parcel of that stability in urban economies could, and probably has to be, the gradual development of manufacturing and other urban T sectors that help absorb labour.

One large obstacle to this process in the past has been import protection in developed countries, for example, in textiles, leather, electronics and other typically labour-intensive industries where developing countries can obtain a comparative advantage. A country like Gabon would probably have too high labour costs at present, among other impediments, to become competitive in these industries. But other countries with a more devalued currency are either already embarking strongly on this path (Indonesia, Mexico) or could do so potentially if a number of additional preconditions were altered (Cameroon, PNG, Ecuador).

This leads to the second general win–win option. This book pointed to trade protectionism in the cattle, food-crop and timber sectors as significant domestic drivers of deforestation in most of the study countries. Conversely, another claim entertained in this book has been that trade liberalisation within these sectors where land demand directly competes with forests would have reduced overall deforestation. Criticism of this claim would probably be centred around two points. On the one hand, trade is a prime motor of global economic growth, which, one could argue, also promotes excessive consumption. This is not the place to enter into a discussion of this factor. Suffice it to observe that periods of crisis with declining trade in our study countries mostly led to more rather than less deforestation, because people were forced to rely on primary-production activities that required extensive land areas. A second counter-argument would be that our national focus of analysis implies that increased food imports as such are bound to be good for forests, simply because this moves land-use pressures to another country, which 'does not count' in our balance sheet. In principle, this is a valid argument, but still trade and competing production options provided *de facto* an additional stabiliser that also generally economised on land use. In other words, to feed an African family with a bowl of rice from a Thai high-yield-variety ricefield is much less land-demanding than to feed the same family with plantains from a shifting cultivation plot in Central Africa. And to fill the barbecue of a meat-loving middle-class family in Quito, it is much less land-demanding to provide that meat from the Argentinean pampas than from an erosion-prone, nutrient-poor plot on the Ecuadorian flanks of the Andes. In other words, trade can also provide important environmental protection against the most perverse types of land-use expansion; it is a crucial lever to influence land-use and forest conditions. This is certainly a different environmental perspective on trade liberalisation than the one we see when NGO activists protest against new initiatives under the WTO. [32]

Thinking beyond oil

Throughout this final chapter, a number of policies have already been discussed which go beyond the issues narrowly relating to the forests of oil-exporting tropical countries. As closing remarks, there are three issues of broader interest we may want to address more explicitly: the environmental Kuznets curve (EKC), structural adjustment and debt relief.

The environmental Kuznets curve

In the mid-1950s, Simon Kuznets put forward the hypothesis that, over the course of a nation's economic development, income inequality would first tend to increase and then decrease at higher average income levels (Kuznets 1955). Much later, other authors also applied this idea to a variety of environmental factors. Tropical deforestation rates, like other environmental indicators, were supposed to follow an EKC, that is, a similar inverted U-curve over growing GDP per-capita levels: things would first get worse (loss rates would increase), before they could eventually improve (slow-down in forest loss) as average incomes grew and 'cleaner' technologies were introduced. As there are normally no single-country long-run time-series available, the alternative has been to use cross-country regression analysis to test this hypothesis, thus assuming that temporal phases of economic and environmental development can be mimicked by country differences at a single point in time. See Chapter 2 for a discussion of the caveats of this method. In general, we can summarise some main findings from these tropics-wide deforestation studies:[33]

- some studies confirm the inverted U-curve, but at low significance levels (much 'white noise');
- results differ greatly according to the sources and measures of deforestation that were employed;
- the curve is more often confirmed for Latin America and Africa than for Asian countries; and
- the estimated 'turning points' occur at very high GDP per-capita levels (mostly US$4,000–6,000).

Note that this literature has normally dealt with tropical developing countries only, and that it does not predict deforestation to stop but only to *slow down* – and to do so only at an advanced stage of economic development. However, another branch of the literature dealing with 'forest transitions' has also studied countries in temperate countries, including high-income developed countries.[34] Several of these studies find an *absolute expansion* in forest cover at higher income levels, although the forests that 'come back' are often plantations rather than biodiversity-rich natural forests. The main structural factors supposed to bring about this land-use change in high-income countries are the abandonment of marginal agricultural areas, a demographic transition reducing population growth, the shifting emphasis from commodity production to tertiary sectors, and the higher demand for forest-provided services (e.g. hydrology, recreation, etc.).

How does oil wealth affect the EKC picture? In both the cross-country studies by Mainardi (1998) and Sunderlin and Wunder (2000), a high level of a specialisation on mineral exports was found to significantly reduce deforestation levels, even when income effects had actually been controlled for. In simplified terms, we could visualise this as a parallel U-curve estimated for the mineral-exporting countries, with a lower deforestation at any given income level than for non-mineral countries. In other words, higher oil wealth does not any makes countries slide up or down the U-curve through income effects, it also partially *delinks* deforestation from GDP per capita. The structural factors responsible for that de-linking have clearly been identified in this book: growing tertiary sectors, higher urbanisation, stagnant agriculture and sluggish expansion in logging. In this way, some of the spatial development and land-use structures in specialised oil countries come to resemble those of developed countries with much higher income levels. The case of a long-term oil-exporter like Venezuela showed this quite clearly. However, the Venezuela case also made it clear that the degree of delinking may be reduced when mineral wealth decreases or proves insufficient to avoid economic crisis. In this sense, delinking countries permanently from their common stage-pattern of deforestation by means of oil wealth alone may not be a fully trustworthy strategy.

Structural adjustment

Structural adjustment programmes are basically designed to improve the macroeconomic performance of countries that are experiencing urgent problems of external and internal imbalances (trade, balance-of-payment and fiscal deficits, debt-service problems, etc.). They typically include measures such as devaluation, cuts in government spending (wages, employment, transfers), tax increases, trade and market liberalisation, etc. The SAPs have been much criticised in the past for selling out on the environment in the name of short-term goals of economic recovery, hence calling for alternative designs that would explicitly take environmental dimensions into account (e.g. Cruz and Repetto 1992; Reed 1992, 1996). Certainly, policy effects always depend heavily on what environmental targets we include – basically no policy safeguard can be expected to benefit *all* environmental dimensions at the same time, as the links to the macroeconomic variables targeted by the SAPs are often contradictory. But the country experiences in this book should at least allow us to say something about the specific impact of SAPs on forests.

One of the crucial and yet most controversial structural adjustment elements is (the degree of) currency devaluation. This book generally confirms that devaluation has a strong impact on forest degradation – to the extent that it strongly promotes accelerated timber extraction for exports – and also a fairly strong effect on deforestation. Devaluation is a *sine qua non* for productive diversification. For many countries, that means diversifying into new land- and forest-using sectors that become profitable at the margin. Indeed, price realignment and diversification may be necessary conditions for successful economic adjustment. But the experience of some of our study countries (e.g. Indonesia or Ecuador) also shows that a sharp devaluation in response to crisis and capital outflows can cause a temporary real *undervaluation* of the currency, inducing an over-shoot in the expansion of tradable sectors and excessive forest loss.

A second controversial element of SAPs has been trade liberalisation and export incentives that create a more open economy aiming at a more efficient use of resources. As discussed in the previous section, there often seem to be clear forest benefits in taking a firm stance on the import liberalisation of some home-market-oriented sectors in agriculture (food crops, livestock) and forestry (timber for construction, pulp and paper, etc.). At the same time, while vested interests in the domestic sectors that are affected are likely to resist such a policy strongly, there seem to be no clear economic-wide arguments against this type of liberalisation. On the other hand, perhaps the Indonesian experience in particular showed that an extreme emphasis on export incentives can cause much forest loss and degradation.

Finally, a debatable element has been the substantial, rapid cutbacks on government spending, both in terms of generally reducing the size of the public sector and in particular in diminishing the efficiency of forestry and national-park agencies that are hit by reduced budgets (Reed 1996). This book confirms that the first mechanism – large increases in unemployment and reduced urban labour absorption – can provide incentives to re-engage in rural-based activities that rely on forest conversion. Indeed, any policy that increases total labour absorption – preferably outside agriculture[35] – is likely to be good for forests. On the other hand, the book also places serious question marks over the latter assumption of specific agencies' capacity to change. Efficiency in forestry or conservation may already have been very low beforehand. Budgets may also have been largely donor-financed already over longer periods, as with conservation efforts in Central Africa. Our findings are thus much in line with those of Kaimowitz *et al.* (1998: 63) for Bolivia and Indonesia: 'government capacity to carry out such [forest regulation and protection] policies was already weak before the SAP and, if anything, improved as a result of greater public concern for the environment'.

On the whole, is it thus a good idea to screen SAPs for their alleged forest impacts? Moreover, is it actually feasible to design alternatives? After all, the *raison d'être* of the SAPs is to improve the economy and produce more foreign exchange, and most available pathways of adjustment seemed to involve some trade-offs with land-use extensification and forests. Also, on the list of ten partial macroeconomic and sectoral factors with potentially high forest impacts, the weight and intensity of factors varies greatly across countries. As the partial effects have contradictory links to forests – some reduce deforestation while others accelerate it – the net effect of a certain set of identical policies would be highly variable in, say, Cameroon, Venezuela and PNG. Consequently, it seems impossible to provide universal recipes or ready-made checklists of criteria that an SAP must meet to qualify for a 'forest-sustainable' green policy label; a careful country-specific design would be necessary as a minimum requirement.

On the other hand, we have also seen that certain policy changes do have predictable and generalisable impacts on forests, so forest concerns may be one justification among several to fine-tune SAPs accordingly. It has already been emphasised that the avoidance of large structural disruptions and of abrupt policy changes would probably be good for forests. Import liberalisation in forest-wasteful industries would be another recommendation. But the design of a forest-friendly SAP is not only about a checklist of what things *not to do*. On the contrary, one would probably want to make sure that some of the most forest-damaging spending is also included on the lists of fiscal cuts. Whereas items such as transport subsidies and government-directed settlement programmes have often been retrenched, subsidised credits for cattle and other extensive land uses, and especially for road-construction projects, have been less likely to suffer cutbacks (see Kaimowitz *et al* 1998: 62, and later). The latter seems particularly important because few large road projects in the tropics are implemented without the participation of international development banks or external credit agencies. The SAPs can probably seldom avoid causing *any* deforestation or forest degradation impact at all. But they can maintain *a sound policy balance* that counterbalances these impacts.

Debt relief

Environmental concerns have also featured among justifications in the international call for the large-scale reduction or write-off of foreign debt in Heavily Indebted Poor Countries (HIPCs). As we saw in Chapter 1, in analytical terms many effects of international financial transfers like foreign aid or debt reduction are identical to those of an oil windfall. Other things remaining equal, such transfers tend to reduce foreign-exchange scarcity, strengthen public-sector budgets, cause higher national spending, RER appreciation and promote higher urbanisation.[36] This book has shown that in most cases, this development pattern is good news for forests. There can thus be little doubt that the implementation of a comprehensive global programme of debt relief in HIPCs would also help reduce global deforestation and forest degradation. But the analysis also indicated that this outcome is not guaranteed in every country, as pointed out in the 'improved Ecuadorian recipe for forest destruction' from above. Speaking boldly, we would expect the forest-protecting effect of debt reduction to be much more certain in a poor African country dominated by shifting cultivation than in a Latin American country where most cleared land is used for pasture. In the latter, deforestation may actually be stimulated. If forest conservation is a principal justification for debt reduction, this outcome may not be acceptable. If so, more direct tools might increase efficiency. As in the SAPs, one could aim to tie debt relief to policy change through 'green conditionalities' – though the past has also demonstrated the strength of political resistance to this concept.

Another avenue might lead more into the conservation concession direction, that is, direct payments to indebted governments for spatially specific forest-protecting action – or sometimes for *non-action*. For instance, it has sometimes been proposed that governments and/or local stakeholders be paid an annuity in order not to build planned roads through certain biodiversity-rich forests, thus compensating them for forgone development benefits. Of course, such proposals immediately raise questions about national sovereignty, additionality, perverse incentives for new road-planning, the lack of benefit 'trickle down' to local people, etc. On the other hand, the comparison of our country studies has powerfully underlined both the spatially specific importance of road-building and the general importance of wealth transfers. Furthermore, at least the PNG case showed that a 'near-closure of the frontier' (due to PNG-specific land-tenure obstacles) need not necessarily impede agricultural growth or cause disastrous livelihood impacts. Perhaps it is also time to experiment with a range of novel, direct instruments to find more promising ways of promoting macroeconomic development, while reducing the pace of land-use expansion and safeguarding the global values of tropical forests.

Notes

1 We recall from the Cameroon case study (Chapter 6) that, due to secret extra-budgetary operations, the size of oil exports for certain years has been underestimated in the official statistics. Also, the economic impact of the boom was greatly exacerbated by the coffee and cocoa price boom – the latter still being more important export commodities at the time – and by foreign borrowing. Hence, the *macroeconomic* boom in Cameroon was larger than what the official oil exports would seem to indicate.

2 For Ecuador, an upper boundary rather than a range was estimated; thus the point estimate represents a best guess rather than the midpoint of the range.

3 The word 'currently' is emphasised in italics because previously Venezuela was unanimously considered to be a case of low deforestation, which is important *vis-à-vis* the hypothesis being tested.

4 Wherever there were differences across chapters in the scale of the variables' units (e.g. 'millions of US$' versus 'thousands of US$'), the coefficients have been scaled up or down accordingly, in order to allow the elasticities to be compared.

5 This occurs because the Dutch Disease is a model of the real side of the economy's medium-term adjustment. This means that, at least in its basic version, it is not concerned with either with monetary factors and exchange rate dynamics or short run rigidities and lags.

6 For Gabon, Venezuela and even Ecuador, absolute production value-added was used, for Cameroon agriculture's share of GDP, and for PNG the per-capita value-added.

7 When excluding the time dummy, the variable is estimated with the unexpected positive sign in PNG.

8 The Cameroon, Venezuela and Ecuador coefficients are comparable (all use absolute agricultural value added). For a small country like Gabon, a 1 per cent fall in the RER is expected to trigger a surprisingly large rise in agricultural value added of US$2.29 million, which is more impressive than the coefficient for the much larger Venezuela (US$6.66 million).

9 This statement includes a judgement about the comparative size of deforestation in the rice-exporting country, which I shall return to the section on 'The role of the structure of consumption'.

10 Exceptions were Gabon and PNG, where trade policies did little to hamper food-crop imports. See also the section 'What policies counted most?' on changing consumption structure.

11 For Gabon, we could distinguish between different price elasticities for the *okoumé* species versus other timbers (see Chapter 4).

12 No econometric tests were made for Indonesia, so this claim is based rather on a comparison with the situation in the primary case countries, and more circumstantial evidence from Indonesia.

13 This includes related infrastructure investments, notably bridges and railways.

14 Unfortunately, we cannot really generalise on the impact of railway-building, as this only was significant in one single country, Gabon. The indirect impact here was very limited, but this may have more to do with the framework conditions in that country than with the specific impact of the type of infrastructure.

15 The cautionary term 'net impact' refers to the possibility that resettled people might still have cleared some forest in their place of origin. It also takes into account that some components of a large programme like *transmigrasi* may actually have *reduced* forest-clearing (see Scenario 3).

16 As in Scenario 2, the term 'net reduction in forest-clearing' refers to the fact that this type of resettlement could also cause new forest-clearing in the resettlement sites that are established. But this would be inferior to the size of the abandoned original sites growing back into forest.

17 Pourtier (1989a: 237), my translation from the French.

18 For instance, these 'other things' refer to the structure of consumption (see section on 'The role of the structure of consumption').

19 It is interesting that out of the 6.75 million ha study area, purely residential areas only increased from 9.4 to 14 per cent. The largest change was that cultivated area decreased from 57.8 to 41.4 per cent, while a mosaic of residential and cultivated area grew from 17.9 to 33.4 per cent. This indicates a more mixed and scattered pattern of land use (Prasetyo 1992).

20 In Cameroon, rice imports rose by a factor of ten between 1981 and 1990. Yet rice was also produced in small quantities in the non-forested north.

21 This generalisation is abstracted from what was called Scenario 2, namely large rents with incipient road-building, which tends to be a transitory phase.

22 I am grateful to Dr Stuart White for this illustration of the effects of fallow-shortening on deforestation, as elaborated in his comments on an earlier draft of this book.

23 This is not to say that population-dense developing countries in general cannot stabilise or increase their forest cover – India and China seem to be good examples of countries that are achieving that. A conjecture is that the difference to Nigeria is, on the one hand, the dominance of a land-intensive food crop like rice in these countries and, on the other, a more developed urban industrial sector that has increasingly absorbed the labour surplus.

24 One might well argue for each of the policy elements that they had been set in a deterministic fashion, as explained by a combination of internal and external pressures. For instance, one might claim, as has been done elsewhere, that road-building decisions are responses to spatially specific political pressures from pre-established settlers and economic interest groups, making them an endogenous policy variable (Andersen *et al.* 2002). While there is doubtless some truth in this approach regarding some countries, this does not at all seem the logical conclusion to be drawn from this book, because autonomy in policy-making was pronounced.

25 Atilla Sönmez, interview 3 January 1990, cited in Winters (1996: 147).

26 Interview with Mr Placido Miko Abogo, Secretary-General of the Convergence for Social Democracy opposition party in Equatorial Guinea, cited in Onishi 2000: 3–4.

27 Collier (2000) analyses seventy-two cases of civil war, and finds that low-income countries with a high share of primary commodity exports are significantly more at risk of conflict. His '[e]conomic analysis sees rebellion as more like a form of organised crime' (ibid: 1).

28 I owe this observation to Tom Rudel, one of the reviewers of this book.

29 This assumes that a Ministry of the Environment exists – in its absence, it could be the representative of another central public agency that is in charge of environmental issues.

30 Again, I am indebted to Dr Stuart White for both the idea of formulating the policy recommendations in this manner, and for some of the wording regarding the policy objectives.

31 In theory, one could imagine agreements that are pegged to the development in world-market oil-prices. In practice, this would run up against significant domestic spending and rent-seeking pressures, and the problem is how to design a binding mechanism involving a sovereign government.

32 Let us takes the last point a little further. Let us imagine for a second that we are stepping into the shoes of a global land-use planner who, independently of present production patterns, could redistribute production structures among countries. A number of strategic questions emerge if we think of expanding food crops and cattle-ranching as two major sources of land-use change. Is it perhaps undesirable for single countries or mega-regions to attain self-sufficiency in food staples as populations continue to grow? Is it true, as Nobel Laureate Norman Borlaug expresses it, that 'low-yield farming is only sustainable for people with high death rates', and that safeguarding the expansion of these diverse but low-return systems is 'locking up half of the planet's arable land as a low-yield gene museum' (Borlaug 2002)? Are the tropics in general perhaps not well-suited to be massive food providers? Is it more promising for them to focus on tropical speciality crops for export, except for prime agricultural areas that can sustain higher-yield farming techniques? These are difficult questions, and answering them definitely goes beyond the scope of this book. But, if the global land-use planner is inclined to answer these questions in the negative, and if he implements his planning accordingly by giving priority to national food-security and sectoral import-substitution objectives in sheltered land-extensive sectors with static technologies, his choice is bound to have a high cost for forests.

33 For résumés of the EKC deforestation literature, see, for example, Barbier (1997) and Culas and Dutta (2002: 7–9).

34 See, for example, Mather and Needle (1998) or Rudel *et al.* (2002: 88–9).

35 Higher labour intensity in agriculture could also reduce deforestation pressures, but if the technological or product-mix change is associated with higher agricultural returns, the scale and land use of these activities could actually expand, potentially to the detriment of forests.

36 Notably, this assumes that under the comparative scenario prior to the debt reduction, the debt was actually serviced so that significant capital-outflows occurred. Even if debt was not serviced, the reduction could still promote some capital inflows by encouraging new private investment flows.

Bibliography

Abril-Ojeda, G. 1991. 'Export booms and development in Ecuador', in M. Blomström and P. Meller (eds), *Diverging Paths. Comparing a Century of Scandinavian and Latin American Economic Development*. Baltimore and London: John Hopkins University Press.

Achard, F., Eva, H., Glinni, A., Mayaux, P., Richards, T. and Stibig, H. J. (eds). 1998. 'Identification of deforestation: hot spot areas in the humid tropics'. *Trees Publications Series B*. Research Report No. 4, Ispra: European Commission Joint Research Centre.

Achard, F., Eva, H. D., Stibig, H.-J., Mayaux, P., Gallego, J., Richards, T. and Malingreau, J.-P. 2002. 'Determination of deforestation rates of the world's humid tropical forests', *Science*, 297(9), August.

Adams, J. S. and McShane, T. O. 1996. *The Myth of Africa. Conservation Without Illusion*. Berkeley: University of California Press.

Adjaye, A. 1996. 'Revitalising growth in Papua New Guinea's agricultural sector', *Agricultural Systems* 51(3): 299–316.

Ahmed, S. and Chhibber, A. 1992. 'Successful adjustment to oil shocks: the rare case of Indonesia', *The Bangladesh Development Studies* 20(2/3): 185–213.

Aicher, C., Grimmig, M. and Müller, B. 1998. 'The Imataca forest reserve: golden future for Venezuela's south-east?', *SEFUT Working Paper* No. 2. Freiburg: DFG Graduiertenkolleg, Socioeconomics of Forest Use in the Tropics and Sub-tropics, January.

Ajamil, C. and Carrillo, P. 1999. *Descripción del sector petrolero ecuatoriano*, Quito: MoE and Project Petramaz.

Allen, B. 1999. 'Rural poverty in Papua New Guinea'. *Development Bulletin* 49: 42–4.

Allen, B. J., Bourke, R. M. and Hanson, L. 2001. 'Dimensions of Papua New Guinea village agriculture', in R. M. Bourke, M. G. Allen and J. Salisbury (eds), *Proceedings of the Papua New Guinea Food and Nutrition 2000 Conference: Food Security in Papua New Guinea*. Canberra: Australian Centre for International Agricultural Research.

Allen, B. J., Bourke, R. M. and Hide, R. L. 1995. 'The sustainability of Papua New Guinea agricultural systems: the conceptual background', *Global Environmental Change* 5(4): 297–312.

Amelung, T. and Diehl, M. 1992. *Deforestation of Tropical Rainforests. Economic Causes and Impact on Development*, J. C. B. Mohr: Tübingen.

Amend, T. 1990. *Marine und litorale Nationalparks in Venezuela. Anspruch, Wirklichkeit und Zukunftsperspektiven*. Mainzer Geographische Studien, Heft 32, Johannes-Gutenberg Universität.

Amigransa, E. B., de Vries, T., Flores, M., Garzón, P., Granda, A., Jocknick, C., la Torre, L., Martínez, E. and Viteri, C. 1997. *Les voix de la resistance: L'exploitation pétrolière dans les pays tropicaux*. Quito: Oilwatch.

Amin, A. A. 1995. 'The effects of structural adjustment programme in Cameroon's major exports cacao, coffee and cotton', *Les Cahiers d'OCISCA* No. 17. Yaoundé: ORSTOM.

Amin, A. A. 1996. 'The effects of exchange rate policy on Cameroon's agricultural competitiveness', *AERC Research Paper* No. 42. Nairobi: African Economic Research Consortium, March.

Amin, S. 1972. 'Underdevelopment and dependence in black Africa', *Journal of Modern African Studies* 10(4): 503–24.

Amuzegar, J. 1982. 'Oil wealth: A very mixed blessing', *Foreign Affairs* 60.

Andersen, L. E. 1996. 'The causes of deforestation in the Brazilian Amazon', *Journal of Environment & Development*, (5)3: 309–28.

Andersen, L. E., Granger, C. W. J., Hung, L. L., Reis, E. and Weinhold, D. 1996. 'Report on Amazon deforestation', University of California, Department of Economics, *Discussion Paper 96–40*. San Diego: UCL , November.

Andersen, L. E., Granger, C. W. J., Reis, E. J., Weinhold, D. and Wunder, S. (2002). *The Dynamics of Deforestation and Economic Growth in the Brazilian Amazon*. Cambridge University Press, Cambridge (UK), 200.

Angelsen, A. 1997. 'Tropical agriculture and deforestation: economic theories and a study from Indonesia'. A dissertation submitted for the PhD degree at the Norwegian School of Economics and Business Administration, Bergen.

Angelsen, A. and Culas, R. 1996. 'Debt and deforestation: a tenuous link', *Working Paper* 1996: 10. Bergen: Christian Michelsen Institute.

Angelsen, A. and Kaimowitz, D. (eds). 2001. *Agricultural Technologies and Tropical Deforestation*. Wallingford (UK): CABI Publishing, 422.

Ansari, M. I. 1990. 'The Dutch Disease: the Canadian experience', *Weltwirtschaftliches Archiv*.

Aranda, S. 1984. *La economía venezolana. Una interpretación de su modo de funcionamento*, Editorial Pomaire, Caracas.

Araujo, M. C. 1998. 'Gobernabilidad durante la crisis y política de ajuste', *Documento de Trabajo* No. 6, Quito: Cordes.

Arif, M., Madiadipura, H. and Zulfikar, H. 1996. 'CGPRT crops in Indonesia: a statistical profile', Part I Economic Indicators, Annual Crop Production, National, Provincial. *CGPRT Working Paper Series* No. 18, Bogor: CGPRT Centre.

Aron, J. 2000. 'The Zambian copper boom and crash, 1964–80', in P. Collier and J. W. Gunning (eds), *Trade Shocks in Developing Countries, Volume 1: Africa*. Oxford: Oxford University Press, 259–302.

Arvelo-Jiménez, N. 1984. 'The politics of cultural survival in Venezuela: beyond indigenismo', in M. Schmink and C. H. Wood (eds), *Frontier Expansion in Amazonia*. Fla.: Gainesville.

Asafu-Adjaye, J. 1996. 'Revitalising growth in Papua New Guinea's agricultural sector', *Agricultural Systems* 51: 299–316.

Ascher, W. 1998. 'From oil to timber: the political economy of off-budget development financing in Indonesia', *Indonesia* 65: 37–61.

ATIBT. 1999. *La lettre de 'ATIBT'*. July – No. 10.

ATIBT. 2000. Statistics and markets. *ATIBT Newsletter*. Summer.

Attiga, A. A. 1981. 'How oil revenues can destroy a country', *Petroleum Intelligence Weekly, Special Supplement*.

Aubreville, A. 1948. 'Étude sur les fôrets de l'AEF et du Cameroun', *Bulletin Scientifique STAT*, No. 2.

Aune, J. B., Glomsrod, S., Iversen, V. and Wiig, H. 1996. 'Structural adjustment and land degradation: a CGE analysis for Tanzania', in *Structural Adjustment Policies and Environmental Degradation in Tanzania, Zambia, and Ethiopia*. Preliminary Seminar Report, 9–10 January. Ås: Agricultural University of Norway.

Australian Agency for International Development (AusAID). 2000. *The Economy of Papua New Guinea. Macroeconomics Policies: Implications for Growth and Development in the Informal Sector*. 1999 Report. International Development Issues No. 53. Canberra: AusAID.

Auty, R. M. 2001. 'The political economy of resource-driven growth', *European Economic Review* 45: 839–46.

Aweto, A. O. 1990. 'Plantation forestry and forest conservation in Nigeria', *Environmentalist* 10(2): 127–34.

Azam, J.-P. 2000. 'The uranium boom in Niger, 1975–82', in P. Collier and J.W. Gunning (eds), *Trade Shocks in Developing Countries, Volume 1: Africa*. Oxford: Oxford University Press, 351–98.

Azam, J.-P. and Chambas, G. 2000. 'The groundnut and phosphates boom in Senegal, 1974–77', in P. Collier and J. W. Gunning (eds), *Trade Shocks in Developing Countries, Volume 1: Africa*. Oxford: Oxford University Press, 226–58.

Azam, J.-P. and Shahabuddin, Q. 2000. 'The remittance boom in Bangladesh, 1978–86', in P. Collier and J. W. Gunning (eds), *Trade Shocks in Developing Countries, Volume 2: Asia and Latin America*. Oxford University Press, 285–307.

Baardsen, S. 1987. 'The influence of varying domestic petroleum income on the Norwegian pulp and paper sector', *Scandinavian Forest Economics* 87(29): 171–82.

Bakoup, F. and Tarr, D. 1998. 'How integration into the Central African economic and monetary community affects Cameroon's economy. General equilibrium estimates', *Policy Research Working Paper* No. 1872. Washington, DC: World Bank.

Banco Central de Venezuela. 1977. *La economía venezolana en los últimos treinta y cinco años*. Caracas: BCV.

Bank of Papua New Guinea (BPNG). 2001. *Bank of Papua New Guinea Quarterly Economic Bulletin* 29(2), 43 + appendix.

Baptista, A. 1997. *Bases cuantitativas de la economía venezolana 1830–1995*. Caracas: Fundación Polar.

Barbier, E. 1997. 'Introduction to the environmental Kuznets curve', *Environment and Development Economics*' 2(4), 369–79.

Barbier, E. 2001. 'The economics of tropical deforestation and land use', *Land Economics* 77(2): 155–72.

Barbier, E. and Burgess, J. C. 1996. 'Economic analysis of deforestation in Mexico', *Environment and Development Economics* 1: 203–39.

Barbier, E., Bockstael, N., Burgess, J. and Strand, L. 1994. 'The timber trade and tropical deforestation in Indonesia', in K. Brown and D. Pearce (eds), *The Causes of Tropical Deforestation. The Economic and Statistical Analysis of Factors giving Rise to the Loss of Tropical Forests*, London: UCL Press, 242–70.

Barham, B. L. and Coomes, O. T. 1994. 'Reinterpreting the Amazon rubber boom: investment, the state, and Dutch disease', *Latin American Research Review* 29(2):73–109.

Barker, T. and Brailovsky, V. (eds). 1981. *Oil or Industry? Energy, Industrialisation and Economic Policy*. London and Mexico City: Academic Publishers.

Barnes, J. F. 1992. *Gabon: Beyond the Colonial Legacy*. Boulder: Westview Press.

Barnett, T. E. 1990. 'A summary of the Report of the Commission of Inquiry into aspects of the timber industry in Papua New Guinea', Hobart, Tasmania: The Asia-Pacific Action Group, November, 59.

Barr, C. 2001. *Banking on Sustainability: Structural Adjustment and Forestry Reform in Post-Suharto Indonesia*. Bogor: CIFOR and Washington, DC: WWF.

Barret-Lefeuvre, G. and Dufoulon, G. 1979. *La forêt gabonaise*. Libreville: Ministère des Eaux et Forêts.

Barreto, A, and Pérez-Puelles, S. 1991. 'Estudio integral de la actividad minera en la cuenca hidrográfica del Alto Caroní', unpublished paper presented at the II Jornadas Profesionales de EDELCA, October.

Basquin, P., van Beek, G., Christy, P., Clist, B., Guicherit, R., Lahm, R., Moungazi, A., Reitsma, J., Waardenburg, H., White, L. and Wilks, C. 1991. 'Maguelou. An environmental study of the Ofoubou area for Dupont E & PN°8 BV', unpublished report, Libreville: Africa Forest.

Bassey, N. 1999. 'The mangroves in the Niger river delta', in *Oilwatch: The Oil Flows, the Forest Bleeds*. Quito. Electronic version at www.oilwatch.org.ec (accessed April 2002).

Bawa, K. S. and Dayanandan, S. 1997. 'Socio-economic factors and tropical deforestation', *Nature* 386: 562–3.

BDPA. 1998. *Études préalables à la définition d'une politique d'aménagement et de développement de la Province de l'Estuaire. Rapport de synthèse*. Libreville: Ministère de l'Agriculture, de l'Elevage et du Développement Rural.

Bellamy, J. A. and McAlpine, J. R. 1995. *Papua New Guinea Inventory of Natural Resources, Population Distribution and Land Use Handbook*, 2nd edn., *PNGRIS Publication* No. 6. Canberra: Australian Agency for International Development.

Benalcázar, R. R. 1989. *Análisis del desarrollo económico del Ecuador*. Quito: Ediciones Banco Central del Ecuador.

Benjamin, N. 1996. 'Adjustment and income distribution in an agricultural economy: a general equilibrium analysis of Cameroon', *World Development* 24(6): 1003–13.

Benjamin, N., Devarajan, S. and Weiner, R. J. 1989. 'The Dutch disease in a developing country: oil reserves in Cameroon', *Journal of Development Economics* 30: 71–92.

Bennett, E. L. and Robinson, J. G. 2000. 'Hunting of wildlife in tropical forests: implications for biodiversity and forest peoples', *Environment Department Papers, Biodiversity Series*. Washington, DC: World Bank.

Bevan, D., Collier, P. and Gunning, J. W. 1987. 'Consequences of a commodity boom in a controlled economy: accumulation and redistribution in Kenya', *World Bank Economic Review* 1(3).

Bevan, D., Collier, P. and Gunning, J. W. 1989. 'Fiscal response to a temporary trade shock: the aftermath of the Kenyan coffee boom', *Applied Economics Discussion Paper* No. 68. Oxford: Institute of Economics and Statistics.

Bevan, D., Collier, P. and Gunning, J. W. 1999. *Nigeria and Indonesia: The Political Economy of Poverty, Equity, and Growth*. A World Bank comparative study. Washington, DC: World Bank.

Bevan, D., Collier, P. and Gunning, J. W. 2000. 'Anatomy of a temporary trade shock: the Kenyan coffee boom, 1976–79', in P. Collier and J. W. Gunning (eds), *Trade Shocks in Developing Countries*, Oxford: Oxford University Press.

Bevilacqua, M., Cárdenas, L., Flores, A. L., Hernández, L., Lares, E., Mansutti, A., Miranda, M., Ochoa, J., Rodríguez, M. and Selig, E. 2001. 'The state of Venezuela's forests: a case study of the Guayana region', A Global Forest Watch Report, review version (final version published in April 2002).

Bhagwati, J. 1983. 'Directly Unproductive, Profit-seeking (DUP) activities', *Journal of Political Economy* 5.

Bikié, H, Ndoye, O. and Sunderlin, W. D. 2000b. 'L'impact de la crise économique sur les systèmes agricoles et le changement du couvert forestier dans la zone forestière humide du Cameroun', *Occasional Paper* No. 27. Bogor, Indonesia: Center for International Forestry Research.

Bikié, H., Collomb, J. G., Djomo, L., Minnemeyer, S., Nguofo, R. and Nguiffo, R. 2000a. *An Overview of Logging in Cameroon. A Global Forest Watch Cameroon Report*. Washington, DC: World Resources Institute.

Bisbal, F. J. 1988. 'Impacto humano sobre los habitat de Venezuela', *Interciencia* 13(5), 225–32.

Blandford, D., Friedman, D., Lynch, S., Mukherjee, N. and Sahn, D. E. 1994. 'Oil boom and bust: the harsh realities of adjustment in Cameroon', in D. E. Sahn (ed.), *Adjusting to Policy Failure in African Economies*. Ithaca and London: Cornell University Press.

Blaney, S., Mbouity, S., Nzamba, P. M., Nkombé, J. M., Thibault, M., Mavoungu, O. M. and Bayé, J. P. 1998. *Complexe d'aires protégées de Gamba*. WWF. Unpublished report.

Blarcom, B. van, Knudsen, O. and Nash, J. 1993. 'The reform of public expenditures for agriculture', *World Bank Discussion Papers* No. 216. Washington, DC: World Bank.

Bleaney, M. and Greenaway, D. 1993. 'Long-run trends in relative prices of primary commodities and in the terms of trade of developing countries', *Oxford Economic Papers*, 45: 349–63.

Blyth, C. 1988. 'Government expenditure in Papua New Guinea', *INA Discussion Paper* No. 34. Port Moresby: Institute of National Affairs.

Borcherdt, C. 1985. *Geographische Untersuchungen in Venezuela (II)*, Stuttgart: Geographisches Institut der Universität Stuttgart.

Borcherdt, C. 1988. 'Über neuere Veränderungen in unterschiedlichen Wirtschaftsräumen Venezuelas unter den Vorzeichen konjunktureller Schwankungen und wachsender Umweltprobleme', in R. Sevilla (ed.), *Venezuela – Kultur-und Entwicklungsprobleme eines OPEC-Landes in Südamerika*. Tübingen: Institut für Wissenschaftliche Zusammenarbeit.

Borlaug, N. 2002. 'We can feed the world: Here's how', *Wall Street Journal*, 13 May.

Bosworth, M. and Anderson, K. 2000. *Reforming Trade Policy in Papua New Guinea and the Pacific Islands. What Roles for WTO and APEC?* Adelaide: Centre for International Economic Studies and Port Moresby: Institute of National Affairs.

Boumah, A. 1975. *Le chemin de fer Transgabonais: Facteur essentiel de développement économique et d'aménagement du territoire national*. Ministry of Planning and Development, Libreville. Unpublished report.

Bourguignon, F. 1980. 'Oil and income distribution in Venezuela 1968–76', in J. De Bandt, P. Mándi and D.Seers (eds), *European Studies in Development*. New Trends in European Development Studies, The European Association of Development Institutes, London: Macmillan.

Bourke, M. 2001. 'Intensification of agricultural systems in Papua New Guinea', *Asia Pacific Viewpoint*, 42(2/3), August/December, 219–35.

Bourke, R. M. 2000. *Intensification of Food Production and Land Use in Papua New Guinea*. Paper presented at the Monbusho International Symposium 'Vegeculture in Eastern Asia and Oceania', Osaka, Japan, 7–9 November.

Bourke, R. M. 1997. 'Management of fallow species composition with tree planting in Papua New Guinea', *Resource Management in Asia-Pacific Seminar Series*, WP 1997/5, Human Geography. Canberra: ANU.

Bravo, E. 1999. 'Derechos humanos y petroleo en países tropicales', in *Oilwatch: The Oil Flows, the Forest Bleeds*. Quito; electronic version at www.oilwatch.org.ec (accessed April 2002).

Brito, F. F. 1996. *Historia económica y social de Venezuela*, Vol. II. Caracas: UCV-Ediciones de la Biblioteca.

Bromley, R. 1981. 'The colonisation of humid tropical areas in Ecuador', *Singapore Journal of Tropical Geography*, 2(1).

Brown, K. and Ekoko, F. 2001. 'Forest encounters: synergy among agents of forest change in southern Cameroon', *Society and Natural Resources* 14, 269–90.

Brown, K. and Pierce, D. W. (eds) 1994. *The Causes of Tropical Deforestation. The Economic and Statistical Analysis of Factors Giving Rise to the Loss of Tropical Forests*. London: UCL Press, 338 pp.

Brownrigg, L. A. 1981. 'Economic and ecological strategies of Lojano migrants to El Oro', in N. E. Whitten (ed.), *Cultural Transformation and Ethnicity in Modern Ecuador*. Chicago: Urbana.

Brugière, D. 1999. *Aires protégées et diversité biologique au Gabon*. Montpellier: CIRAD Forêt.

Brunck, F., Grison, F. and Maitre, H. F. 1990. *L'Okoumé: Aucoumea klaineana Pierre*. Nogent-sur-Marne: Centre Technique Forestier Tropical.

Brunner, J. and Ekoko, F. 2000. 'Cameroon', in F. J. Seymour and N. K. Dubash (eds), *The Right Conditions. The World Bank, Structural Adjustment, and Forest Policy Reform*. Washington, DC: World Resources Institute, 59–77.

Bryant, D., Nielsen, D. and Tangley, L. 1997. 'The last frontier forests: ecosystems and economies on the edge – What is the status of the world's remaining large, natural forest ecosystems?', Washington, DC: World Resources Institute, Forest Frontiers Initiative.

Buiter, W. H. and Purvis, D. D. 1984. 'Oil, disinflation and export competitiveness: a model of the "The Dutch Disease" ', in J. S. Bhandari and B. H. Putnam (eds), *Economic Interdependence and Flexible Exchange Rates*. Cambridge, Mass: University of Chicago Press.

Burgess, J. C. 1991. 'Economic analysis of frontier agricultural expansion and tropical deforestation', MSc Thesis, Environmental and Resource Economics. London: University College London.

Burritt, R. 1997. 'Environmental disclosures in annual reports of Australian gold and copper mining companies with activities in Papua New Guinea and/or Indonesia', *Working Paper* 1997/13. Resource Management in Asia Pacific. Canberra, Australia: Australian National University.

Buschbacher, R. J. 1987. 'Deforestation for sovereignty over remote frontiers: case study no. 4: government-sponsored pastures in Venezuela near the Brazilian border', in C. F. Jordan (ed.), *Amazonian Rain Forests: Ecosystem Disturbance and Recovery*. New York: Springer-Verlag, 46–57.

Cabarle, B. J., Crespi, M., Dodson, C. H., Luzuriaga, C., Rose, C. D. and Shores, J. N. 1989. *An Assessment of Biological Diversity and Tropical Forests for Ecuador*. Quito: USAID.

Campbell, D. and J. C. Riddell 1982. 'Population growth and land use change in the Mandara Mountains of North Cameroon', *The East Lakes Geographer* 17: 5–16.

Capistrano, A. D. 1990. 'Macroeconomic influences on tropical forest depletion, a cross country analysis', PhD dissertation. Gainesville: University of Florida.

Capistrano, A. D. and Kiker, C. F. 1995. 'Macro-scale economic influences on tropical forest depletion', *Ecological Economics* 14: 21–9.

Carrero, N. O. 1995. 'Antecedentes de las estadísticas forestales en Venezuela', *Seforven*, year 6, No. 12, June, Caracas: MARNR.

Carret, J. C. and Clément, J. 1993. *La compétitivité des bois d'œuvre Africains*. Paris: Ministère de la Cooperation.

Carrière, S. 1999. 'Le dileme de la route dans la vallée du Ntem', *APFT Working Paper* No. 6, *La route en forêt tropicale: Porte ouverte sur l'avenir?*, 38–41.

Carvallo, G. and de Hernández, J. R. 1984. *Temas de la Venezuela agroexportadora*. Caracas: Fondo Editorial Tropykos.

Casagrande, J. B., Thompson, S. I. and Young, P. D. 1964. 'Colonization as a research frontier: the Ecuadorian case', in R. A. Manners (ed.), *Process and Pattern in Culture: Essays in Honor of Julian H. Steward*. Chicago: Aldine.

Casson, A. 2000. 'The hesitant boom: Indonesia's oil palm sub-sector in an era of economic crisis and political change', *CIFOR Occasional Paper* No. 29. Bogor: CIFOR.

Catalán, A. 1993. 'El proceso de deforestación en Venezuela entre 1975 y 1988', *Serie de Informes Técnicos*, DGSIASV/IT/338, Caracas: MARNR, October.

Cattaneo, A. 2001. 'Deforestation in the Brazilian Amazon', *Land Economics* 77(2): 219–40.

CENDES 1995. 'Structural adjustment and the environment. Venezuela country study', Summary report, Centro de Estudios del Desarrollo. Caracas: UCV, January.

Centeno, J. C. 1990. *El desarrollo forestal de Venezuela*. Mérida: IFLA, January.

Centeno, J. C. 1995. 'Forest concession policy in Venezuela', Paper commissioned by World Resources Institute, Washington, DC (unpublished).

Centeno, J. C. 1997a. 'Deforestation out of control in Venezuela', available online at http://www.ciens.ula.ve/~ jcenteno/DEFOR-EN.html.

Centeno, J. C. 1997b. 'The assault on Imataca', available online at http://www.ciens.ula.ve/~ jcenteno.

Central Intelligence Agency (CIA). 1999a. 'The world factbook 1999 – Gabon' available online at http://www.adci.gov/cia/publications/factbook/gb.html.

Central Intelligence Agency (CIA). 1999b. 'The world factbook 1999 – Cameroon', available online at http://www.adci.gov/cia/publications/factbook/cm.html.

Chomitz, K. M. and Gray, D. A. 1996. 'Roads, lands, markets, and deforestation: a spatial model of land use in Belize', *World Bank Economic Review* 10: 487–512.

Christy, P., Lahm, S., Louis, A., Moungazi, A., Reitsma, J. and Wilks, C. 1991a. 'Alonha. An environmental study in the lake Ezanga area for Conoco Onango (Gabon) Ltd'. Libreville: Africa Forest.

Christy, P., Lahm, S., Louis, A., Moungazi, A., Reitsma, J. and Wilks, C. 1991b. Atem. An environmental study of the Atem area for Conoco Ndombo (Gabon) Ltd. Libreville: Africa Forest.

Christy, P., Moungazi, A. and Wilks, C. 1990a. An environmental study of the Nyonyie area for the Sun Gabon oil company. Libreville: Africa Forest.

Christy, P., Moungazi, A. and Wilks, C. 1990b. Étude environmentale de la zone d'Oyan pour Sun Gabon Oil Company. Libreville: Africa Forest.

Christy, P., Jaffré, R. Ntougou, O. and Wilks, C. 2002. (draft) *La forêt gabonaise*, Libreville: French Development Cooperation and Multipress.

Chrystal, K. A. 1984. 'Dutch Disease or Monetarist medicine? The British economy under Mrs Thatcher', *Federal Bank of St. Louis Review* 66.

Clist, B. 1995. *Gabon: 100,000 ans d'histoire*. Libreville: Centre Culturel Français Saint-Exupéry.

Cochrane, M. A., Alencar, A , Schulze, M. D., Souza, C. M., Nepstad, D. C., Lefebvre, P. and Davidson, E. A. 1999. 'Positive feedbacks in the fire dynamic of closed canopy tropical forests', *Science* 284, 1834–6.

Collier, G. A. 1998. *!Basta! Tierra y rebelión zapatista en Chiapas*. Mexico City: Food First Books.

Collier, G. A., Mountjoy, D. C. and Nigh, R. B. 1994. 'Peasant agriculture and global change: a Maya response to energy development in southeastern Mexico', *BioScience* 44(6): 398–407.

Collier, P. 1987. 'Macroeconomic effects of oil on poverty in Nigeria', *IDS Bulletin*, No. 18. Sussex: Institute of Development Studies.

Collier, P. 1988. 'Oil shocks and food security in Nigeria', *International Labour Review* 127(6): 761–82.

Collier, P. 2000. 'Economic causes of civil conflict and their implications for policy', Global Policy Forum – World Bank/UN Security Council: Diamonds in Conflict, at: http://www.globalpolicy.org/security/issues/diamond/wb.htm, accessed on 26 September 2000.

Collier, P. and Gunning, J. W. 2000a. *Trade Shocks in Developing Countries, Volume 1: Africa*. Oxford: Oxford University Press.

Collier, P. and Gunning, J. W. 2000b. *Trade Shocks in Developing Countries, Volume 2: Asia and Latin America*. Oxford: Oxford University Press.

Collier, P. and Gunning, J. W. 2000c. 'Trade shocks: theory and evidence', in P. Collier and J. W. Gunning (eds), *Trade Shocks in Developing Countries, Volume 1: Africa*. Oxford: Oxford University Press, 1–63.

Collins, N. M., Sayer, J. A. and Whitmore, T. C. (eds). 1991. *The Conservation Atlas of Tropical Forests: Asia and the Pacific*. New York: Simon & Schuster and London: IUCN.

Collomb, J. G., Mikissa, J. B., Minnemeyer, S., Mundunga, S., Nzao, H. N., Madouma, J., Mapaga, J. D., Mikolo, C., Rabenkogo, N., Akagah, S., Bayani-Ngoye, E. and Mofouma, A. 2000. 'A first look at logging in Gabon', A Global Forest Watch report. Washington, DC: World Resources Institute.

Comafors and IPS. 2001. *El bosque en el Ecuador. Una visión transformada para el desarrollo y la conservación*. Informe final, Quito: GTZ.

Commander, S. and Peek, P. 1986. 'Oil exports, agrarian change and the rural labour process: the Ecuadorian Sierra in the 1970s', *World Development* 14(1).

Connell, J. 1997. *Papua New Guinea. The Struggle for Development*. London and New York: Routledge.

Contreras-Hermosilla, A. 2000. 'The underlying causes of forest decline', *CIFOR Occasional Paper* No. 30. Bogor: CIFOR, June.

Cook, C. D. 1996. 'The divided island of New Guinea: people, development and deforestation', in E. Sponsel, Leslie, T. N. Headland and R. C. Bailey (eds), *Tropical Deforestation: The Human Dimension*. New York: Columbia University Press.

Corden, W. M. 1983. 'The economic effects of a booming sector', *International Social Science Journal* 35(3): 441–54.

Corden, W. M. 1984. 'Booming sector and Dutch disease economics: survey and consolidation', *Oxford Economic Papers* 36(3): 359–80.

Corden, W. M. and Neary, J. P. 1982. Booming sector and de-industrialisation in a small open economy', *Economic Journal* 92: 825–48.

Coronil, F. 1997. *The Magical State: Nature, Money and Modernity in Venezuela*. Chicago and London: The University of Chicago Press.

Corrales, L. and Kleinn, C. 2000. *FRA 2000. Bibliografía comentada. Cambios en la cobertura forestal. Ecuador*. Rome: Food and Agricultural Organization of the United Nations (unpublished).

Costanza, R., Cumberland, J., Daly, H., Goodland, R. and Norgaard, R. 1997. *An Introduction to Ecological Economics*. Florida: St Lucie Press.

Cruz, W. and Repetto, R. 1992. *The Environmental Effects of Stabilization and Structural Adjustment Programs: The Philippine Case*. Washington, DC: World Resource Institute.

Cuddington, J. T. 1988. 'Fiscal policy in commodity exporting LDCs', *World Bank Policy, Planning and Research Working Paper* (unpublished background paper for World Development Report 1988).

Culas, R. and Dutta, D. 2002. 'The underlying causes of deforestation and environmental Kuznets curve. A cross-country analysis', University of Sydney, School of Economics and Political Science, available online at www.bus.qut.edu.au/esam02/program/papers/Culas_Richard.pdf, accessed on 6 September 2002.

Dahanayake, P. A. S. 1982. 'An analysis of effective exchange rate changes in Papua New Guinea', *IASER Discussion Paper* No. 43. Boroko, Papua New Guinea: Institute of Applied Social and Economic Research.

Dauthuille, C. 1996. *L'Économie du Gabon, par les timbres*. Libreville: Office des Postes et Telecommunications.

Davidson, J. n.d. 'Forestry in Papua New Guinea: a case study of the Gogol woodchip project near Madang', unpublished report. Canberra: CSIRO.

Davis, G. 1995. 'Learning to love the Dutch Disease: evidence from the mineral economies', *World Development* 23(10), 1765–79.

Davis, J. M. 1983. 'The economic effects of windfall gains in export earnings 1975–78', *World Development* 11.

De Macedo, J. B. 1982. 'Currency diversification and export competitiveness: a model of the Dutch Disease in Egypt', *Journal of Development Economics* 11.

De' Ath, C. and Michalenko, G. 1980. 'High technology and original peoples: the case of deforestation in Papua New Guinea and Canada', *Impact of Science on Society* 30(3): 197–209.

Debroux, L. and Karsenty, A. 1997. 'L'implantation des sociétés forestières Asiatiques en Afrique Centrale: Rimbunan Hijau au Cameroun'. *Bois et Forêts des Tropiques* 254(4).

Defries, R. S., Hansen, M. C., Townshend, J. R. G., Janetos, A. C. and Lovelands, T. R. 2000. 'A new global 1-km dataset of percentage tree cover derived from remote sensing', *Global Change Biology* 6.

Deininger, K. W. and Minten, B. 1999. 'Poverty, policies, and deforestation: the case of Mexico', *Economic Development and Cultural Change* 47(2): 313–44.

DeJanvry, A., Sadoulet, E. and Fargeix, A. 1991. 'Politically feasible and equitable adjustment: some alternatives for Ecuador', *World Development* 19(11).

DeLancey, M. W. 1989. 'Cameroon: dependence and independence', in Bowman, L. W. (ed.), *Profiles: Nations of Contemporary Africa*. San Fransisco: Westview Press.

Delvingt, W., Vermeulen, C., Auzel, P. and Doucet, J. L. 2000. 'For a more realistic policy of sustainable management of tropical African rainforest', *The ATIBT Newsletter*, Summer.

Denevan, W. M. 1976. *The Native Population of the Americas in 1492*. Madison: University of Wisconsin Press.

Densley, B. (eds) 1977. *Agriculture in the Economy. A Series of Review Papers. Subsistence Agriculture*. J. M. Macewan/DPI.

Department of Primary Industry. 1979. *Handbook on Agricultural Statistics 1978*. Papua New Guinea: Department of Primary Industry. Policy, Planning and Coordination Branch.

Department of Primary Industry. 1983. *Handbook on Agricultural Statistics 1981*. Papua New Guinea: Department of Primary Industry. Policy, Planning and Coordination Branch.

Department of Primary Industry. 1986. *Handbook on Agricultural Statistics 1984*. Papua New Guinea: Department of Primary Industry, Policy and Planning Division.

Department of Territories (Australia). 1967. *Compendium of Statistics for Papua and New Guinea*. Canberra: Department of Territories, Statistics Section, May.

Depierre, D. and Ole, Z. 1976. Parcs nationaux et réserves de faune du Cameroun. *Bois et Forêts des Tropiques* No. 170.

Devarajan, S. 2000. 'Cameroon', in P. Collier and J. W. Gunning (eds), *Trade Shocks in Developing Countries, Volume 1: Africa*. Oxford: Oxford University Press, 399–419.

Devarajan, S. and de Melo, J. 1987. 'Adjustment with a fixed exchange rate: Cameroon, Côte d'Ivoire, and Senegal', *World Bank Economic Review* 1(3): 447–87.

Diaw, C. M. 1997. 'Si, nda bot and ayong: shifting cultivation, land uses and property rights in southern Cameroon', *Rural Development Forestry Network Paper*. UK: Overseas Development Institute.

Dick, J. 1991. 'Forest land use, forest use zonation, and deforestation in Indonesia: a summary and interpretation of existing information', Background paper to UNCED for the State Ministry for Population and Environment (KLH) and the Environmental Impact Management Agency (BAPEDAL), Jakarta.

Direction Générale de la Statistique et des Études Économiques (DGSEE). 1993. *Recensement général de la population et de l'habitat. Principaux résultats*. Libreville: DGSEE.

Direction Générale de la Statistique et des Études Économiques (DGSEE). 1994a. *Enquête budget consommation. Les conditions de vie des populations Africaines. A Libreville et Port-Gentil, vol. I: Analyse*. Ministère de la Planification, de l'Environnement et du Tourisme. Libreville: DGSEE.

Direction Générale de la Statistique et des Études Économiques (DGSEE). 1994b. *Enquête budget consommation. Les conditions de vie des populations Africaines. A Libreville et Port-Gentil. Vol. II: Resultats*. Ministère de la Planification, de l'Environnement et du Tourisme. Libreville: DGSEE.

Direction Générale de la Statistique et des Études Économiques (DGSEE). 1997. *Les comptes de la nation. Comptes provisoires 1995*. Libreville: DGSEE.

Direction Générale de l'Économie (DGE). 1998. *Tableau de bord de l'économie*. Libreville: DGE.

Direction Générale de l'Économie (DGE). 1999. *Tableau de bord de l'économie*. Libreville: DGE.

Direction Générale de l'Économie (DGE). 2001. *Tableau de bord de l'économie*. Libreville: DGE.

Dorney, S. 2000. *Papua New Guinea. People, Politics and History since 1975*. Sydney: ABC Books.

Dounias, E. 1999. 'Le paradoxe de la route en Afrique forestière', *APFT Working Paper* No. 6, *La route en forêt tropicale: porte ouverte sur l'avenir?*, 25–33.

Doward, J. 2000. *Mineral Riches Fuel War, not the Poor*. Global Policy Forum – The World Bank/UN Security Council: Diamonds in Conflict, *The Observer*, 18 June, available online at http://www.globalpolicy.org/security/docs/resourc1.htm, accessed on 26 September 2000.

Downton, M. W. 1995. 'Measuring tropical deforestation: development of the methods', *Environmental Conservation* 22(3), Autumn, 229–40.

Drouineau, S. and Nasi, R. 1999. *L'aménagement forestier au Gabon. Historique, bilan, perspectives*. Project FORAFRI.

Duncan, R. C., Chand, S., Graham, B., Lawson, T. and Duncan, R. 1998. *Exchange Rate Policy in Papua New Guinea*. Canberra: Australian National University.

Duncan, R. C. and Xu, X. 2000. 'Should Papua New Guinea adopt a stronger exchange rate regime?' *Pacific Economic Bulletin* 15(2). Canberra, Australia: Asia Pacific Press, 36–45.

Dupuy, B., Gérard, C., Maître, H. F., Marti, A. and Nasi, R. 1998. *Gestion des écosystèmes forestiers denses d'Afrique tropicale humide*. Montpellier: CIRAD.

Durieu, L., Avelling, C. and Vives, M. 2000. 'Un exploitant forestier au Gabon pourrait démarrer ses activités dans une aire protégée, et ce, en apparante toute legalité', *CANOPEE – Bulletin sur l'Environnement en Afrique Central*, No. 16, February.

Eastwood, R. K. and Venables, A. J. 1982. 'The macroeconomic implications of a resource discovery in an open economy' *Economic Journal* 92.

Eaton, P. 1983. 'The plantation redistribution scheme in Papua New Guinea', *Land Reform, Land Settlement and Cooperatives*, 1–2: 52–61.

Eba'a-Atyi, R. 1998. 'Cameroon's logging industry: structure, economic importance and effects of devaluation', *Occasional Paper* No. 14. Bogor, Indonesia: Center for International Forestry Research.

ECAM 1997. *Enquête Camerounaise Auprès des Ménages*, Vol. 2 *(Résultats)*. Yaoundé: Ministère de l'Économie et des Finances, December.

Economist Intelligence Unit (EIU). 1995. 'Country profile: Cameroon, Central African Republic, Chad 1994–95'. London: The Economist Intelligence Unit.

Economist Intelligence Unit (EIU). 1997. 'Country profile: Cameroon, Central African Republic, Chad, 1997–98'. London: The Economist Intelligence Unit.

Economist Intelligence Unit (EIU). 1999a. 'Country profile: Gabon 1999–00'. London: The Economist Intelligence Unit.

Economist Intelligence Unit (EIU). 1999b. 'Country profile: Venezuela 1999–00'. London: The Economist Intelligence Unit.

Economist Intelligence Unit (EIU). 1999c. 'Country profile: Cameroon, Central African Republic, Chad, 1999–00'. London: The Economist Intelligence Unit.

Economist Intelligence Unit (EIU). 1999d. 'Country profile: PNG 1999–00'. London: The Economist Intelligence Unit.

Economist Intelligence Unit (EIU). 1999e. 'Country profile: Indonesia 1999–2000'. London: The Economist Intelligence Unit.

Economist Intelligence Unit (EIU). 2001. 'Country Profile 2001: Gabon, Equatorial Guinea'. London: The Economist Intelligence Unit.

Eden, M. J. 1985. 'Forest cultivation and derived savanna and grassland: comparative studies from Southern Papua and South West Guyana' in J. C. Tothill and J. J. Mott (eds), *Ecology and Management of the World's Savannas*. Canberra: Australian Academy of Science.

EDIAFRIC. 1985. *L'économie Gabonaise*. EDIAFRIC – la documentation Africaine, Paris.

Edwards, C. R. 1986. 'The human impact on the forest in Quintana Roo, Mexico', *Journal of Forest History* 30(3): 120–7.

Edwards, S. 1986. 'A commodity export boom and the real exchange rate: the money-inflation link', in J. P. Neary and S. van Wijnbergen (eds), *Natural Resources and the Macroeconomy*. Oxford, 229–48.

Eide, E. 1973. 'Virkninger af statens oljeinntekter på norsk økonomi', *Sosialøkonomen* 10.

Ekanade, O. 1987. 'Small-scale cocoa farmers and environmental change in the tropical rain forest regions of southwestern Nigeria', *Journal of Environmental Management* 25: 61–70.

Ekstrom, J. P. 1981. 'Colonist strategies of verticality in an Eastern valley', in N. E. Whitten (ed.), *Cultural Transformation and Ethnicity in Modern Ecuador*. Chicago: Urbana.

Ella Nguema Rolly, E. 1979. '*Les 182 kilomètres du Transgabonais*', Bachelor's thesis (unpublished). Libreville: Université Nationale Omar Bongo.

Ellis, D. M. 1999. 'Where there are no roads', in *La route en forêt tropicale: porte ouverte sur l'avenir? APFT Working Paper* No. 6.

Ellman, M. 1981. 'Natural gas, restructuring and de-industrialisation: the Dutch experience of industrial policy', in T. Barker, and V. Brailovsky (eds), *Oil or Industry? Energy, Industrialisation and Economic Policy*. London and Mexico City: Academic Publishers.

Enders, K. and Herberg, H. 1983. 'The Dutch Disease: causes, consequences, cures and calmatives', *Weltwirtschaftliches Archiv* 119.

Energy Information Administration (EIA) 2000. 'Cameroon energy oil information. USA': Energy Information Administration, available online at http://www.eia.doe.gov.

Enright, M. and Saavedra, E. S. 1995. *Venezuela: El reto de la competitividad – Resumen ejecutivo*. Caracas: CONAPRI.

Erwidodo and Hadi, P. U. 1999. 'Effects of trade liberalisation on agriculture in Indonesia: commodity aspects', CGPRT Centre, *CGPRT Working Papers* No. 48. Bogor, October.

Essama-Nssah, B. and Gockowski, J. J. 2000. 'Forest sector development in a difficult political economy: an evaluation of Cameroon's forest development and World Bank assistance', Washington: Operations Evaluation Department, World Bank (unpublished).

Estreguil, C. and Lambin, E. F. 1996. 'Mapping forest-cover disturbances in Papua New Guinea with AVHRR data', *Journal of Biogeography* 23: 757–73.

Eva, H. D., Glinni, A., Janvier, P. and Blair-Myers, C. 1999. 'Vegetation map of tropical South America at 1:5,000,000', *TREES publication Series D: Thematic Outputs* No. 2. Ispra, Italy: TREES JRC.

Ezeala-Harrison, F. 1993. 'Structural readjustment in Nigeria: diagnosis of a severe Dutch disease syndrome', *American Journal of Economics and Sociology* 52(2):193–208.

Fabella, R. and Jayasuriya, S. 2000. 'The impact of temporary trade shocks on an economy in disequilibrium: the Philippines, 1986–89' in P. Collier and J. W. Gunning (eds), *Trade Shocks in Developing Countries, Volume 2: Asia and Latin America*. Oxford, 260–84.

Fallon, J., King, T. and Zeitsch, J. 1995. 'Exchange rate policy in Papua New Guinea', *INA Discussion Paper* No. 64. Port Moresby: Institute of National Affairs.

FAO. 1993. Forest Resources Assessment 1990. Tropical countries. *FAO Forestry Paper 112*. Rome: Food and Agricultural Organization of the United Nations.

FAO. 1996. *FAO Production Yearbook 1995*, Vol. 49, Rome: FAO.

FAO. 1997a. *State of the World's Forests*. Rome: FAO.

FAO. 1997b. Available online at http://www.fao.org; gopher://faov02.FAO.ORG...try.latin%Decuador.txt

FAO. 1998. Available online at http://www.fao.org; gopher://faov02.FAO.ORG...try.latin%Dvenezuel.txt.

FAO. 2000a. FRA 2000. 'On definitions of forest and forest change', *FRA Working Paper* No. 33. Rome: FAO.

FAO. 2000b. FAOSTAT, available online at http://apps.fao.org.

FAO. 2001a. 'Results of the Global Forest Resources Assessment 2000', Committee on Forestry, Item 8(b) of the Provisional Agenda, Fifteenth Session, Secretariat Note Rome, Italy, 12–16 March 2001.

FAO. 2001b. FAOSTAT, available online at http://apps.fao.org.

FAO. 2001c. 'Global forest resources assessment 2000. Main report', *FAO Forestry Paper* No. 140. Rome: FAO.

FAO. 2002a. Forest Resources Assessment 2000 – Papua New Guinea. Available online at http://www.fao.org/forestry/fo/country/is.jsp?geo_id = 118&lang_id = 1&page_id = 507, accessed on 23 January 2002.

FAO. 2002b. Forest Resources Assessment 2000 – Nigeria, available online at http://www.fao.org/forestry/fo/country/index.jsp?geo_id = 1&lang_id = 1, accessed on May 2002.

Fardmanesh, M. 1991. 'Dutch Disease economics and the oil syndrome: an empirical study', *World Development* 19(6): 711–17.

Feltenstein, A. 1992. 'Oil prices and rural migration: the Dutch Disease goes south', *Journal of International Money and Finance*, 11(3): 273–92, June.

Fernando, N. A. 1992. *Provincial Human Development in Papua New Guinea: Its Level and Inequality 1972 and 1980*. Port Moresby: National Research Institute.

Figueroa, S. 1995. 'The Ecuadorian system of protected natural areas' in S. Amend and T. Amend (eds), *National Parks without People? The South American Experience*. Quito: IUCN.

Filer, C. (ed). 1997. *The Political Economy of Forest Management in Papua New Guinea*. The National Research Institute and International Institute for Environment and Development. Papua New Guinea: Boroko.

Filer, C. 2001. 'Between a rock and a hard place: mining projects, "indigenous communities" and Melanesian states', in B.Y. Imbun and P. A. McGavin (eds), *Mining in Papua New Guinea: Analysis and Policy Implications*. Port Moresby: University of Papua New Guinea Press.

Filer, C. with Sekhran, N. 1998. 'Loggers, donors and resource owners', *Policy that Works for Forests and People Series*, No. 2. London: International Institute for Environment and Development.

Filer, C., Dubash, N. K. and Kalit, K. 2000. *The Thin Green Line: World Bank Leverage and Forest Policy Reform in Papua New Guinea*. The National Research Institute and Australian National University, Boroko, Papua New Guinea and Canberra, Australia.

Filer, C., Henton, D., Jackson, R. 2000. *Landowner Compensation in Papua New Guinea's Mining and Petroleum Sectors*. Port Moresby, Papua New Guinea: PNG Chamber of Mines and Petroleum.

Finger. 1998. *Metals from the Forest: Mining and Forest Degradation*. Switzerland: Gland.

Forsyth, P. J. 1986. 'Booming sectors and structural change in Australia and Britain: A comparison', in J. P. Neary and S. van Wijnbergen (eds), *Natural Resources and the Macroeconomy*. Oxford: Blackwell, 251–83.

Forsyth, P. J. and Kay, J. A. 1980. 'The economic implications of North Sea oil revenues', *Fiscal Studies* 1.

Forsyth, P. J. and Nicholas, S. J. 1983. 'The decline of Spanish industry and the price revolution: A neoclassical analysis', *Journal of European Economic History* 12.

Franqueville, A. 1984. 'La population rurale africaine face à la pénétration de l'économie moderne: le cas du Sud-Cameroun', in C., Blanc-Pamard, J. Bonnemaison, J. Boutrais, V. Lassailly-Jacob and A. Lericollais (eds), *Le développement rural en questions. Paysages, espaces ruraux, systèmes agraires. Maghreb, Afrique noire, Mélanésie*. Paris: ORSTOM.

Franqueville, A. 1987. *Une Afrique entre le village et la ville. Les migrations dans le sud du Cameroun*. Paris: ORSTOM.

French, B. 1978. *Food Crops of Papua New Guinea*. PNG: Vudal College of East New Britain, October, 55 pp.

Freyne, D. F. and McAlpine, J. R. 1985. 'Land clearing and development in Papua New Guinea', *Proceedings of CRAM Workshop, Bangkok, Thailand, 27 August – 3 September 1985*, 153–63.

Fuhr, M., Delègue, M. A., Nasi, R. and Minkoué, J. M. 1998. *Dynamique et croissance d l'okoumé en zone côtière du Gabon*. Montpellier: CIRAD.

FWI/GFW 2002. *The State of the Forest: Indonesia*. Bogor, Indonesia: Forest Watch Indonesia and Washington, DC: Global Forest Watch.

Gabaldón, M. 1995. 'Inhabitants in Venezuelan national parks: policies and management', in S. Amend and T. Amend (eds), *National Parks without People? The South American Experience*. Quito: IUCN.

Gami, M. and Nasi, R. 2001. 'Sustainability and security of intergenerational access to resources. Participatory mapping studies in Gabon', in C. Colfer and Y. Byron (eds), *People Managing Forests. The Links between Human Well-Being and Sustainability*. Washington, DC: RFF.

García, G., Rodríguez, R., Marcano, B. L., Penfold, R. and Sánchez, G. 1997. *La sostenibilidad de la política fiscal en Venezuela*. Revista Banco Central de Venezuela 11(2).

Garnaut, R. and Baxter, P. with Krueger, A.O. 1984. 'Exchange rate and macro-economic policy in independent Papua New Guinea', *Pacific Research Monograph*, Development Studies Centre. Canberra: The Australian National University, 173 pp.

Gartlan, S. 1989. *La conservation des écosystèmes forestiers du Cameroun*. Gland: IUCN.

Gavin, M. 2000. 'The Mexican oil boom, 1977–85', in P. Collier and J. W. Gunning (eds), *Trade Shocks in Developing Countries, Volume 2: Asia and Latin America*. Oxford: Oxford University Press, 164–200.

Geist, H. J. and Lambin, E. F. 2001. 'What drives tropical deforestation? A meta-analysis of proximate and underlying causes of deforestation based on subnational case study evidence', *LUCC Report Series*, No. 4. Louvain-la-Neuve: Ciaco.

Gelb, A. and Associates. 1988. *Oil Windfalls: Blessings or Curse?* New York: Oxford University Press.

Gelb, A. and Marshall-Silva, J. 1988. 'Ecuador: Windfalls of a new exporter' in A. Gelb and associates (eds), *Oil Windfalls – Blessing or Curse?* Oxford: Oxford University Press and New York: The World Bank.

Gelb, A. with Auty, R. 1988. 'Trinidad and Tobago: Windfalls in a small parliamentary democracy', in A. Gelb and Associates (eds), *Oil Windfalls: Blessings or Curse?* New York: Oxford University Press, 262–88.

Gelb, A. with Bienen, H. 1988. 'Nigeria: from windfall gains to welfare losses?' in A. Gelb and Associates (eds), *Oil Windfalls: Blessings or Curse?* New York: Oxford University Press, 227–61.

Gelb, A. with Bourguignon, F. 1988. 'Venezuela: absorption without growth', in A. Gelb and Associates (eds), *Oil Windfalls: Blessings or Curse?* New York: Oxford University Press, 288–326.

Gelb, A. with Bourguignon, F. 1988. 'Venezuela: absorption without growth', in A. Gelb and Associates (eds), *Oil Windfalls – Blessing or Curse?* Oxford: Oxford University Press and New York: World Bank.

Gelb, A. with Conway, P. 1988. 'Algeria: windfalls in a socialist economy', in A. Gelb and Associates (eds), *Oil Windfalls: Blessings or Curse?* New York: Oxford University Press, 147–69.

Gelb, A. with Glassburner, B. 1988. 'Indonesia: Windfalls in a poor rural economy', in A. Gelb and Associates (eds), *Oil Windfalls: Blessings or Curse?* New York: Oxford University Press, 197–226.

Gelb, A. with Marshall-Silva, J. 1988. 'Ecuador: windfalls of a new exporter', in A. Gelb and Associates (eds), *Oil Windfalls: Blessings or Curse?* New York: Oxford University Press, 170–96.

Gelb, A., Knight, J. B. and Sabot, R. H. 1988. 'Lewis through a looking glass: public sector employment, rent-seeking and economic growth', *World Bank Policy, Planning and Research Working Paper*.

Ghanem, H. 2000. 'The Ivorian cocoa and coffee boom of 1976–79: the end of a miracle?', in P. Collier and J. W. Gunning (eds), *Trade Shocks in Developing Countries, Volume 1: Africa*. Oxford University Press, 142–62.

Gibson, J. 1995. 'Food consumption and food policy in Papua New Guinea', *INA Discussion Paper* No. 65. Port Moresby: Institute of National Affairs, September.

Gibson, J. 2001. 'Food security and food policy in Papua New Guinea', *INA Discussion Paper* No. 83, Port Moresby: Institute of National Affairs, June.

Gill, T. 1931. *Tropical Forests of the Caribbean*. Baltimore: Tropical Plant Research Foundation.

Gockowski, J. J. 1994. 'Supply responses among small holder producers of arabica and robusta coffee in Cameroon', PhD thesis. Florida: University of Florida (unpublished).

Gómez, H. 1995. 'Deforestaciones y reservas forestales', *Seforven*, year 6, No. 12, June. Caracas: MARNR.

Gómez, H. and Malavé, J. 1998. 'Un país cada vez más pobre', *Debates IESA* 4(1), July/September.

González-Vega, C. 2000. 'Costa Rica: mismanagement of the coffee boom', in P. Collier and J. W. Gunning (eds), *Trade Shocks in Developing Countries, Volume 2: Asia and Latin America*. Oxford: Oxford University Press, 42–81.

Goodman, R., Lepani, C. and Morawetz, D. 1987. 'The economy of Papua New Guinea. An independent review', *Pacific Policy Papers* No. 1. Canberra, Australia: Australian National University.

Government of Papua New Guinea, 1989. *White Paper on Agriculture: National Policies for the Development of the Agricultural Sub-sector*. Port Moresby, Papua New Guinea.

Grainger, A. 1996. 'An evaluation of the FAO tropical forest resource assessment, 1990', *Geographical Journal* 162(1).

Grau, P. C. 1997. 'El paisaje venezolano', in G. Morón (coord.), *Venezuela quinientos años*. Caracas: Citibank.

Greenaway, D., and Lamusse, R. 2000. 'Private and public sector responses to the 1972–75 sugar boom in Mauritius', in P. Collier and J. W. Gunning (eds), *Trade Shocks in Developing Countries, Volume 1: Africa*. Oxford: Oxford University Press, 207–25.

Greenaway, D., and Pillay, S. S. 2000. 'An evaluation of the 1979–85 petroleum boom in Malaysia', in P. Collier and J. W. Gunning (eds), *Trade Shocks in Developing countries, Volume 2: Asia and Latin America*. Oxford: Oxford University Press, 308–25.

Gregory, R. G. 1976. 'Some implications of the growth of the mineral sector', *Australian Journal of Agricultural Economics* 20.

Grilli, E. R. and Yang, M. C. 1988. 'Primary commodity prices, manufactured goods prices, and the terms of trade of developing countries: what the long run shows', *World Bank Economic Review* 2(1), 1–47.

Grynberg, R., Wigston, D. and Vemuri, S. R. 1988. 'Transfer pricing and trade malpractice in the Papua New Guinea (PNG) log export industry', *Journal of Interdisciplinary Economics* 2: 215–43.

Gubry, P., Lamlenn, S. B., Ngwé, E., Tchégho, J.-M., Timnou, J.-P. and Veron, J. 1996. *Le retour au village. Une solution à la crise économique au Cameroun?* Paris: Éditions l'Harmattan.

Gullison, R. E. and Losos, E. C. 1993. 'The role of foreign debt in deforestation in Latin America', *Conservation Biology* 7: 140–7.

Gumoi, M. 1994. 'Floating the kina, a historical decision by the Government', *The Times (13 October)*, p. 31.

Gupta, D. 1995. 'The exchange rate issue and the cash flow problem in Papua New Guinea: some self-inflicted wounds', *NRI Special Publication* No. 21. Papua New Guinea: National Research Institute.

Gutiérrez, A. 1992. *Las relaciones entre el ingreso, su distribución y el consumo de alimentos: el caso de Venezuela en la década de los ochenta*. Cuadernos del CENDES, segunda época, September/December.

Gutiérrez, O. R. 1996. 'La empresa forestal en función de la reforma agraria (El caso de EMIFOCA, Barinas-Venezuela)', *Cuadernos Geográficos*, Vol. 10. Universidad de Los Andes, Facultad de Ciencias Forestales, Mérida.

Gylfason, T. 2001. 'Natural resources, education, and economic development', *European Economic Review* 45: 847–59.

Haase, A. H. V. 1997. 'Racionalidad des uso del recurso bosque por pequeños productores agrícolas, Estado Sucre, Venezuela', *Compilación de los resultados de los trabajos de investigación realizados en Venezuela entre septiembre de 1996 y octubre de 1997*. Carúpano, October.

Haque, N. U., Husain, A. M. and Montiel, P. J. 1994, 'An empirical "dependent economy" model for Pakistan', *World Development* 22(10):1585–97.

Harcourt, C. S. and Sayer, J. A. 1996. *The Conservation Atlas of Tropical Forests. The Americas*, IUCN, CIFOR, WCMC, BP. New York: Simon and Schuster.

Harden, R. and Sugden, C. 1999. 'An update of recent events: the Papua New Guinea economy', *Pacific Economic Bulletin* 14(2), 1–11.

Harper, A. and Israel, M. 1999. 'The killing of the fly: state-corporate victimization in Papua New Guinea', *Working Paper* 1999/22. Resource Management in Asia Pacific. Canberra, Australia: Australian National University.

Harrigan, J. 2000. 'Malawi's positive 1977–79 trade shock', in P. Collier and J. W. Gunning (eds), *Trade Shocks in Developing Countries, Volume* 1: *Africa*. Oxford: Oxford University Press, 163–206.

Harris, G. T. 1982. 'Subsistence agriculture and nutrition in Papua New Guinea: a research review', *IASER Discussion Paper* No. 42. Papua New Guinea: IASER.

Hartley, A. C. 1996. 'Rainforest regeneration: Kutubu petroleum development project, PNG', in P. G. Buchanan (ed.), *Petroleum Exploration, Development and Production in PNG*. Proceedings of the 3rd PNG Petroleum Convention, Port Moresby.

Hasanuddin, L. (ed.) 1996. 'Mitos-mitos pengelolaan hutan di Indonesia [Forest management myths in Indonesia]', *NGO Position Paper* No. 2, Akatiga, KPA, LATIN, LBBT, SHK, WALHI, YABSHI, YLBHI, Pelangi, Sejati, Walhi (publisher), Jakarta.

Hausman, R. 2000. 'Dealing with negative oil shocks: the Venezuelan experience in the 1980s', in P. Collier and J. W. Gunning (eds), *Trade Shocks in Developing countries, Volume* 2: *Asia and Latin America*. Oxford: Oxford University Press, 120–63.

Hausmann, R. 1990. *Dealing with Negative Oil Shocks: The Venezuelan Experience in the Eighties*. Paper for the Conference on Temporary Trade Shocks. Oxford: Oxford University (unpublished).

Hecht, S. and Cockburn, A. 1990. *The Fate of the Forest: Developers, Destroyers and Defenders of the Amazon*. New York: Harper Perennial, 357.

Heimberger, B. 1994. *Geldwirtschaftliche Analyse eines Rohstofflandes. Beitrag zur Erklärung der ökonomischen Entwicklung Venezuelas (1958–1993)*, MA thesis. Berlin: Freie Universität, July.

Hill, C. and Knight, J. 2000. 'The diamond boom, expectations, and economic management in Botswana', in P. Collier and J. W. Gunning (eds), *Trade shocks in developing Countries, Volume 1: Africa*. Oxford: Oxford University Press, 303–50.

Hill, H. 1992. 'Regional development in a boom and bust petroleum economy: Indonesia since 1970', *Economic Development and Cultural Change* 40(2): 351–79.

Hiroaka, M. and Yamamoto, S. 1980. 'Agricultural development in the Upper Amazon of Ecuador', *Geographical Review* 70(4): 423–446.

Hoel, M. 1981. 'Employment effects of an increased oil price in an economy with short-run labour immobility', *Scandinavian Journal of Economics* 83.

Holmes, D. 2000. 'Deforestation in Indonesia: a review of the situation in Sumatra, Kalimantan and Sulawesi', Draft. Jakarta: World Bank (unpublished).

Holzknecht, H. 1996. 'Policy reform, customary tenure and stakeholder clashes in Papua New Guinea's rainforests', *Rural Development Forestry Network Paper*, No. 19c. UK: Overseas Development Institute.

Hoogeveen, J. A. M. and van Soest, D. P. 1993. 'Tropical rainforest degradation in Cameroon and Ecuador: a socio-economic approach', *Series Development and Security* No. 41. Centre for Development Studies, University of Groningen.

Houghton, R. A., Lefkowitz, D. S. and Skole, D. L. 1991. 'Changes in the landscape of Latin America between 1850 and 1985. I. Progressive loss of forests, *Forest Ecology and Management* 38: 143–72.

Huber, O. and Alarcón, C. 1988. *Mapa de vegetación de Venezuela*. Caracas: MARNR and TNC.

Hulme, D. 1982. 'Land settlement schemes in Papua New Guinea. An overview', *Land Reform, Land Settlement and Cooperatives Development Studies*, Vol. 1–2, 21–42, Administrative Staff College, Port Moresby.

Hulme, D. 1983. 'Papua New Guinea scene no. 24. An economic appraisal of the Hoskins Oil Palm Scheme', *Australian Geographer* 15(5): 330–4.

Hutchison, M. M. 1994. 'Manufacturing sector resiliency to energy booms: empirical evidence from Norway, the Netherlands, and the United Kingdom', *Oxford Economic Papers* 46: 311–29.

Hyndman, D. 1997. 'Changing relations of production in the creation of the Ok Tedi mining enclave in Papua New Guinea', *Working Paper* 1997/11. Resource Management in Asia Pacific. Canberra, Australia: Australian National University.

ICRA and IGAD. 1996. 'Agriculture itinerante et enjeux fonciers dans la Province de l'Estuaire, Gabon', *Serie de Documents de Travail* No. 55. Libreville: Centre International pour la Recherche Agricole orientée vers le développement and Institut Gabonais d'Appui au Développement.

IDA and IMF. 2000. *Cameroon*. Decision point document for the enhanced Heavily Indebted Poor Countries (HIPC) Initiative. Washington, DC: International Development Association and the International Monetary Fund, 15 September.

IGAD. 1997. *L'approvisionnement vivrier de Franceville*. Institut Gabonais d'Appui au Développement (IGAD), Libreville.

Igua, P. B. K. 2001. 'Food security strategies for Papua New Guinea', *CGPRT Working Paper* No. 56, Bogor, Indonesia, 104.

IFAD, IDRC, CIIFAD, ICRAF and IIRR. 2001. *Shifting Cultivation: Towards Sustainability and Resource Conservation in Asia*. Cavite, Philippines: International Institute of Rural Reconstruction, 421 pp.

IMF. 1990. *International Financial Statistics Yearbook*. Washington, DC: International Monetary Fund.

IMF. 1991. *International Financial Statistics Yearbook*. Washington, DC: International Monetary Fund.

IMF. 1992. *International Financial Statistics Yearbook*. Washington, DC: International Monetary Fund.

IMF. 1996. *International Financial Statistics Yearbook*. Washington, DC: International Monetary Fund.

IMF. 1999a. *International Financial Statistics Yearbook*. Washington, DC: International Monetary Fund.

IMF. 1999b. 'Gabon: Statistical annex', *Staff Country Report* No. 99/12. Washington, DC: International Monetary Fund.

IMF. 1999c. 'Cameroon: enhanced structural adjustment facility medium-term economic and financial policy framework paper (1999/2000–2001/02),' available online at http://www.imf.org/external/np/pfp/ 1999/cameroon/INDEX.htm.

IMF. 2000. 'Cameroon: statistical annex', *Staff Country Report* No. 00/81. Washington, DC: International Monetary Fund.

IMF. 2001. 'Nigeria: selected issues and statistical appendix', *IMF Country Report* No. 01/132, August, Washington, DC: International Monetary Fund, available online at http://www.imf.org/external/pubs/ft/scr/2001/cr01132.pdf, accessed on April 2001.

Indjieley, M. 1998. *Analyse socioculturelle et économique de la consommation de la viande de brousse a Libreville*. Synthèse des travaux. Libreville: Seminaire FORAFRI de Libreville.

INEFAN. 1995. *Plan de acción forestal para el Ecuador*. Final document. Quito: Ministry of Agriculture.

Infante, A. 1995. 'Las instituciones forestales venezolanas ante la deforestación', *Seforven*, year 6, No. 12, June. Caracas: MARNR.

Inman, K. 1990. 'Social causes of deforestation in third world countries', Master's thesis. Athens: University of Georgia.

Inman, K. 1993. 'Fueling expansion in the third world: population, development, debt, and the global decline of forests', *Society and Natural Resources* 6: 17–39.

IPCC 2001. 'Climate change 2001: impacts, adaptation and vulnerability. Summary for policy-makers', Intergovernmental Panel on Climate Change (WGII), available online at http://www.ipcc.ch.

IPIECA. 2002a. TOTAL in Mahakam Delta, Indonesia. International Petroleum Industry Environmental Conservation Association (IPIECA)/UNEP, available online at http://www.ipieca.org/publications/sens_envir_case_studs/total_mahakam/total_mahakam.html, accessed on April 2002

IPIECA. 2002b. Case study nine: 'PT Caltex Pacific Indonesia', International Petroleum Industry Environmental Conservation Association (IPIECA)/UNEP, available online at http://www.ipieca.org/downloads/TechCoopPDFs/PT%20Caltex%20in%20Indonesia.pdf, accessed on April 2002.

IRF. 1988. World Road Statistics. Washington, DC: International Road Federation.

IRF. 1994. World Road Statistics. Washington, DC: International Road Federation.

IRF. 2000. World Road Statistics. Vernier: International Road Federation.

ITTO and INEFAN. 1994a. *Estrategia para la industria sostenida de la madera en el Ecuador*, Proyecto PD 137/91, Resumen del informe final. Quito, May.

ITTO and INEFAN. 1994b. *Desarrollo integrado de la estructura de comercialización de la madera y productos de la madera en el Ecuador*, Proyecto PD 154/91 Rev 2(1), Resumen ejecutivo y propuestas. Quito, May.

IUCN. 1990. *La conservation des écosystèmes forestiers du Gabon. Basé sur le travail de Chris Wilks*. Le programme de l'UICN pour les forêts tropicales.

IUCN 1993. *Oil and Gas Exploration and Production in Mangrove Areas*, Gland (Switzerland) and Cambridge (UK): IUCN (The World Conservation Union) with E&P Forum, London (47 pp).

James, N. 1997. 'An overview of Papua New Guinea's mineral policy', *Resources Policy* 23 (1/2), 97–101.

Janz, K. 1993. 'World forest resources assessment 1990: an overview', *Unasylva* 44(3).

Jarrett, F. O. and Anderson, K. 1989. 'Growth, structural change and economic policy in Papua New Guinea: implications for agriculture', *Pacific Policy Paper* No. 5. National Centre for Development Studies. Canberra, Australia: Australian National University.

Jayasuriya, S. 2000. 'Temporary shocks, consumption smoothing and economic adjustment: Sri Lanka, 1973–76', in P. Collier and J. W. Gunning (eds), *Trade Shocks in Developing Countries, Volume 2: Asia and Latin America*. Oxford: Oxford University Press, 233–59.

Jones, J. R. 1990. *Colonization and Environment: Land Settlement Projects in Central America*. Tokyo: United Nations University.

Jones, L. T. and McGavin, P. A. 2001. *Land Mobilisation in Papua New Guinea*. Canberra, Australia: Asia Pacific Press, Australian National University.

Josling, T. 1997. 'Agricultural trade policies in the Andean Group. Issues and options', *World Bank Technical Paper* No. 364. Washington, DC: World Bank.

Jua, N. 1993. 'State, oil and accummulation', in P. Geschiere, and P. Konings (eds), *Itinéraires d'accumulation au Cameroun*. Paris: Karthala.

Kahn, J. R. and McDonald, J. A. 1994. 'International debt and deforestation', in K. Brown, and D. Pearce (eds), *The Causes of Tropical Deforestation. The Economic and Statistical Analysis of Factors Giving Rise to the Loss of Tropical Forests*, London: University College London Press, 57–67.

Kahn, J. R. and McDonald, J. A. 1995. 'Third-world debt and tropical deforestation', *Ecological Economics* 12(2): 107–23.

Kaimowitz, D. 1992. 'Aid and development in Latin America', *Latin American Research Review* 27(2), 202–11.

Kaimowitz, D. and Angelsen, A. 1998. *Economic Models of Tropical Deforestation: A Review*. Bogor: Center for International Forestry Research, 139 pp.

Kaimowitz, D. and Smith, J. 2001. 'Soybean technology and loss of natural vegetation in Brazil and Bolivia', in A. Angelsen and D. Kaimowitz (eds), *Agricultural Technologies and Tropical Deforestation*. Wallingford (UK): CAB International with CIFOR, 195–211.

Kaimowitz, D., Erwidodo, Ndoye, O., Pacheco, P. and Sunderlin, W. 1998. 'Considering the impact of structural adjustment policies on forests in Bolivia, Cameroon, and Indonesia', *Unasylva* 49(149), 57–64.

Kalulumia, P. 1997. 'External debt, fiscal policy and terms-of-trade disturbances in the dynamics of an open economy: the case of Cameroon', *African Development Review* 9(2).

Kamas, L. 1986. 'Dutch disease economics and the Colombian export boom', *World Development* 14(9):1177–98.

Kammesheidt, L. A., Torres, L. and Franco, W. 1995. 'Efecto de la explotación selectiva sobre la estructura y sostenibilidad del bosque tropical: um caso de los Llanos occidentales venezolanas', *Revista Forestal Venezolana* 39(1). Mérida: Universidad de Los Andes.

Kant, S. and Redantz, A. 1997. 'An econometric model of tropical deforestation', *Journal of Forestry Economics* 3: 51–86.

Karl, T. L. 1986. 'Petroleum and political pacts', in G. O'Donnell, P. C. Schmitter and L. Whitehead (eds), *Transition from Authoritarian Rule: Latin America*. Baltimore/London: Johns Hopkins University Press.

Karl, T. L. 1995. 'The Venezuelan petro-state and the crisis of "its" democracy', in J. McCoy, A. Serbin, W. C. Smith and A. Stambouli (eds), *Venezuelan Democracy under Stress*. Miami: University of Miami, North-South Center.

Karl, T. L. 1997. *The Paradox of Plenty. Oil Booms and Petro-states*. Berkely, Los Angeles and London: University of California Press, 342 pp.

Karsenty, A. 2000. *Étude sur la fiscalité forestière en Afrique Centrale. En vue de favoriser une développement forestier durable dans la sous région (Gabon, Cameroun, Congo-Brazzaville, Centrafrique, Guinée Équatoriale)*. Libreville: WWF (unpublished).

Kayizzy-Mugerwa, S. 1989. 'External shocks and adjustment in a mineral-dependent economy: a short-run model for Zambia', Memo 123. School of Economics and Legal Science, Gothenburg University.

Kimerling, J. (with five secondary authors). 1991. *Amazon Crude*. Washington, DC: Natural Resources Defence Council.

Kimsey, M. 1991. 'An analysis of spatial and temporal variation in tropical deforestation', PhD dissertation. Athens: University of Georgia.

Klappa, S. 1999. 'The road to development: Krisa', in La route en forêt tropicale: porte ouverte sur l'avenir? *APFT Working Paper* No. 6. Bruxelles: Commission Européenne/DG VIII.

Kleinn, C. and Corrales, L. 2000. *FRA 2000. Bibliografía comentada. Cambios en la cobertura forestal. Venezuela*. Rome: Food and Agricultural Organization of the United Nations.

Knudsen, O. and Parnes, A. 1975. *Trade Instability and Economic Development*. Lexington: Lexington Books.

Kocher-Schmid, C. 1999. 'Roads to nowhere: Vanimo – Kilimeri', in: 'La route en forêt tropicale: porte ouverte sur l'avenir? *APFT Working Paper* No. 6. Bruxelles: Commission Européenne/DG VIII.

Konings, P. 1996. 'The post-colonial state and economic and political reforms in Cameroon', in A. E. Fernández Jilberto and A. Mommen (eds), *Liberalization in the Developing World: Institutional and Economic Changes in Latin America, Africa and Asia*. London and New York: Routledge, 244–65.

Kremers, J. J. M. 1986. 'The Dutch disease in the Netherlands', in J. P. Neary and S. van Wijnbergen (eds), *Natural Resources and the Macroeconomy*. Oxford: Blackwell, 96–135.

Krueger, A. O. 1978. 'The political economy of the rent-seeking society', *American Economic Review* 64.

Kuntz, S. and Siegert, F. 1999. 'Monitoring of deforestation and land use in Indonesia with multi-temporal ERS data', *International Journal of Remote Sensing* 20(14).

Kuznets, S. 1955. 'Economic growth and income inequality', *American Economic Review* 45(1), 1–28.

Kwapena, N. 1985. 'Tropical rain forests and plantation forestry in Papua New Guinea', *The Environmentalist* 5(10): 91–100.

Lam, N. V. 1984. 'The commodity export sector in Papua New Guinea', *Monograph 22*. Boroko, Papua New Guinea: Institute of Applied Social and Economic Research.

Lamb, D. 1990. 'Exploiting the tropical rain forest: an account of pulpwood logging in Papua New Guinea', *Man and the Biosphere Series* 3: 259.

Lane-Poole, C. E. 1925. 'The forest resources of the territories of Papua and New Guinea', *Report*. Melbourne: Government of the Commonwealth of Australia.

Lanly, J.-P. 1982. 'Tropical forest resources', *FAO Forestry Paper* 30. Rome, Italy: FAO.

Laporte, N., Goetz, S. J., Justice, C. O., and Heinecke, M. 1998. 'A new land cover map of Central Africa derived from multi-resolution, multi-temporal AVHRR data', *International Journal of Remote Sensing* 18: 3537–50.

Laporte, N., Justice, C. and Kendall, J. 1995. 'Mapping the dense humid forest of Cameroon and Zaire using AVHRR satellite data', *International Journal of Remote Sensing* 1(6).

Larivrière, J. 1996. *L'Atlas pour la conservation des forêts tropicales d'Afrique*. IUCN, Paris: Jean-Pierre de Monza.

Larrea, C. 1992. 'The mirage of development: oil, employment, and poverty in Ecuador', PhD thesis, Graduate Programme in Social and Political Thought. Ontario: York University.

Levang, P. 2001. 'Transmigration in Indonesia: the land beyond – tanah sabrang. *Indonesian French Review* 40: 19–23.

Levantis, T. 2000. *Papua New Guinea: Employment, Wages and Economic Development*. Canberra: Australian National University and Port Moresby: Institute of National Affairs.

Liverman, D., Wilder, M., Patterson M. and Vásquez, M. 1997. 'Land use change in Mexico', Prepared for delivery at the 1997 meeting of the Latin American Studies Association, Continental Plaza Hotel, Guadalajara, Mexico, 17–19 April 1997.

Looney, R. E. 1992. 'Real or illusory growth in an oil-based economy: government expenditure and private sector investment in Saudi Arabia', *World Development* 20(9), 1367–75.

López, J., Pernía, E. and Pozzobón, E. 1996. 'Avance de las deforestaciones en la Reserva Forestal de Caparo, Estado Barinas, Venezuela, entre 1987 y 1994', *Revista Forestal Venezolana* 40(2).

López, R. 1993. 'Economy-wide policies, agricultural productivity, and environmental factors: the case of Ghana', Workshop on Economy-wide Policies and the Environment. Washington, DC: World Bank.

Loraas, K. T. and Eraker, H. 1999. 'Saga's environmental status report is criticised', *NorWatch Newsletter* No. 7, April, Oslo, available online at www.fivh.no/norwatch/english.asp?artID = 612, accessed on April 2002.

Louman, B. and Siaguru, P. 1995. 'Land tenure arrangements and forest based development in Papua New Guinea', *BOS NiEuwsletter* 32(3): 51–8.

Love, R. 1994. 'Drought, Dutch disease and controlled transition in Botswana agriculture', *Journal of Southern African Studies* 20(1): 71–83.

Lux, M. 1997. *Integración actual y potencial del sector forestal en la economía regional del estado Sucre, con vista a un desarrollo regional sostenible*. Compilación de los resultados de los trabajos de investigación realizados en Venezuela entre septiembre de 1996 y octubre de 1997, Carúpano, October.

McAlpine, J. and Quigley, J. 1998. 'Forest Resources of Papua New Guinea. Summary statistics from the Forest Inventory Mapping (FIM) System', Papua New Guinea: Coffey MPW Ltd for the Australian Agency for International Development and the Papua New Guinea National Forest Service.

McAlpine, J. and Quigley, J. n.d. 'Natural resources, land use and population distribution of Papua New Guinea. Summary statistics from PNGRIS', *PNGRIS Report* No. 7. Canberra: Australian Agency for International Development.

McAlpine, J. R., Freyne, D. F. and Keig, G. 2001. 'Land use and rural population change in PNG 1975–1995', in R. M. Bourke, M. G. Allen, J. Salisbury (eds), *Proceedings of the Papua New Guinea Food and Nutrition 2000 Conference: Food Security in Papua New Guinea*. Canberra: Australian Centre for International Agricultural Research.

Macbean, A. J. 1966. Export instability and economic development. Cambridge, MA: Harvard University Press, 364 pp.

Mackie, C. 1984. The lessons behind East Kalimantan's forest fires. *Borneo Research Bulletin* 16: 63–74.

Maddock, R. and McLean, J. 1984. 'Supply-side shocks: the case of Australian gold', *Journal of Economic History* 44.

MAEDR 1975. *Recensement agricole 1974–1975. Résultat du dépouillement manuel*. Libreville: Ministère de L'Agriculture de L'Elevage er du Développement Rural.

Magrin, G. 1994. 'L'approvisionnement de Libreville en produits vivriers: vers une dialectique positive de la crise?', MA thesis (Geography), Université de Paris.

Mainardi, S. 1998. 'An econometric analysis of factors affecting tropical and subtropical deforestation', *Agrekon* 37(1): 23–45.

Maizels, A. 1988. 'Commodity instability and developing countries: the debate', *WIDER Working Paper* No. 34, Helsinki.

Maley, J. 1998. *Variation à long terme des écosystèmes forestiers du sud Cameroun et du Gabon au cours des derniers millénaires*. Libreville: Séminaire FORAFRI de Libreville.

Malingreau, J.-P. 1993. 'Satellite monitoring of the world's forests: a review', *Unasylva* 44(3).

Mamingi, N., Chomitz, K. M., Gray, D. A. and Lambin, E. 1996. 'Spatial patterns of deforestation in Cameroon and Zaire', Washington, DC: World Bank, Policy Research Department available online at http://www.worldbank.org/html/prddr/prdhome/peg/wps08/index.htm, accessed on 28 August 2001.

Manning, M. 1999. 'Factors contributing to the lack of investment in Papua New Guinea. A private sector survey', *Discussion Paper* No. 74. Port Moresby, Papua New Guinea: Institute of National Affairs.

Manning, M. 2000. 'Privatisation in Papua New Guinea – where at and where to?' *Pacific Economic Bulletin* 15(2): 91–105.

Mansoorian, A. 1991. 'Resource discovery and "excessive" external borrowing', *The Economic Journal*, 101: 1497–509.

Marchés Tropicaux. 1994. *L'hébdomadaire de l'Afrique: Cameroun 1994*. Paris, April.

Marchés Tropicaux. 1998. *L'hébdomadaire de l'Afrique: Gabon*. Paris, June.

Marchés Tropicaux. 1999. *L'hébdomadaire de l'Afrique: Cameroun*. Paris, December.

Marchés Tropicaux. 2000. *L'hébdomadaire de l'Afrique: Gabon*. Paris, July.

MARNR. 1982. *Mapa de la vegetación actual de Venezuela, Sistemas Ambientales Venezolanos*. Caracas: Ministerio del Ambiente y de los Recursos Naturales Renovables.

MARNR. 1995. *Balance ambiental de Venezuela 1994–95*. Caracas: Ministerio del Ambiente y de los Recursos Naturales Renovables.

MARNR 1996. *Balance ambiental de Venezuela. Apéndice 1996*. Caracas: Ministerio del Ambiente y de los Recursos Naturales Renovables.

MARNR/MEM/UNEP/US Country Study Program 1996. *Inventario de emisiones de gases de efecto invernadero – Venezuela*. Caracas: Ministerio del Ambiente y de los Recursos Naturales Renovables.

Marsden, K. 1993. 'Employment strategy in Papua New Guinea: an international perspective', *International Labour Review* 132(2): 229–44.

Marshall, A. R. and Rau, M.T. 1999. Lower Ok Tedi and Middle Fly – estimate of current vegetation dieback and classification of flood plain vegetation. Summary version, August, available at http://www.oktedi.com/environment/papers.htm.

Martin, R. and van Wijnbergen, S. 1986. 'Shadow prices and intertemporal aspects of remittances and oil revenues in Egypt', in J. P. Neary and S. van Wijnbergen (eds), *Natural Resources and the Macroeconomy*. Oxford: Blackwell, 142–67.

Mather, A. and Needle, C. L. 1998. 'Trends in global forest cover: issues in explanation and modelling', in Y. Himiyama and L. Crissman (eds), *Information Bases for Land Use/Cover Change Research*, Proceedings of IGU-LUCC'97. Brisbane, Australia, July, pp. 84–91.

Matthews, E. 2001. Understanding the FRA 2000. *Forest Briefing* No. 1. Washington, DC: World Resources Institute

Matute, D. 1984. *Las deforestaciones con fines agropecuarios*. MARNR, División de Vegetación, Serie Informes Técnicos DGSIIA/IT/188, Caracas.

Mawuli, A. 1998. 'The accelerating depreciation of the kina – a recipe for economic disaster', *The PNG Business (April)* 10.

Mawuli, A. 2001. 'Poverty perspective: creating, perpetuating or eradicating', in B. Boeha and J. Robins (eds), *Issues and Trends in National Development. Searchlight on Papua New Guinea. Special Publication* No. 30, Vol. 2, July–December 2000, Boroko, Papua New Guinea: National Research Institute 11–15.

Mawuli, A. and Sanida, O. 2000. 'Landowners' mineral rent-quest and use in Papua New Guinea', *National Research Institute (NRI) Discussion Paper* No. 95. Papua New Guinea: National Research Institute.

Mayaux, P., Achard, F. and Malingreau, J.-P. 1998. 'Global tropical forest area measurements derived from coarse resolution satellite imagery: a comparison with other approaches', *Environmental Conservation* 25(1).

Mayaux, P., Janodet, E., Blair-Myers, C. M. and Legeay-Janvier, P. 1997. *Vegetation Map of Central Africa*, Scale 1/5M. *TREES Publications Series D*, N°1. European Commission, Luxembourg.

Mbendi 2000. 'Information for Africa', available online at http://www.mbendi.co.za/a_sndmsg/news_view.asp?I = 11239, accessed on July 2000.

Mbendi. 2002. 'Nigeria: oil and gas industry', *Mbendi Information for Africa*, available online at www.mbendi.co.za/indy/oilg/af/ng/p0005.htm.

Mertens, B. and Lambin, E. F. 1997. 'Spatial modeling of deforestation in Southern Cameroon: spatial disaggregation of diverse deforestation processes', *Applied Geography* 17: 143–68.

Mertens, B. and Lambin, E. F. 2000. 'Land-cover-change trajectories in southern Cameroon', *Annals of the Association of American Geographers* 90(3): 467–94.

Mertens, B., Sunderlin, W. D., Ndoye, O. and Lambin, E. F. 2000. 'Impact of macroeconomic change on deforestation in south Cameroon: integration of household survey and remotely-sensed data', *World Development* 28(6): 983–99.

Milner, C. R. 1990. 'Identifying and quantifying anti-export bias: the case of Cameroon', *Weltwirtschaftliches Archiv* 126(1): 142–55.

MINEFI 1998a. *Annuaire statistique du Cameroun 1997*. Yaoundé: Ministère de l'Économie et des Finances.

MINEFI 1998b. *Cameroon in figures: 1997*. Yaoundé: Ministry of Economy and Finance, Department of Statistics and National Accounts, August.

MINEFI 2000. *Audit économique et financier du secteur forestier du Cameroun. Rapport Final*. Yaoundé: Ministère de l'Économie et des Finances and CIRAD-Forêt.

Ministry of Environment 1999. *Informe de Avance del Ecuador, Hacia la 'Meta Año 2000' de la organización internacional de maderas tropicales*. Quito: Ministry of Environment and ITTO.

Ministry of Equipment 1984. *Informations de base sur le réseau routier camerounais*. Yaoundé: Ministry of Equipment, Department of Highways.

Ministry of Forestry/FAO 1990. *Situation Outlook of the Forestry Sector in Indonesia*, Vol. 2. Jakarta: Forest resource base.

Miranda, M. (with A. Blanco-Uribe Q., L. Hernández, J. Ochoa G. and E.Yerena). 1998. 'All that glitters is not gold. Balancing conservation and development in Venezuela's frontier forests'. Forest Frontier Initiative, Washington, DC: World Resources Institute.

MML 2001. 'Sustainability plan (mine closure plan)' year 2000 consultation document. Port Moresby: Misima Mines Limited.

Moari, G. M. and Era, S. K. 1996. 'Safety and the environment and Kutubu – a success story', in P. G. Buchanan (ed.), *Petroleum Exploration, Development and Production in PNG*. Proceedings of the 3rd PNG Petroleum Convention, Port Moresby.

Montenegro, S. 1989. 'External shocks and macroeconomic policy in a small open developing economy' PhD thesis. Oxford University.

Montenegro, S. 2000. 'One decade of external coffee shocks in Colombia, 1975–85' in P. Collier and J. W. Gunning (eds), *Trade Shocks in Developing Countries, Volume 2: Asia and Latin America*. Oxford: Oxford University Press, 1–41.

Morales, J. A. 2000. 'Bolivia's tin and natural gas crises of 1985–89' in P. Collier and J. W. Gunning (eds), *Trade Shocks in Developing Countries, Volume 2: Asia and Latin America*. Oxford: Oxford University Press, 82–119.

Mosley, P. 1991. 'Ecuador', in P. Mosley, J. Harrigan and J. Toye (eds), *Aid and Power. The World Bank and Policy-based Lending*, Vol. 2, case studies. London: Routledge.

Mourata, M. Hon Sir 2001. *Economic and Development Policies*, Vol. 1. Papua New Guinea.

Muhamad, N. Z. 2002. 'Rates of deforestation in Indonesia. An annotated bibliography', unpublished consultancy report. Bogor, Indonesia: CIFOR.

Müller, B. 1997. *'Esto es mi ambiente, aquí tengo vida'. La colonización de los bosques del estado Bolívar (Venezuela) por los buscadores de oro y diamantes en el contexto del desarrollo regional*. Compilación de los resultados de los trabajos de investigación realizados en Venezuela entre septiembre de 1996 y octubre de 1997. Carúpano. October.

Murray Li, T. 2000. 'Migrants, locals and government powers on Sulawesi's cocoa/forest frontier' unpublished seminar notes, 14 July, Bogor, Indonesia: CIFOR.

Mwanawina, I. and Sankhayan, P. L. 1996. 'A computable general equilibrium model to measure the effects of structural adjustment programmes on ecology and development in Zambia', in *Structural Adjustment Policies and Environmental Degradation in Tanzania, Zambia, and Ethiopia*. Preliminary Seminar Report, 9–10 January. Agricultural University of Norway, Ås.

Myers, N. 1994. 'Tropical deforestation: rates and patterns', in K. Brown and D. W. Pearce (eds), *The Causes of Tropical Deforestation*. London: UCL Press Limited.

Nadarajah, T. 1994. *The Sustainability of Papua New Guinea's Forest Resource. Discussion Paper* No. 76. Papua New Guinea: National Research Institute.

Nankani, G. 1979. 'Development problems of mineral-exporting countries', *World Bank Staff Working Paper* No. 354. Washington, DC.

Ndongko, W. A. and Vivekananda, F. 1989. *Economic Development of Cameroon. Africa – The Awakening Giant,* Vol. 2. Stockholm: Bethany Books.

Ndoye, O. and Kaimowitz, D. 2000. 'Macro-economics, markets, and the humid forests of Cameroon, 1967–1997', *Journal of Modern African Studies* 38(2): 225–53.

Ndoye, O., Ruiz-Pérez, M., Mamoun, A. D. and Lema Ngono, D. 1998. Les effets de la crise économique et de la dévaluation sur l'utilisation des plantes médicinales au Cameroun. Implications pour la gestion durable des forêts. *Seminaire FORAFRI de Libreville – Session 1: états populations et forêts.*

NEA 1994. *Un Homme, un Pays. El Hadj Omar Bongo. Le Gabon.* Dakar: Les Nouvelles Éditions Africaines.

Neary, P. and van Wijnbergen, S. 1986. 'Natural resources and the macroeconomy: a theoretical framework', in J. P. Neary and S. van Wijnbergen (eds), *Natural Resources and the Macroeconomy.* Oxford: Blackwell, 13–45.

Neba, A. S. 1987. *Géographie moderne de la Rèpublique du Cameroun.* Camden, New Jersey: Neba Edition.

Nepstad, D. C., Veríssimo, A., Alencar, A., Nobre, C., Lima, E., Lefebvre, P., Schlesinger, P., Potter, C,. Moutinho, P., Mendoza, E., Cochrane, M., and Brooks, V. 1999. 'Large-scale impoverishment of Amazonian forests by logging and fire', *Nature* 398, 8 April.

Newbery, D. M. G. and Stiglitz, J. E. 1981. *The Theory of Commodity Price Stabilization. A Study in the Economics of Risk.* Oxford: Clarendon Press.

Newbold, R. A., Lockaby, B. G. and Thomas, C. E. 1988. Conversion of forestland to oil and gas production in northern Louisiana: An evaluation', *Journal of Soil and Water Conservation* 43: 331–4.

Newton, A. 1994. *Central Africa: A Travel Survival Kit.* Hongkong: Lonely Planet.

Noreng, Ø. 1981. 'Petroleum revenues and industrial income', in T. Barker, and V. Brailovsky (eds), *Oil or industry? Energy, Industrialisation and Economic Policy.* London and Mexico City: Academic Publishers.

O'Brien, K. L. 1998. *Sacrificing the Forest. Environmental and Social Struggles in Chiapas.* Boulder: Westview Press.

O'Collins, M. 1999. 'Reflections on poverty assessments in Papua New Guinea, Fiji and Vanuatu', *Pacific Economic Bulletin*: 33–46.

O'Faircheallaigh, C. 1984. *Mining and Development: Foreign-Financed Mines in Australia, Ireland, Papua New Guinea and Zambia.* London and Sydney: Croom Helm and New York: St Martin's Press.

OCEI 1997. *Venezuela. Síntesis estadística 1996.* Caracas: Oficina Central de Esadística e Informática.

Ochoa-Gaona, S. and González-Espinosa, M. 2000. 'Land use and deforestation in the highlands of Chiapas, Mexico', *Applied Geography* 20: 17–42.

Ohtsuka, R., Inaoka, T., Umezaki, M., Nakada, N. and Abe, T. 1995. 'Long-term subsistence adaptation to the diversified Papua New Guinea environment human ecological assessments and prospects', *Global Environmental Change* 5(4), 347–53.

Ok Tedi 2002. available online at http://www.oktedi.com/environment/impacts.htm, accessed on 1 February 2002.

Onishi, N. 2000. 'Oil riches, and risks, in tiny African nation', Global Policy Forum – World Bank/UN Security Council: Diamonds in Conflict, *New York Times*, 23 July, available online at http://www.globalpolicy.org/security/docs/oil5.htm, accessed on 26 September 2000.

Ortiz-Chour, H. 1999. FRA 2000. 'Annotated bibliography forest cover change, Nepal', *FRA Working Paper* No. 12. Rome: Forestry Department, Food and Agriculture Organization of the United Nations.

Osemeobo, G. J. 1988. 'The human causes of forest depletion in Nigeria', *Environmental Conservation* 15(1): 17–28.

Osemeobo, G. J. 1992. 'Fuelwood exploitation from natural ecosystems in Nigeria: socio-economic and ecological implications', *Journal of Rural Development* (Hyderabad, India) 11(2): 141–55.

Osgood, D. 1994. 'Government failure and deforestation in Indonesia', in K. Brown and D. Pearce (ed.), *The Causes of Tropical Deforestation the Economic and Statistical Analysis of Factors giving Rise to the Loss of Tropical Forests*, London: UCL Press, 217–25.

Ovono-Edzang, N. 2001. 'Innovations et mutations agricoles dans les provinces du Woleu-Ntem et de l'Estuaire (Gabon)', PhD thesis. Tropical Geography, Bordeaux: University of Michel de Montaigne, June.

Oyejide, T. A. 2000. 'Trade shock, oil boom, and the Nigerian economy, 1973–83', in P. Collier and J. W. Gunning (eds), *Trade Shocks in Developing Countries, Volume 1: Africa*. Oxford: Oxford University Press, 420–47.

Päivinen, R., Gillespie, A. J. R., Davis, R. and Holmgren, P. FRA 2000. *Assessing State and Change in Global Forest Cover: 2000 and Beyond*. Rome: Forestry Department, Food and Agriculture Organization of the United Nations.

Paldam, M. 1997. 'Dutch disease and rent seeking: the Greenland model', *European Journal of Political Economy* 13: 591–614.

Parks, P. J. and Bonifaz, M. 1995. 'Nonsustainable use of renewable resources: mangrove deforestation and mariculture in Ecuador', in S. Hanna and M. Munasinghe (eds), *Property Rights in a Social and Ecological Context. Case Studies and Design Applications*. Washington: Beijer and The World Bank.

Parvin, M. and Dezhbakhsh, H 1988. 'Trade, technology transfer and hyper-Dutch disease in OPEC: theory and evidence', *International Journal of Middle East Studies* 20: 469–77.

Paus, E. 1995. 'Exports, economic growth and the consolidation of peace in El Salvador', *World Development* 23(12): 2173–93.

Pearce, D., Barbier, E. and Markandya, A. 2000. *Sustainable Development: Economics and Environment in the Third World*. London: Earthscan Publications Limited.

Peters, C. M. 1996. 'The ecology and management of non-timber forest resources', *Technical Paper* No. 322. Washington, DC: World Bank.

Pichón, F. J. 1997. 'Settler households and land-use patterns in the Amazon frontier: farm-level evidence from Ecuador', *World Development* 25(1).

Pinto, B. 1987. 'Nigeria during and after the oil boom: a policy comparison with Indonesia', *World Bank Economic Review* 1(3): 419–55.

Pittier, H. 1948. *Trabajos escogidos*. Buenos Aires: Imprenta López.

Placer Dome. 2001. *Sustainability Report 2000 – Porgera Mine*. Port Moresby: Placer Dome.

Plonczak, M. 1997. 'Die Nutzung der Naturwälder in Venezuela', *Allgemeine Forst-und Jagdzeitung*, year 168, no. 3/4, March/April, Frankfurt a.M.: Sauerländer's Verlag.

Pokam, J. and Sunderlin, W. D. 1999. 'L'impacte de la crise économique sur les populations, les migrations, et le couvert forestier du Sud-Cameroun', *CIFOR Occasional Paper* No. 25. Bogor, Indonesia: Center for International Forestry Research.

Polier, N. 1996. 'Of mines and min: modernity and its malcontents in Papua New Guinea', *Ethonology* 35(1): 1–16.

Pollard, H. L. 1985. 'The erosion of agriculture in an oil economy: the case of export crop production in Trinidad', *World Development*, 13(7), 819–36, July.

Polume, S. 1997. 'Export instability, investment risk and mineral taxation in Papua New Guinea', PhD thesis. Canberra: Australian National University.

Poupart, N. and Pilichowski, E. 1997. *Republic of Gabon. Poverty in a Rent-Based Economy*. Washington, DC: World Bank.

Pourtier, R. 1982. 'Stratégie ferroviaire et politique de l'espace le Transgabonais', *Hérodote*, No. 25.

Pourtier, R. 1984. 'Agro-industrie et développement rural au Gabon: une contradiction?' in C. Blanc-Pamard, J. Bonnemaison, J. Boutrais, V. Lassailly-Jacob and A. Lericollais (eds), *Le développement rural en questions. Paysages, espaces ruraux, systèmes agraires*. Paris: Editions de l'ORSTOM.

Pourtier, R. 1989a. *Le Gabon. Tome 1: Espace-histoire-société*. Paris: L. Harmattan.

Pourtier, R. 1989b. *Le Gabon. Tome 2: État et développement*. Paris: L. Harmattan.

Prasetyo, L. B. 1992. 'Application of a Geographical Information System (GIS) for land-use changes study towards better understanding of deforestation processes: case studies in Jabotabek and South Sumatra, Indonesia' PhD thesis. Japan: University of Tsukuba.

Prebisch, R. 1950. *The Economic Development of Latin America and its Principal Problems*. New York: United Nations.

Quane, D. 2002. 'Indonesia: a major cocoa supplier to the US market. Market AG', Market Asia and RAP Archives, available online at http://www.marketag.com/ma/news/archive/v43/ indo.stm, accessed on May 2002.

Ramos, H. and Robison, L. 1990. 'Credit and credit policies', in M. D. Whitaker and D. Colyer (eds), *Agriculture and Economic Survival. The Role of Agriculture in Ecuador's development*. Boulder, San Francisco and Oxford: Westview Press, 225–46.

RAN 2002. 'Oil in Venezuela's Orinoco delta – a Citigroup case study', Rainforest Action Network at http://www.ran.org/ran_campaigns/citigroup/cs_venezoil.htm, accessed 1 February 2002.

RAN and Project Underground. 1998. 'Drilling to the ends of the earth. The ecological, social and climate imperative for ending petroleum exploration'. San Francisco: Rainforest Action Network, September.

RAN 2002. 'Rainforest Action Network campaign, accessed on 1 February 2002, available online at www.ran.org/ran_campaigns/citigroup/cs_pnggobe.html.

Raponda-Walker, A. and Sillans, R. 1961. *Les plantes utiles du Gabon*. Paris: Éditions Paul Lechevalier.

Reed, D. (ed.). 1992. *Structural Adjustment and the Environment*. WWF International, Boulder: Westview Press.

Reed, D. (ed.). 1996. *Structural Adjustment, the Environment, and Sustainable Development*. London: Earthscan Publications/World Wildlife Fund.

Reed, D. and Sheng, F. 1998. 'Macroeconomic policies, poverty and the environment', *Discussion Paper*, Macroeconomics Program Office (MPO), WWF.

Reinhart, C. M. and Wickham, P. 1993. 'Commodity prices: cyclical weakness or secular decline?' *IMF Staff Papers* No.41, June.

Rémy, M. 1996. *Le Gabon aujourd'hui*. Libreville: Jaguar.

République Gabonaise. 2000. *Loi de finances pour l'an 2000. Rapport économique, social et financier*. Libreville.

Richard, A. and Léonard, G. 1993. *Le Gabon: Géographie active*. Libreville: Institut Pédagogique National.

Richards, D. G. 1994. 'Booming-sector economic activity in Paraguay 1973–86: a case of Dutch disease?' *Journal of Development Studies* 31:310–33.

Richards, T., Gallego, J. and Achard, F. 1999. 'Sampling for forest change assessment at the Pan Topical Scale', in Y. Laumonier (eds), *Data Management and Modelling Using Remote Sensing and GIS for Tropical Forest Land Inventory*. Proceedings of the EU-FIMP/INTAG International Conference, 26–9 October 1998, Jakarta, Indonesia: 83–92.

Rippert, G. 1997. *Étude des systèmes de production dans le bassin d'approvisionnement de Franceville, Gabon*. Libreville: Institut Gabonais d'Appui au Développement (IGAD).

Robson, P. 1985. 'Trade and trade policy issues in Papua New Guinea', *INA Discussion Paper* No. 21. Port Moresby: Institute of National Affairs.

Rodríguez, A. E. 1991. *The Deforestation Problem in Oil-importing Developing Countries: A Capital Theory Approach to a Renewable Resource*. Austin: University of Texas.

Rodríguez, B. C. 1994. *El oro y su impacto en la economía venezolana*, in III Simposio Internacional del Oro en Venezuela, 2–4 October. Caracas: Asociación Venezolana del Oro.

Roemer, M. 1984. 'Dutch disease in developing countries: swallowing bitter medicine', *Development Discussion Paper*, No. 156. Harvard Institute for International Development.

Roemer, M. 1994. 'Dutch disease and economic growth: the legacy of Indonesia', *Development Discussion Paper*, No. 489. Harvard Institute for International Development.

Rojas L. J. 1993. La colonización agraria de las reservas forestales: un proceso sin solución? *Cuadernos Geográficas*, Vol. 10. Universidad de Los Andes, Facultad de Ciencias Forestales, Mérida.

Romero, D. 1994. 'Economic contributions of Venezuelan protected areas: the tragedy of the commons and perspectives', in M.Munasinghe and J.McNeely (eds), *Protected Area Economics and Policy. Linking Conservation and Sustainable Development*. Washington, DC: IUCN and World Bank.

Ropivia, M. L. and Djeki, J. 1995. *Atlas de la formation territoriale du Gabon. Frontières et unités administratives des origines à nos jours*. Libreville: Institut National de Cartographie.

Ros, J. 1987. 'Mexico from the oil boom to the debt crisis: an analysis of policy responses to external shocks 1978–85', in R. Thorp and H. Whitehead (eds), *Latin American Debt and the Adjustment Crisis*. London: Macmillan.

Rosero, J. A. 1992. *Evaluación económica del impacto ambiental del proyecto de desarrollo rural integral Pichincha I*, IDEA Technical Document No. 46. Quito.

Ross, M. L. 1999. 'The political economy of the resource curse', *World Politics* 51(2): 297–322.

Ross, M. L. 2000. 'Does resource wealth lead to authoritarian rule? Explaining the "Midas touch," '. Paper presented at the World Bank Research Group workshop 'The economics of political violence', Princeton University, March 18–19.

Ross, M. L. 2001. *Timber Booms and Institutional Breakdown in Southeast Asia*. Cambridge: Cambridge University Press.

Rowthorn, R. E. and Wells, J. R. 1987. *Deindustrialization and Foreign Trade*. Oxford: Oxford University Press.

Rudel, T. K. 1993. *Tropical Deforestation: Small Farmers and Land Clearing in the Ecuadorian Amazon* (with B. Horowitz). New York: Columbia University Press.

Rudel, T. K. and Roper, J. 1996. 'Regional patterns and historical trends in tropical deforestation, 1976–1990: A qualitative comparative analysis', *Ambio* 25: 160–6.

Rudel, T. K. and Roper, J. 1997a. 'The paths to rain forest destruction: cross-national patterns of tropical deforestation, 1975–1990', *World Development* 25(1): 53–65.

Rudel, T. K. and Roper, J. 1997b. 'Forest fragmentation in the humid tropics: a cross-national analysis', *Singapore Journal of Tropical Geography* 18.

Rudel, T. K., Bates, D. and Machinguiashi, R. 2002. 'A tropical forest transition? Agricultural change, out-migration, and secondary forests in the Ecuadorian Amazon', *Annals of the Association of American Geographers* 92(1), 87–102.

Ruiz, L. 2000. *Amazonía ecuatoriana: escenario y actores del 2000*. Quito: EcoCienca and IUCN.

Sachs, J. D. and Warner, A. M. 2001. 'The curse of natural resources', *European Economic Review* 45: 827–838.

Salazar-Carrillo, J. 1976. *Oil in the Economic Development of Venezuela*. New York: Praeger.

Salter, W. G. 1953. 'Internal and external balance', *The Economic Record* 36: 226–38.

Saulei, S. M. and Aruga, J. A. 1994. 'The status and prospects of non-timber forest products development in Papua New Guinea', *Commonwealth Forestry Review* 73(2): 97–105.

Saunders, J. C. 1993a. 'Agricultural land use of Papua New Guinea. Explanatory notes to map', *PNGRIS Publication* No. 1. Canberra: Australian Agency for International Development.

Saunders, J. C. 1993b. 'Forest resources of Papua New Guinea. Explanatory notes to map', *PNGRIS Publication* No. 2. Canberra: Australian Agency for International Development.

Sayer, J., Hartcourt, C. and Collins, N. M. (eds) 1992. *The Conservation Atlas of Tropical Forests. Africa*. New York: IUCN and Simon & Schuster.

SBC Warburg 1996. *Republic of Gabon*. A Division of Swiss Bank Corporation.

Scherr, S. J. 1989. 'Agriculture in an export boom economy: a comparative analysis of policy and performance in Indonesia, Mexico and Nigeria', *World Development* 17(4): 543–60.

Scotland, N., Fraser, A. and Jewell, N. 1999. Indonesian Forest Inventory Data. Indonesia-UK Tropical Forest Management Program/ITFMP, Jakarta.

Sekhran, N. and Miller, S. (eds). 1994. *Papua New Guinea Country Study on Biological Diversity*. Port Moresby: Department of Environment and Conservation.

Seymour, F. J. and Dubash, N. K. (eds). 2000. *The Right Conditions. The World Bank, Structural Adjustment, and Forest Policy Reform*. Washington, DC: World Resources Institute, 59–77.

Shafik, N. 1994a. 'Macroeconomic causes of deforestation: barking up the wrong tree?' in K. Brown, and D. Pearce (eds), *The Causes of Tropical Deforestation. The Economic and Statistical Analysis of Factors giving Rise to the Loss of Tropical Forests*, 86–95. London: University College London Press.

Shafik, N. 1994b. 'Economic development and environmental quality: an econometric analysis', *Oxford Economic Papers* 46: 757–73.

Shafik, N. 2000. 'Multiple trade shocks and partial liberalisation: Dutch disease and the Egyptian economy', in P. Collier and J. W. Gunning (eds), *Trade shocks in developing countries, Volume* 1: *Africa*. Oxford: Oxford University Press, 448–84.

Shearman, P. 2001. 'Giving away another river: an analysis of the impacts of the Porgera mine on the Strickland River system', in B. Y. Imbun and P. A. McGavin (eds), *Mining in Papua New Guinea: Analysis and Policy Implications*. Port Moresby: University of Papua New Guinea Press.

Shell Gabon. 1996. *Rabi Phase II*. Environmental statement. Libreville.

SICA. 2001. *Ecuador: Uso de la tierra*, available online at www.sica.gov.ec. Quito: World Bank and Ministry of Agriculture.

Siegert, F. and A. Hoffmann A. 2000. 'The 1998 forest fires in East Kalimantan (Indonesia): a quantitative evaluation using high resolution, multitemporal ERS-2 SAR images and NOAA-AVHRR hotspot data', *Remote Sensing of Environment* 72, 64–77.

Siegert, F., Ruecker, G., Hinrichs, A. and Hoffmann, A. A. 2001. 'Increased damage from fires in logged forests during droughts caused by El Nino', *Nature* 414: 437–40.

Sierra, R. 1996. *El uso del recurso forestal y la deforestación en el Noroccidente del Ecuador, 1983–1993*, Quito: EcoCiencia.

Sierra, R. 2000. 'Dynamics and patterns of deforestation in the western Amazon: the Napo deforestation front, 1986–1996', *Applied Geography* 20.

Sierra, R. 2001. 'The role of domestic timber markets in tropical deforestation and forest degradation in Ecuador: implications for conservation planning and policy', *Ecological Economics* 36.

Sierra, R. and Brown, L. A. 1994. 'Frontier migration as a multi-stage phenomenon reflecting the interplay of macroforces and local conditions: the Ecuador Amazon', Papers in Regional Science, *Journal of the RSAI* 73(3), 267–88.

Sierra, R. and Stallings, J. 1998. 'The dynamics and social organization of tropical deforestation in Northwest Ecuador, 1983–1995', *Human Ecology* 26(1).

Simons, K. 1996. *Lessons from the Rainforest: Policy and Perceptions of Logging in Gabon*. Libreville. (Unpublished Fulbright Research draft).

Simpson, G., Finlaysin, M., Brooksbank, J., Goie, A. and Goldman, L. 1998. 'Lost opportunity? The management of oil revenues in Papua New Guinea', unpublished paper.

Singer, H. 1950. 'The distribution of trade between investing and borrowing countries', *American Economic Review* 21.

Singh, K. D. 1993. 'The 1990 tropical forest resources assessment', *Unasylva* 44(3).

Sizer, N. and Plouvier, D. 2000. *Increased investment and trade by transnational logging companies in Africa, the Caribbean and the Pacific: Implications for the sustainable management and conservation of tropical forests*. Bruxelles: World Resources Institute & World Wide Fund for Nature, 121 pp.

Sizer, N. and Rice, R. 1995. *Backs to the Wall in Suriname: Forest Policy in a Country in Crisis*. Washington, DC: World Resources Institute.

Skeldon, R. 1978. 'Recent urban growth in Papua New Guinea', *IASER Discussion Paper*, No. 21. Papua New Guinea: IASER.

Smeraldi, R. and Veríssimo, J. A. 1999. *Hitting the Target: Timber Consumption in the Brazilian Domestic Market and Promotion of Forest Certification*. São Paulo: Amigos da Terra and IMAFLORA and Belém: IMAZON.

Smith, W. C. and McCoy, J. 1995. 'Venezuelan democracy under stress', in J. McCoy, A. Serbin, W. C. Smith and A. Stambouli (eds), *Venezuelan Democracy Under Stress*. Miami: University of Miami, North-South Center.

Snape, R. H. 1977. 'Effects of mineral development on the economy', *Australian Journal of Agricultural Economics* 21.

Söderling L. 2002. 'Escaping the curse of oil? The case of Gabon', *IMF Working Paper* No. 02–93. Washington, DC: International Monetary Fund.

Soladoye, M. O. and Ola-Adams, B. A. 1990. 'The present status and future of Nigerian forests', in *Proceedings of the Twelfth Plenary Meeting of AETFAT*. Hamburg, September 4–10, 1988. Mitteilungen aus dem Institut für Allgemeine Botanik Hamburg, 23a, 189–95.

Sournia, G. 1998. *Les aires protegées d'Afrique francophone*. Paris: Éditions Jean-Paul de Monza, ACCT, UICN.

Southgate, D., Sierra, R. and Brown, L. 1991. 'The causes of tropical deforestation in Ecuador: A statistical analysis', *World Development* 19(9).

Southgate, D. and Whitaker, M. 1992. *Development and the Environment: Ecuador's Policy Crisis*. Quito: IDEA, June.

Southgate, D. and Whitaker, M. 1994. *Economic Progress and the Environment. One Developing Country's Policy Crisis*. New York and Oxford: Oxford University Press, 150.

Spencer, I. T. and Heywood, P. 1982. 'Staple foods in Papua New Guinea: their relative supply in urban areas'. *Food and Nutrition Bulletin* 5: 40–6.

Stacey, M. and Lucas, J. 1989. *Investigating Geography: Highlands Communities in Papua New Guinea*. Melbourne: Longman Cheshire.

Statistisches Bundesamt 1994. *Länderbericht Gabun 1994*. Wiesbaden.

Steel, E. 1994. *Étude sur le volume et la valeur du commerce de la viande de brousse au Gabon*. Libreville: WWF (unpublished).

Steigum, E. Jr. 1989. 'Wealth, structural adjustment and optimal recovery from the Dutch disease', Bergen: Norwegian School of Economics and Business Administration, *Discussion Paper* 06/89.

Stibig, H.-J., Janvier, P. and Beuchle, R. 2002. *Forest Map of Insular Southeast Asia at Scale 1/5M. TREES Publications Series D*, No. 3. Luxembourg: European Commission.

Stibig, H.-J. and Achard, F. 1999. 'TREES II – Present activities for global and regional tropical forest cover monitoring: example of South East Asia. Data management and modelling using remote sensing and GIS for tropical forest land inventory', in Y. Laumonier (ed.) *Proceedings of the EU-FIMP/ INTAG International Conference*. 26–9 October 1998. Jakarta, Indonesia, 73–81.

Strauss K. R. A. 1993. Un perfil cronológico del tiempo prehispánico venezolano. in L. Cipriano Rodríguez (coord.), *Los grandes períodos y temas de la historia venezolana*. Instituto de Estudios Hispanoamericanos, Caracas: Universidad Central de Venezuela.

Struthers, J. J. 1990. 'Nigerian oil and exchange rates: indicators of "Dutch disease" ', *Development and Change* 21: 309–41.

SUFOREN 1991. *Diagnóstico: Plan de Acción Forestal, 1991–95*. Quito: MAG.

Sumahadi, Sarbini, B. and Prihatno, K., B. 1997. 'Forest resource monitoring and sssessment at national level: An Indonesian experience', in *A State-of-the-Art Report on Some Recent Forestry Policies, Initiatives and Achievements in Indonesia: Concepts, Strategies and Actions for Sustainable Forest Management and Forestry Development Towards 21st Century*. Jakarta: Ministry of Forestry.

Sunderlin, W. D. 1993. 'Environment, equity and the state: A perspective through the Java Social Forestry Program', PhD thesis. Graduate School, Cornell University, January.

Sunderlin, W. D. 1997. 'Shifting cultivation and deforestation in Indonesia: steps toward overcoming confusion in the debate', *ODI Network Paper* No. 21b. London: Overseas Development Institute.

Sunderlin, W. D. and Rodríguez, J. A. 1996. 'Cattle, broadleaf forests and the agricultural modernization law of Honduras: the case of Olancho', *CIFOR Occasional Paper* No. 7(E). Bogor: CIFOR, March.

Sunderlin, W. D. and Wunder, S. 2000. 'The influence of mineral exports on the variability of tropical deforestation', *Environmental and Development Economics* 5: 309–32.

Sunderlin, W. D., Ndoye, O., Bikié, H., Laporte, N., Mertens, B. and Pokam, J. 2000. Economic crisis, small-scale agriculture, and forest cover change in southern Cameroon', *Environmental Conservation* 27(3): 284–90.

Sunderlin, W. D. Resosudarmo, I. A. P. 1996. 'Rates and causes of deforestation in Indonesia: towards a resolution of the ambiguities', *CIFOR Occasional Paper* No. 9. Bogor: CIFOR.

Sunderlin, W. D., Resosudarmo, I. A. P., and Ndoye, O. 2001. 'The effect of economic crises on small farmers and forest cover: a comparison of Cameroon and Indonesia', in M. Palo and J. Uusivuori (eds), *World Forests, Society, and Environment*. Dordrecht: Kluwer, 115–25.

Sunderlin, W. D., Resosudarmo, I. A. P., Rianto, E. and Angelsen, A. 2000. 'The effect of Indonesia's economic crisis on small farmers and natural forest cover in the outer islands', *CIFOR Occasional Paper* No. 28(E). Bogor: CIFOR.

Sunderlin, W. D. and Pokam, J. 2002. 'Economic crisis and forest cover change in Cameroon: the roles of migration, crop diversification, and gender division of labor', *Economic Development and Cultural Change* 50(3): 581–606.

Tabatabai, H. 1996. *Statistics on Poverty and Income Distribution: An ILO Compendium of Data*. Geneva: International Labour Office.

Tandjeu, J. B. 1998. Défi de la conservation de la faune dans un contexte de crise économique: Cas du Cameroun. *Seminaire FORAFRI de Libreville – Session 3: produits de la forêt*.

Taylor, L., Yurukoglu, K. T. and Chaudhry, S. 1986. 'A macro model of an oil exporter: Nigeria', in J. P. Neary and S. van Wijnbergen (eds), *Natural Resources and the Macroeconomy*. Oxford: Blackwell, 201–24.

Taylor, R. 1997. 'The state versus custom – regulating Papua New Guinea's timber industry', *Working Paper* 1997/9. Resource Management in Asia Pacific. Canberra, Australia: Australian National University.

Tchoungui, R., Gartlan, S., Simo, J. A., Sikod, F., Youmbi, A., Ndjatsana, M. and Winpenny, J. 1995. 'Structural adjustment and sustainable development in Cameroon: a World Wide Fund for Nature study', *ODI Working Paper* No. 83. UK: Overseas Development Institute.

Temu, I. 1995. 'Price policy analysis: the case of coffee in Papua New Guinea', *NRI Discussion Paper* No. 79. Papua New Guinea: National Research Institute.

Thapa, K., Bilsborrow, R. E. and Murphy, L. 1996. 'Deforestation, land use and women's agricultural activities in the Ecuadorian Amazon', *World Development* 24(8).

The Economist 1977. 'The Dutch Disease', 26 November.

Thiele, R. and Wiebelt, M. 1994. 'Policies to reduce tropical deforestation and degradation: a computable general equilibrium analysis for Cameroon', *Quarterly Journal of International Agriculture* 33(2).

Thompson, H. 1995. 'Environment and development: the forests of Papua New Guinea', *Journal of Interdisciplinary Economics* 6(2): 133–49.

Timmer, C. P. 1994. 'Dutch disease and agriculture in Indonesia: the policy approach', *Development Discussion Paper* No. 490. Harvard: HIID.

Toft, S. (ed.). 1997. 'Compensation for resource development in Papua New Guinea', *Monograph No. 6*. Port Moresby: Law Reform Commission of Papua New Guinea.

Toornstra, F. H., Person, G. A. and Youmbi, A. 1994. *Deforestation in Context: The Southern Forest Region in Cameroon*. Centre for Environmental Science Leiden: Leiden University.

Topik, S. C. 1997. 'How to turn nothing into something', *World Trade Magazine*, 10(7): 56–66.
Torres, C. 2000. *Análisis de los patrones locales en el proceso de deforestación del nororiente ecuatoriano*, MSc thesis, Universidad Central del Ecuador. Quito: FIGEMPA.
Transparency International 1999. 1999 Bribe Payers Index. 1999 Corruption Perceptions Index, available online at www.transparency.de/documents/cpi/index.htm.
Transparency International 2000. The Corruption Perceptions Index 1999, available online at http://www.transparency.de/documents/cpi/index.html, accessed on 7 November 2000.
Transparency International 2001. 'The Corruption Perceptions Index 2000, available online at http://www.transparency.de/documents/cpi/2000/cpi2000.html, accessed 11 November 2001.
Transparency International 2002. 'The Corruption Perceptions Index 2001, available online at http://www.transparency.org/cpi/2001/cpi2001.html, accessed on 29 May 2002.
Trefon, T. 1999. 'Libreville et son appétence opiniâtre de forêt', *Afrique contemporaine*, No. 190, 2nd trimestre.
Tsalefac, M. 1994. 'Sécheresse, déforestation et érosion sur les montagnes de l'Ouest du Cameroun', *Cahiers d'Outre-Mer* 47(185).
Umeazaki, M., Kuchikura, Y., Yamauchi, T. and Ohtsuka, R. 2000. 'Impact of population pressure on food production: an analysis of land use change and subsistence pattern in the Tari Basin in Papua New Guinea highlands', *Human Ecology* 28(3), 359–81.
UNCTAD. 1999. *1997 International Trade Statistics Yearbook*, Vol. 1 (trade by country). New York: United Nations.
UNDP. 1996. *Cooperation au développement. Gabon. Rapport 1995*. Libreville.
UNDP. 1999a. *Cooperation au développement. Gabon. Rapport 1998*. Libreville.
UNDP. 1999b. *Rapport national sur le développement humain. Rapport detaillé*. Ministère de la Planification de la Progammation du Développement et de l'Amènagement du Territoire, Libreville.
UNDP. 1999c. Cameroon. *Human Development Report 1998*. Poverty in Cameroon. Yaoundé: United Nations Development Program.
UNDP. 1999d. *Coopération pour le développement. Cameroun. Rapport 1998*. Yaoundé: United Nations Development Program.
UNDP. 1999e. *Papua New Guinea. Human Development Report 1998*. Port Moresby: Government of PNG and United Nations Development Programme.
UNEP. 2001. 'An assessment of the status of the world's remaining closed forests', UNEP/DEWA/TR-01–2, Nairobi: UNEP/NASA/USGS/USFS.
Uquillas, J. 1984. 'Colonization and spontaneous settlement in the Ecuadoran Amazon', in M. Schmink and C. H. Wood (eds), *Frontier Expansion in Amazonia*. Gainesville: University of Florida Press.
US Geological Survey. 2000. Cameroon metals. http://www/informatik.tu-muenchen.de/~kouamwot/cameroun/metaux/met.html.
Usui, N. 1996. 'Policy adjustments to the oil boom and their evaluation: the Dutch disease in Indonesia', *World Development* 24(5): 887–900.
Vaez-Zadeh, R. 1988. 'Oil wealth and economic behavior: the case of Venezuela 1965–81', *IMF Working Paper* No. 56, Washington DC: IMF.
Van Soest, D. 1995. *The Role of the Agricultural Sector in the Deforestation Process: A Case Study of Cameroon*. Department of General Economics, Groningen: University of Groningen.
Veillon, J.-P. 1976. 'Deforestaciones en los Llanos Occidentales de Venezuela desde 1950 hasta 1975', in L. S. Hamilton *et al.* (eds), *Conservación de los bosques húmedos de Venezuela*. Sierra Club-Consejo de Bienestar Rural.
Veillon, J.-P. 1989. *Los bosques naturales de Venezuela. Parte I: El medio ambiente*. Universidad de los Andes, Instituto de Silvicultura, Mérida.
Veillon, J.-P. 1997. *Los bosques naturales de Venezuela. Parte III: Los bosques tropóficos, o veraneros de la zona de vida: bosque seco tropical*. Universidad de los Andes, Instituto Forestal Latinoamericano (IFLA), Mérida.

Venturini, O. L. 1969. *La carretera Panamericana y la colonización del Piedemonte nor-occidental andino (zona de El Vigía)*. Mérida: ULA, Instituo de Geografía, Facultad de Ciencias Forestales.

Vivas R. F. 1993. 'Una periodificación de Venezuela colonial', in L. Cipriano Rodríguez (coord.), *Los grandes períodos y temas de la historia Venezolana*. Instituto de Estudios Hispanoamericanos, Universidad Central de Venezuela, Caracas.

Warr, P. G. 1986. 'Indonesia's other Dutch disease: economic effects of the petroleum boom', in J. P. Neary and S. Van Wijnbergen (eds), *Natural Resources and the Macroeconomy*. Cambridge, Mass.: MIT Press.

Warr, P. G. 2000. 'Indonesia: trade shocks and construction booms', in P. Collier and J. W. Gunning (eds), *Trade Shocks in Developing Countries, Volume 2: Asia and Latin America*. Oxford: Oxford University Press, 326–54.

Warr, P. G. and Chaiyindeepum, S. 2000. 'Thailand: trade shocks and domestic responses', in P. Collier and J. W. Gunning (eds), *Trade Shocks in Developing Countries, Volume 2: Asia and Latin America*. Oxford: Oxford University Press, 201–32.

Waters, W. 1997. 'The road of many returns. Rural bases of the informal urban economy in Ecuador', *Latin American Perspectives*, issue 94, 24(3): 50–64.

Wesley-Smith, T. 1990. 'Papua New Guinea in 1990: a year of crisis', *Asian Survey* 31(2): 188–95.

Wetzel, D. 2000. 'Ghana's management of a temporary windfall: the cocoa boom of 1976–77', in P. Collier and J. W. Gunning (eds), *Trade Shocks in Developing Countries, Volume 1: Africa*. Oxford: Oxford University Press, 94–141.

Whalley, J. 1982. 'Foreign trade in Papua New Guinea', *Institute of National Affairs Discussion Paper* No. 10. Port Moresby, Papua New Guinea: Institute of National Affairs.

Whitaker, M. D. and Greene, D. 1990. 'Development policy and agriculture', in M. D. Whitaker and D. Colyer (eds), *Agriculture and Economic Survival. The Role of Agriculture in Ecuador's Development*. Boulder, San Francisco and Oxford: Westview Press, 21–41.

Whitaker, M. D. and J. Alzamora 1990a. 'Production agriculture: nature and characteristics', in M. D. Whitaker and D. Colyer (eds), *Agriculture and economic survival. The Role of Agriculture in Ecuador's Development*. Boulder, San Francisco and Oxford: Westview Press, 131–60.

Whitaker, M. D. and J. Alzamora 1990b. 'Irrigation and agricultural development', in M. D. Whitaker and D. Colyer (eds), *Agriculture and Economic Survival. The Role of Agriculture in Ecuador's Development*. Boulder, San Francisco and Oxford: Westview Press, 161–92.

White, H. 1992. 'The macroeconomic impact of development aid: a critical survey', *The Journal of Development Studies* 28(2): 163–83.

White, L. and Oslisly, R. 1998. 'Lopé: a window on the history of the central African rain forests', *Séminaire FORAFRI de Libreville*, Libreville.

Whitworth, A. 1993. 'Public enterprise policy: independence to 1991', *INA Discussion Paper* No. 57. Port Moresby: Institute of National Affairs.

Wiebelt, M. 1994. 'Stopping deforestation in the Amazon: trade-off between ecological and economic targets?' *Weltwirtschaftliches Archiv* 131: 542–68.

Wigston, D. L. 1984. 'Historical aspects of the forestry problem in Papua New Guinea', in H. K. Steen, (ed.), *History of Sustained-Yield Forestry: A Symposium*. Port Moresby: University of Technology.

Wilkie, D. S. and Carpenter, J. 1998. The impact of bushmeat hunting on forest fauna and local economies in the Congo Basin: a review of the literature, *CARPE-WCS Paper*, Libreville.

Wilks, C. 1992. 'État de l'environment', in *Étude environmentale de la zone d'Avocette*. Report prepared for EU (unpublished). Libreville.

Wilks, C. 2000. *La forêt gabonaise*. Libreville (unpublished draft).

Wilks, C. 2002. 'Evolution des forêts du Gabon' and 'Conversions en cours', in P. Christy, R. Jaffré, O. Ntougou and C. Wilks (eds), 2002 (draft), *La forêt gabonaise*. Libreville: French Development Cooperation and Multipress, 19–72.

Williams, S. 1982. 'New approaches to agriculture and rural development. Commonwealth Development Corporation', *Agribusiness Worldwide* 3(8): 26–9.

Winters, J. A. 1996. *Power in Motion: Capital Mobility and the Indonesian State*. Ithaca and London: Cornell University Press.

Wolff, E., Trefon, T., Sintzoff, M. and Mwanza, H. n.d. *Géographie des relations ville-forêt*. Rapport final. Libreville: CARPE.

Wood, A. 1988. 'Global trends in real exchange rates 1960–84', *World Bank Discussion Paper* No. 35.

Wood, A., Stedman-Edwards, P. and Mang, J. (eds). 2000. *The Root Causes of Biodiversity Loss*. London: Earthscan and WWF.

Wood, H. A. 1972. 'Spontaneous agricultural colonization in Ecuador', *Annals of the Association of American Geographers* 62(4), December.

Woods, P. 1989. 'Effects of logging, drought and fire on structure and composition of tropical forests in Sabah, Malaysia', *Biotropica* 21, 290–8.

World Bank. 1992. *World Tables 1992*. London/Baltimore: Johns Hopkins University Press.

World Bank 1997. *Papua New Guinea: Accelerating Agricultural Growth – An Action Plan*. Report No. 16737-PNG. Washington, DC: World Bank.

World Bank. 1999a. World Development Indicators. CD-Rom version (WinSTARS), Washington, DC: World Bank.

World Bank. 1999b. *World Development Report 1998/99*. Washington, DC: World Bank.

World Bank. 2000a. 'Gabon at a glance', available online at www.worldbank.org/data/countrydata/ang/cmr_aag.pdf.

World Bank. 2000b. 'PNG at a glance', available online at www.worldbank.org/data/countrydata/.

World Bank. 2001a. World Development Indicators, CD-Rom version (WinSTARS), Washington, DC: World Bank.

World Bank. 2001b. *World Development Report 2000/2001: Attacking poverty*, Washington, DC: World Bank, 333 pp.

World Conservation Monitoring Centre (WCMC). 1996. 'Cameroon: forest statistics', available online at http://www.wcmc.org.uk/cgi-bin/fp_countryquery.

WRI. 1992. *World Resources 1992–3. A Guide to the Global Environment*. World Resources Institute, New York/Oxford: Oxford University Press.

WRI. 1994. *World Resources 1994–95. A Guide to the Global Environment*. World Resources Institute, New York/Oxford: Oxford University Press.

WRI. 1995. *World Resources 1995–96. A Guide to the Global Environment*. World Resources Institute, New York/Oxford: Oxford University Press.

WRI. 1998. *World Resources 1998–99. A Guide to the Global Environment*. World Resources Institute, New York/Oxford: Oxford University Press.

Wunder, S. 1991. 'Dutch Disease theory and the case of Colombia', PhD thesis, part I ('The Dutch Disease – a theory of trade shocks'), part II ('An empricial application of the Dutch Disease model: Colombia 1967–88', Institute of Economics, University of Copenhagen, *Red Mimeo Series* No. 41, 550 pp.

Wunder, S. 1995. 'Conservation status of Sangay National Park. With special emphasis on the Guamote-Macas road', Report to the UNESCO World Heritage Committee, IUCN Regional Office for South America (unpublished), Quito, October.

Wunder, S. 1996a. 'Deforestation and the uses of wood in the Ecuadorian Andes', *Mountain Research and Development* 16(4).

Wunder, S. 1996b. *Los caminos de la madera*. Programa Regional Bosques Nativos Andinos (PROBONA)/IUCN/DDA. Quito: Argudo Hnos.

Wunder, S. 1997. 'From Dutch disease to deforestation – a macroeconomic link? A case study from Ecuador', *CDR Working Paper 97(6)*. Copenhagen: Centre for Development Reseach.

Wunder, S. 2000. *The Economics of Deforestation: the Example of Ecuador*. Houndmills: Macmillan Press and New York: St Martin's Press, June.

Wunder, S. 2001a. 'Ecuador goes bananas: Incremental tecnnological change and forest loss', in A. Angelsen and D. Kaimowitz (eds), *Agricultural Technology and Tropical Deforestation*. Wallingford (UK): CAB International, April.

Wunder, S. 2001b. 'Poverty alleviation and tropical forests – what scopes for synergies?', *World Development* 29(11): 1817–33, November.

Wunder, S. 2003 (forthcoming). 'When the Dutch Disease met the French connection. Oil, macroeconomics and forests in Gabon', Report prepared for the CARPE-CIFOR-USAID project 'The impact of macroeconomic and agricultural policies on forest conditions in Gabon'.

WWF. 2003. 'To dig or not to dig? Criteria for determining the suitability or acceptability of mineral exploration, extraction and transport from ecological and social perspectives. A discussion paper for WWF'. Gland (Switzerland), February (pp. 19).

WWF and IUCN. 1996. *Forests for life*. The WWF/IUCN forest policy book. World Wide Fund for Nature International and the World Conservation Union, Godalming, February.

Yañez, L. 1995. 'Deforestación, degradación y desertificación en Venezuela', *Seforven*, year 6, no. 12, June. Caracas: MARNR.

Yates, D. A. 1996. *The Rentier State in Africa: Oil Rent Dependency and Neocolonialism in the Republic of Gabon*. Asmara: Africa World Press Inc.

Young, C. E. F. 1995. 'Public policy and deforestation in the Brazilian Amazon', Report prepared to CREED/IIED. London: International Institute for Environment and Development, July.

Younger, S. D. 1992. 'Aid and the Dutch disease: macroeconomic management when everybody loves you', *World Development* 20(11):1587–97.

Zambrano S., L. 1995. 'What we have done and what we can still do in economic policy', in J. McCoy, A. Serbin, W. C. Smith and A. Stambouli (eds), *Venezuelan Democracy under Stress*. Miami: University of Miami, North-South Center.

Zomo Yebe, G. 1993. *Compendre la crise de l'économie Gabonaise*. Paris: Éditions L'Harmattan.

Name index

Abogo, P. M. 382
Abril-Ojeda, G. 226
Achard, F. 63, 81, 89, 136, 219, 323
Adams, J. S. 85, 90, 109, 110, 129
Ahidjo, President 177, 179
Aicher, C. 11, 153
Ajamil, C. 12, 220
Alarcón, C. 227
Alfonso, J. P. P. 168
Allen, B. J. 249, 250, 251, 285, 292, 294, 295
Alzamora, J. 235, 236, 243
Amelung, T. 217, 218, 253
Amend, T. 155, 169
Amigransa, E. B. 50, 91, 176, 220, 222, 310
Amin, A. A. 28, 178, 182, 191, 200, 211, 212
Amos, G. 12, 280
Amuzegar, J. 15
Andersen, L. E. 55, 62, 381
Anderson, K. 262, 270, 276, 277, 281, 289
Angelsen, A. 11, 37, 39, 40, 55, 83, 324
Ansari, M. I. 14
Aranda, S. 132, 150, 154, 157, 161
Araujo, M. C. 226, 227
Arif, M. 316
Aruga, J. A. 294
Arvelo-Jiménez, N. 132, 155, 157
Asafu-Adjaye, J. 268
Ascher, W. 312, 317, 323, 366
Attiga, A. A. 15
Aubreville, A. 127
Auty, R. M. 25, 27
Aweto, A. O. 208

Baardsen, S. 14, 31
Bakoup, F. 191
Banks, G. 261
Baptista, A. 160, 161, 168
Barbier, E. 81, 299, 300, 302, 317, 382
Barham, B. L. 14
Barker, P. 12
Barnes, J. F. 97
Barnett, T. E. 274
Barr, C. 49, 314, 315, 317, 319, 324
Barret-Lefeuvre, G. 102
Barreto, A. 140
Barro-Chambrier, A. 127
Basquin, P. 126, 127
Bassey, N. 304
Baxter, P. 262
Bellamy, J. A. 254, 293
Benalcázar, R. R. 215, 246
Benjamin, N. 183, 192, 197
Bennett, E. L. 203, 213
Betancourt, J. 11
Bevan, D. 21, 25, 311, 312, 313
Bevilacqua, M. 135, 140
Bhagwati, J. 21
Bickerton, J. 11, 92, 127, 129
Bienen, H. 305, 306
Bikié, H. 11, 12, 170, 173, 174, 175, 185, 187, 190, 196, 202, 205, 212
Bille, M. 11
Bisbal, F. J. 139, 156
Biya, P. 177, 179
Blaise, N. G. 12
Blandford, D. 178, 179, 181, 192, 195, 204, 211
Blaney, S. 92, 127
Blarcom, B. van 154, 155
Blyth, C. 265, 280, 282
Bond, A. 12
Bongo, O. 96, 97, 98, 118, 127
Bonifaz, M. 31, 247
Bonnell, S. 293
Borcherdt, C. 139, 144, 145, 154, 156, 162
Borja, R. 227
Borlaug, N. 382
Bosworth, M. 276, 277, 289

Boumali, A. 112
Bourguignon, F. 140, 141, 144, 157, 168
Bourke, M. 12, 250, 252, 261, 292, 294
Bravo, E. 310
Brito, F. F. 132
Bromley, R. 215, 216, 237
Brown, K. 54, 190, 198, 222, 251
Brugière, D. 109
Brunck, F. 103
Brunner, J. 196, 212
Bryant, D. 75, 167
Bucaram, A. 227
Burgess, J. C. 299, 300, 302
Burritt, R. 256
Burton, J. 259, 293
Buschbacher, R. J. 158

Caballé, G. 87
Cabarle, B. J. 214, 217, 218
Cabeza, M. 11
Cabrera, E. 11, 150, 156
Caldera 144, 150, 155, 157
Campbell, D. 201
Capistrano, A. D. 39
Carillo, P. 220
Carpenter, J. 121, 129, 197, 213
Carrero, N. O. 132
Carrero, O. 133
Carret, J.-C. 11, 188, 212
Carrière, S. 198
Carvallo, G. 132
Casagrande, J. B. 239
Casson, A. 318, 319
Catalán, A. 134, 135, 139, 144, 168, 169
Cattaneo, A. 40
Centeno, J. C. 11, 133, 136, 139, 145, 149, 157, 158, 159, 168, 169
Chan, Sir J. 264
Chávez, H. 144, 150–1, 153, 371
Chomitz, K. M. 55
Christy, P. 11, 90, 126, 127
Clément, J. 188, 212
Cochrane, M. A. 70
Cockburn, A. 55
Collier, G. A. 54, 297, 298, 300, 301, 302
Collier, P. 21, 23, 304, 306, 307, 322, 364, 382
Collins, N. M. 74, 248, 250, 251, 253, 254, 274, 275
Collomb, J. G. 87, 88, 101, 102, 103, 109, 126, 128
Commander, S. 30, 236, 240
Connell, J. 248, 250, 256, 257, 259, 262, 263, 264, 267, 268, 269, 270, 281, 294
Contreras-Hermosilla, A. 37, 42, 55
Cook, C. D. 251, 253
Coomes, O. T. 14
Corden, M. 14, 19, 20, 29
Coronil, F. 141, 143, 154, 168, 364
Corrales, L. 83
Costanza, R. 83
Cruz, W. 378
Culas, R. 39, 382
Culverwell, J. 197

Dahanayake, P. A. S. 262
Dauthuille, C. 93, 100, 120
Davidson, J. 275
Davis, G. 27
Davis, J. M. 22
Davis, R. 11
De'Ath, C. 275
Debroux, L. 187, 190
Defries, R. S. 59
Deininger, K. W. 299, 300
DeJanvry, A. 227
DeLancey, M. W. 177, 180, 198
de la Torre 235, 241
Delfín, P. 11
Delvingt, W. 190, 205
de Melo, J. 179, 183
Denevan, W. M. 130, 246
Dennis, R. 11, 83, 318, 323
Densley, B. 252
Depierre, D. 197
Dermawan, A. 10
Descoing, B. 126
de Silva 267
Devarajan, S. 179, 183
de Wachter, P. 11, 111
Dezhbaksh, H. 20
Diaw, C. 12, 172, 191, 201, 212
Dick, J. 321
Diehl, M. 217, 218, 253
Djeki, J. 85
Domingo, C. 11, 168
Dorney, S. 264, 265, 281, 287, 294
Dounias, E. 198
Doward, J. 364
Downs 262
Downton, M. W. 62, 83
Droineau, S. 86
Duba, D. 12
Dubash, N. K. 275, 365
Dufoulon, G. 11, 102, 109, 110, 112
Duncan, R. C. 262, 264, 266, 267, 294
Durán Ballen 227
Durieu, L. 110
Dutta, D. 382

Eaton, P. 283
Eba'a Atyi, R. 187, 190, 210
Ebiang-Ebang, V. 11
Eden, M. J. 250
Edwards, C. R. 298
Eide, E. 14
Ekoko, F. 190, 196, 198, 212
Ekstrom, J. P. 247
Ekuma, E. 11
Ella Nguema Rolly, E. 111, 112
Ellis, D. M. 282
Enders, K. 20, 29
Enn, D. 12, 295
Enright, M. 153
Eraker, H. 311
Erwidodo 316
España, L. P. 11, 161
Essama-Nssah, B. 187, 190, 195, 200, 202, 203, 204, 205, 212, 213
Esteves, C. 145
Estreguil, C. 275
Eva, H. D. 59, 74
Ezeala-Harrison, F. 305, 306

Fallon, J. 262, 266, 267, 276
Fardmanesh, M. 30
Faure, J. J. 78
Febres-Cordero 226, 227, 364
Feltenstein, A. 30, 298, 300, 322
Fernando, N. A. 295
Fiebach, J. 11, 305, 309, 323
Figueroa, S. 236
Filer, C. 11, 12, 249, 250, 252, 257, 258, 259, 261, 265, 274, 275
Fontes, J. 126
Forni, E. 11
Forsyth, P. J. 14
Franqueville, A. 200, 201, 203–4
French, B. 269
Freyne, D. F. 251, 255, 292, 294
Fuhr, M. 127

Gabaldón, M. 11, 155
Gaethe, R. 12, 246
Gami, N. 11, 102
García, G. 141, 144
Garnaut, R. 262
Gartlan, S. 170
Gavin, M. 296, 297, 298, 322
Geist, H. J. 63, 75, 81, 83
Gelb, A. 21, 22, 140, 141, 144, 168, 224, 226, 236, 247, 305, 306, 310, 311, 312, 314
Gibson, J. 269, 271, 287–8
Gill, T. 130, 139, 147, 167
Glassburner, B. 310, 312, 314
Glynn, T. 323
Gockowski, J. 12, 182, 187, 190, 191, 195, 200, 202, 203, 204, 205, 212, 213
Gómez, H. 144, 149, 159
González-Espinosa, M. 297, 301
Goodman, R. 262, 264, 280
Grainger, A. 73, 76, 83
Grau, P. C. 167
Gray, D. A. 55
Greene, D. 226
Grogen, D. 12
Grynberg, R. 274
Gubry, P. 213
Gumoi, T. M. 11, 12, 262, 277, 292, 294, 295
Gunning, J. W. 21, 23
Gupta, D. 262, 263, 264, 266
Gutiérrez, A. 149, 156, 159, 162, 169
Gylfason, T. 27, 29

Haase, A. H. V. 145
Hadi, P. U. 316
Hannibal, L. W. 323
Harcourt, C. S. 74, 130, 133, 134, 138, 140, 156, 158, 159, 216, 217, 238, 298, 299, 300
Harden, R. 264, 267
Harper, A. 257
Harris, G. T. 251
Hasanuddin, L. 316
Hauser, S. 12, 197, 213
Hausmann, R. 132, 141, 143
Hecht, S. 55
Heimberger, B. 140
Henner, H. F. 99
Herberg, H. 20, 29
Herrera Campins, L. 143, 363
Heywood, P. 287, 295
Hide, R. 295
Hill, H. 311
Hiraoka, M. 220, 238
Hoel, M. 14
Hoffman, F. 68
Holmes, D. 315, 318, 323
Holzknecht, H. 12, 251, 252
Hoogeveen, J. A. M. 191, 213
Houghton, R. A. 81
Huber, O. 11
Hulme, D. 283, 284
Hurst, P. 253
Hurtado, O. 224, 227
Huyizzy, E. 11
Hyndman, D. 259, 293

Igua, P. B. K. 285, 287, 288
Imine, S. 12
Indjieley, M. 121
Infante, A. 144–5, 169
Israel, M. 257
Izko, X. 12

James, N. 259, 293
Janz, K. 73, 79
Jarrett, F. O. 262, 270, 276, 277, 281
Jones, L. T. 252
Jorez, J.-P. 11
Josling, T. 150, 232, 242
Jua, N. 177, 179, 180, 181, 195, 198, 211

Kahn, J. R. 38
Kaimowitz, D. 9, 11, 37, 40, 54, 55, 181, 182, 183–4, 185, 186, 192, 204, 212, 320, 324, 379
Kalit, K. 12, 275
Kalulumia, P. 183
Kamas, L. 29
Kammesheidt, L. A. 169
Kanawi, W. 12, 257, 280
Kapanamur, E. 12
Karl, T. L. 28, 132, 140, 141, 297, 300, 305
Karo, R. 12, 292
Karsenty, A. 11, 107, 187, 188, 190, 196, 212
Kavanamur, E. 295
Kennedy, R. F. Jr. 246
Kiele-Sapak, R. 12, 293
Kiker, C. F. 39
Kimerling, J. 220, 221, 222, 223, 246
King, T. 213
Kizer, B. 11
Klappa, S. 283
Kleinn, C. 83
Knudsen, O. 28
Kocher-Schmid, C. 283
Konings, P. 180, 181
Krueger, A. O. 21, 262
Kuntz, S. 62, 68
Kuznets, S. 376–7
Kwapena, N. 275

Lam, N. V. 262, 266, 268, 269
Lamb, D. 275, 295
Lambin, E. F. 55, 63, 75, 81, 83, 275
Lane-Poole, C. E. 250, 274
Lanly, J.-P. 73
Laporte, N. 74, 170, 173, 174
Larivière, J. 101, 126
Larrea, C. 225, 227, 228, 231, 234, 235, 237, 239, 240–2, 247
Lelang, J. 12
Léonard, G. 85, 86, 97, 104, 113, 118, 127, 128
Levang, P. 323, 353
Levantis, T. 266
Lezama, A. T. 11
Liano, A. 10
Liverman, D. 299
Lombardo, S. 11
López, J. 169
López-Portillo 296
Lorass, K. T. 311
Louman, B. 252
Lucas, J. 269, 294
Lusinchi 143, 150, 155, 363
Lux, M. 11
Lynch, U. 12

McAlpine, J. 11, 251, 253, 254, 255, 273, 276, 283, 292, 293, 294
MacBean, A. J. 28
McCoy, J. 144
McCrea, P. 12, 270, 295
McDonald, J. A. 38
Mackie, C. 70
McRea 213
McShane, T. O. 85, 90, 102, 109, 110, 129
Magrin, G. 89
Mahuad 227
Mainardi, S. 36, 55, 377
Maizels, A. 28
Malavé, J. 159
Malingreau, J.-P. 62, 83
Mamingi, N. 200
Mandeakali, L. 12, 294
Manning, M. 12, 195, 266, 282, 285, 286, 292
Mansoorian, A. 21
Mansutti, A. 11, 162
Marsden, K. 265, 286, 294, 295
Marshall, A. R. 257
Marshall-Silva, J. 224, 226, 236, 247
Masera, O. 299
Mather, A. 382
Matthews, E. 63, 72, 79, 83
Matute, D. 133, 134, 135
Mawuli, A. 259, 260, 261, 262, 285, 293
Mayaux, P. 62, 74, 75, 87, 133, 134, 173, 253, 254, 308, 328, 329
Mbendi 176, 303, 304
Mborou, C. 11
Medjo, F. R. 12
Mengué, M. A. 11
Mertens, B. 11, 55, 62, 174, 200, 211
Meye M'eya, S. 11, 98

Meza, J. 247
Michalenko, G. 275
Michaud, P. 128
Mika, A. 12
Mikesell 266
Miller, S. 275, 280, 294
Milne, M. 12
Milner, C. R. 191, 212
Minten, B. 299, 300
Miranda, M. 11, 31, 133, 139, 140, 149, 150, 155, 158, 168, 169
Moari, G. M. 258
Moncada, S. 11
Mondiai, K. 12
Montenegro, S. 22
Moriconi-Ebrard 118, 203
Mosley, P. 227, 231, 247
Mourata, M. 275, 294
Muhamad, N. Z. 315, 318, 323, 324
Müller, B. 11, 139, 140
Murray Li, T. 318
Musacchio, M. 11, 322
Myers, N. 59, 87, 167, 170, 173, 174, 253, 254

Nadarajah, T. 294
Nankani, G. 23
Nasi, R. 11, 62, 86, 101, 102, 113
Ndong, E. 11
Ndongko, W. A. 177
Ndong-Obiang, L. S. 11
Ndoye, O. 11, 182, 183–6, 201, 202, 204, 212, 213
Neary, P. 14, 22, 29, 311
Needle, C. L. 382
Nehru, V. 323
Nepstad, D. C. 70, 72
Newbery, D. M. G. 28
Newbold, R. A. 47
Ngoye, A. 11, 111
Nguéma Meye, P.-H. 11
Nguépi, J. 78
Nguinguiri, J.-C. 12
Nicholas, S. J. 14
Noreng, Ø. 25

Obiang, M. M. 11, 127
O'Brien, K. L. 297, 299, 300, 301
Ochoa-Gaona, S. 297, 301
Ochoa, O. 11
O'Collins, M. 284, 285
O'Faircheallaigh, C. 262
Ohtsuka, R. 292, 294
Ojara, R. 12, 246
Ola-Adams, B. A. 308
Ole, Z. 197
Ondo, P. O. 11, 128
Ondo, R. 11
Onishi, N. 364, 382
Ortiz-Chour, H. 83
Osemeobo, G. J. 308
Osgood, D. 313, 314
Oslisy, R. 110
Ovono-Edzang, N. 128
Oyejide, T. A. 303, 307, 322

Päivinen, R. 62, 63, 75
Paolilla, A. 11, 138
Parkin, R. 11
Parks, P. J. 31, 247
Parnes, A. 28
Parvin, M. 20
Pavarotti, L. 15
Peek, P. 30, 240
Peñate, A. 11, 168
Peng, P. 12, 292, 295
Pérez, C. A. 141, 144, 150, 155, 168
Pérez-Puelles, S. 140
Peters, C. M. 70
Pichón, F. J. 222, 247
Pierce, D. W. 54, 83
Pietracci, B. 10
Pilchowski, E. 97, 111, 115, 116, 117, 128
Pinto, B. 25, 305, 311
Pittier, H. 167
Plonczak, M. 133, 156, 167
Plouvier, D. 31
Podolefski 251
Pokam, J. 204, 213, 357
Polier, N. 259
Pollard, H. L. 30
Polume, S. 264
Poupart, N. 97, 111, 115, 116, 117, 128
Pourtier, R. 11, 86, 89–90, 96, 97, 101, 104, 108, 112, 113, 115, 116, 117, 118, 119, 125, 126, 128, 129, 355, 381
Prajanthi, W. 10
Prasetyo, L. B. 381
Prebisch, R. 28
Puntodewo, A. 10

Quane, D. 318
Quigley, J. 251, 253, 254, 255, 276, 292, 293

Raikes, P. 11
Ramos, H. 235
Raponda-Walker, A. 86, 113, 126
Rau, M. T. 257

Recalde, O. 12, 246
Reed, D. 39, 155, 159, 169, 378
Reinhart, C. M. 28
Repetto, R. 378
Resosudarmo, I. A. P. 315, 316, 317, 321, 323
Rice, R. 31
Richard, A. 85, 86, 97, 104, 113, 118, 127, 128
Richards, D. G. 31
Richards, T. 316
Riddell, J. C. 201
Rios de Hernández, J. R. 132
Rippert, G. 89, 126
Riutort, M. 11
Robinson, J. G. 203, 213
Robison, L. 235
Robson, P. 276
Rodner, C. 168
Rodríguez, D. 11, 31, 55, 139, 140
Roemer, M. 30
Rojas, L. J. 133, 144, 145, 149, 156, 158, 159, 163, 168, 169
Romero, D. 155
Roper, J. 36, 39, 55, 73, 75, 299, 309
Ropivia, M. L. 85
Rosero, J. A. 247
Ross, M. L. 27, 317, 364, 366
Rowell, A. 304
Rowland, I. 12, 294
Rowthorn, R. E. 15
Rudel, T. K. 11, 36, 39, 55, 73, 75, 81, 83, 216, 222, 223, 230, 235, 237, 246, 247, 299, 309, 382
Ruiz, L. 220, 221, 246

Saavedra, E. S. 153
Sachs, J. D. 27, 29
Salazar-Carillo, J. 162
Salgado, V. 12, 246
Salter, W. G. 29
Sánchez, J. M. 11, 217
Sanida, O. 259, 260, 261, 293
Saracco, F. 11
Saro-Wiwa, K. 304
Saulei, S. M. 294
Saunders, J. C. 254, 255, 293
Sayer, J. 74, 87, 102, 126, 130, 133, 134, 138, 140, 156, 158, 174, 211, 216, 217, 298, 300, 306, 307, 308, 309
Scherr, S. J. 25, 30, 298, 300, 302, 306, 307, 311, 312, 313, 314
Schmidt, R. 27, 219, 246
Scotland, N. 315
Segovia, E. 11
Sekhran, N. 249, 250, 274, 275, 280, 294
Senat, P. 12, 283
Seymour, F. J. 275, 365
Shearman, P. 257
Sheng, F. 159
Siaguru, P. 251
Siegert, F. 62, 68, 70
Sierra, R. 12, 217, 222, 223, 230, 232, 247
Sillans, R. 86, 113, 126
Simons, K. 93, 102
Simpson, G. 260
Singer, H. 28
Singh, K. D. 73
Siriak, E. 12, 212
Sissoko, O. 11
Sizer, N. 31
Skeldon, R. 286, 287
Skole, D. 83
Skov, F. 246
Smeraldi, R. 346
Smith, J. 55
Smith, W. C. 144
Snape, R. H. 14
Snook, L. 11, 297, 298
Söderling, L. 95, 125, 127
van Soest, D. 191, 201, 213
Soladoye, M. O. 308
Solem, R. 11
Sönmez, A. 381
Sournia, G. 109, 197
Southgate, D. 11, 81, 219, 220, 221, 222, 223, 231, 232, 241, 246, 247
Spencer, I. T. 287
Sperandío, M. H. 11
Stacey, M. 269, 294
Stallings, J. 230
Steel, E. 121, 122, 129
Steigum, E. 14
Steininger, M. 83
Stibig, H.-J. 75, 323
Stiglitz, J. E. 28
Strauss, K. R. A. 167
Sugden, C. 264, 267
Suharto, President 310, 312, 313, 314, 317, 321, 363
Sukarno, President 310
Sumahadi 316, 323
Sunderlin, W. D. 9, 11, 12, 29, 36, 54, 55, 64, 204, 209, 213, 296, 297, 315, 316, 317, 318, 319, 321, 323, 324, 357, 363, 364, 377
Sutowo, I. 312

Tabatabai, H. 238, 295
Tacconi, L. 11, 12, 277, 281
Tandjeu, J. B. 202, 204

Tarr, D. 191
Taylor, L. 307
Taylor, R. 251
Tchamou, N. 11, 12
Tchinou, D. 12
Tchoungui, R. 181, 182, 186
Temu, I. 268, 276
Thaopile, M. 129
Thapa, K. 223, 247
Thiel, H. 12, 247
Thiele, R. 184, 191
Thompson, H. 274
Timmer, C. P. 314
Tobar, A. 12
Toft, S. 259, 293
Toledo 299
Toornstra, F. H. 190, 201, 211, 213
Topik, S. C. 14
Toro 217
Torres, C. 221, 222
Touré, S. 11, 126
Trefon, T. 112, 117, 121
Tsalefac, M. 202, 213
Tucker, C. 83

Umeazaki, M. 252, 294
Uquillas, J. 216, 238, 247

Veillon, J.-P. 132, 133, 134, 135, 149, 167, 168
Venturini, O. L. 139, 156, 158, 161, 169
Veríssimo, J. A. 346
Vinchent, R. 11, 108, 126, 128
Vivas, R. F. 131
Vivekananda, F. 177
Vogel, J. 12
Vos, R. 234
Voubou, B.-H. 11, 111

Warner, A. M. 27, 29
Warr, P. G. 310, 311, 313, 314, 319
Waters, W. 236, 240, 351
Wells, J. R. 15
Wesley-Smith, T. 264
Whalley, J. 277
Whitaker, M. D. 219, 220, 221, 226, 231, 232, 235, 236, 241, 243, 246, 247
White, K. J. 251
White, L. 110
White, S. 11, 12, 29, 83, 381, 382
Whitworth, A. 295
Wickham, P. 28
Wiebelt, M. 184, 191
Wigston, D. L. 251
van Wijnbergen, S. 22, 29, 311
Wilkie, D. S. 121, 129, 197, 213
Wilks, C. 11, 49, 83, 87, 88, 90, 92, 93, 100, 101, 109, 126, 127, 128
Williams, S. 283
Wingti, P. 264, 281
Winters, J. A. 28, 312, 323, 363, 381
Wolff, E. 89, 111, 117, 118, 119
Wood, A. 29, 40, 42
Wood, H. A. 247
Woods, P. 70
Wunder, S. 15, 29, 36, 54, 59, 78, 129, 215, 217, 228, 230, 234, 246, 247, 296, 297, 377

Xu, X. 262

Yamamoto, S. 220, 238
Yañez, L. 168
Yang, M. C. 28
Yates, D. A. 86, 96, 97, 98, 108, 121, 127, 128
Young, C. E. F. 39
Young, D. 11, 104

Zambrano, S. L. 141, 143, 167
Zamora, R. 11
Ziza, S. 11, 127
Zomo Yebe, G. 94, 96, 97, 98, 99, 100, 104, 105, 108, 117, 121

Subject index

Abuja 305
Aceh 310, 364
Adamaoua Province 172
afforestation 58, 65
Africa Forest 83, 90, 91, 127, 128
afub awondo fields 212
Agricultural Bank of Papua New Guinea 281
agriculture 30–1, 342–5, 325–82; in Cameroon 183–7, 195–7, in Ecuador 227–8, 235–6; in Gabon 99–101, 107–10; in Indonesia 311–15; in Mexico 297–8; in Nigeria 304–7; in Papua New Guinea 266–74, 280–1; in Venezuela 144–7, 154–5; *see also* shifting cultivation, slash-and-burn cultivation, subsistence farming
Agripog project 128
Agrolandia 330
aid xiv
Air Niugini 295
alang-alang grasslands 321
Alberta 55
Algeria 24, 310
aluminium 141
Amazon 1, 51, 55, 63, 65, 71, 157, 214, 216, 219–23, 235, 237, 238, 241–3, 245, 247, 339, 350, 353
Amazonas State 136, 164
Amerada Hess 129
Andean pact 232
Andes 130, 133, 161, 164, 214, 216, 219, 230, 235, 237, 239, 243, 246, 247, 352, 376
Angola 35, 54, 129, 322
Anzoátegui State 138
ARCO concession 246
Argentina 344, 373, 376
ASEAN 314
Australia 14, 15, 59, 258, 266, 294
Australian Centre for International Agricultural Research (ACIAR) 10
Australian National University (ANU) 12
Avocette 90
Ayos 200
ayous 127, 187
azobe 187

Baka 200
Bakassi peninsula 176, 303
Bakola pygmies 176
Bali 310, 313
balsa wood 230
bamboo 70
Bamboutou mountains 213
Bamiléké 201
bananas 66, 67, 287, 358, 373; in Cameroon 182, 183, 186, 206; in Ecuador 214–15, 226, 227, 232, 234; in Gabon 99, 108, 129
Bangladesh 24
Bantu 84, 201
BAPPENAS, Indonesian National Development Planning Agency 312
Barinas State 156
barley 241, 358
Barnett Enquiry 274
bauxite 211
Baygone Plain Programme 201
beef 119–22, 129, 163, 241, 276, 277, 288, 291, 300, 344, 357–9, 361, 366; *see also* cattle, pastureland
beeswax 86
Bélinga 127
Benoue 201
Beti 180
Bewani 283
Biafra 303, 364
biodiversity 40, 64, 72, 102, 110, 150, 172, 230, 280, 347, 371, 380
Bitam 100
bivariate correlation coefficients 39
Bogor 10

Bolívar State 133, 136, 139, 149, 156, 169
Bolivia 24, 35, 379
Booué 127
Borneo 68, 316; *see also* Kalimantan
Botswana 24
Bougainville Island 256, 257, 262, 264, 289, 294, 364
Brasília 305
Brazil 2, 34, 35, 36, 38, 39, 55, 62, 63, 132, 136, 247, 344, 346, 353
Brazil nuts 70
brazilwood 131
Brunei Darussalam 35
bushmeat 48, 169, 348, 352, 358, 359; in Cameroon 190, 202–3, 205, 206, 373; in Gabon 93, 102, 117, 121–2, 337
bushpigs 205

cadmium 257
Cairns 294
Caisse de stabilisation et de péréquation 104
Calabar 323
Caltex-Indonesia 11, 310, 311
Cameroon 170–213; agriculture in 183–7, 195–7; deforestation in 170–6; forestry in 187–90, 195–7; mentioned xiii, xvi, 2, 3, 4, 9, 11, 24, 29, 31, 33, 35, 38, 44, 56, 61–3, 75, 77–9, 82, 84, 86, 95, 96, 100, 104, 105, 107, 110, 111, 117, 120, 126–9, 227, 303, 325–82; oil boom in 177–82; poverty in 201–3; recommendations for 372–3; roads in 197–200; trade policy in 191–2; urbanisation in 203–5
Cameroonian Development Corporation (CDC) 195
Campeche 296, 297
Canada 14, 55
Canaima National Park 155
Canberra 305
canopy cover 58–60
Caparo Forest Reserve 149, 169
Capital District 286
Cap Lopez 91
Caracas 11, 143, 162, 356
Caroní river 140
cassava 99, 100, 212, 241, 250, 307, 314, 316, 358
cattle 5, 55, 66, 81, 277, 281, 288, 340, 344, 345, 347, 349, 355, 357, 359–62, 366–8, 376, 379, 382; in Cameroon 172, 205, 209, 213; in Ecuador 214, 215, 219, 220, 222, 223, 228, 231, 235, 239, 242, 245, 331, 334–5, 373; in Gabon 100, 108, 115, 121; in Mexico 300, 303, 335–6; in Venezuela 130, 132, 133, 136, 144–4, 151, 153, 156, 159, 162–5, 167–9, 332–4
cedar 132
Central African Economic and Customs Union (UDEAC) 191
Central African Economic and Monetary Community (CEMAC) 191
Central African Regional Program for Environment (CARPE) 11, 12
Central African Republic 3, 120
Centre for Development Research (CDR) 9, 10
Centre for International Forestry Research (CIFOR) xv, 9, 10, 12, 74, 76, 190
Centre for Research on the Cultural and Biological Diversity of Andean Rainforests 246
Centre Province 172, 202, 204, 212
CEPAL 147
Chad 176, 188
Chad, Lake 170
Chanal 301
charcoal 68, 117, 136
Chevron Niugini 260
Chiapas 54, 297, 299–302, 336
chicken 108, 162
China 102, 274, 381
Chocó 214, 373
Ciudad Bolívar 154, 162
Ciudad Guayana 162
CLIRSEN 216, 217
coca 24, 223
cocoa 16, 21, 23, 24, 29, 81, 306, 307, 309, 318, 319, 323, 334, 336, 352; in Cameroon 172, 177, 179, 181–4, 191, 194, 201, 203, 204, 206–8, 210, 211, 351, 355, 374, 380; in Ecuador 214, 226, 227, 232, 234; in Gabon 104, 108, 123, 128; in Papua New Guinea 267, 268–9, 281, 284, 290; in Venezuela 131, 150–1, 153, 154, 164
coconuts 250, 267–8
cocoyams 212, 307
coffee 10, 16, 23, 24, 26, 46, 54, 81, 314, 318, 334; in Cameroon 172, 177, 179, 181–4, 191, 192, 194, 195, 201–4, 206–8, 210, 211, 212, 351, 380; in Ecuador 215, 219, 222, 226, 227, 232, 234; in Gabon 104, 108, 123, 128; in Papua New Guinea 262, 267–9, 281–3, 290, 292, 301; in Venezuela 132, 145, 150–1, 154, 164
Colombia 3, 10, 22, 24, 29, 54, 147, 223, 247
Commonwealth Development Corporation 283

computable general equilibrium (CGE) 37, 39–41
CONASUPO 298
Congo 3, 35, 54, 84, 97, 100, 105, 110, 117, 127, 322
Conquest of the South (CODESUR) 155, 158
conservation 1, 45, 209, 254, 280, 284, 348, 367, 368–9
Consultative Group on International Agricultural Research (CGIAR) xv
convergence of evidence method 8, 75, 80
Convergence for Social Democracy 382
COPEI 141, 155
Copenhagen 9, 10
copper 5, 24, 256, 257, 262, 264, 335
copra 250, 257, 267, 268, 269, 284, 290, 318
Cordillera del Condor 353
cork 58
corn 133
correlation analysis 8
corruption 6, 21, 28, 97, 141, 154, 305, 312, 349, 350, 364, 366, 373; in Cameroon 180, 181, 196, 210; in Papua New Guinea 260, 265, 289, 292–3
Costa 214, 216, 218, 219, 226, 230, 235, 237, 238, 243, 246
Costa Rica 24
cotton 24, 131, 132, 172, 177, 181, 234
CREA 235, 238
Cuenca 235
Cuyabeno Wildlife Reserve 223
cyanide 256, 257, 311

Danida 10, 246
de-agriculturalisation 26, 27, 30, 40, 105
debt 8, 21, 36–9, 42, 95, 159, 363, 376, 378, 379–80, 382
defaunation *see* bushmeat
deforestation *passim;* discussed 56–83, 325–82; in Cameroon 170–6; in Ecuador 214–20; in Gabon 84–91; in Indonesia 315–22; in Mexico 299–303; in Nigeria 307–9; in Papua New Guinea 248–56; in Venezuela 130–8
deforestation theories 3
Delphi technique 75, 80
Denmark 10, 364
Department of Environment and Conservation 280
devaluation 37, 39, 182, 227, 265, 267, 268, 269, 274, 275, 287, 288, 294, 334, 335, 342, 362, 371–3, 375, 378
diamonds 24, 139, 140, 167
dipterocarp 70, 248, 314, 317
DIVA programme 246
Dja forest 190
dollarisation 227, 246, 373
donors (aid) xiv, 6, 75, 82, 110, 280, 281, 312, 363, 370
Douala 170, 172, 176, 188, 191, 197, 198, 200, 201, 204, 207
drugs 54
duikers 205
Dutch Disease xvi, 3, 4, 7, 9, 13–29, 30, 32, 41, 51, 297, 315, 319, 334, 336, 342, 355, 371, 380; in Cameroon 179, 182, 187, 192, 194, 207, 208; in Ecuador 227, 230, 231, 235, 241, 242, 245; in Gabon 104, 123; in Nigeria 304–7; in Papua New Guinea 266, 267, 289; in Venezuela 132, 145, 147, 151, 164, 165

Eastern Highlands Province 267
East Province 172, 182, 188, 190, 198, 200, 202, 211, 212, 351–2, 373
East Sepik Province 286
Ebolowa 12, 212
Echira 91
Ecociencia 246
ECOFAC project 110, 127
Ecuador 214–47, 291, 325–82; agriculture in 227–8, 235–6; deforestation in 214–20; forestry in 228–31, 236; mentioned xiii, xvi, xvii, 2–4, 9–12, 19, 22, 24, 33, 35, 45, 50, 51, 61, 63, 65, 67, 74, 76–9, 81, 83, 158, 209, 252, 258, 291, 300, 310; oil in 220–7; poverty in 238–40; recommendations for 373; roads in 236–7; settlement in 237–8; trade policy in 231–2; urbanisation in 240–1
Ecuadorian Institute for Agrarian Reform and Colonisation (IERAC) 238
Ecuadorian Protected Area System 236
Egypt 22, 24
Elf Aquitaine 91, 176
Elf Gabon 128
El Niño 230, 293
El Oro Province 247
El Salvador 24
Enga Province 293
Engel effect 241
environmentalism 1, 2, 6, 64, 365, 369, 375, 376, 379
Equatorial Guinea 84, 100, 105, 129, 303, 364, 382
esep fields 344
Esmeraldas Province 214, 219, 352, 373
Esso 303
Estuary Province 85, 90, 93, 109, 113, 126–8

eucalyptus 128, 167, 230
European Union 100, 110, 120
explosive source method 47
Exxon Mobil 176, 303

Falcón 167
Far North Province 201
ficus 248
Financial Funds Mechanism 235
Finland 364
firewood 31, 44, 45, 47, 69, 121, 239, 252, 348
Fly river 256, 257, 259
Food and Agricultural Organisation, United Nations (FAO) 36, 60, 64, 65, 67, 68, 69, 70, 75, 77, 78, 79, 82, 83, 87, 88, 89, 90, 133–5, 147, 164, 168, 173, 174, 188, 209, 216, 217, 218, 231, 234, 249, 253, 254, 255, 256, 271, 279, 293, 299, 301, 302, 308, 316, 317, 323, 324, 328, 362
forest degradation *passim;* discussed 69–71, 345–8
Forest Inventory and Mapping System 78, 253, 254, 255, 292, 293
Forest Resources Information System (FORIS) 73, 83, 173
forestry *passim;* discussed 30–1, 325–82; in Cameroon 187–90, 195–7; in Ecuador 228–31, 236; in Gabon 101–4, 109–10; in Papua New Guinea 250, 254–6, 267, 274, 280, 286, 287, 290, 291; in Venezuela 147–50, 155
Foumban 201
France 102, 105
Franceville 89, 93, 97, 104, 110, 111, 113, 118, 119, 126, 127
Freiburg 11
Frontier Forest Project 75, 88

Gabon 84–129; agriculture in 99–101, 107–10; deforestation in 84–90, 107–10; forestry in 101–4, 107–10; mentioned xiii, xvi, xvii, 2, 3, 4, 7, 11, 24, 29, 32, 35, 49, 56, 60, 61, 73–9, 81, 83, 130, 136, 141, 144, 149, 154, 157, 164, 170, 176, 179, 184, 196, 198, 200, 201, 205, 208, 241, 248, 300, 325–82; mining in 93–9; oil in 91–9; railways in 111–12; recommendations for 368–9, 371; roads in 110–11; settlement in 113–14; trade policy in 104–5; urbanisation in 117–19
Galápagos Islands 247
Gambá-Ivinga 91
gas *see* natural gas
Geomatics International Inc. 323
Ghana 3, 24, 38
Giffen goods 55
Gini index 159
Global Forest Watch 75, 173, 321
globalisation 28
global warming 6
Gobe 258, 260, 264
Gogol Valley 274, 275
gold 5, 14, 69, 70, 125, 136, 139, 140, 167, 211, 256, 261, 264, 335, 339, 372
Goroka 286
greenhouse gases 6–7
Greenland 24
Gregory effect 14
Groningen field 13
groundnuts 24, 100, 133, 172, 212, 306
GSFC 83
Guamote 247
guano 14
Guatemala 298, 299, 301
guayacán 131
Guayaquil 242, 364
Guayas 215, 358
guided democracy 310
Guinea 35
Gulf of Papua 257, 258
Gulf Province 260, 294
Guri dam 143
Gusap 277
Guyana 35, 130, 139, 157, 164, 167

Hagen, Mount 261, 283, 294, 295
hard kina policy 262, 265, 278
Haus Poroman 12
Haut-Ogooué Province 119, 128
Hévégab 128
Hides field 258, 289
Higaturu 283
Higaturu Oil Palm 268
Highlands Highway 258, 282, 283
Holocene 56
homesteading 158, 223, 243, 372, 373
honey 85
Hoskins Oil Palm Scheme 284
'H' theory 167
Huistán 301
Humid Forest Zone (HFZ) 172, 182–4, 186, 191, 192, 194, 195, 198–201, 203–10, 212–13, 372
hunter-gatherers 56, 84, 250
hunting xvii, 5, 102, 109, 121, 126, 169, 251, 304
hydroelectricity 31, 55, 342
hypotheses 4–5

IDB 247
idle lands 132
IERAC 238, 247
Ife 323
igname 206
Ikonos satellite 62
Imataca Forest Reserve 139, 150, 169
immerseration school 3
import substitution 18, 26, 129, 147, 151, 167, 191, 204, 271, 297
INC 238
INDA 238
India 59, 381
Indonesia 310–22, 325–82; agriculture in 311–19; forestry in 314–19; mentioned xiii, xvi, xvii, 2, 3, 5, 7, 8, 9, 10, 22, 24, 25, 28, 35, 45, 64, 213, 226, 227, 256, 322; oil in 310–15
INEC 218, 246
INEFAN 216–18, 247
INPARQUES 155
Institut Gabonais d'Appui au Dévelopement (IGAD) 89, 128
insurance 16
Integrated Operational Zones (OZI) 128
International Geosphere Biosphere Program (IGBP) 59
International Institute for Environment and Development (IIED) 275
International Institute for Tropical Agriculture (IITA) 12
International Monetary Fund (IMF) xiv, xvii, 96, 98, 143, 144, 157, 181, 202, 211, 262, 264, 293, 363–5
International Road Federation (IRF) 281
International Tropical Timber Organisation (ITTO) 72, 217, 247
Ipassa 109
Irian Jaya 248, 310; *see also* West Papua
iron 5, 20, 97, 141, 211
Iron Age 84
ITIC (Technical Institute for Immigration and Colonisation) 157–8
IUCN (World Conservation Union) 10, 12, 42, 74, 76, 87, 88, 133, 173, 254, 308
ivory 85, 99
Ivory Coast 3, 24, 39, 212

Jaba river 257
Jakarta 356
Jamaica 35
Jambi 311
JANT mill 275
Japan 15, 274
Jaringan Advokasi Tambang (JATAM) 323
Java 310, 313, 316, 323, 324, 328

Kalimantan 310, 311, 315; *see also* Borneo
Kare, Mount 259
Kavieng 261
Kenya 3, 24
Kikori 258
Kiunga 261
Kole 176
Korea 274
Korup National park 197
Kribi 176
Kribi-Campo basin 176
kunai grasses 250
Kutubu, Lake 258
Kutubu oilfield 258–9, 337
Kuwait 5
Kuznets curve 376–8

Lae 258, 282, 286
Lagos 305
La Guaira 132
La Lopé 11, 109
Lambaréné 111, 126
La Mémé 172
Landsat satellite 62, 72, 83, 135, 173, 253, 254, 257, 292, 301, 316, 323
Lastoursville 90, 126
Law of Agricultural Orientation 129
Law of Fallow Lands 247
Law of Idle Lands and Colonisation 247
Lekabi 120
Libreville 83, 89, 90, 93, 97, 104, 110, 111, 112, 117, 118, 119, 121, 125, 128, 129
Lihir 260, 261, 262
linear regression 168
Littoral Province 170, 172, 182, 201, 202
llanos 130, 132–3, 135, 136, 138, 156, 164, 167, 168, 337, 352
logging xvii, 5, 6, 51, 54, 59, 69, 70, 71, 347, 348, 351–2, 359–60, 368, 369, 375, 377; in Cameroon 173, 187, 188, 198, 200, 206, 209–10; in Ecuador 222, 230, 242, 373; in Gabon 86, 88, 92, 102, 107, 112, 123, 127, 371; in Indonesia 315, 320; in Mexico 301; in Nigeria 308; in Papua New Guinea 252, 253, 255, 256, 267, 274, 275, 280, 283, 291, 374; in Venezuela 138, 164, 166, 167, 169, 372; *see also* deforestation, forestry, timber
Lombok 310

Macas 247
Madang 12, 256, 273, 275, 283, 286, 295

Madura 310
Mahakam delta 311
mahogany 132
maize 121, 130, 131, 151, 206, 212, 219, 241, 300, 301, 303, 307, 314, 316, 344, 358, 361
Makokou 90, 109
malaria 169, 216, 251
Malaysia 24, 35, 274, 316, 345
Manabí Province 219
Mandara mountains 201
Mandji 95
manganese 86, 93, 97, 98, 99, 107, 113, 118, 127, 339
mangroves 31, 47, 48, 68, 91, 138, 219, 228, 248, 254, 337, 342; in Cameroon 170, 172, 176; in Nigeria 303, 304, 307, 309, 311, 320, 339
manioc 105, 118, 121, 127, 129, 130
Maracaibo 132, 133, 139, 156, 161, 164, 167, 168, 169, 337
Margarita Islands 167
Markham valley 277, 283, 288, 293, 358
MARN 149, 155, 168
MARNR 133, 134, 135, 139, 147, 168
Marqués de Comillas 299–300
Maryland 242
Mauritius 24
Mbalmayo 212
Mendi 258, 261
Mengomo 212
Merang 311
meranti 127
mercury 69, 70, 140, 257
Mérida 11
methods 8–9, 40–2
Mexico 296–303, 325–82; agriculture in 297–8; forests in 299–303; mentioned xiii, xvii, 2, 3, 7, 24, 35, 38, 39, 54, 158, 224, 309, 311, 319, 320, 322, 324; oil in 296–8
millet 172, 307
Milne Bay Province 256
milpa 297, 302, 344
MINEFI 203
Mineral Resources Stabilisation Fund (MRSF) 262, 264, 266, 288
mining 5, 14, 30, 31, 33–7, 46, 61, 69, 70, 81, 337, 339–40, 342, 348, 360, 373; in Cameroon 197, 206, 209, 211; in Gabon 86, 93–9, 118; in Papua New Guinea 248, 252, 255, 256–7, 258, 259, 260, 261, 262, 264, 271, 280, 285, 289, 291; in Venezuela 140–4, 164, 166, 167, 339, 372
Minkebé 109
Minvoul 100
Misima 256, 261, 262, 264
Mitzic 128
moabi 190
Moanda 93, 127
Morobe Province 277
Morona-Santiago Province 223, 237, 246
Morrocoy 169
Moungo 202
multiple regression models 127
multivariate regression models 39

Napo river 219, 221, 222, 223
National Aeronautics and Space Administration (NASA) 74, 83
National Forest Authority 254, 270, 275, 279, 280, 292, 295, 350
National Forest Inventory 315
National Forestry School 109
natural gas xvi, 3, 13, 24, 125, 303, 310
Ndélélé 211
Ndian 172
N'djolé 111
neoclassical school 3
Nepal 83
Netherlands, the xvi, 13–14, 15, 22
New Britain 256, 268, 276, 286
New Order government 310, 314
ngon melons 212
Ngounié 119, 120
nickel 211
Niger 24
Niger Delta 304, 309
Nigeria 303–10, 325–82; agriculture in 304–7; forests in 307–10; mentioned xiii, xvii, 2, 3, 7, 22, 24, 25, 35, 138, 176, 179, 184, 188, 208, 258, 310, 311, 319, 320, 322, 323; oil in 303–4
Nigeria disease 304
Nigerian National Oil Corporation 303
Nioungou 129
NOAA-AVHRR 62, 76, 87, 133, 135, 173, 292, 323
North Sea xvi
Norway 22, 31
nothofagus 248
Nusa Tenggara Province 323
Nyanga Province 119, 120
Nyonyie 90

OAU 97
Oaxaca 298
OCTRA 112
Office National du Café et du Cacao (ONCC) 191

Office National de Commercialisation des Produits de Base (ONCPB) 191
Ofoubou 91
Ogoni 304
Ogooué-Ivindo Province 119
Ogooué-Lolo Province 119
Ogooué-Maritime Province 128
Ogooué Province 85
Ogooué river 113
oil *passim*; also 19, 21, 24, 30–55, 334, 351; in Cameroon 177–82; in Ecuador 220–7; in Gabon 91–9; in Indonesia 310–11;in Mexico 296–7; in Nigeria 303–4;in Papua New Guinea 256–66; in Venezuela 138–9
oil palm 71, 151, 306, 309, 323, 335, 336, 340, 352; in Cameroon 172, 183, 186; in Ecuador 215, 223, 231, 234; in Gabon 85, 86, 100, 108; in Indonesia 314, 317, 318, 319, 321; in Papua New Guinea 267, 268, 281, 283, 284, 286
Oilwatch 50, 91
okoumé 86, 93, 99, 101, 102, 103, 105, 106, 107, 110, 113, 127, 381
Ok Tedi mine 256, 257, 259, 261, 262, 264, 293
Omar Bongo University 111
OPEC 20, 168, 310
Oriente 214, 216, 218, 219, 221, 222, 223, 224, 230, 237, 238, 246, 247
Orinoco 130, 132, 135, 136, 138, 145, 147, 149, 153, 156, 158, 164, 334, 372
ORSTOM 12
Owendo 93, 96, 97, 111, 112, 127, 129
Oyem 11, 89, 100, 111
ozigo 101, 102, 127

Pacto de Punto Fijo 141
padouk 190
Pakistan 3, 24
palm oil *see* oil palm
Pan-American Highway 156
Pandora field 257
Panguna mine 256, 257, 259, 261, 262, 264, 288
paper 31, 153
Papua, Gulf of 257, 258
Papua New Guinea 248–95, 325–82; agriculture in 250–1, 254–7, 260–1, 267, 277–80, 286–7, 290–3; deforestation in 248, 251–8, 262, 268–9, 274–6, 280–1, 288–91, 293; forestry in 250, 254–6, 267, 280, 286–7, 290–1; mentioned xiii, xvi, 1, 2, 3, 4, 7, 8, 11, 12, 19, 33, 35, 44, 55, 77, 78; oil in 256–66; poverty in 284–7, 290–1, 295; recommendations for 374; roads in 248, 266, 276, 281–3, 284, 290–1, 293, 295; settlement in 250–1, 258–9, 283–4, 290–1, 293–5; trade policy in 267, 276–7, 290–1; urbanisation in 248, 267, 271, 285–7, 291
Papua New Guinea Resource Information System (PNGRIS) 253–5, 271, 273, 275, 293
Papua, West *see* Irian Jaya
Paraguay 24, 216
páramo 130, 214
Pastaza Province 223
pastureland 6, 81, 299, 335, 340, 358, 359, 361, 380; in Ecuador 214, 218, 219, 222, 231, 234, 239, 243, 245; in Venezuela 136, 139, 145, 165, 168; *see also* beef, cattle
Pathfinder project 74–5
patronage 28
peanuts *see* groundnuts
Pearson coefficient 127, 211
pepper 319
Pertamina 312, 323, 340
Peru 14, 35, 45, 219, 223, 238, 247, 353
Petroecuador 220, 340
petroleum 20
Philippines 38, 250, 294
phosphates 24, 211
Pichincha Province 240, 242
pine 230
pinus caribea 167
pixel sizes 60–1
plantains 139, 234, 334, 340; in Cameroon 172, 186, 206, 209, 212; in Gabon 100, 105, 108, 111, 121
plywood 231
poaching 109, 200
Pointe-Noire 93
Poland 15
policy recommendations 367–74
political ecology school 3
pollution 48, 70, 221, 257, 258, 304, 339; *see also* cadmium, copper, cyanide, mercury, mining, zinc
pometia 248
population 36, 37, 41, 42, 73, 327–8, 357, 362, 368, 370, 377; in Cameroon 186, 188, 204–5, 209, 210–11; in Ecuador 214, 215–16, 223, 238; in Nigeria 306–9, 336; in Papua New Guinea 250, 251, 259, 267, 279, 283, 284, 287, 288, 289, 293, 294, 295, 335
Poreporena Highway 282
Porgera mine 256, 257, 261, 262, 264, 293
Port Gentil 91, 104, 119, 121, 126, 128, 129
Port Moresby 250, 257, 273, 277, 280, 282, 284, 285, 286, 292, 293, 294, 295

poverty 45, 354–5, 360, 361; in Cameroon 201–3, 207; in Ecuador 238–40, 245; in Gabon 115–17; in Papua New Guinea 261, 284–5, 286, 287, 290, 291, 295; in Venezuela 159–61, 165–6, 169
PREDESSUR 158
pribumi 312
price mechanisms 16–19
privatisation 181
Public Officers Superannuation Fund 294
Puerto Ayacucho 136, 156
Puerto Cabello 132

Queensland 258, 266
Quickbird satellite 62
Quintana Roo State 298
Quito 12, 237, 240, 242, 364, 376

Rabaul 264, 286
Rabi I 92, 337
Rabi II 92, 337
Rabi-Kounga 91, 95
railways 122, 172, 180, 198, 200, 215, 381; *see also* Transgabonese Railway
Rainforest Action Network 258
Ramu 277, 288, 293, 358
ranching *see* cattle, pastureland
rattan 70
reforestation 58, 65, 123
regression analysis 33, 39, 54, 55, 105, 151, 168, 192, 230, 232
relative prices *passim*; discussed 16–18
remote sensing 67, 69
resource curse 27, 29
Riau 310
rice 16, 20, 121, 145, 153, 162, 336, 344, 358, 361, 372, 374, 381; in Cameroon 180, 186, 191, 206, 207, 209; in Ecuador 228, 231, 234, 235, 241; in Indonesia 312–13, 314, 316, 319, 320; in Papua New Guinea 267, 269, 271, 287, 307
Rimbunan Hijau 187
Rio del Rey 176
roads xvii, 5, 36, 37, 39, 44, 49–50, 53, 54, 55, 66, 67, 315, 317, 320, 324, 350–2, 359, 360, 368, 374, 379, 380, 381; in Cameroon 172, 180, 190, 197–200, 201, 205, 336–7, 373; in Ecuador 215, 219, 223, 228, 230, 236–7, 239, 243, 244–5, 246, 334, 339, 373; in Gabon 92, 99, 101, 102, 122, 332; in Mexico 298, 300, 302, 335–6; in Nigeria 304, 305; in Papua New Guinea 248, 256, 258, 259, 266, 276, 279, 281–3, 284, 290, 291, 293, 295, 374; in Venezuela 132, 138–9, 164, 165, 332–4, 372
round wood equivalents (RWE) 72
Royal Dutch 310
rubber 14, 58, 100, 212, 283, 306, 309, 314, 318, 319, 352
rural–urban migration *see* urbanisation
Rutgers University 81

Sabah 316
Saga 311
sago 294
San Carlos 158
Sangay National Park 247
Santa Elena Province 220
Santiago 223, 237, 247
Santo Domingo 237, 247
sapelli 187
Sarawak 316
Saudi Arabia 5, 24
savannah 56, 59, 77, 138, 250, 305, 308, 330, 337, 350, 358, 359, 367; in Cameroon 170, 172, 201, 206; in Gabon 84–5, 88, 91, 92, 93, 100, 104, 109, 120
Seberida 324
SEFORVEN 149, 155
Selva Lacandona 300, 301
Senegal 24
Sepik 282, 286, 294
Setté Cama 92
settlement, directed 4, 39, 45, 302, 352–4; in Cameroon 200–1; in Ecuador 237–8; in Gabon 112, 113–14; in Papua New Guinea 283; in Venezuela 157–9, 166; *see also* Transmigrasi
Shell 91, 92, 126, 220, 303, 304, 311, 340
shifting cultivation 56, 61, 65, 67, 69, 70, 81, 158, 297, 308, 309, 321, 336, 337, 340, 344, 348, 353, 361, 362, 376, 380; in Cameroon 170, 172, 184, 204, 206; in Papua New Guinea 250, 254, 271; *see also* slash-and-burn cultivation
shrimp-farming 31, 46, 68, 219, 228, 245, 342, 373
Shushifindi project 238
SICA 218, 219, 223
Sierra (Ecuador) 214, 216, 219, 223, 226, 227, 230, 234, 235, 237, 240, 243, 247
silver 256
Siwai 269
slash-and-burn cultivation 45, 64, 67, 182, 211, 247, 354, 359, 361; in Gabon 89, 90, 92, 108, 126, 127; in Venezuela 130, 132, 145; *see also* shifting cultivation

slash-and-mulch cultivation 222
slavery 85, 99
Société de Développement du Cacao (SODECAO) 195
Société National des Bois du Gabon (SNBG) 107
Société Sucrière du Haut-Ogcoué (SOSUHU) 104, 128
Sogacel 128
Soloman Islands 269
sorghum 133, 172, 307
sources 8–9
South Africa 100, 105, 120
Southern Cone countries 161
Southern Province 260
South Province 172, 202, 204
South-West Province 172
soybeans 24, 55, 215, 234, 316, 319, 336
Spain 14
SPOT satellite 62
Sri Lanka 294
State of the World's Forests (SOFO) 73, 83
steel 141, 305
Steppelandia 330
Strickland, river 257
structural adjustment programmes (SAP) 7, 8, 39, 98, 264, 275, 348, 375, 376, 378–9, 380; in Cameroon 177, 180–2, 188, 191, 196, 210; in Venezuela 143, 144, 150, 159
subsistence farming 64, 92, 99, 132, 207, 255, 261, 269, 280, 284, 293
Sucre 145
Sucumbíos 221, 222, 223, 246
Suez 141
SUFOREN 217, 218, 223, 247
sugar 24, 55, 100, 104, 108, 131, 132, 151, 172, 183, 186, 216, 276, 277, 291, 300, 345
Sulawesi 318
Sumatra 311, 324
Suriname 35, 219
sustainability 348, 379
sweet potatoes 250, 287, 291, 358
swiddening *see* shifting cultivation
Switzerland 15
Sylvania 330
synthetic aperture radar (SAR) 62

Tabasco State 297, 299, 301
Tanzania 38
Tarapoa 222
taro, Chinese 250
tea 172, 177, 183, 186, 206, 283, 352
Team Ten 312
Technical Institute for Immigration and Colonisation (ITIC) 157–8
terminalia 248
Texaco 220, 222, 340
textiles 212
Thailand 39, 294
theories 3–4
Ticoporo 149, 169
timber 30, 31, 37, 38, 39, 42, 46, 47, 48, 50, 53, 72, 74, 299, 302, 305–6, 307, 314–15, 317, 320, 321, 324, 340, 345, 346–7, 348, 350, 355, 357, 359, 360, 361, 364, 366, 376, 378; in Cameroon 180, 182, 187, 188, 190, 192, 196, 198, 200, 206, 208, 210; in Ecuador 221, 228–31, 234, 236, 242, 245, 246, 352, 373; in Gabon 85, 103, 105, 106, 107, 113, 117, 118, 121, 122, 371; in Papua New Guinea 248, 258, 267, 274, 275, 276, 277, 278, 280, 282, 284, 288, 294; in Venezuela 132, 136, 149, 153, 156, 157, 162, 167, 372; *see also* deforestation, forestry, logging
tin 24, 211
Tipishca 222
tobacco 131, 132, 172, 177, 206
Togo 35
tontine system 195
Torres Strait 258
TOTAL 311
trade liberalisation 37, 39
trade policy 37–9; in Cameroon 191–2; in Ecuador 231–2; in Gabon 104–5; in Papua New Guinea 267, 276–7; in Venezuela 150–1
Transecuadorian pipeline 221, 226
Transgabonese Railway 96, 97, 98, 99, 104, 111–13, 123, 350, 351
Transmigrasi 45, 55, 317, 320, 323, 352, 353, 354, 381
Transparency International 364
Trans-PNG Highway 283
transport *see* railways, roads
Trans-Sylvania 330
Trinidad and Tobago 24, 35, 310
Tropenbos 190
Tropical Ecosystem Environment Observations by Satellite (TREES) 62, 63, 76, 83, 87–8, 89, 109, 126, 135, 173, 174, 253, 254, 292, 308, 316, 323, 328
tse-tse fly 213
'*T*' test 127
Tumbes 219
Turén 149
two-sector trade model 29

UNESCO 83
United Nations Development Program (UNDP) 111, 116, 127, 246, 295
United Nations Environment Program (UNEP) 59, 74, 147, 253, 254, 292
Upata 162
uranium 24, 86, 92, 98, 99, 107, 118, 127, 339
urbanisation 4, 26, 36, 37, 41, 46, 48, 52, 54, 55, 68, 302, 309, 319, 327, 331, 346, 351, 354, 355–7, 358, 360–1, 367, 375, 377, 379; in Cameroon 173, 186, 188, 192, 200, 201, 203–5, 207, 209, 332; in Ecuador 227, 231, 240–1, 245; in Gabon 90, 97, 101, 113, 116–19, 122, 123, 332; in Papua New Guinea 248, 267, 271, 285–7, 288, 291; in Venezuela 132, 157, 161–2, 165
USAID 221
USGS 253
Uxpanapa 298

Venezuela 130–69, 325–82; agriculture in 144–7, 154–5; deforestation in 130–44; forestry in 147–50, 155; mentioned xvii, 2, 3, 4, 8, 9, 10, 11, 22, 24, 28, 32, 33, 35, 65, 74, 76, 77, 78, 79, 80, 81, 83, 98, 170, 209, 214, 241, 252, 291, 300, 303, 310, 324; mining in 139–40; oil in 138–9; poverty in 159–61, 165–6; recommendations for 371–2; roads in 156–7; settlement in 157–9; trade policy in 150–1; urbanisation in 161–2
Venezuelan Forest Service *see* SEFORVEN
Venezuelan Investment Fund (FIV) 141
Veracruz 298
Vibroseis method 47

Waigani 295
wantok system 261, 284–5, 289, 291, 355
West Africa 1
Western Highlands 272, 286
Western Province 256
West Papua 248, 251, 310; *see also* Irian Jaya
West Province 182, 202
Wewak 289
wheat 16, 131, 232, 241, 267, 269–71, 287, 307, 345, 358, 361, 374
Woleu 100, 113, 119, 128
wood-processing 126
Woods Hole Research Centre 74
World Bank xiv, xv, xvi, 12, 95, 97, 98, 115, 116, 118, 127, 143, 161, 176, 179, 181, 186, 187, 196, 203, 211, 238, 251, 254, 263, 264, 268–9, 271, 274–6, 284–6, 292, 294, 317, 323, 363, 364
World Conservation Monitoring Centre (WCMC) 74, 76
World Conservation Union 10; *see also* IUCN
World Forest Survey 63
World Resources Institute (WRI) 72, 75, 87, 88, 133, 134, 135, 168, 173, 174, 217, 218, 253, 254, 274
World Trade Organisation (WTO) 277, 373, 376
World Wide Fund for Nature (WWF) 42, 92, 121

Yabassi-Bafang initiative 201
yams 99, 100, 212, 250, 251, 307
Yaoundé 11, 180, 191, 197, 198, 200, 201, 204, 207, 212
yellow fever 216

Zafiro field 303
Zaire 200
Zambia 24, 35
Zamora-Chinchipe Province 223
Zapatistas 302
zinc 211, 257, 311
Zulia State 337